ENUMERATIVE COMBINATORICS

DISCRETE
MATHEMATICS
AND
ITS APPLICATIONS

Series Editor

Kenneth H. Rosen, Ph.D.

AT&T Bell Laboratories

Abstract Algebra Applications with Maple,
Richard E. Klima, Ernest Stitzinger, and Neil P. Sigmon

Algebraic Number Theory, *Richard A. Mollin*

An Atlas of The Smaller Maps in Orientable and Nonorientable Surfaces,
David M. Jackson and Terry I. Visentin

An Introduction to Crytography, *Richard A. Mollin*

Combinatorial Algorithms: Generation Enumeration and Search,
Donald L. Kreher and Douglas R. Stinson

Cryptography: Theory and Practice, Second Edition, *Douglas R. Stinson*

Design Theory, *Charles C. Lindner and Christopher A. Rodgers*

Frames and Resolvable Designs: Uses, Constructions, and Existence,
Steven Furino, Ying Miao, and Jianxing Yin

Fundamental Number Theory with Applications, *Richard A. Mollin*

Graph Theory and Its Applications, *Jonathan Gross and Jay Yellen*

Handbook of Applied Cryptography,
Alfred J. Menezes, Paul C. van Oorschot, and Scott A. Vanstone

Handbook of Discrete and Computational Geometry,
Jacob E. Goodman and Joseph O'Rourke

Handbook of Discrete and Combinatorial Mathematics, *Kenneth H. Rosen*

Introduction to Information Theory and Data Compression,
Darrel R. Hankerson, Greg A. Harris, and Peter D. Johnson

Network Reliability: Experiments with a Symbolic Algebra Environment,
Daryl D. Harms, Miroslav Kraetzl, Charles J. Colbourn, and John S. Devitt

Quadratics, *Richard A. Mollin*

The CRC Handbook of Combinatorial Designs,
Charles J. Colbourn and Jeffrey H. Dinitz

Handbook of Constrained Optimization,
Herbert B. Shulman and Venkat Venkateswaran

ENUMERATIVE COMBINATORICS

CHARALAMBOS A. CHARALAMBIDES

CHAPMAN & HALL/CRC

A CRC Press Company
Boca Raton   London   New York   Washington, D.C.

Library of Congress Cataloging-in-Publication Data

Charalambides, Ch. A.
  Enumerative combinatorics / Charalambos A. Charalambides.
      p. cm. — (The CRC Press series on discrete mathematics and its applications)
  Includes bibliographical references and index.
  ISBN 1-58488-290-5
  1. Combinatorial enumeration problems. I. Title. II. Series.

QA164.8 .C48 2002
511′.62—dc21   2002019775

Visit the CRC Press Web site at www.crcpress.com

© 2002 by Chapman & Hall/CRC

No claim to original U.S. Government works
International Standard Book Number 1-58488-290-5
Library of Congress Card Number 2002019775
Printed in the United States of America 1 2 3 4 5 6 7 8 9 0
Printed on acid-free paper

Preface

Combinatorics, at the early stages of its emergence, treated the enumeration and properties of permutations, combinations and partitions of a finite set under various conditions. Its advent coincides with that of discrete probabilities in the 17th century. The advance of probability theory and statistics, with an ever-increasing demand for more general configurations, as well as the appearance and growth of computer science, have undoubtedly contributed to the rapid development of combinatorics during the last decades. This subject was then expanded to cover enumeration and examination of properties as well as investigation of existence and construction of configurations with specified properties.

This book provides a systematic coverage of the subject of enumeration of configurations with specified properties. It is designed to serve as a textbook for introductory or intermediate level enumerative combinatorics courses usually given to undergraduate or first year graduate students in mathematics, computer science, combinatorics, or mathematical statistics. The broad field of applications of combinatorial methods renders this book useful to anyone interested in operational research, physical and social sciences.

In Chapter 1, the two basic counting principles of addition and multiplication are introduced after a brief presentation of the necessary elements of set theory. Along with these principles, the notion of a recurrence (recursive) relation is introduced and its connection with a difference equation is pointed out. Also, discrete probability is briefly introduced for use in the subsequent chapters. In the last section, the symbols of summation and product are also presented and some of their properties are discussed.

Chapter 2 is devoted to the enumeration and properties of permutations, combinations, divisions and partitions of a finite set and also the enumeration of integer solutions of a linear equation. Further, some basic elements of enumeration of lattice paths are presented. This chapter concludes with several classical applications in discrete probability theory and statistics.

Vandermonde's factorial formula, Newton's binomial formula and the multinomial formula are presented in Chapter 3, after a suitable extension of factorials and binomials. Stirling's approximation formula is also given.

In Chapter 4, in continuation to the counting principles of addition and multiplication, the principle of inclusion and exclusion is extensively treated. In addition, the Bonferroni inequalities are derived.

The famous problem of coincidences (probleme des recontres), which perhaps constitute the first application of the inclusion and exclusion principle, is examined in Chapter 5 in the general framework of enumeration of permutations with fixed points. The related problem of enumeration of permutations with successions is examined in the same chapter.

Chapter 6 is devoted to a thorough presentation of generating functions, which constitute an important means of unifying the treatment of enumerative combinatorial problems.

In several enumeration problems, the number of configurations satisfying specified conditions can only be expressed recursively. In Chapter 7 we present the basic methods of solving linear recurrence relations.

Chapter 8 is devoted to an extensive treatment of the Stirling numbers of the first and second kind, which are the coefficients of the expansion of factorials into powers and of powers into factorials, respectively. In addition, the coefficients of the expansion of generalized factorials into usual factorials are examined.

The enumeration of distributions (of balls into urns) and occupancy (of urns by balls) is closely related to the enumeration of permutations and combinations. A formulation of this problem in a general framework and its treatment from a different point of view warrants a separate chapter. This treatment is the subject of Chapter 9.

A few elementary aspects of the combinatorial theory of partitions of integers are discussed in Chapter 10. Specifically, after the introduction of the basic concepts, recurrence relations and generating functions of the numbers of partitions with summands of specified values are obtained. Also, relations between the numbers of various partitions are concluded. The last section of this chapter includes some interesting classical q-identities.

Chapter 11 deals with the partition polynomials in n variables. The coefficient of the general term of these polynomials is the number of partitions of a finite set of n elements in specified numbers of subsets of the same cardinality and the summation is extended over all partitions of the number n. As particular cases, the partition polynomials include the exponential, logarithmic and potential polynomials, which owe their particular names to the form of their generating functions. The inversion of a power series by using the potential polynomials is presented in the last section of this chapter.

Enumeration problems emerging from the representation of a permutation as a product of cycles are treated in Chapter 12.

The problem of counting the number of equivalence classes of a finite set under a group of its permutations is the subject of Chapter 13.

Finally, Chapter 14 considers the Eulerian and the Carlitz numbers. In the last two sections of this chapter, these numbers are used to express the number of permutations with a given number of ascending runs (or rises) and the number of permutations with repetitions with a given number of non-descending runs (or rises).

A distinctive feature of the presentation of the material covered in this book is the comments (remarks) following most of the definitions and theorems. In these remarks the particular concept or result presented is discussed and extensions or generalizations of it are pointed out. In concluding each chapter, brief bibliographic notes, mainly of historical interest, are included.

At the end of each chapter, a rich collection of exercises is provided. Most of these exercises, which are of varying difficulty, aim to the consolidation of the concepts and results presented, while others complement, extend or generalize some of the results. So, working these exercises must be considered an integral part of this text. A few of the exercises are marked with an asterisk, indicating that they are more challenging. Hints and answers to the exercises are included at the end of the book. Before trying to solve an exercise, the less experienced reader may first look up the hint to its solution.

The material of this book has been presented several times to classes at the department of mathematics of the University of Athens, Greece, since 1972. Its first Greek edition, containing only Chapters 1 to 7, was published in 1984. Since then, the comments and suggestions communicated to me by students and colleagues who used it as a textbook for an introductory course in combinatorics contributed to improvements of certain points. The need for a textbook for a second, more advanced course in combinatorics led to its substantial expansion. The revised second Greek edition, expanded by the addition of Chapters 8 to 14 was published in 1990. Thanks are due to my colleagues Dr. M. Koutras and Dr. A. Kyriakoussis for their comments and suggestions while I prepared this revision.

The preparation of this English edition was an opportunity for me to clarify and improve several points. The comments of the reviewers and the series editor were of great help and are gratefully acknowledged. Special thanks are also due to Mrs. Rosa Garderi for the excellent typesetting of the book.

Charalambos A. Charalambides

Athens, September 2001

The Author

Charalambos A. Charalambides, is professor of mathematics at the University of Athens, Greece. Dr. Charalambides received a diploma in mathematics (1969) and a Ph.D. in mathematical statistics (1972) from the University of Athens. He was a visiting assistant professor at McGill University, Canada (1972-73), a visiting associate professor at Temple University, Philadelphia (1985-86) and a visiting professor at the University of Cyprus (1995-96). Since 1979 he has been an elected member of the International Statistical Institute (ISI). Professor Charalambides' research interests include enumerative combinatorics, discrete probability and parametric statistical estimation. He is an associate editor of *Communications in Statistics* and co-edited *Probability and Statistical Models with Applications*, Chapman Hall/CRC Press.

ccharal@math.uoa.gr

To Angelos and Cassandra

Contents

Chapter 1

BASIC COUNTING PRINCIPLES

1.1 INTRODUCTION

Combinatorics, at the early stages of its emergence, was basically concerned with the enumeration of permutations, combinations and partitions of a finite set under various conditions. The demand for construction and study of more general configurations expanded its subject. Combinatorics, nowadays established as a branch of discrete mathematics, deals with the existence, construction, enumeration and examination of properties of configurations satisfying specified conditions.

The appearance of combinatorial problems may be traced back far into time. In a letter considered to have been addressed by Archimedes to Eratosthenes, it is proposed, subject to certain conditions, to "compute the number of cattle of the Sun." This problem is one of the rare allusions in antiquity to combinatorics and its confrontation depends on consideration of the polygonal numbers of Pythagoras, Nicomachos and Diophantos.

Investigation of the existence or nonexistence of a configuration with certain specified properties constitutes the famous problem of "magic squares." This is the problem of placing positive integers in a square of n rows and n columns in such a way that the sum of the numbers in any row, column or diagonal is the same. Two simple examples of magic squares for $n = 3$ and $n = 5$ are given in Figure 1.1. The magic squares were known to Chinese in antiquity. The "Grant Plan," which is described in one of the oldest divinatory books in China, constitutes one such configuration, and legend claims it was decorated upon the back of a divine tortoise that emerged from the river Lo. Substituting the various sets of marks by positive integers, we obtain the left magic square of Figure 1.1. The magic squares were also known to ancient Greeks, according to a reference to them by Theona Smyrneou. The Indians and later the Arabs worked on these squares. The Greek monk Emmanuel Moschopoulos gave the first method of constructing magic squares in the 14th century. Euler and other great mathematicians

FIGURE 1.1
Magic squares

4	9	2
3	5	7
8	1	6

11	24	7	20	3
4	12	25	8	16
17	5	13	21	9
10	18	1	14	22
23	6	19	2	15

also showed interest in the same problem, in a more abstract setting, in the construction of finite geometries.

For certain configurations that can be easily obtained, such as the combinations of n objects taken k at a time, it is natural to ask for their multitude. In this way a new development in the evolution of combinatorics began with deriving formulas for the number of configurations satisfying specified properties. Of course, combinatorics has been greatly developed in this direction as a result of the powerful influence of probability (with the classical definition of Laplace) and statistics.

For a long time combinatorics was considered as the art of counting. From this point of view, elements of combinatorics may be traced almost everywhere. The majority of the classical formulas has been discovered and rediscovered several times. Perhaps the oldest example is the binomial coefficients which, in the 12th century, were known to the students of the Indian mathematician Bhakra; according to a recent discovery, the recursive computation of these coefficients was taught in 1265 by the Persian philosopher Nasir-Ad-Din. Pascal and Fermat rediscovered the binomial coefficients as a by-product of their study of games of chance. It is also known that Cardan, in 1560, proved that the number of subsets of a set of n elements is 2^n. In 1666, Leibniz published the first treatise in combinatorics "Dissertatio de Arte Combinatoria." As the configurations became more complex, great effort was put in counting techniques. In this direction, the most celebrated discovery was Laplace's generating function technique.

In the 20th century several new applications of combinatorics appeared in statistics (design of experiments), in coding theory (the problem of capacity of a set of signals), in operational research (the traveling salesman problem), etc. Pólya proved his famous theorem on counting.

When the configurations under consideration become too complex and the derivation of their exact number very difficult, the effort is focused on asymptotic values, bounds, inequalities, etc. A particularly curious case in this direction are the Ramsey numbers, which are very close to the binomial coefficients; it is not even known how to calculate them when their parameter values are greater than 7.

In certain quite difficult problems the only method of dealing with them remains that of listing all possible configurations. This method of reasoning by exhaustion has been used in proving important results and in other branches of mathematics.

This book concentrates on counting configurations satisfying specified conditions. In this chapter, some necessary basic elements of set theory are presented, in the next section, to pave the way for the presentation of the basic counting principles of addition and multiplication. Combinatorics is tightly connected with discrete probability theory; this connection greatly influenced its rapid development. It is thus reasonable that a brief introduction of the notion of probability and its connection to counting configurations follows the presentation of the basic counting principles. In order to make the book self-contained, the symbols of summation and product are presented and their basic properties are discussed.

1.2 SETS, RELATIONS AND MAPS

1.2.1 Basic notions

The concept of a set is a primitive notion in set theory like the concepts of a point and a straight line in Euclidean geometry. For the presentation of the basic concepts of combinatorics, only a few elements of set theory are needed and not its axiomatic foundation. In this respect, it is sufficient to consider a *set* as a (well-defined) collection of distinct objects. Sets are usually designated by capital letters of the alphabet with or without subscripts and their elements by lower case letters. The fact that the element a *belongs to* (or *is a member of*) the set A is expressed by writing $a \in A$. The negation of this statement is expressed by writing $a \notin A$. In describing which objects are contained in a set A we use the notation

$$A = \{a_1, a_2, \ldots, a_n, \ldots\},$$

which requires that the list of elements of the set A is known, or the notation

$$A = \{a : a \text{ has property } P\},$$

where P is a characteristic property of the elements of the set A.

Two special sets are often of interest: the *universal* set, designated by Ω, which is the set of all objects under consideration, and the *empty* or *null* set, designated by $\emptyset$, which does not contain any of the objects under consideration. It must be noted that these two sets may vary from case to case. For example, in studying the real roots of polynomials, the set of real numbers R is taken as the universal set. In this framework the set

$$\{x \in R : x^2 - 2x + 2 = 0\}$$

is an empty set as it is known that the quadratic equation $x^2 - 2x + 2 = 0$ has only complex roots and, more specifically, the roots $x_1 = 1 - i$ and $x_2 = 1 + i$, where $i = \sqrt{-1}$ is the imaginary unit. Further, if the study is extended to the complex roots of polynomials, the set of complex numbers C is taken as the universal set. In this case the set of roots of the polynomial $x^2 - 2x + 2$,

$$\{x \in C : x^2 - 2x + 2 = 0\} = \{1 - i, 1 + i\},$$

is not empty.

A set B is called a *subset* of a set A if and only if for every $b \in B$, we have $b \in A$ (every element of B is also an element of A). This is indicated by writing $B \subseteq A$ (and is read as B is a subset of A or B is included in A) or equivalently by writing $A \supseteq B$ (and is read as A is a superset of A or A includes B). If $B \subseteq A$ and there exists $a \in A$ such that $a \notin B$, then B is said to be a *proper subset* of A and this is indicated by $B \subset A$ or equivalently by $A \supset B$.

The fact that $B \subseteq A$ does not exclude $A \subseteq B$; when both relations hold, the sets A and B, consisting of the same elements, are called *equal* and this is indicated by $A = B$.

The set of all subsets of a set A is called the *power set* of A, and is denoted by $\mathcal{P}(A)$. In the sequel, all sets under consideration are considered as subsets of a universal set Ω, that is, all are elements of $\mathcal{P}(\Omega)$.

Schematic figures are frequently used for illustrating pictorially various sets. The *Venn diagrams* are such figures in which the universal set Ω is defined by a closed area of the plane containing its elements; the elements of Ω are defined by geometrical points of this plane. The subsets of Ω are defined by subareas. Figure 1.2 represents the fact that $B \subset A$.

FIGURE 1.2
Subset of a set

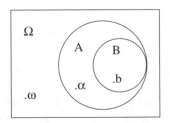

Note that these schematic figures are useful in verifying the validity of theorems of the theory of sets and also in indicating their proofs. Naturally, the proofs must exclusively be based only on the definitions of the notions.

1.2.2 Cartesian product

The concept of an ordered pair and, generally, of an ordered n-tuple is needed for the definition of the Cartesian product of two and generally of n sets. According to the definition of the equality of sets, the pair (two-element set) $\{a, b\}$ is equal to the pair $\{b, a\}$. Further, there are cases where it matters which element is first and which is second; for example in analytic geometry the pair of coordinates (a, b) of a point on the plane designates its abscissa and ordinate, respectively. The necessity of this distinction of the elements of a pair leads to the introduction of the concept of an ordered pair.

A pair of elements a and b (not necessarily different) in which a is considered as the first and b as the second element is called an *ordered pair* and is denoted by (a, b). According to this definition, two pairs (a, b) and (c, d) are equal if and only if $a = c$ and $b = d$.

The concept of an ordered n-tuple $(a_1, a_2, \dots, a_n)$ can be inductively defined as follows. Thus, for $n = 3$, an ordered triple (a_1, a_2, a_3) is defined as

$$(a_1, a_2, a_3) = ((a_1, a_2), a_3),$$

an ordered pair with first element the ordered pair (a_1, a_2) and second the element a_3. Generally, an *ordered n-tuple* $(a_1, a_2, \dots, a_n)$ is defined as

$$(a_1, a_2, \dots, a_{n-1}, a_n) = ((a_1, a_2, \dots, a_{n-1}), a_n),$$

an ordered pair with first element the ordered $(n-1)$-tuple $(a_1, a_2, \dots, a_{n-1})$ and second the element a_n. In several cases it is convenient to adopt the vector terminology where the first element a_1 is called first coordinate, the second element a_2 is called second coordinate, etc.

After the introduction of the concept of an ordered pair and, generally, of an ordered n-tuple, the Cartesian product can by defined as follows:

The *Cartesian product* of the sets A and B, denoted by $A \times B$, is defined as the set of ordered pairs in which the first coordinate is an element of the set A and the second coordinate is an element of the set B, that is

$$A \times B = \{(a, b) : a \in A, b \in B\}.$$

This definition can be extended to n sets $A_1, A_2, \dots, A_n$ as follows:

$$A_1 \times A_2 \times \cdots \times A_n = \{(a_1, a_2, \dots, a_n) : a_1 \in A_1, a_2 \in A_2, \dots, a_n \in A_n\}.$$

Particularly, if $A_1 = A_2 = \cdots = A_n = A$, the Cartesian product is denoted by A^n.

The Cartesian product $A \times B$ is geometrically represented by the points (a, b) of the plane, with abscissa $x = a$ taking values from the set A ($x - axis$) and ordinate $y = b$ taking values from the set B ($y - axis$). Generally, the Cartesian product $A_1 \times A_2 \times \cdots \times A_n$ is represented by the points $(a_1, a_2, \ldots, a_n)$ of the n-dimensional space with the k-coordinate taking values from the set A_k ($x_k - axis$), $k = 1, 2, \ldots, n$.

Example 1.1 Routes

Suppose that there are three different roads a_1, a_2 and a_3 from town O to town A, and four different roads b_1, b_2, b_3 and b_4 from town A to town B. Find the different routes from town O to town B.

A route from town O to town B may be represented by an ordered pair (a, b), where a is a road from the set $A = \{a_1, a_2, a_3\}$ of roads from town O to town A, and b is a road from the set $B = \{b_1, b_2, b_3, b_4\}$ of roads from town A to town B. Thus, the set of routes from town O to town B is the Cartesian product:

$$A \times B = \{(a_1, b_1), (a_1, b_2), (a_1, b_3), (a_1, b_4), (a_2, b_1), (a_2, b_2),$$
$$(a_2, b_3), (a_2, b_4), (a_3, b_1), (a_3, b_2), (a_3, b_3), (a_3, b_4)\}.$$

This is geometrically represented in Figure 1.3.   ▯

FIGURE 1.3
Set of routes

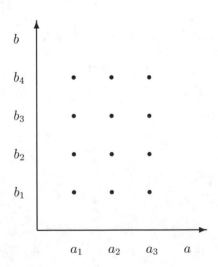

1.2.3 Relations

A binary *relation* from the set A to the set B is a subset $\mathcal{R}$ of the Cartesian product $A \times B$. The ordered pair (a, b) satisfies the relation $\mathcal{R}$ if and only if $(a, b) \in \mathcal{R}$. This is usually denoted by $a\mathcal{R}b$. If $B = A$, then $\mathcal{R}$ is called a relation on A. Such relations are for example the equality relation and the inclusion relation, which have already been introduced in Section 1.2.1. These relations are defined on the set $\mathcal{P}(\Omega)$ of the subsets of Ω. Note that the equality relation on a set A is the diagonal D_A of the Cartesian product A^2.

The above definition is very general. The most interesting relations are those satisfying certain desirable properties. Such properties are the following:

A binary relation $\mathcal{R}$ on a set A is called:

(a) *Reflexive*, if and only if for every $a \in A$, it holds $a\mathcal{R}a$

(b) *Symmetric*, if and only if $a\mathcal{R}b$ implies $b\mathcal{R}a$

(c) *Antisymmetric*, if and only if $a\mathcal{R}b$ and $b\mathcal{R}a$ imply $a = b$

(d) *Transitive*, if and only if $a\mathcal{R}b$ and $b\mathcal{R}c$ imply $a\mathcal{R}c$

The *inverse* (or *reciprocal*) of a binary relation $\mathcal{R}$ on A denoted by $\mathcal{R}^{-1}$ is defined as follows: $a\mathcal{R}^{-1}b$, if and only if $b\mathcal{R}a$.

An *equivalence relation* is a binary relation that is reflexive, symmetric and transitive. Such relation is, for example, the equality relation. An *order relation* is a binary relation that is reflexive, antisymmetric and transitive. Such relation is, for example, the inclusion relation $\subseteq$. The inverse of this relation is the relation $\supseteq$.

1.2.4 Maps

A subset F of the Cartesian product $A \times B$ is called a *map* (or *function* or *correspondence*) of the set A into the set B if and only if, for every $a \in A$, there exists only one b such that $(a, b) \in F$. Thus, if $(a, b_1) \in F$ and $(a, b_2) \in F$, then $b_1 = b_2$. In the ordered pair $(a, b) \in F$, the element $a \in A$ is called *archetype* and the element $b \in B$ is called the *image* of a by F and is usually denoted by $b = F(a)$. Consequently, a map F of the set A into the set B associates to each element $a \in A$ only one element $b \in B$, the image of a by F. If, in addition, for every element $b \in B$, there exists at least one element $a \in A$ such that $(a, b) \in F$, then F is called *surjective (onto)*. A map F of the set A into the set B is called *injective (one-to-one)* if and only if there exist at most one element $a \in A$ such that $(a, b) \in F$. Thus, if $(a_1, b) \in F$ and $(a_2, b) \in F$, then $a_1 = a_2$. Finally, a map F is called *bijective (one-to-one and onto)* if F is both surjective and injective.

The diagonal T_A of the Cartesian product A^2,

$$T_A = \{(a,b) \in A^2 : a = b\},$$

is called *identity map* of the set A. Thus, T_A associated to each element $a \in A$ the same element $T_A(a) = a$.

If $F \subseteq A \times B$ is a *bijective map* of the set A into the set B, then its *inverse map*, denoted by $F^{-1} \subseteq B \times A$, is defined as follows: $(b,a) \in F^{-1}$ if and only if $(a,b) \in F$ or, equivalently, $a = F^{-1}(b)$ if and only if $b = F(a)$ for every $a \in A$ and $b \in B$.

Consider the set $N = \{1, 2, \ldots, n, \ldots\}$ of natural numbers. A map of the set N into a set A,

$$\{(n, a_n) : n \in N, a_n \in A\},$$

which corresponds an element $a_n \in A$ to each natural number $n \in N$, is particularly called a *sequence* of elements of A. This sequence is usually denoted as $a_n \in A$, $n = 1, 2, \ldots$. The element a_n is called the n-th term of the sequence.

Consider, in general, an index set I. A map of the set I into a set A,

$$\{(i, a_i) : i \in I, a_i \in A\},$$

which corresponds an element $a_i \in A$ to each $i \in I$, is called a *family* of elements of A. This family is usually denoted as $a_i \in A$, $i \in I$. The term family is used instead of the term map when the interest is focused on the elements $a_i \in A$ and not on the map itself. Note that a sequence is a particular case, $I \equiv N$, of a family.

Example 1.2 Maps
Consider the sets $A = \{0, 1, \ldots, 9\}$, $B = \{0, 1, 2\}$ and their Cartesian product

$$A \times B = \{(a,b) : a \in \{0, 1, \ldots, 9\}, \ b \in \{0, 1, 2\}\}.$$

(a) The subset

$$\{(0,0), (2,0), (4,0), (6,0), (8,0), (1,1), (3,1), (5,1), (7,1), (9,1)\},$$

of $A \times B$, is a map F from the set A into the set B such that

$$F(0) = 0, \ F(2) = 0, \ F(4) = 0, \ F(6) = 0, \ F(8) = 0$$

and

$$F(1) = 1, \ F(3) = 1, \ F(5) = 1, \ F(7) = 1, \ F(9) = 1.$$

Note that the image $b = F(a)$ of an element $a \in A$ is the remainder of the division of a by 2. Further, this map F is not surjective since there is no element $a \in A$ such that $F(a) = 2$.

(b) The subset

$$\{(0,0),(3,0),(6,0),(9,0),(1,1),(4,1),(7,1),(2,2),(5,2),(8,2)\},$$

of $A \times B$, is a surjective map G from the set A into the set B such that

$$G(0) = 0, \quad G(3) = 0, \quad G(6) = 0, \quad G(9) = 0, \quad G(1) = 1,$$

$$G(4) = 1, \quad G(7) = 1, \quad G(2) = 2, \quad G(5) = 2, \quad G(8) = 2.$$

This map corresponds to each element $a \in A$ the remainder $b = G(a)$ of the division of a by 3. $\quad \square$

1.2.5 Countable and uncountable sets

Two sets A and B are called *equivalent*, and this is denoted by $A \sim B$, if and only if there exists a bijective map of the set A into the set B. For example, the set $A = \{2, 4, \ldots, 2n\}$ is equivalent to the set $B = \{1, 2, \ldots, n\}$ since the map $F(a) = a/2$ for every $a \in A$ is bijective. More generally, the set $A = \{a_1, a_2, \ldots, a_n\}$ is equivalent to the set $B = \{1, 2, \ldots, n\}$, with $F(a_k) = k$ for every $k = 1, 2, \ldots, n$, is a bijective map from A into B.

A set A is called *finite*, with n elements, if and only if it is equivalent to the subset $\{1, 2, \ldots, n\}$ of natural numbers. The empty set $\emptyset$ is considered finite with 0 elements. A set that is not finite is called *infinite*. A set A is called *infinitely countable* if and only if it is equivalent to the set $N = \{1, 2, \ldots, n, \ldots\}$ of natural numbers. Denoting by a_k the element of A that corresponds to the natural number k, for $k = 1, 2, \ldots$, the set A can be represented as

$$A = \{a_1, a_2, \ldots, a_n\},$$

if it is finite with n elements, or as

$$A = \{a_1, a_2, \ldots, a_k, \ldots\},$$

if it is infinitely countable. A set A is called *countable* if it is finite or infinitely countable. A set that is not countable is called *uncountable*.

1.2.6 Set operations

The *union* of two sets A and B is defined as the set that includes the elements of Ω belonging to A or B (the conjunction or being not exclusive) and is denoted by $A \cup B$, that is

$$A \cup B = \{\omega \in \Omega : \omega \in A \text{ or } \omega \in B\}.$$

This definition is extended to n sets $A_1, A_2, \ldots, A_n$:

$$A_1 \cup A_2 \cup \cdots \cup A_n = \{\omega \in \Omega : \omega \in A_k \text{ for at least one } k \in \{1, 2, \ldots, n\}\}$$

and more generally to a family of sets $\{A_i, i \in I\}$:

$$\bigcup_{i \in I} A_i = \{\omega \in \Omega : \omega \in A_i \text{ for at least one subscript } i \in I\}.$$

The *intersection* of two sets A and B is defined as the set that includes the common elements of the two sets and is denoted by $A \cap B$, that is

$$A \cap B = \{w \in \Omega : \omega \in A \text{ and } \omega \in B\}.$$

For reasons of economy the notation AB instead of $A \cap B$ is frequently used. This definition is extended to n sets $A_1, A_2, \ldots, A_n$:

$$A_1 \cap A_2 \cap \cdots \cap A_n = \{\omega \in \Omega : \omega \in A_k \text{ for all subscripts } k \in \{1, 2, \ldots, n\}\}$$

and more generally to a family of sets $\{A_i, i \in I\}$:

$$\bigcap_{i \in I} A_i = \{\omega \in \Omega : \omega \in A_i \text{ for all subscripts } i \in I\}.$$

The union and intersection are pictorially illustrated by Venn diagrams in Figure 1.4; the shaded area represents the set under consideration.

FIGURE 1.4
Union and intersection

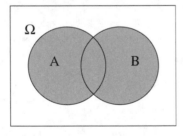 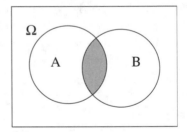

The *complement* (with respect to the universal set Ω) of a set A is defined as the set that includes the elements of Ω not belonging in A and is denoted by A' or A^c or CA, that is

$$A' = \{\omega \in \Omega : \omega \notin A\}.$$

The (set theoretic) *difference* of a set B from a set A is defined as the set that includes the elements of A not belonging in B and is denoted by $A - B$, that is

$$A - B = \{\omega \in \Omega : \omega \in A, \omega \notin B\}.$$

Note that

$$A' = \Omega - A, \ A - B = A \cap B'.$$

FIGURE 1.5
Complement and difference

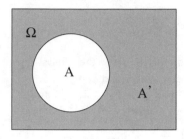 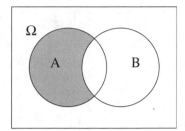

The Venn diagrams of the difference of two sets and the complement of a set are given in Figure 1.5.

The operations of union and intersection of sets share many similarities and differences, with the operations of summation and multiplication of real numbers, as follows from the next theorem.

THEOREM 1.1
For any sets A, B and C the following properties of the union, intersection and complementation hold:
(a) *Associativity of the union and intersection:*

$$(A \cup B) \cup C = A \cup (B \cup C), \quad (A \cap B) \cap C = A \cap (B \cap C).$$

(b) *Distributivity of the intersection with respect to the union and of the union with respect to the intersection:*

$$A \cap (B \cup C) = (A \cap B) \cup (A \cap C), \quad A \cup (B \cap C) = (A \cup B) \cap (A \cup C).$$

(c) *Commutativity of the union and intersection:*

$$A \cup B = B \cup A, \quad A \cap B = B \cap A.$$

(d) *The null set $\emptyset$ is the neutral element for the union:*

$$A \cup \emptyset = \emptyset \cup A = A,$$

while the universal set Ω is the neutral element for the intersection

$$A \cap \Omega = \Omega \cap A = A.$$

(e) *The complementation assigns to each set A, the set A' such that*

$$A \cap A' = \emptyset, \quad A \cup A' = \Omega.$$

(f)
$$\Omega' = \emptyset, \ \emptyset' = \Omega, \ (A')' = A$$

(g)
$$A \cup A = A, \ A \cup \Omega = \Omega, \ A \cap A = A, \ A \cap \emptyset = \emptyset.$$

PROOF   It can be easily verified that these properties are almost direct consequences of the definitions of the union, intersection and complementation. The proof of (b), which requires a somewhat longer series of steps than the others, may be carried out as follows. Consider an element $\omega \in A \cap (B \cup C)$. Then ω belongs to both A and $B \cup C$ and thus it belongs to A and to at least one of B and C. This implies that ω belongs to both A and B or to both A and C and hence it belongs to $A \cap B$ or to $A \cap C$. Therefore $\omega \in (A \cap B) \cup (A \cap C)$ and $A \cap (B \cup C) \subseteq (A \cap B) \cup (A \cap C)$. Similarly, it can be shown that $(A \cap B) \cup (A \cap C) \subseteq A \cap (B \cup C)$, whence $A \cap (B \cup C) = (A \cap B) \cup (A \cap C)$. The proof of $A \cup (B \cap C) = (A \cup B) \cap (A \cup C)$ is quite similar.   ∎

REMARK 1.1   The set $\mathcal{P}(\Omega)$ of all subsets of Ω, furnished with the operations of union $\cup$, intersection $\cap$ and the map $C : \mathcal{P}(\Omega) \to \mathcal{P}(\Omega)$, which to every element $A \in \mathcal{P}(\Omega)$ assigns its complement $CA = A' \in \mathcal{P}(\Omega)$, is made a *Boolean algebra* since, according to Theorem 1.1, properties (a) to (e), constituting the definition of such an algebra, are satisfied.   ∎

Two important interrelations between the operations of union, intersection and complementation are given in the next theorem.

THEOREM 1.2 De Morgan's formulae
Let A and B be subsets of a universal set Ω. Then

$$(A \cup B)' = A' \cap B', \ \ (A \cap B)' = A' \cup B'.$$

PROOF   Consider an element $\omega \in (A \cup B)'$. Then $\omega \notin A \cup B$ and hence $\omega \notin A$ and $\omega \notin B$. This implies $\omega \in A'$ and $\omega \in B'$. Thus $\omega \in A' \cap B'$ and $(A \cup B)' \subseteq A' \cap B'$. Similarly it can be shown that $A' \cap B' \subseteq (A \cup B)'$, whence $(A \cup B)' = A' \cap B'$. The proof of the second formula can be similarly carried out.   ∎

REMARK 1.2   De Morgan's formulae can be extended to n subsets $A_1, A_2,$ $\dots, A_n$ of Ω:

$$(A_1 \cup A_2 \cup \cdots \cup A_n)' = A'_1 \cap A'_2 \cap \cdots \cap A'_n,$$

$$(A_1 \cap A_2 \cap \cdots \cap A_n)' = A'_1 \cup A'_2 \cup \cdots \cup A'_n,$$

and to a family of subsets $\{A_i, i \in I\}$ of Ω:

$$\left(\bigcup_{i \in I} A_i\right)' = \bigcap_{i \in I} A_i', \quad \left(\bigcap_{i \in I} A_i\right)' = \bigcup_{i \in I} A_i'.$$

The derivation of these formulae follows verbatim the lines of the proof of Theorem 1.2. ∎

The distinction of two sets according to whether they have elements in common or not will be useful in the sequel. In this respect, the next definition is introduced.

Two sets A and B are said to be *disjoint* if they do not have elements in common, that is, if $A \cap B = \emptyset$. More generally, the sets $A_1, A_2, \dots, A_n$ are said to be *pairwise* or *mutually disjoint* if $A_i \cap A_j = \emptyset$ for all pairs of subscripts $\{i, j\}$ with $i \neq j$, from the set of indices $\{1, 2, \dots, n\}$. In such a case, the operation of the union is designated by $+$ or $\sum$ instead of $\cup$.

Some useful properties of the Cartesian product are shown in the next theorem.

THEOREM 1.3
For any sets A, B, C and D the following relations hold:

$$(A \cup B) \times C = (A \times C) \cup (B \times C),$$

$$(A \cap B) \times (C \cap D) = (A \times C) \cap (B \times D).$$

PROOF   Consider an element $u \in (A \cup B) \times C$. Then $u = (a, c)$ with $a \in A \cup B$, $c \in C$ and hence $a \in A$ or $a \in B$ and $c \in C$, which implies $(a, c) \in A \times C$ or $(a, c) \in B \times C$. Therefore $u = (a, c) \in (A \times C) \cup (B \times C)$ and $(A \cup B) \times C \subseteq (A \times C) \cup (B \times C)$. It can be shown, following the inverse procedure, that $(A \times C) \cup (B \times C) \subseteq (A \cup B) \times C$ and thus $(A \cup B) \times C = (A \times C) \cup (B \times C)$. The proof of the second formula is quite similar.   ∎

1.2.7 Divisions and partitions of a set

An *n-division* of a set W (or a division of a set W in n subsets) is an ordered n-tuple of sets $(A_1, A_2, \dots, A_n)$ that are pairwise disjoint subsets of W and their union is W, that is:

$$A_i \subseteq W, \ i = 1, 2, \dots, n, \ A_i \cap A_j = \emptyset, \ i, j = 1, 2, \dots, n, \ i \neq j,$$

$$A_1 + A_2 + \cdots + A_n = W.$$

Note that, in a division of a set, the inclusion of one or more empty sets is not excluded. For example, the ordered sequence (A_1, A_2, A_3, A_4), with $A_1 = \{w_1\}$, $A_2 = \{w_2, w_3, w_4\}$, $A_3 = \{w_5\}$ and $A_4 = \emptyset$, is a 4-division of the set $W = \{w_1, w_2, w_3, w_4, w_5\}$.

An *n-partition* of a set W (or a partition of a set W in n subsets) is a set of n sets $\{A_1, A_2, \dots, A_n\}$ that are pairwise disjoint and not null subsets of W and their union is W, that is:

$$A_i \subseteq W, \ A_i = \emptyset, \ i = 1, 2, \dots, n, \ A_i \cap A_j = \emptyset, \ i, j = 1, 2, \dots, n, \ i \neq j,$$

$$A_1 + A_2 + \cdots + A_n = W.$$

Note that, in a partition, as opposed to a division of a set, no empty sets are included and the sets constituting it are not ordered.

1.3 THE PRINCIPLES OF ADDITION AND MULTIPLICATION

As has been noted in the introduction, counting configurations constitutes a major part of combinatorics. The set of configurations is in any case finite and so the problem of counting them is a problem of counting the elements of a finite set.

The number of elements of a finite set A is denoted by $N(A)$ or $|A|$ and is called the *cardinal* of it. In the case of a finite universal set Ω, its cardinality is taken as $N(\Omega) \equiv N$. At this point, though clear from the relevant definitions of the preceding section, it is worth noting explicitly the following lemma.

LEMMA 1.1
If A and B are finite and equivalent sets, then

$$N(A) = N(B).$$

Thus, the cardinal of a finite set A may be deduced by determining a finite set B, equivalent to A, with known cardinality.

Some basic properties of cardinality are proved in the next theorem.

THEOREM 1.4
   (a) *If A and B are finite and disjoint sets, then*

$$N(A + B) = N(A) + N(B).$$

(b) *If A is a subset of a finite universal set Ω and A' its complement, then*

$$N(A') = N - N(A).$$

(c) *If A and B are finite sets, then*

$$N(A - B) = N(A) - N(A \cap B)$$

and particularly for $B \subseteq A$,

$$N(A - B) = N(A) - N(B).$$

PROOF   (a) Since $A \cap B = \emptyset$, any element of $A + B$ belongs either to A only or to B only and thus

$$N(A + B) = N(A) + N(B).$$

(b) Note that the sets A and A' are disjoint and according to part (a),

$$N(A + A') = N(A) + N(A').$$

Further, $A + A' = \Omega$, whence

$$N \equiv N(\Omega) = N(A) + N(A')$$

and

$$N(A') = N - N(A).$$

(c) Since

$$(A \cap B') \cap (A \cap B) = (A \cap A) \cap (B' \cap B) = A \cap \emptyset = \emptyset,$$

the sets $A \cap B' = A - B$ and $A \cap B$ are disjoint and further

$$(A \cap B') + (A \cap B) = A \cap (B' + B) = A \cap \Omega = A.$$

Hence

$$N(A) = N((A \cap B') + (A \cap B)) = N(A \cap B') + N(A \cap B)$$

and

$$N(A - B) = N(A \cap B') = N(A) - N(A \cap B).$$

In particular, for $B \subseteq A$, whence $A \cap B = B$, it follows that

$$N(A - B) = N(A) - N(B)$$

and the proof of the theorem is completed.   ∎

REMARK 1.3   The set function $N(\cdot)$, which corresponds to every set $A \in \mathcal{P}(\Omega)$ its cardinal $N(A)$, is (a) *nonnegative*: $N(A) \geq 0$ for every set $A \in \mathcal{P}(\Omega)$ and (b) *finitely additive*: $N(A + B) = N(A) + N(B)$ for any disjoint sets $A, B \in \mathcal{P}(\Omega)$, according to part (a) of Theorem 1.4. These properties made $N(\cdot)$ a *finitely additive measure* (or simply *measure*) on $\mathcal{P}(\Omega)$. The most known measures are also the length, area and volume in geometry, the mass in physics and the probability (measure) in the theory of probability. This last measure is introduced in the next section for the needs of the probabilistic applications of combinatorics.   ∎

As regards the cardinal of the union of more than two finite and pairwise disjoint sets, the next corollary of the first part of Theorem 1.4 is shown.

COROLLARY 1.1
If $A_1, A_2, \dots, A_n$ are finite and pairwise disjoint sets, then

$$N(A_1 + A_2 + \cdots + A_n) = N(A_1) + N(A_2) + \cdots + N(A_n). \qquad (1.1)$$

PROOF   Note first that, according to part (a) of Theorem 1.4, relation (1.1) holds for $n = 2$. Suppose that (1.1) holds for $n - 1$, that is

$$N(A_1 + A_2 + \cdots + A_{n-1}) = N(A_1) + N(A_2) + \cdots + N(A_{n-1}).$$

It will be shown that (1.1) holds also for n. For this reason, set $A = A_1 + A_2 + \cdots + A_{n-1}$ and $B = A_n$, whence

$$A \cap B = (A_1 + A_2 + \cdots + A_{n-1}) \cap A_n = A_1 \cap A_n + A_2 \cap A_n + \cdots + A_{n-1} \cap A_n = \emptyset,$$

$$A + B = (A_1 + A_2 + \cdots + A_{n-1}) + A_n = A_1 + A_2 + \cdots + A_{n-1} + A_n.$$

Thus, the sets A and B are finite and disjoint and according to part (a) of Theorem 1.4, $N(A + B) = N(A) + N(B)$, and the hypothesis that (1.1) holds for $n - 1$, it follows that

$$N(A_1 + A_2 + \cdots + A_n) = N(A_1 + A_2 + \cdots + A_{n-1}) + N(A_n)$$
$$= N(A_1) + N(A_2) + \cdots + N(A_{n-1}) + N(A_n).$$

Hence, according to the principle of mathematical induction, (1.1) holds for every integer $n \geq 2$.   ∎

REMARK 1.4   Relation (1.1) is often referred to as the *addition principle* and can also be stated as follows: if an element (object) ω_i can be selected in k_i different ways, for $i = 1, 2, \dots, n$, and the selection of ω_i excludes the simultaneous selection of ω_j, $i, j = 1, 2, \dots, n$, $i \neq j$, then any of the elements (objects) ω_1 or ω_2 or $\cdots$ or ω_n can be selected in $k_1 + k_2 + \cdots + k_n$ ways.   ∎

REMARK 1.5 For each division $(A_1, A_2, \ldots, A_n)$, as well as for each partition $\{A_1, A_2, \ldots, A_n\}$, of a finite set W, the sets $A_1, A_2, \ldots, A_n$ are finite and pairwise disjoint. Thus, by Corollary 1.1, the following relation holds:

$$N(W) = N(A_1 + A_2 + \cdots + A_n) = N(A_1) + N(A_2) + \cdots + N(A_n),$$

for both a division and a partition of a finite set. ∎

The next theorem is concerned with the cardinality of the Cartesian product of finite sets.

THEOREM 1.5
If A and B are finite sets, then

$$N(A \times B) = N(A)N(B). \tag{1.2}$$

PROOF Let $A = \{a_1, a_2, \ldots, a_k\}$ and $B = \{b_1, b_2, \ldots, b_r\}$. Then the set A may be written in the form

$$A = A_1 + A_2 + \cdots + A_k, \quad A_i = \{a_i\}, \quad i = 1, 2, \ldots, k$$

and the Cartesian product $A \times B$, according to Theorem 1.3, in the form

$$A \times B = (A_1 + A_2 + \cdots + A_k) \times B = (A_1 \times B) + (A_2 \times B) + \cdots + (A_k \times B),$$

where the sets (Cartesian products) $A_1 \times B, A_2 \times B, \ldots, A_k \times B$ are pairwise disjoint. Hence

$$N(A \times B) = N(A_1 \times B) + N(A_2 \times B) + \cdots + N(A_k \times B).$$

Noting that, for any $i \in \{1, 2, \ldots, k\}$, the Cartesian product $A_i \times B$ is the set of the ordered pairs (a_i, b_j) with first element the only element a_i of the set $A_i = \{a_i\}$ and second element any of the elements $b_j, j = 1, 2, \ldots, r$, of the set B, it follows that

$$N(A_i \times B) = N(B), \quad i = 1, 2, \ldots, k.$$

Thus

$$N(A \times B) = kN(B) = N(A)N(B)$$

and the proof of the theorem is completed. ∎

Example 1.3 Routes revisited
Suppose that there are three different roads a_1, a_2 and a_3 from town O to town A, four different roads b_1, b_2, b_3 and b_4 from town A to town B, and two different roads c_1 and c_2 from town O directly to town B. Calculate the number of different routes from town O to town B.

A route from town O to town B through town A may be represented by an ordered pair (a, b), where a is a road from the set $A = \{a_1, a_2, a_3\}$ and b is a road from the set $B = \{b_1, b_2, b_3, b_4\}$. Thus the set of routes from town O to town B through town A is the Cartesian product $A \times B = \{(a, b) : a \in A, b \in B\}$. Also, the set of routes from town O to town B directly is $C = \{c_1, c_2\}$. Therefore, by the addition principle and expression (1.2), the number of different routes from town O to town B equals

$$N(A)N(B) + N(C) = 14. \quad \square$$

A subset S_2 of the Cartesian product Ω^2, of a finite universal set Ω with itself, cannot always be written as a Cartesian product $A \times B$, with $A \subseteq \Omega$ and $B \subseteq \Omega$. Nevertheless, an expression similar to (1.2) may be obtained for the number of elements of S_2 when the number of selections for the first coordinate and, for each of these selections, the number of selections for the second coordinate, are known. Specifically, the next corollary, which is readily deduced from Corollary 1.1 and Theorem 1.5, is concerned with the number of elements of S_2.

COROLLARY 1.2
If $S_2 = (A_1 \times B_1) + (A_2 \times B_2) + \cdots + (A_k \times B_k)$, where $A_1, A_2, \ldots, A_k$ are finite and pairwise disjoint sets and $B_1, B_2, \ldots, B_k$ are finite sets, then

$$N(S_2) = N(A_1)N(B_1) + N(A_2)N(B_2) + \cdots + N(A_k)N(B_k). \quad (1.3)$$

In particular, if $A_i = \{a_i\}$ and $B_i = \{b_{i,1}, b_{i,2}, \ldots, b_{i,r}\}$, $i = 1, 2, \ldots, k$, whence $S_2 = \{(a_i, b_{i,j}),\ i = 1, 2, \ldots, k,\ j = 1, 2, \ldots, r\}$, and introducing $A = \{a_1, a_2, \ldots, a_k\}$, then

$$N(S_2) = N(A)N(B_i) = kr. \quad (1.4)$$

The cardinality of the Cartesian product of more than two finite sets is inductively deduced from Theorem 1.5.

COROLLARY 1.3
If $A_1, A_2, \ldots, A_n$ are finite sets, then

$$N(A_1 \times A_2 \times \cdots \times A_n) = N(A_1)N(A_2) \cdots N(A_n). \quad (1.5)$$

PROOF   Note first that, according to Theorem 1.5, relation (1.5) holds for $n = 2$. Suppose that (1.5) holds for $n - 1$, that is,

$$N(A_1 \times A_2 \times \cdots \times A_{n-1}) = N(A_1)N(A_2) \cdots N(A_{n-1}).$$

It will be shown that (1.5) holds also for n. For this reason put $A = A_1 \times A_2 \times \cdots \times A_{n-1}$ and $B = A_n$, whence

$$A \times B = (A_1 \times A_2 \times \cdots \times A_{n-1}) \times A_n = A_1 \times A_2 \times \cdots \times A_{n-1} \times A_n.$$

Thus, according to Theorem 1.5, $N(A \times B) = N(A)N(B)$, and the hypothesis that (1.5) holds for $n - 1$, it follows that

$$N(A_1 \times A_2 \times \cdots \times A_n) = N(A_1 \times A_2 \times \cdots \times A_{n-1})N(A_n)$$
$$= N(A_1)N(A_2) \cdots N(A_{n-1})N(A_n)$$

and according to the principle of mathematical induction, (1.5) holds for every integer $n \geq 2$. ∎

Example 1.4 Binary number system

In the binary number system each number is represented by a binary sequence of 0s and 1s. For example, the numbers 5 and 11, which are expressed in terms of powers of 2 as $5 = 1 \cdot 2^2 + 0 \cdot 2^1 + 1 \cdot 2^0$ and $11 = 1 \cdot 2^3 + 0 \cdot 2^2 + 1 \cdot 2^1 + 1 \cdot 2^0$, are represented by the binary sequences $(1, 0, 1)$ and $(1, 0, 1, 1)$, respectively. Note that, with the exception of the number 0, which is represented by the one digit sequence (0), all the other binary sequences start with digit 1. Calculate the number of four-digit binary sequences.

The first digit of a four-digit binary sequence is necessarily 1. Further, a four-digit binary sequence $(1, a_1, a_2, a_3)$ uniquely corresponds to an ordered triple (a_1, a_2, a_3), with $a_i \in A_i = \{0, 1\}$, $i = 1, 2, 3$. Thus, the set B_4 of four-digit sequences is equivalent to the set $A_1 \times A_2 \times A_3$, of ordered triples (a_1, a_2, a_3), with $a_i \in A_i = \{0, 1\}$, $i = 1, 2, 3$ and by Lemma 1.1 and Corollary 1.3,

$$N(B_4) = N(A_1 \times A_2 \times A_3) = N(A_1)N(A_2)N(A_3) = 2^3.$$

Clearly, the 8 four-digit binary sequences are the following

$$(1, 0, 0, 0), \ (1, 0, 0, 1), \ (1, 0, 1, 0), \ (1, 0, 1, 1),$$

$$(1, 1, 0, 0), \ (1, 1, 0, 1), \ (1, 1, 1, 0), \ (1, 1, 1, 1)$$

and represent the integers 8, 9, 10, 12, 13, 14 and 15. ⬚

Example 1.5 Factors of a positive integer

Evaluate the number of positive integers that are factors of the number 300.

Note first that the number 300 is expressed as a product of prime numbers as

$$300 = 2^2 \cdot 3 \cdot 5^2.$$

Each factor of this number is of the form

$$2^{i_1} \cdot 3^{i_2} \cdot 5^{i_3},$$

where $i_1 \in A_1 = \{0, 1, 2\}, i_2 \in A_2 = \{0, 1\}$ and $i_3 \in A_3 = \{0, 1, 2\}$. Therefore, by (1.5), the number of different factors of the number 300 equals

$$N(A_1 \times A_2 \times A_3) = N(A_1)N(A_2)N(A_3) = 3 \cdot 2 \cdot 3 = 18.$$

The set of these 18 factors is

$$\{1, 2, 3, 4, 5, 6, 10, 12, 15, 20, 25, 30, 50, 60, 75, 100, 150, 300\}. \qquad \square$$

An extension of expression (1.4) to a subset S_n of the n-fold Cartesian product Ω^n is given in the following corollary.

COROLLARY 1.4

Let S_n be a subset of elements $(\omega_1, \omega_2, \dots, \omega_n)$ of the n-fold Cartesian product Ω^n, of a finite universal set Ω with itself. Specifically, assume that the first coordinate ω_1 can be selected from a set $A = \{a_1, a_2, \dots, a_{k_1}\}$ of k_1 elements and, for each selection $\omega_1 = a_{i_1}$, the second coordinate ω_2 can be selected from a set $A_{i_1} = \{a_{i_1,1}, a_{i_1,2}, \dots, a_{i_1,k_2}\}$ of k_2 elements, $i_1 = 1, 2, \dots, k_1$, and so on. Finally, assume that for each selection $\omega_1 = a_{i_1}, \omega_2 = a_{i_1,i_2}, \dots, \omega_{n-1} = a_{i_1,i_2,\dots,i_{n-1}}$, the last coordinate ω_n can be selected from a set $A_{i_1,i_2,\dots,i_{n-1}} = \{a_{i_1,i_2,\dots,i_{n-1},1}, a_{i_1,i_2,\dots,i_{n-1},2}, \dots, a_{i_1,i_2,\dots,i_{n-1},k_n}\}$ of k_n elements, $i_r = 1, 2, \dots, k_r$, $r = 1, 2, \dots, n-1$. Then

$$N(S_n) = N(A)N(A_{i_1}) \cdots N(A_{i_1,i_2,\dots,i_{n-1}}) = k_1 k_2 \cdots k_n. \qquad (1.6)$$

REMARK 1.6   Relation (1.6) is often referred to as the *multiplication principle* and can be restated as follows: if an element (object) ω_1 can be selected in k_1 ways and for each of these ways an element ω_2 can be selected in k_2 ways and so on and for each of these ways an element ω_n can be selected in k_n ways, then all the elements (objects) ω_1 and ω_2 and $\cdots$ and ω_n can be selected (sequentially) in $k_1 k_2 \cdots k_n$ ways.

In many applications the set of different selections of ω_i cannot be identified in advance, but only after the selections of $\omega_1, \omega_2, \dots, \omega_{i-1}$. This does not cause any difficulty in the application of the multiplication principle since the only requirement is the cardinality of the set of selections of ω_i. This point is further clarified in the next example.   ∎

Example 1.6 Booking tickets

Suppose that there are n stations on a railway line. How many different kinds of tickets have to be provided so that booking is possible from any station to any other station?

Let us represent the route from station s_i to station s_j by the pair $(s_i, s_j), i \neq j$. The set of different kinds of tickets to be provided is equal to the set S_2 of ordered pairs (a, b) of different stations. Note that the set of different selections of a is $A = \{s_1, s_2, \dots, s_n\}$, while the set of different selections of b is not known in advance, since b has to be different from a. After the selection $a = s_i$ from the set A, the set of different selections of b is $B_i = A - \{s_i\} = \{s_1, s_2, \dots, s_{i-1}, s_{i+1}, \dots, s_n\}$. Irrespective of the possibility of identifying the set B_i, its cardinality is in any case equal to $N(B_i) = n - 1$ and hence formula (1.4) can be applied. Therefore

$$N(S_2) = N(A)N(B_i) = n(n-1),$$

which is the required number of different kinds of tickets.   □

Example 1.7   The number of subsets of a finite set

Consider a finite set $W_n = \{w_1, w_2, \dots, w_n\}$ and let $\mathcal{A}_n = \mathcal{P}(W_n)$ be the set of subsets of W_n. The number $s_n = N(\mathcal{A}_n)$ may be determined as follows.

Note first that, to any subset U of W_n there corresponds an ordered n-tuple $(a_1, a_2, \dots, a_n)$ such that $a_j = 0$ if $w_j \notin U$, and $a_j = 1$ if $w_j \in U$ for $j = 1, 2, \dots, n$. In this way, the null set $\emptyset \subseteq W_n$ corresponds to the ordered n-tuple $(0, 0, \dots, 0)$, with all components zero; the one-element set $\{w_1\} \subseteq W_n$ corresponds to the ordered n-tuple $(1, 0, \dots, 0)$, with the first component equal to one and the rest components zero; and the set $W_n \subseteq W_n$ corresponds to the ordered n-tuple $(1, 1, \dots, 1)$, with all components equal to one. This correspondence is one to one and, according to Lemma 1.1, the number s_n of subsets of W_n is equal to the number of ordered n-tuples $(a_1, a_2, \dots, a_n)$, with $a_j \in A_j = \{0, 1\}$, which, in turn, is equal to the number of elements of the Cartesian product $A_1 \times A_2 \times \dots \times A_n$, with $A_j = \{0, 1\}, j = 1, 2, \dots, n$. Then, according to formula (1.5), this number is equal to $s_n = 2^n$.   □

Example 1.8   Sum of terms of a geometric progression

The sum of the first n terms of a geometric progression with the first term equal to 1 and proportion equal to 2 may be derived combinatorially as follows :

Let $\mathcal{E}_n$ be the set of nonempty subsets of a finite set $W_n = \{w_1, w_2, \dots, w_n\}$ and C_k the set of subsets of the finite set $W_k = \{w_1, w_2, \dots, w_k\}$, each of which contains w_k, $k = 1, 2, \dots, n$. Then $C_1, C_2, \dots, C_n$ are pairwise disjoint and $C_1 + C_2 + \dots + C_n = \mathcal{E}_n$. Note that, to each subset of W_k containing the element w_k, corresponds a subset of W_{k-1} (without any restriction), from which it is obtained by adding w_k, $k = 2, 3, \dots, n$. This correspondence is one to one and hence the number $N(C_k)$ is equal to the number of subsets of W_{k-1}, which, in turn, according to Example 1.7, is equal to $N(C_k) = 2^{k-1}$, $k = 2, 3, \dots, n$. Especially, the number of subsets of $W_1 = \{w_1\}$ containing the element w_1 is equal to $N(C_1) = 1$. Further, the number of nonempty subsets of W_n is equal to $N(\mathcal{E}_n) = 2^n - 1$. Therefore, on using (1.1), it follows that

$$1 + 2 + 2^2 + \dots + 2^{n-1} = 2^n - 1,$$

which is the required expression.   ∎

Before concluding this section on the basic counting principles, it is worth noting that, in several enumeration problems, the number of configurations satisfying specified conditions can only be expressed recursively. In addition, even when the direct expression of this number in a closed form is possible, a recurrence relation is useful at least for tabulation purposes. The notion of a *recurrence relation*, which is extensively used in the sequel, is briefly introduced here. (The basic methods of solving recurrence relations are presented in Chapter 7.)

Consider a sequence of numbers a_n, $n = 0, 1, \dots$. In combinatorics a_n may represent the number of subsets of a finite set $W_n = \{w_1, w_2, \dots, w_n\}$ that satisfy a set of specified conditions. In general, we may assume that

$$a_{n+r} = F(n, a_n, a_{n+1}, \dots, a_{n+r-1}), \quad n = 0, 1, \dots.$$

This equation, in which the term a_{n+r} is expressed as a function of the preceding r terms, $a_n, a_{n+1}, \dots, a_{n+r-1}$, of the sequence, is called *recurrence relation of order r*. The notion of a recurrence relation is also introduced in the case of a double-index sequence $a_{n,k}$, $n = 0, 1, \dots$, $k = 0, 1, \dots$. So, the equation

$$a_{n+r,k+s} = F(n, k, a_{n,k}, a_{n,k+1}, a_{n+1,k}, \dots, a_{n+r-1,k+s}, a_{n+r,k+s-1})$$

is called *recurrence relation of order (r, s)*. Notice that r is the order of this recurrence relation with respect to the first index (variable) and s is its order with respect to the second index (variable). In this case the term $a_{n+r,k+s}$ is expressed as a function of the $(n+1)(k+1) - 1 = nk + n + k$ terms $a_{n,k}, a_{n,k+1}, a_{n+1,k}, \dots, a_{n+r-1,k+s}, a_{n+r,k+s-1}$ of the double-index sequence. From the computation point of view, the tabulation of the number a_n, for $n = 0, 1, \dots$, or the number $a_{n,k}$, for $n = 0, 1, \dots$, $k = 0, 1, \dots$, can be done more easily step-by-step by using the corresponding recurrence relation. For this purpose, the knowledge of the r *initial conditions* (values) $a_0, a_1, \dots, a_{r-1}$, in the first case, and of the $r + s$ *initial conditions* (sequences) $a_{0,k}, a_{1,k}, \dots, a_{r-1,k}$ and $a_{n,0}, a_{n,1}, \dots, a_{n,s-1}$, in the second case, are required. The initial conditions guarantee the uniqueness of the solution of the recurrence relation. In this section only a very simple example of a recurrence relation is discussed.

Closely connected with a recurrence relation is the (*finite*) *difference equation*, which is defined as follows. The first order finite difference, with increment h, of a function $y = f(x)$ denoted by $\Delta_h f(x)$, is defined by $\Delta_h f(x) = f(x + h) - f(x)$. Recursively, the r-th order finite difference of $f(x)$ is defined by $\Delta_h^r f(x) = \Delta_h[\Delta_h^{r-1} f(x)]$, $r = 2, 3, \dots$. Introducing the displacement (shift) of $f(x)$, which is defined by $E_h f(x) = f(x + h)$ and recursively by $E_h^r f(x) = E_h[E_h^{r-1} f(x)] = f(x + rh)$, $r = 2, 3, \dots$, it follows that $\Delta_h f(x) = E_h f(x) - If(x)$, where $If(x) = f(x)$. An equation of the form

$$E_h^r f(x) = F(x, f(x), E_h f(x), \dots, E_h^{r-1} f(x)), \quad x \in R$$

or equivalently of the form

$$\Delta_h^r f(x) = G(x, f(x), \Delta_h f(x), \dots, \Delta_h^{r-1} f(x)) = 0, \quad x \in R$$

is called *difference equation of order r*. In the particular case where the function $y = f(x)$ is defined only on a countable set of points $\{x_0, x_1, \dots, x_n, \dots\}$ which, in the applications of the calculus of finite differences, are usually equidistant, $x_n = x_0 + nh$, $n = 0, 1, \dots$, the transformation $z_n = (x_n - x_0)/h$, $n = 0, 1, \dots$ is used. So, the function f is transformed to the function g, with $g(n) = f(x_0 + nh)$, which is defined on the set $\{0, 1, \dots\}$ of nonnegative integers. In this particular case, using the sequence

$$y_n = g(n) = f(x_0 + nh), \quad n = 0, 1, \dots,$$

and since $E_h^j(x_n) = f(x_n + jh) = g(n + j) = y_{n+j}$, $j = 1, 2, \dots$, the difference equation, with $x = x_n$, may be written as

$$y_{n+r} = F(n, y_n, y_{n+1}, \dots, y_{n+r-1}), \quad n = 0, 1, \dots,$$

which is a recurrence relation of order r. Note that the method of solving a finite difference equation is the same as that of solving the recurrence relation even when the function $y = f(x)$ is defined for every $x \in R$. In the last case an investigation of the solution is required.

Example 1.9 **A recurrence relation for the number of subsets of a finite set**

As in Example 1.7, consider a finite set $W_n = \{w_1, w_2, \dots, w_n\}$ and let $A_n = P(W_n)$ be the set of subsets of W_n. The number $s_n = N(A_n)$ may be deduced recursively as follows.

The set $A_{n+1} = P(W_{n+1})$ of subsets of $W_{n+1} = \{w_1, w_2, \dots, w_n, w_{n+1}\}$ can be divided into the following two disjoint subsets: the set A of subsets of W_{n+1}, each of which does not include the element w_{n+1} and the set B of subsets of W_{n+1}, each of which includes the element w_{n+1}. Hence $A_{n+1} = A + B$ and, by the addition principle, $N(A_{n+1}) = N(A) + N(B)$. Clearly, $A = A_n$ and so $N(A) = N(A_n)$. Further, to each subset B that belongs to B there uniquely corresponds the subset $A = B - \{w_{n+1}\}$ that belongs to A_n and so, by Lemma 1.1, $N(B) = N(A_n)$. Consequently $N(A_{n+1}) = 2N(A_n)$ and

$$s_{n+1} = 2s_n, \quad n = 0, 1, \dots,$$

with $s_0 = 1$, which is the number of subsets of the null set. This is a homogenous recurrence relation of the first order with constant coefficients. Iterating (applying repeatedly) it, we find

$$s_{n+1} = 2s_n = 2(2s_{n-1}) = 2^2 s_{n-1} = \cdots = 2^{n+1} s_0$$

and, since $s_0 = 1$, we conclude that $s_n = 2^n$. □

1.4 DISCRETE PROBABILITY

The theory of probability is concerned with the study of mathematical models, known as *stochastic models*, which are used in explaining random or stochastic phenomena or experiments. The basic characteristic of these experiments is that the conditions under which they are performed and the values of various quantities appearing in them do not predetermine the outcome, but do predetermine the set of possible outcomes. The element of randomness lies in the inability of predetermining the outcome of random phenomena or experiments.

The set Ω of the possible outcomes of a random phenomenon or experiment is called *sample space* and the elements ω of Ω are called *sample points*. It should be noted that it is possible to define more than one set of possible outcomes for each random phenomenon or experiment and, according to the requirements of the specific problem, the more appropriate of these is chosen as the sample space. The inappropriate choice of the sample space leads to many paradoxes. The sample space may be finite or countably infinite or uncountable. For the cases of finite or more generally countable sample spaces, on which this presentation concentrates, every subset A of Ω is called an event. An event $A = \{\omega\}$ containing only one element of Ω is called an elementary or simple event.

The *classical definition of probability* was first expressed by De Moivre (1711) as follows: the probability of the occurrence of an event is the fraction whose numerator is the number of possibilities *favorable* for the occurrence of the event and the denominator is the number of all possibilities, provided that all possibilities are equally probable. This condition is essential because, otherwise, by considering the two possibilities of the occurrence and the non-occurrence of an event, it can be concluded that its probability is one half. This is not true in general since these two cases are not always equally probable.

The concept of equally probable cases should be defined independently of the notion of (the measure) of probability; otherwise, in the classical definition of probability, there would be a vicious circle. This is achieved by the adoption of *the principle of the want of sufficient reason*. So if, according to all available data, no reason is known for regarding any of the possibilities more or less probable than any other, then all possibilities are regarded as equally probable. It should be noted that the classical definition of probability applies only on finite sample spaces.

The foundation of the theory of probability based on the classical definition of probability is attributed to Laplace (1812). Consider a finite sample space $\Omega = \{\omega_1, \omega_2 \ldots, \omega_N\}$, whose elements (sample points, possibilities) are, according to the principle of the want of sufficient reason, equally probable and an event $A \in \mathcal{P}(\Omega)$. The *probability* of A, denoted by $P(A)$, is given by the expression

$$P(A) = \frac{N(A)}{N},$$

where $N(A)$ is the number of elements of A and $N \equiv N(\Omega)$ is the number of elements of the sample space Ω. Note that the condition of equally probable sample points (possibilities) is then expressed by

$$P(\{\omega_1\}) = P(\{\omega_2\}) = \cdots = P(\{\omega_N\}) = \frac{1}{N}.$$

According to the definition of the classical (uniform) probability, the calculation of the probability $P(A)$ of an event A in a finite sample space Ω whose elements are equally probable is a purely combinatorial problem of counting the numbers $N(A)$ and N of certain configurations.

REMARK 1.7   It is worth presenting the most important properties of the classical probability that subsequently inspired the suitable choice of axioms in the axiomatic foundation of the theory of probability.

The set function $P(\cdot)$, which, in the case of a finite sample space Ω whose elements are equally probable, assigns the number $P(A) = N(A)/N$ to each event $A \in \mathcal{P}(\Omega)$ is (a) *nonnegative*: $P(A) \geq 0$ for every event $A \in \mathcal{P}(\Omega)$, (b) *normalized*: $P(\Omega) = 1$ and (c) *finitely additive*: $P(A + B) = P(A) + P(B)$ for any disjoint (mutually exclusive) events $A, B \in \mathcal{P}(\Omega)$.

Properties (a) and (c), which follow directly from the definition of classical probability and the corresponding properties of the set function $N(\cdot)$, made $P(\cdot)$ a finitely additive measure on $\mathcal{P}(\Omega)$ (see Remark 1.3). Property (b), which is a direct consequence of the definition of classical probability, distinguishes the probability from other measures.

Note that, from (c) and the principle of mathematical induction, the next expression is deduced:

$$P(A_1 + A_2 + \cdots + A_n) = P(A_1) + P(A_2) + \cdots + P(A_n),$$

for any pairwise disjoint events $A_1, A_2, \ldots, A_n \in \mathcal{P}(\Omega)$. Also, if W_i, $i = 1, 2, \ldots, n$ are finite sample spaces, each with equally probable elements, and $A_i \in \mathcal{P}(W_i)$, $i = 1, 2, \ldots, n$, then from (1.5) it follows that

$$P(A_1 \times A_2 \times \cdots \times A_n) = P(A_1)P(A_2) \cdots P(A_n),$$

provided the elements of the sample space $\Omega = W_1 \times W_2 \times \cdots \times W_n$ are also equally probable. ∎

An extension of the classical definition of probability, in the case of a finite sample space Ω whose elements are not necessarily equally probable, constitutes the next definition of probability:

A set function $P(\cdot)$ defined on the set of events $\mathcal{P}(\Omega)$, assuming real values and satisfying properties (a), (b) and (c) is called *probability* (or *measure of probability*). These properties are quite general and, for the calculation of the probability of any event $A \in \mathcal{P}(\Omega)$, the knowledge of the probabilities $p_i = P(\{\omega_i\})$ of the elementary events $\{\omega_i\}$, $i = 1, 2, \ldots, N$ is required. Indeed, if $A = \{\omega_{i_1}, \omega_{i_2}, \ldots, \omega_{i_k}\}$, then, with $A_{i_r} = \{\omega_{i_r}\}$, $r = 1, 2, \ldots, k$, it follows that

$$A = A_{i_1} + A_{i_2} + \cdots + A_{i_k}, \quad A_{i_r} \cap A_{i_s} = \emptyset, \quad r \neq s,$$

and so

$$P(A) = \sum_{r=0}^{k} P(\{\omega_{i_r}\}).$$

In the special case in which $P(\{\omega_1\}) = P(\{\omega_2\}) = \cdots = P(\{\omega_N\}) = 1/N$, it reduces to

$$P(A) = \frac{k}{N},$$

with $k = N(A)$, which is the classical definition of probability.

Three useful properties of the probability that follow directly from the classical definition of probability and Theorem 1.4 are quoted here for easy reference in the following chapters.

The probability of the complementary event A' of A may be expressed as

$$P(A') = 1 - P(A).$$

For any events A and B, the next relation holds

$$P(A - B) = P(A) - P(A \cap B)$$

and particularly for $B \subseteq A$,

$$P(A - B) = P(A) - P(B).$$

Examples with applications of the classical definition of probability are given in the following chapters after the introduction of the combinatorial concepts necessary for their presentation.

1.5 SUMS AND PRODUCTS

A finite sum of n terms, $a_1, a_2, \ldots, a_n$, is denoted by

$$s_n \equiv a_1 + a_2 + \cdots + a_n$$

or briefly by

$$s_n = \sum_{j=1}^{n} a_j.$$

The number 1 is called the lower limit and the number n the upper limit of the sum. Note that the lower limit can be any integer $r < n$. Clearly, the value of the sum

$$s_{r,n} = \sum_{j=r}^{n} a_j$$

is completely determined by the general term a_j $(j = r, r + 1, \ldots, n)$ and the limits r and n and does not depend on the (bound) variable j even though it occurs in its expression. Thus, the variable j can be replaced by another variable i without any effect on the value of the sum:

$$s_{r,n} = \sum_{j=r}^{n} a_j = \sum_{i=r}^{n} a_i = a_r + a_{r+1} + \cdots + a_n.$$

More generally, the transformation $i = j + m$, with inverse $j = i - m$, where m is a given integer, may be used. In this case, the general term a_j $(j = r, r + 1, \ldots, n)$ becomes $b_i = a_{i-m}$ $(i = r + m, r + m + 1, \ldots, n + m)$ and the sum $s_{r,n}$ is transformed to

$$s_{r,n} = \sum_{j=r}^{n} a_j = \sum_{i=r+m}^{n+m} a_{i-m} = \sum_{i=r+m}^{n+m} b_i$$

and particularly for $m = -r$

$$s_{r,n} = \sum_{j=r}^{n} a_j = \sum_{i=0}^{n-r} a_{i+r} = \sum_{i=0}^{n-r} b_i.$$

This brief notation of simple (with respect to one variable) sums is also used to represent double (with respect to two variables) sums and generally multiple sums. So, the sum of the terms $a_{i,j}$, $j = 1, 2, \ldots, k$, $j = 1, 2, \ldots, n$, is denoted by

$$S_{k,n} = \sum_{j=1}^{n} \sum_{i=1}^{k} a_{i,j}$$

and, more generally, the sum of the terms $a_{j_1,j_2,\ldots,j_r}$, $j_k = 1,2,\ldots,n_k$, $k = 1,2,\ldots,r$, is denoted by

$$S_{n_1,n_2\ldots,n_r} = \sum_{j_r=1}^{n_r} \cdots \sum_{j_2=1}^{n_2} \sum_{j_1=1}^{n_1} a_{j_1,j_2,\ldots,j_r}.$$

REMARK 1.8   If the set J of the values of the variable j and, more generally, if the sets $J_1, J_2, \ldots, J_r$ of the values of the variables $j_1, j_2, \ldots, j_r$, respectively, are not sets of consecutive integers, the following notation of the sums is adopted:

$$s = \sum_{j \in J} a_j, \quad S = \sum_{j_r \in J_r} \cdots \sum_{j_2 \in J_2} \sum_{j_1 \in J_1} a_{j_1,j_2,\ldots,j_r}.$$

When the sets J and $J_1, J_2, \ldots, J_r$ are defined by the conditions C and $C_1, C_2, \ldots, C_r$, respectively, the following notation of the sums may be used:

$$s = \sum_{C} a_j, \quad S = \sum_{C_r} \cdots \sum_{C_2} \sum_{C_r} a_{j_1,j_2,\ldots,j_r}.$$

According to the last notation, the sums

$$s_n = \sum_{j=1}^{n} a_j, \quad S_{k,n} = \sum_{j=1}^{n} \sum_{i=1}^{k} a_{i,j}$$

may, equivalently, be written as:

$$s_n = \sum_{1 \leq j \leq n} a_j, \quad S_{k,n} = \sum_{1 \leq j \leq n} \sum_{1 \leq i \leq k} a_j.$$

If the expression of the conditions C and $C_1, C_2, \ldots, C_r$ is not very simple, then it is preferable to avoid writing it underneath the summation sign. In this case the conditions are written after the sum in the form of phrases like "where the summation is extended over all j or $j_1, j_2, \ldots, j_r$ such that the conditions C or $C_1, C_2, \ldots, C_r$ are satisfied."   ∎

REMARK 1.9   If in the sum

$$s_n = \sum_{j=1}^{n} a_j$$

the general term is constant (independent of the subscript j): $a_j = a$, $j = 1, 2, \ldots, n$, then

$$s_n = \sum_{j=1}^{n} a = a + a + \cdots + a = na.$$

The special case $a = 1$ is particularly noted:

$$\sum_{j=1}^{n} 1 = n.$$

In general,

$$\sum_{j \in J} 1 = N(J)$$

and

$$\sum_{j_r \in J_r} \cdots \sum_{j_2 \in J_2} \sum_{j_1 \in J_1} 1 = N(J_1) N(J_2) \cdots N(J_r),$$

for any finite sets $J, J_1, J_2, \ldots, J_r$. ∎

Some basic properties of finite sums, which constitute simple extensions of the corresponding properties of the sum and product of two real numbers, are presented in the next theorem.

THEOREM 1.6

$$\sum_{j=1}^{n} b a_j = b \sum_{j=1}^{n} a_j,$$

$$\sum_{j=1}^{n} a_j + \sum_{i=1}^{n} b_i = \sum_{j=1}^{n} (a_j + b_j),$$

$$\sum_{j=1}^{n} a_j \sum_{i=1}^{k} b_i = \sum_{j=1}^{n} \sum_{i=1}^{k} a_j b_i = \sum_{i=1}^{k} \sum_{j=1}^{n} a_j b_i,$$

$$\sum_{j=1}^{n} \sum_{i=1}^{k} a_{i,j} = \sum_{i=1}^{k} \sum_{j=1}^{n} a_{i,j}.$$

REMARK 1.10 As regards the possibility of changing the order of summation in double (or multiple) sums, special care should be given when the set of values of one variable depends on the set of values of the other variable. A characteristic example of this nature constitutes the sum

$$\sum_{j=1}^{n} \sum_{i=1}^{j} a_{i,j},$$

where, for a given value of the variable $j \in J_1 = \{j : 1 \leq j \leq n\}$, the set $I_1 = \{i : 1 \leq i \leq j\}$ of values of the variable i depends on the value $j \in J_1$. This dependence can be reversed without altering the set $K = \{(i,j) : 1 \leq i \leq j \leq n\}$, of values of the pair (i,j) of variables in the double sum. Indeed, for a given value of the variable $i \in I_2 = \{i : 1 \leq i \leq n\}$, the set of values of the variable j is $J_2 = \{j : i \leq j \leq n\}$. Hence

$$\sum_{j=1}^{n}\sum_{i=1}^{j} a_{i,j} = \sum_{(i,j)\in K} a_{i,j} = \sum_{i=1}^{n}\sum_{j=i}^{n} a_{i,j}.$$

As a second example, the relation

$$\sum_{j=1}^{n}\sum_{i=1}^{n-j+1} a_{i,j} = \sum_{i=1}^{n}\sum_{j=1}^{n-i+1} a_{i,j}$$

can be similarly established. ∎

A finite product of n terms, $a_1, a_2, \ldots, a_n$, is denoted by

$$p_n \equiv a_1 a_2 \cdots a_n$$

or briefly by

$$p_n = \prod_{j=1}^{n} a_j.$$

Clearly, all notational and other remarks previously made on finite sums are applied, with the necessary changes, to finite products as well. The brief notation of simple products is also used for the presentation of double and more generally multiple products. So, the product of the terms $a_{i,j}$, $i = 1, 2, \ldots, k$, $j = 1, 2, \ldots, n$, is denoted by

$$P_{k,n} = \prod_{j=1}^{n}\prod_{i=1}^{k} a_{i,j}$$

and more generally, the product of the terms $a_{j_1,j_2,\ldots,j_r}$, $j_k = 1, 2, \ldots, n_k$, $k = 1, 2, \ldots, r$, is denoted by

$$P_{n_1,n_2,\ldots,n_r} = \prod_{j_r=1}^{n_r} \cdots \prod_{j_2=1}^{n_2}\prod_{j_1=1}^{n_1} a_{j_1,j_2,\ldots,j_r}.$$

It can be easily shown that

$$\prod_{j=1}^{n}\prod_{i=1}^{k} a_{i,j} = \prod_{i=1}^{k}\prod_{j=1}^{n} a_{i,j},$$

$$\prod_{j=1}^{n}\prod_{i=1}^{j} a_{i,j} = \prod_{i=1}^{n}\prod_{j=i}^{n} a_{i,j},$$

$$\prod_{j=1}^{n}\prod_{i=1}^{n-j+1} a_{i,j} = \prod_{i=1}^{n}\prod_{j=1}^{n-i+1} a_{i,j}.$$

The union of n sets, $A_1, A_2, \ldots, A_n$, is briefly denoted by

$$\bigcup_{i=1}^{n} A_i \equiv A_1 \cup A_2 \cup \cdots \cup A_n$$

and the intersection of these sets by

$$\bigcap_{i=1}^{n} A_i \equiv A_1 \cap A_2 \cap \cdots \cap A_n$$

and both share properties analogous to those of finite sums and products.

Example 1.10 Sum of terms of an arithmetic progression
The general term of an arithmetic progression is $a_j = a + jb$, $j = 1, 2, \ldots, n$.
The sum

$$s_n(a, b) = \sum_{j=1}^{n} a_j, \quad a_j = a + jb, \quad j = 1, 2, \ldots, n,$$

may be evaluated as follows. Using the transformation $i = n - j + 1$, with inverse
$j = n - i + 1$, the general term $a_j = a + jb$ $(j = 1, 2, \ldots, n)$ takes the form
$b_i \equiv a_{n-i+1} = \{a + (n+1)b\} - ib$ $(i = 1, 2, \ldots, n)$. Therefore

$$s_n(a, b) = \sum_{i=1}^{n} b_i, \quad b_i = \{a + (n+1)b\} - ib, \quad i = 1, 2, \ldots, n$$

and

$$2s_n(a, b) = \sum_{j=1}^{n} a_j + \sum_{i=1}^{n} b_i = \sum_{j=1}^{n}(a_j + b_j).$$

Since $a_j + b_j = 2a + (n+1)b$, $j = 1, 2, \ldots, n$, it follows that

$$2s_n(a, b) = n\{2a + (n+1)b\}$$

and

$$s_n(a, b) = \sum_{j=1}^{n}(a + jb) = \frac{n\{2a + (n+1)b\}}{2}.$$

In particular, for $a = 0$ and $b = 1$, the last expression reduces to

$$s_n = \sum_{j=1}^{n} j = \frac{n(n+1)}{2}. \quad \square$$

Example 1.11
Evaluate the sum

$$c_{n,2} = \sum_{j=1}^{n} j^2.$$

Using the identity

$$(j+1)^3 - j^3 = 3j^2 + 3j + 1,$$

we get

$$3c_{n,2} = 3\sum_{j=1}^{n} j^2 = \sum_{j=1}^{n}(j+1)^3 - \sum_{j=1}^{n} j^3 - 3\sum_{j=1}^{n} j - \sum_{j=1}^{n} 1.$$

Putting in the first sum of the right-hand side $i = j + 1$ and, since (see Example 1.10 and Remark 1.9)

$$\sum_{j=1}^{n} j = \frac{n(n+1)}{2}, \quad \sum_{j=1}^{n} 1 = n,$$

we find

$$3c_{n,2} = \sum_{i=2}^{n+1} i^3 - \sum_{j=1}^{n} j^3 - \frac{3n(n+1)}{2} - n = (n+1)^3 - 1 - \frac{3n(n+1)}{2} - n$$

$$= \frac{n(2n^2 + 3n^2 + 1)}{2} = \frac{n(n+1)(2n+1)}{2}.$$

Hence

$$c_{n,2} = \sum_{j=1}^{n} j^2 = \frac{n(n+1)(2n+1)}{6}. \quad \square$$

Example 1.12
Evaluate the sum

$$C_n = \sum_{\substack{j=1 \\ i+j=n}}^{n} \sum_{i=1}^{n} ij.$$

Introducing in the inner sum the transformation $k = n-i$, with inverse $i = n-k$, and taking into account the restriction $i + j = n$, it follows that, for a given $j \in \{1, 2, \ldots, n\}$, $k = j$. Then the sum C_n reduces to

$$C_n = \sum_{j=1}^{n} j(n-j) = n \sum_{j=1}^{n} j - \sum_{j=1}^{n} j^2.$$

Since (see Examples 1.10 and 1.11)

$$\sum_{j=1}^{n} j = \frac{n(n+1)}{2}, \quad \sum_{j=1}^{n} j^2 = \frac{n(n+1)(2n+1)}{6},$$

it follows that

$$C_n = \frac{n^2(n+1)}{2} - \frac{n(n+1)(2n+1)}{6} = \frac{n(n^2-1)}{6}. \quad \Box$$

Example 1.13 Sum of terms of a geometric progression

The general term of a geometric progression is $a_j = ab^{j-1}$, $j = 1, 2, \ldots, n$. The sum

$$g_n(a, b) = \sum_{j=1}^{n} a_j = a \sum_{j=1}^{n} b^{j-1}$$

may be evaluated as follows: multiplying both sides of this expression by $(1 - b)$, we get

$$(1 - b)g_n(a, b) = a(1 - b) \sum_{j=1}^{n} b^{j-1} = a \left(\sum_{j=1}^{n} b^{j-1} - \sum_{j=1}^{n} b^j \right).$$

Putting in the second sum of the last equality $i = j+1$, whence $j = i - 1$, we find

$$(1 - b)g_n(a, b) = a \left(\sum_{j=1}^{n} b^{j-1} - \sum_{i=2}^{n+1} b^{i-1} \right) = a(1 - b^n)$$

and so

$$g_n(a, b) = \sum_{j=1}^{n} ab^{j-1} = \frac{a(1 - b^n)}{1 - b}.$$

Note that the sum

$$g(a, b) = \sum_{j=1}^{\infty} ab^{j-1},$$

of an infinite number of terms of a geometric progression $a_j = ab^{j-1}, j = 1, 2, \ldots,$
with $|b| < 1$, may be calculated by using the last result. Indeed, from

$$\sum_{j=1}^{\infty} ab^{j-1} = \lim_{n \to \infty} \sum_{j=1}^{n} ab^{j-1} = \lim_{n \to \infty} \frac{a(1-b^n)}{1-b}$$

and since $\lim_{n \to \infty} b^n = 0$ for $|b| < 1$, it follows that

$$g(a, b) = \sum_{j=1}^{\infty} ab^{j-1} = \frac{a}{1-b}. \qquad \square$$

Example 1.14

Evaluate the double sum

$$S_n(a, b) = \sum\sum_{1 \le i \le j \le n} a^{i-1} b^{j-1}.$$

The set of values of the pair (i, j) of (bound) variables of this sum,

$$K = \{(i, j) : 1 \le i \le j \le n\},$$

may be expressed as follows: for a given value of the variable $j \in J = \{j : 1 \le j \le n\}$, the set of values of the variable i is $I = \{i : 1 \le i \le j\}$ and hence

$$S_n(a, b) = \sum_{(i,j) \in K} a^{i-1} b^{j-1} = \sum_{j \in J} \sum_{i \in I} a^{i-1} b^{j-1} = \sum_{j=1}^{n} \sum_{i=1}^{j} a^{i-1} b^{j-1}.$$

Upon using the formula for the sum of a finite number of terms of a geometric progression (see Example 1.13), we get

$$S_n(a, b) = \sum_{j=1}^{n} \left(b^{j-1} \sum_{i=1}^{j} a^{i-1} \right) = \sum_{j=1}^{n} b^{j-1} \frac{1 - a^j}{1 - a}$$

$$= \sum_{j=1}^{n} \frac{b^{j-1}}{1-a} - \sum_{j=1}^{n} \frac{a(ab)^{j-1}}{1-a}$$

and so

$$S_n(a, b) = \frac{1 - b^n}{(1-a)(1-b)} - \frac{a(1 - a^n b^n)}{(1-a)(1-ab)}. \qquad \square$$

Example 1.15

Evaluate the products

$$p_n = \prod_{j=0}^{n} a^j, \quad P_n = \prod_{j=0}^{n} \prod_{i=0}^{j} a^i b^{j-i}.$$

The first product, on using the formula for the sum of a finite number of terms of an arithmetic progression

$$\sum_{j=1}^{n} j = \frac{n(n+1)}{2},$$

is evaluated as

$$p_n = \prod_{j=0}^{n} a^j = a^{1+2+\cdots+n} = a^{n(n+1)/2}.$$

In a similar way the second product is expressed as

$$P_n = \prod_{j=0}^{n} \prod_{i=0}^{j} a^i b^{j-i} = \prod_{j=0}^{n} a^{j(j+1)/2} b^{j(j+1)/2} = \prod_{j=0}^{n} (ab)^{j(j+1)/2}$$

and, since

$$\sum_{j=1}^{n} \frac{j(j+1)}{2} = \frac{1}{2} \sum_{j=1}^{n} j^2 + \frac{1}{2} \sum_{j=1}^{n} j$$

$$= \frac{n(n+1)(2n+1)}{12} + \frac{n(n+1)}{4} = \frac{n(n+1)(n+2)}{6},$$

is deduced as

$$P_n = \prod_{j=0}^{n} \prod_{i=0}^{j} a^i b^{j-1} = (ab)^{n(n+1)(n+2)/6}. \quad \square$$

1.6 BIBLIOGRAPHIC NOTES

The basic counting principles of addition and multiplication have been presented, along with the necessary elements of set theory. Also, discrete probability was introduced and properties of sums and products have been discussed. Much of this classical material may be found in calculus and probability books. The first two chapters of the book by T. Apostol (1962) are devoted to this subject. The classical book of W. Feller (1968) may also be used for further reading. The book by F. N. David (1962) contains a lot of information on counting and its connection with the probability of games of chance.

1.7 EXERCISES

1. Suppose that three distinguishable (different) coins are tossed. Write down the elements $s = (a, b, c)$ of the set S_3 of possible outcomes and verify that $N(S_3) = 8$. A map F from the set S_3 to the set $N_3 = \{0, 1, 2, 3\}$, which, to each element (outcome) of S_3 corresponds the number of heads registered in it, is useful in probability theory. For each element $s = (a, b, c)$ from the set S_3, write down its image $F(s) = n$ from the set N_3.

2. Suppose that two distinguishable dice (one white and one black) are rolled and let $S_2 = \{(w, b) : w = 1, 2, \ldots, 6, \ b = 1, 2, \ldots, 6\}$ be the set of possible results. (a) Find the number $N(S_2)$ of possible results. Let A_k be the subset of possible results (w, b) with sum $w + b = k$, for $k = 2, 3, \ldots, 12$. (b) Write down the elements of each of the sets $A_2, A_3, \ldots, A_{12}$.

3. Suppose that three distinguishable dice (one white, one black and one red) are rolled and let $S_3 = \{(w, b, r) : w = 1, 2, \ldots, 6, \ b = 1, 2, \ldots, 6, \ r = 1, 2, \ldots, 6\}$ be the set of possible results. (a) Find the number $N(S_3)$ of possible results. Further, let A be the subset of possible results (w, b, r) with sum $w + b + r > 10$ and A' its complement (with respect to S_3). Show that the sets A and A' are equivalent, and conclude that $N(A) = 108$.

4. Suppose that the 5 left-hand gloves $L_5 = \{l_1, l_2, \ldots, l_5\}$ of 5 pairs of gloves $G_5 = \{(l_1, r_1), (l_2, r_2), \ldots, (l_5, r_5)\}$ are kept in drawer D_1, while the corresponding 5 right-hand gloves $R_5 = \{r_1, r_2, \ldots, r_5\}$ are kept in a second drawer D_2. Calculate the total number of ways of choosing one left-hand glove and one right-hand glove. Also, find the number of ways of choosing one left-hand glove and one right-hand glove that do not form a pair.

5. Find the number $N(B_n)$ of n-digit binary sequences and the (total) number $N(T_n)$ of binary sequences of at most n digits.

6. *The Morse code.* A letter of the alphabet is represented by a sequence of telegraphic "dot" and "dash" signals with repetitions allowed. Find the number of letters that can be represented by sequences of at most n symbols.

7. Calculate the number of positive integers with exactly one digit equal to 3 that are (a) greater than or equal to 10 and less than 100 and (b) greater than or equal to 100 and less than 1000. Combining these results, conclude (c) the total number of positive integers with exactly one digit equal to 3 that are less than 1000.

8. Calculate the number of odd positive integers that are (a) greater than 10 and less than or equal to 100, (b) greater than 100 and less than or equal to 1000 and conclude (c) the total number of odd positive integers that are less than or equal to 1000.

9. An alphabet is, in general, a set $L = \{l_1, l_2, \ldots, l_n\}$, of n different letters. An ordered k-tuple of letters from the alphabet L (repetitions allowed) is a k-letter word. Calculate (a) the total number of three-letter words from the alphabet L, (b) the number of three- letter words that do not have repeated letters and (c) the number of three-letter words that contain the letter l_n.

10. Evaluate the number of positive integers that are factors of the following numbers: (a) $2^3 \cdot 3^2 \cdot 5^1$, (b) 900 and (c) 9800.

11. (*Continuation*) Let $N = p_1^{k_1} p_2^{k_2} \cdots p_n^{k_n}$, where p_i, $i = 1, 2, \ldots, n$, are prime numbers and k_i, $i = 1, 2, \ldots, n$, are positive integers. Show that the number of positive integers that are factors of N is equal to $A(k_1, k_2, \ldots, k_n) = (k_1 + 1)(k_2 + 1) \cdots (k_n + 1)$.

12. An examination sheet contains n multiple-choice questions. A list of four alternative answers, among which only one is correct, is provided to each question. A student taking such an exam answers each question by marking one of the given answers. Find the number of different ways of answering the examination.

13. *Soccer's results foresight.* Each card of soccer's results foresight contains 13 matches. The ordered pairs of teams are written on the card in horizontal rows with one pair in each row. The foresights are marked with the symbols 1, X and 2 for win, tie and defeat of the home (first) team. A complete series of foresights consists of 13 symbols put in a column, one for each match. (a) Find the number of different columns that can be written. (b) If one specific symbol is used to mark each of 6 suitably chosen matches, two specific symbols are used to mark each of 5 given matches and all three symbols are used to mark each of the remaining 2 matches, how many different columns can be written?

14. In Greece, for the registration of automobiles, three letters and a four-digit number are used. The 14 letters A, B, E, Z, H, I, K, M, N, O, P, T, Y, X, common in the Greek and Latin alphabets, are used. Find the number of automobiles that can be registered with this system.

15. In a telegraphic station n different telegraphs are to be distributed to k operators for expedition. Find the number of different distributions.

16. Find the number of divisions of a finite set of n elements in (a) two subsets, (b) k subsets.

17. Show that the number of maps f of the set $X = \{x_1, x_2, \ldots, x_n\}$ into the set $Y = \{y_1, y_2, \ldots, y_k\}$ is equal to k^n.

18*. *The pigeonhole principle.* Show that if $n+1$ objects are distributed into n cells, then at least one cell contains at least two objects. Using this principle, prove that, if $K = \{k_1, k_2, \ldots, k_{n+1}\} \subseteq \{1, 2, \ldots, 2n\}$, then two integers k_i and k_j, belonging to K, exist, such that each of these divides the other.

19*. (*Continuation*). Let $k_1, k_2, \ldots, k_n$ be positive integers. Prove that two subscripts i and j, with $1 \leq i < j \leq n$, exist, such that the sum $k_{i+1} + k_{i+2} + \cdots + k_j$ is divisible by n.

20*. (*Continuation*). Show that among any six numbers chosen from the set $\{1, 2, \ldots, 10\}$ there are at least two such that one of them is divisible by the other.

Chapter 2

PERMUTATIONS AND COMBINATIONS

2.1 INTRODUCTION

The notions of permutations, combinations and partitions of finite sets are very basic in the development of enumerative combinatorics; in fact, for a long time the aim of combinatorial analysis was to enumerate such permutations, combinations and partitions. Thus, an introduction to combinatorics naturally begins with a thorough study of these notions. After introducing the notions of permutations and combinations of a finite set in the first two sections of this chapter, a host of emerging enumerative problems is explored. The permutations and combinations of a finite set without and with repetition are enumerated. Also, the number of permutations with a specified number of repetitions, for each element of the finite set, is derived. In addition, recurrence relations for these numbers are deduced. The divisions of a finite set into subsets, which constitute an extension of combinations, are then enumerated. Further, the enumeration of partitions of a finite set, which is related to that of divisions, is treated in the same section.

In a separate section, the problem of counting the number of integer solutions of a linear equation with unit coefficients is reduced to a problem of enumerating combinations. Some basic elements of enumeration of lattice paths, related to the enumeration of certain combinations, are presented. The reflection principle, which facilitates the enumeration of lattice paths, is demonstrated. Moreover, the famous ballot problem that led to the development of lattice paths is treated. The last section of this chapter is devoted to discussion of several applications in discrete probability and statistics. Specifically, the classical probabilistic problems of the most beneficial bet and the distribution of shares, with which a systematic study of enumeration of permutations and combinations began, are discussed. In addition, the Maxwell-Boltzman, Bose-Einstein and Fermi-Dirac stochastic models in statistical mechanics are briefly examined.

2.2 PERMUTATIONS

The concept of an ordered pair and generally of an ordered n-tuple, necessary for the definition of the Cartesian product of sets, has already been introduced in Section 1.2.2. The next definition of a permutation is based on this concept.

DEFINITION 2.1   *Let $W_n = \{w_1, w_2, \dots, w_n\}$, a finite set of n elements. An ordered k-tuple $(a_1, a_2, \dots, a_k)$, with $a_r \in W_n$, $r = 1, 2, \dots, k$, is called k-permutation of the set W_n or simply k-permutation of n.*

Note that the elements (components) of a k-permutation of n may or may not be different elements of W_n. For the first case, the term k-*permutation of n* is preserved, while for the second and when unrestricted repetitions are allowed, the term k-*permutation of n with repetition* is used. When any restrictions on the number of repetitions exist they are explicitly specified. The unqualified term, *permutation,* is used for the particular case of an n-permutation of n.

It is clear from the definition of a k-permutation of n, without or with repetition, that two k-permutations of n are different if at least one element in one of these permutations does not belong to the other. Further, two k-permutations of n containing the same k elements are different if at least one element occupies different positions in these permutations. Also, a k-permutation of n (without repetition) is meaningful when $1 \leq k \leq n$, while a k-permutation of n with (unrestricted) repetition is always meaningful for $k \geq 1$ and $n \geq 1$.

The following examples serve as to illustrate the different kinds of permutations and indicate the method of counting them. The latter will be helpful in understanding the general methods of enumeration of the different kinds of permutations.

Example 2.1
(a) The 2-permutations of the set $W_4 = \{w_1, w_2, w_3, w_4\}$, of 4 elements, are the following:

$$(w_1, w_2), (w_1, w_3), (w_1, w_4), (w_2, w_1), (w_2, w_3), (w_2, w_4),$$

$$(w_3, w_1), (w_3, w_2), (w_3, w_4), (w_4, w_1), (w_4, w_2), (w_4, w_3).$$

Note that, in any 2-permutation (a_1, a_2) of the set $W_4 = \{w_1, w_2, w_3, .w_4\}$, the first element a_1 can be selected from the set $A_1 = W_4$, of 4 elements, while, after the selection of the first element, the second element a_2, which must be different from a_1, can be selected from the set $A_2 = W_4 - \{a_1\}$, of 3 elements. Thus, according to the multiplication principle, the number of 2-permutations of 4 equals $4 \cdot 3 = 12$. A simple counting of these permutations verifies this result.

(b) The permutations of the set $W_3 = \{w_1, w_2, w_3\}$, of 3 elements, are the following:

$$(w_1, w_2, w_3), \ (w_1, w_3, w_2), \ (w_2, w_1, w_3),$$

$$(w_2, w_3, w_1), \ (w_3, w_1, w_2), \ (w_3, w_2, w_1),$$

the number of which, according to the multiplication principle, is equal to $3 \cdot 2 \cdot 1 = 6$. ☐

Example 2.2
The 2-permutations of the set $W_4 = \{w_1, w_2, w_3, w_4\}$, with repetition, are the following:

$(w_1, w_1), (w_1, w_2), (w_1, w_3), (w_1, w_4), (w_2, w_1), (w_2, w_2), (w_2, w_3), (w_2, w_4),$

$(w_3, w_1), (w_3, w_2), (w_3, w_3), (w_3, w_4), (w_4, w_1), (w_4, w_2), (w_4, w_3), (w_4, w_4).$

Note that, in any 2-permutation (a_1, a_2) of the set $W_4 = \{w_1, w_2, w_3, w_4\}$ with repetition, the first element a_1 as well as the second element a_2 can be chosen from the set W_4 of 4 elements. Hence, according to the multiplication principle, the number of 2-permutations of 4, with repetition, equals $4 \cdot 4 = 16$. ☐

Example 2.3
The permutations of the set $W_3 = \{w_1, w_2\}$, of two kinds of elements with $k_1 = 2$ elements w_1 and $k_2 = 1$ element w_2 are the following:

$$(w_1, w_1, w_2), \ (w_1, w_2, w_1), \ (w_2, w_1, w_1).$$

Let us pretend that we do not know the total number of these permutations. The problem of enumerating them may be transformed to an equivalent counting problem, which can be solved by applying basic counting principles. In this particular case, a suitable transformation is carried out in two consecutive actions. Firstly, the two like elements w_1 are transformed to distinct, by assigning to each a second index, $w_{1,1}, w_{1,2}$. Secondly, in each permutation the distinct elements $w_{1,1}$ and $w_{1,2}$ are permuted in all possible ways. Since 2 elements are permuted in only 2 ways, from each permutation 2 new permutations are constructed. The permutations thus constructed are all the permutations of the set $W_{2,3} = \{w_{1,1}, w_{1,2}, w_2\}$, of 3 (distinct) elements, which according to part (b) of Example 2.1 are the following 6:

$$(w_{1,1}, w_{1,2}, w_2), \ (w_{1,2}, w_{1,1}, w_2), \ (w_{1,1}, w_2, w_{1,2}),$$

$$(w_{1,2}, w_2, w_{1,1}), \ (w_2, w_{1,1}, w_{1,2}), \ (w_2, w_{1,2}, w_{1,1}).$$

Thus, if x is the required number of permutations, then $2x = 6$ and so $x = 3$.   ▯

Example 2.4

The 3-permutations of the set $W_3 = \{w_1, w_2, w_3\}$, with restricted repetition and, specifically, with the element w_1 allowed to appear at most $k_1 = 2$ times, the element w_2 at most $k_2 = 1$ time and the element w_3 at most $k_3 = 3$ times, are the following 19:

$$(w_1, w_2, w_3), (w_1, w_3, w_2), (w_2, w_1, w_3), (w_2, w_3, w_1), (w_3, w_1, w_2),$$

$$(w_3, w_2, w_1), (w_1, w_1, w_2), (w_1, w_2, w_1), (w_2, w_1, w_1), (w_1, w_1, w_3),$$

$$(w_1, w_3, w_1), (w_3, w_1, w_1), (w_1, w_3, w_3), (w_3, w_1, w_3), (w_3, w_3, w_1),$$

$$(w_2, w_3, w_3), (w_3, w_2, w_3), (w_3, w_3, w_2), (w_3, w_3, w_3). \quad ▯$$

The following two theorems are concerned with the number of permutations (without repetition).

THEOREM 2.1

The number of k-permutations of n, denoted by $P(n, k)$ or $(n)_k$, is given by

$$P(n, k) \equiv (n)_k = n(n - 1) \cdots (n - k + 1). \tag{2.1}$$

PROOF   Let $\mathcal{P}_k(W_n)$ be the set of k-permutations of the set $W_n = \{w_1, w_2, \dots, w_n\}$. In any k-permutation $(a_1, a_2, \dots, a_k)$ of the set W_n, the first element a_1 can be selected from the set $A_1 = W_n$ of n elements, while after the selection of the first element, the second element a_2, which must be different from a_1, can be selected from the set $A_2 = W_n - \{a_1\}$ of $n - 1$ elements. Finally, after the selection of the elements $a_1, a_2, \dots, a_{k-1}$, the last element a_k, which must be different from the $k - 1$ preceding elements, can be chosen from the set $A_k = W_n - \{a_1, a_2, \dots, a_{k-1}\}$ of $n - (k - 1)$ elements. Thus, according to the multiplication principle,

$$P(n, k) = N(\mathcal{P}_k(W_n)) = N(A_1)N(A_2) \cdots N(A_k) = n(n - 1) \cdots (n - k + 1)$$

and the required expression is established.   ∎

For the particular case, $k = n$, of the permutations of n the next corollary follows.

COROLLARY 2.1

The number of permutations of n, denoted by $P(n)$, is given by

$$P(n) \equiv P(n, n) = 1 \cdot 2 \cdot 3 \cdots (n - 1) \cdot n. \tag{2.2}$$

REMARK 2.1 The number $P(n, k) \equiv (n)_k$, $k = 1, 2, \ldots, n$, $n = 1, 2, \ldots$, is a special case of the factorial of a real number x of order k (see Chapter 3 for details) and is called *falling factorial of n of order k*. In particular, the falling factorial of n of order $k = n$, which is the product of all integers from 1 to n, is called n *factorial* and is denoted by $n!$. Thus, the number $P(n)$ of permutations of n, according to (2.2), is

$$P(n) = n!, \quad n = 1, 2, \ldots .$$

Further, multiplying expression (2.1) by $(n - k) \cdot (n - k - 1) \cdots 3 \cdot 2 \cdot 1$ and then dividing it by $(n - k)! = 1 \cdot 2 \cdot 3 \cdots (n - k - 1) \cdot (n - k)$, the number $P(n, k) \equiv (n)_k$ may be rewritten as

$$P(n, k) \equiv (n)_k = \frac{n!}{(n - k)!}, \quad k = 1, 2, \ldots, n, \ n = 1, 2, \ldots .$$

The numbers $P(n, 0) \equiv (n)_0$, $n = 0, 1, \ldots$ and $P(0) = 0!$, which have no combinatorial meaning, are taken by convention as unity:

$$P(n, 0) \equiv (n)_0 = 1, \ n = 0, 1, \ldots, \ P(0) = 0! = 1.$$

If $k > n$, then $P(n, k) \equiv (n)_k = 0$ and (2.1) is still valid. It is worth noting that

$$P(n, n - 1) = P(n, n) = P(n) = n!,$$

which is a consequence of the fact that, specifying the positions of the $n - 1$ elements, the position of the n-th element is uniquely determined.

The $n!$ increases rapidly, with increasing n: $2! = 2$, $3! = 6$, $4! = 24$, $5! = 120$, $6! = 720$, $7! = 5040, \ldots$. A useful approximation of it is presented in Chapter 3. ∎

Example 2.5

Consider a man who has five different keys, only one of which fits the door of his house. He does not recognize the proper key and tries the keys one after the other, until he opens the door. Let us find the number of ways he may try to open the door so that he succeeds (a) at the third trial, (b) until the third trial and (c) at the last trial.

(a) Let us number the five different keys from 0 to 4 assigning the number 0 to the key that fits the door. Note that, to a sequence of three trials, where in the third trial the door opens, there corresponds an ordered triple $(a_1, a_2, 0)$, where $a_j \in \{1, 2, 3, 4\}$, $j = 1, 2$, $a_1 \neq a_2$. Thus the required number of sequences of three trials equals the number of 2-permutations, (a_1, a_2), of the set $\{1, 2, 3, 4\}$, of four keys, that is

$$P(4, 2) = 4 \cdot 3 = 12.$$

These 12 sequences are the following:

$$(1,2,0), \ (1,3,0), \ (1,4,0), \ (2,1,0), \ (2,3,0), \ (2,4,0),$$

$$(3,1,0), \ (3,2,0), \ (3,4,0), \ (4,1,0), \ (4,2,0), \ (4,3,0).$$

(b) The number of ways he may try to open the door so that he succeeds until the third trial, according to the addition principle, equals the sum of the numbers of ways he may try to open the door so that he succeeds at the first or second or third trial. Since the number of ways he may try to open the door so that he succeeds at the k-th trial equals the number $P(4, k-1) = (4)_{k-1}$ of the $(k-1)$-permutations of the set of $\{1, 2, 3, 4\}$, of four keys, for $k = 1, 2, 3$, the required number is given by the sum

$$S_{4,3} = \sum_{k=1}^{3} (4)_{k-1} = 1 + 4 + 12 = 17.$$

(c) The number of ways he may try to open the door so that he succeeds at the fifth (last) trial, according to the analysis of (a), equals the number of permutations (a_1, a_2, a_3, a_4), of the set $\{1, 2, 3, 4\}$, of four keys, that is

$$4! = 4 \cdot 3 \cdot 2 \cdot 1 = 24. \qquad \Box$$

Example 2.6

Let us consider an urn containing n balls numbered from 1 to n. Assume that k balls are successively drawn from the urn without replacement (without returning the ball drawn to the urn after each drawing). The first number drawn wins a great value gift, while the subsequent numbers win gifts of smaller value. Find the number of possible outcomes of the drawings.

Note that each possible outcome consists of k numbers from the set $\{1, 2, \ldots, n\}$. The order in which the numbers are drawn counts, since it determines the distribution of the gifts. Therefore, to each possible outcome there corresponds one and only one k-permutation of n and, according to Theorem 2.1, the required number is given by

$$(n)_k = n(n-1) \cdots (n - k + 1). \qquad \Box$$

Example 2.7

Suppose that three workers are to be selected among the ten workers, $\{w_1, w_2, \ldots, w_{10}\}$, of a small factory. A different job is to be assigned to the first, second and third workers. Find the number of different selections, which (a) do not include w_{10} (who has not finished a delicate job) and (b) include w_{10} (the most skilled worker).

Note first that the order of selection of the three workers counts, since it determines the assignment of the three different jobs. Thus, to each possible selection there corresponds one and only one 3-permutation (a_1, a_2, a_3) of the set

$\{w_1, w_2, \ldots, w_{10}\}$. Further, (a) since w_{10} is not to be included in the selection, (a_1, a_2, a_3) is a 3-permutation of the set $\{w_1, w_2, \ldots, w_9\}$ of the other nine workers and so the required number of selections equals

$$P(9,3) = 9 \cdot 8 \cdot 7 = 504.$$

(b) Since w_{10} is to be included in the selection, he may be selected first or second or third. To each of the selections (w_{10}, a_1, a_2), (a_1, w_{10}, a_2), (a_1, a_2, w_{10}) there corresponds one and only one 2-permutation (a_1, a_2) of the set $\{w_1, w_2, \ldots, w_9\}$ of nine workers. Hence the required number of selections equals

$$3P(9,2) = 3 \cdot 9 \cdot 8 = 216.$$

It is worth noting that the sum of the number of selections that do not include w_{10} and the number of selections that include w_{10} equals

$$P(9,3) + 3P(9,2) = 504 + 216 = 720,$$

which is the total number of selections (without any condition),

$$P(10,3) = 10 \cdot 9 \cdot 8 = 720.$$

This relation is generalized in the following theorem.   ⬜

THEOREM 2.2
The number $P(n,k) = (n)_k$, of k-permutations of n, satisfies the recurrence relations

$$P(n,k) = P(n-1,k) + kP(n-1,k-1), \qquad (2.3)$$

for $k = 1, 2, \ldots, n$, $n = 1, 2, \ldots$, and

$$P(n,k) = nP(n-1,k-1) = (n-k+1)P(n,k-1), \qquad (2.4)$$

for $k = 1, 2, \ldots, n$, $n = 1, 2, \ldots$, with initial conditions:

$$P(n,0) = 1, \; n = 0, 1, 2, \ldots, \; P(n,k) = 0, \; k > n.$$

PROOF   Let $\mathcal{P}_k(W_n)$ be the set of k-permutations of the set $W_n = \{w_1, w_2, \ldots, w_n\}$. If $\mathcal{Q}$ is the set of the k-permutations of W_n that do not include the element w_n and $\mathcal{S}$ the set of the k-permutations of W_n that include the element w_n, then $\mathcal{Q} \cap \mathcal{S} = \emptyset$ and $\mathcal{P}_k(W_n) = \mathcal{Q} + \mathcal{S}$. Hence, according to the addition principle,

$$N(\mathcal{P}_k(W_n)) = N(\mathcal{Q}) + N(\mathcal{S}).$$

Apparently $\mathcal{Q} = \mathcal{P}_k(W_{n-1})$ and $N(\mathcal{Q}) = N(\mathcal{P}_k(W_{n-1})) = P(n-1,k)$. Further, from each $(k-1)$-permutation of W_{n-1}, by attaching the element w_n in any of the

k possible positions (one before the first element, $k-2$ between the $k-1$ elements and one after the last element), k different k-permutations of W_n that include w_n are constructed. Therefore, $N(\mathcal{S}) = kN(\mathcal{P}_{k-1}(W_{n-1})) = kP(n-1, k-1)$ and

$$P(n, k) = P(n-1, k) + kP(n-1, k-1), \ k = 1, 2, \ldots, n, \ n = 1, 2, \ldots.$$

For the proof of (2.4) note that the first r (ordered) elements, $(a_1, a_2, \ldots, a_r)$, of a k-permutation, $(a_1, a_2, \ldots, a_r, a_{r+1}, \ldots, a_k)$, of W_n, can be chosen from the set $A_1 = W_n$ in $P(n, r)$ ways. After every such selection, the last $k-r$ (ordered) elements, $(a_{r+1}, \ldots, a_k)$, can be chosen from the set $A_2 = W_n - \{a_1, a_2, \ldots, a_r\}$ in $P(n-r, k-r)$ ways. Hence, according to the multiplication principle,

$$P(n, k) = P(n, r)P(n-r, k-r), \ r = 1, 2, \ldots, k, \ k = 1, 2, \ldots, n.$$

In particular, for $r = 1$ and $r = k-1$, since $P(n, 1) = n$ and $P(n-k+1, 1) = n-k+1$, the first and second recurrence relations of (2.4) are deduced.

The initial conditions, which often have no combinatorial meaning, are chosen in such a way that their introduction into the recurrence relation gives the correct values to the subsequent terms that have a combinatorial meaning and known values. Thus, from the known values $P(n, 1) = n, n = 1, 2, \ldots$ and the recurrence relation (2.3), it follows that

$$P(n, 0) = P(n+1, 1) - P(n, 1) = n + 1 - n = 1, \ n = 0, 1, \ldots.$$

The initial condition $P(n, k) = 0$ for $k > n$ is obvious. These are also the initial conditions for the recurrence relation (2.4). ∎

REMARK 2.2   As regards the initial conditions of the recurrence relations (2.3) and (2.4), note that, instead of the relations $P(n, 0) = 1, n = 0, 1, \ldots$, which are indirectly deduced, the relations $P(n, 1) = n, n = 1, 2, \ldots$, following directly from the definition of a permutation, may be used. The preference to the first relations is due to some technical advantages that they provide in the study of recurrence relations, especially through generating functions. ∎

COROLLARY 2.2
The number $P(n) \equiv n!$, of permutations of n, satisfies the recurrence relation

$$P(n) = nP(n-1), \ n = 1, 2, \ldots,$$

with initial condition $P(0) = 1$.

In the next theorem the number of permutations with (unrestricted) repetition is derived.

THEOREM 2.3

The number of k-permutations of n with (unrestricted) repetition, denoted by
$U(n, k)$, *is given by*

$$U(n, k) = n^k.$$

PROOF   Note that, in any k-permutation $(a_1, a_2, \ldots, a_k)$ of the set $W_n = \{w_1, w_2, \ldots, w_n\}$ with repetition, the element a_i can be chosen from the set $A_i = W_n$, of n elements, $i = 1, 2, \ldots, k$. Hence, $\mathcal{U}_k(W_n) \equiv A_1 \times A_2 \times \cdots \times A_k$ is the set of k-permutations of the set W_n with repetition and according to the multiplication principle,

$$U(n, k) = N(\mathcal{U}_k(W_n)) = N(A_1 \times A_2 \times \cdots \times A_k)$$
$$= N(A_1)N(A_2) \cdots N(A_k) = n^k.$$

The proof of the theorem is thus completed.   ∎

Example 2.8 **Ternary number system**

In the ternary number system each number is represented by a ternary sequence of 0s, 1s and 2s. For example, the numbers 5 and 11, which are expressed in terms of powers of 3 as $5 = 1 \cdot 3^1 + 2 \cdot 3^0$ and $11 = 1 \cdot 3^2 + 0 \cdot 3^1 + 2 \cdot 3^0$, are represented by the ternary sequences $(1, 2)$ and $(1, 0, 2)$, respectively. Note that, with the exception of the number 0, which is represented by the one-digit sequence (0), in all the other ternary sequences the first digit is different from 0. Calculate the number of four-digit ternary sequences.

The first digit of a four-digit ternary sequence is either 1 or 2. A four-digit ternary sequence $(1, a_1, a_2, a_3)$ corresponds to a 3-permutation (a_1, a_2, a_3) of the set $\{0, 1, 2\}$ with repetition. Thus, the number of four-digit ternary sequences $(1, a_1, a_2, a_3)$, according to Theorem 2.3, equals

$$U(3, 3) = 3^3 = 27.$$

Similarly, the number of four-digit ternary sequences $(2, a_1, a_2, a_3)$ equals $U(3, 3) = 27$ and so the total number of four-digit ternary sequences is given by

$$2U(3, 3) = 54. \quad \square$$

Example 2.9

(a) Find the number of different outcomes in a series of k tosses of a coin.

An outcome of a series of k tosses of a coin may be represented by an ordered k-tuple $(a_1, a_2, \ldots, a_k)$ of letters from the set $\{h, t\}$, where a_i denotes the outcome of the i-th toss, $i = 1, 2, \ldots, k$, and h and t stand for heads and tails, respectively. Hence, the number of different outcomes in a series of k tosses of a coin equals

$$U(2, k) = 2^k,$$

the number of k-permutations of the set $\{h, t\}$. For $k = 3$, the different outcomes are the following:

$$(h, h, h), (h, h, t), (h, t, h), (t, h, h), (h, t, t), (t, h, t), (t, t, h), (t, t, t),$$

in agreement with $U(2, 3) = 2^3 = 8$.

(b) Consider a series of k throws of a die or equivalently a throw of k distinguishable dice. Find the number of different outcomes.

An outcome of a series of k throws of a die may by represented by an ordered k-tuple $(a_1, a_2, \ldots, a_k)$ of numbers from the set $\{1, 2, 3, 4, 5, 6\}$, where a_i denotes the outcome of the i-th throw, $i = 1, 2, \ldots, k$. Therefore, the number of different outcomes in a series of k throws of a die equals

$$U(6, k) = 6^k,$$

the number of k-permutations of the set $\{1, 2, 3, 4, 5, 6\}$, of 6 numbers (faces of a die), with repetition. For $k = 2$, the different outcomes are the following:

$$(1, 1), (1, 2), (1, 3), (1, 4), (1, 5), (1, 6),$$
$$(2, 1), (2, 2), (2, 3), (2, 4), (2, 5), (2, 6),$$
$$(3, 1), (3, 2), (3, 3), (3, 4), (3, 5), (3, 6),$$
$$(4, 1), (4, 2), (4, 3), (4, 4), (4, 5), (4, 6),$$
$$(5, 1), (5, 2), (5, 3), (5, 4), (5, 5), (5, 6),$$
$$(6, 1), (6, 2), (6, 3), (6, 4), (6, 5), (6, 6),$$

in agreement with $U(6, 2) = 6^2 = 36$.   ☐

Example 2.10 Distributions of distinguishable balls into distinguishable urns

Let k distinguishable balls $\{b_1, b_2, \ldots, b_k\}$ be successively distributed into n distinguishable urns (cells) $\{w_1, w_2, \ldots, w_n\}$ of unlimited capacity. Find the number of different distributions .

Clearly, ball b_r can be placed in any of the n urns, $r = 1, 2, \ldots, k$ and so, according to the multiplication principle, the k balls can be distributed into the n urns in n^k different ways.

It is worth noting the one-to-one correspondence between the set of distributions of k distinguishable balls into n distinguishable urns of unlimited capacity and the set of k-permutations of n with (unrestricted) repetition. Indeed, the distribution of the k balls $\{b_1, b_2, \ldots, b_k\}$ into the n urns $W_n = \{w_1, w_2, \ldots, w_n\}$, according to which the ball b_r is placed into the urn w_{i_r}, $r = 1, 2, \ldots, k$, corresponds to the k-permutation $(w_{i_1}, w_{i_2}, \ldots, w_{i_k})$ of the set W_n, of n urns, with (unrestricted) repetition. Thus, the number of distributions of k distinguishable balls into n distinguishable urns of unlimited capacity equals

$$U(n, k) = n^k,$$

the number of k-permutations of n with repetition. The general problem of distributions of balls of different kinds into urns of different kinds and capacity is examined at length in Chapter 9. $\quad\Box$

In concluding, let us turn the discussion to the enumeration of k-permutations of $W_n = \{w_1, w_2, \ldots, w_n\}$ with repetition, in which the element w_i may appear k_i times, $i = 1, 2, \ldots, n$, where $k_1 + k_2 + \cdots + k_n = r \geq k$. The calculation of their multitude, using the two basic counting principles, is quite difficult. This problem is treated more effectively by the use of generating functions. Constituting a powerful tool for the treatment of many combinatorial problems, these functions are examined in Chapter 6, in which we return to this problem. The next theorem is concerned with the particular case of $k = r$.

THEOREM 2.4
The number of permutations of n kinds of elements with $k_1, k_2, \ldots, k_n$ elements, respectively, denoted by $M(k_1, k_2, \ldots, k_n)$, is given by

$$M(k_1, k_2, \ldots, k_n) = \frac{r!}{k_1! k_2! \cdots k_n!}, \quad r = k_1 + k_2 + \cdots + k_n. \quad (2.5)$$

PROOF　Consider a permutation $(a_1, a_2, \ldots, a_r)$, of the n kinds of elements $W_n = \{w_1, w_2, \ldots, w_n\}$, where k_i of the a's are w_i, $i = 1, 2, \ldots, n$, with $k_1 + k_2 + \cdots + k_n = r$. If the k_1 like elements w_1 are transformed to distinct, by assigning to each a second index from 1 to k_1, $w_{1,1}, w_{1,2}, \ldots, w_{2,k_1}$, and permuted in all possible ways, preserving the positions the element w_1 occupies in this permutation, $k_1!$ permutations are constructed. If, in any of these permutations, the k_2 like elements w_2 are transformed to distinct, $w_{2,1}, w_{2,2}, \ldots, w_{2,k_2}$, and permuted in all possible ways preserving the positions the element w_2 occupies in this permutation, $k_2!$ permutations are constructed. Repeating this procedure until all the n kinds of elements are exhausted and then applying the multiplication principle, it follows that, from the permutation $(a_1, a_2, \ldots, a_r)$, $k_1! k_2! \cdots k_n!$ permutations, $(b_1, b_2, \ldots, b_r)$, of the set $W_{n,r} = \{w_{i,j} : j = 1, 2, \ldots, k_i, i = 1, 2, \ldots, n\}$, of r (distinct) elements, are constructed. Moreover, each permutation $(b_1, b_2, \ldots, b_r)$ of the set $W_{n,r}$, of r elements, by deleting the second index from its elements $w_{i,j}$, $j = 1, 2, \ldots, k_i$, $i = 1, 2, \ldots, n$, is reduced to a unique permutation $(a_1, a_2, \ldots, a_r)$ of the n kinds of elements $W_n = \{w_1, w_2, \ldots, w_n\}$, where k_i of the a's are w_i, $i = 1, 2, \ldots, n$, with $k_1 + k_2 + \cdots + k_n = r$. Therefore, $k_1! k_2! \cdots k_n! M(k_1, k_2, \ldots, k_n)$ equals the number of permutations of the set $W_{n,r}$, of r elements. Further, the number of these permutations, according to Corollary 2.1, is equal to $r!$ and hence we find the relation

$$k_1! k_2! \cdots k_n! M(k_1, k_2, \ldots, k_n) = r!,$$

from which (2.5) follows.   ∎

Example 2.11

Consider r recruits, among which k_i come from the i-th district of the country, $i = 1, 2, \ldots, n$, so that $k_1 + k_2 + \cdots + k_n = r$. Distinguishing them according to district of their origin, the number of ways they can be arranged on a line equals

$$M(k_1, k_2, \ldots, k_n) = \frac{r!}{k_1! k_2! \cdots k_n!},$$

the number of permutations of n kinds of elements (districts), with $k_1, k_2, \ldots, k_n$ elements (recruits), respectively. Distinguishing them according to the number of the recruitment invitation, the number of ways they can be arranged on a line equals

$$P(r) = r!,$$

the number of permutations of r elements (numbers).

In particular, for $n = 2$ and $k_1 = 2$, $k_2 = 2$, the $M(2, 2) = 6$ arrangements of the recruits, according to the district of their origin, are the following:

$$(d_1, d_1, d_2, d_2), \ (d_1, d_2, d_2, d_1), \ (d_1, d_2, d_1, d_2),$$
$$(d_2, d_1, d_2, d_1), \ (d_2, d_1, d_1, d_2), \ (d_2, d_2, d_1, d_1),$$

and the corresponding arrangements of the recruits, according to their recruitment invitation (double-index) number are the following:

$$(d_{11}, d_{12}, d_{21}, d_{22}), \ (d_{12}, d_{11}, d_{21}, d_{22}), \ (d_{11}, d_{12}, d_{22}, d_{21}), \ (d_{12}, d_{11}, d_{22}, d_{21}),$$

$$(d_{11}, d_{21}, d_{22}, d_{12}), \ (d_{12}, d_{21}, d_{22}, d_{11}), \ (d_{11}, d_{22}, d_{21}, d_{12}), \ (d_{12}, d_{22}, d_{21}, d_{11}),$$

$$(d_{11}, d_{21}, d_{12}, d_{22}), \ (d_{12}, d_{21}, d_{11}, d_{22}), \ (d_{11}, d_{22}, d_{12}, d_{21}), \ (d_{12}, d_{22}, d_{11}, d_{21}),$$

$$(d_{21}, d_{11}, d_{22}, d_{12}), \ (d_{21}, d_{12}, d_{22}, d_{11}), \ (d_{22}, d_{11}, d_{21}, d_{12}), \ (d_{22}, d_{12}, d_{21}, d_{11}),$$

$$(d_{21}, d_{11}, d_{12}, d_{22}), \ (d_{21}, d_{12}, d_{11}, d_{22}), \ (d_{22}, d_{11}, d_{12}, d_{21}), \ (d_{22}, d_{12}, d_{11}, d_{21}),$$

$$(d_{21}, d_{22}, d_{11}, d_{12}), \ (d_{21}, d_{22}, d_{12}, d_{11}), \ (d_{22}, d_{21}, d_{11}, d_{12}), \ (d_{22}, d_{21}, d_{12}, d_{11}).$$

The total number of these arrangements is $P(4) = 24$.   ☐

2.3 COMBINATIONS

DEFINITION 2.2   *Let $W_n = \{w_1, w_2, \ldots, w_n\}$, a finite set of n elements. A (non-ordered) collection of k elements $\{a_1, a_2, \ldots, a_k\}$, with $a_r \in W_n$, $r = 1, 2, \ldots, k$, is called k-combination of the set W_n or simply k-combination of n.*

Note that the elements of a k-combination of n may or may not be different elements of W_n. For the first case, the term *k-combination of n* is preserved, while for the second and when unrestricted repetitions are allowed, the term *k-combination of n with repetition* is used. When any restrictions on the number of repetitions exist, they are explicitly specified.

Note also that, according to the definitions of a set and a subset, a k-combination of the set W_n, without repetition, is a subset of W_n. Further, a k-combination of n is meaningful when $1 \leq k \leq n$, while a k-combination of n with (unrestricted) repetition is always meaningful for $k \geq 1$, $n \geq 1$.

The examples that follow illustrate the different kinds of combinations and provide an indication of their enumeration.

Example 2.12
The 2-combinations of the set $W_4 = \{w_1, w_2, w_3, w_4\}$, of 4 elements, are the following:

$$\{w_1, w_2\}, \ \{w_1, w_3\}, \ \{w_1, w_4\}, \ \{w_2, w_3\}, \ \{w_2, w_4\}, \ \{w_3, w_4\}$$

Note that, from each 2-combination $\{a_1, a_2\}$ of the set W_4, by permuting its elements, the 2-permutations (a_1, a_2) and (a_2, a_1) of the set W_4 are deduced. Thus, the number of 2-combinations of 4 equals the number of 2-permutations of 4 divided by the number of permutations of 2, that is $12/2 = 6$ (see Example 2.1).   □

Example 2.13
The 2-combinations of the set $W_4 = \{w_1, w_2, w_3, w_4\}$ with repetition are the following:

$$\{w_1, w_1\}, \ \{w_1, w_2\}, \ \{w_1, w_3\}, \ \{w_1, w_4\}, \ \{w_2, w_2\},$$

$$\{w_2, w_3\}, \ \{w_2, w_4\}, \ \{w_3, w_3\}, \ \{w_3, w_4\}, \ \{w_4, w_4\}.$$

Consider also the 2-combinations of the set $W_5 = \{w_1, w_2, w_3, w_4, w_5\}$ (without repetition):

$$\{w_1, w_2\}, \ \{w_1, w_3\}, \ \{w_1, w_4\}, \ \{w_1, w_5\}, \ \{w_2, w_3\},$$

$$\{w_2, w_4\}, \{w_2, w_5\}, \{w_3, w_4\}, \{w_3, w_5\}, \{w_4, w_5\}.$$

Note that a one-to-one correspondence exists between these two sets of 2-combinations. Specifically, the 2-combination $\{w_{i_1}, w_{i_2}\}$, $1 \leq i_1 \leq i_2 \leq 4$, of the set $W_4 = \{w_1, w_2, w_3, w_4\}$ with repetition, uniquely corresponds to the 2-permutation $\{w_{j_1}, w_{j_2}\}$, $1 \leq j_1 < j_2 \leq 4$, of the set $W_5 = \{w_1, w_2, w_3, w_4, w_5\}$ (without repetition), where $j_1 = i_1$ and $j_2 = i_2 + 1$. Since the number of 2-combinations of the set $W_5 = \{w_1, w_2, w_3, w_4, w_5\}$ equals $(5 \cdot 4)/2 = 10$, it follows, by Lemma 1.1, that the number of 2-combinations of the set $W_4 = \{w_1, w_2, w_3, w_4\}$ with repetition is also equal to 10.   □

Example 2.14

The 3-combinations of the set $W_4 = \{w_1, w_2, w_3, w_4\}$ with repetition and the restriction that each element is allowed to appear at most two times are the following 16:

$$\{w_1, w_1, w_2\}, \{w_1, w_1, w_3\}, \{w_1, w_1, w_4\}, \{w_1, w_2, w_2\},$$

$$\{w_1, w_2, w_3\}, \{w_1, w_2, w_4\}, \{w_1, w_3, w_3\}, \{w_1, w_3, w_4\},$$

$$\{w_1, w_4, w_4\}, \{w_2, w_2, w_3\}, \{w_2, w_2, w_4\}, \{w_2, w_3, w_3\},$$

$$\{w_2, w_3, w_4\}, \{w_2, w_4, w_4\}, \{w_3, w_3, w_4\}, \{w_3, w_4, w_4\}.$$ □

The following two theorems are concerned with the number of combinations (without repetition).

THEOREM 2.5

The number of k-combinations of n, denoted by $C(n, k)$ or $\binom{n}{k}$, is given by

$$C(n, k) \equiv \binom{n}{k} = \frac{n(n-1) \cdots (n-k+1)}{k!} = \frac{n!}{k!(n-k)!}. \qquad (2.6)$$

PROOF   Let $\mathcal{C}_k(W_n)$ be the set of k-combinations and $\mathcal{P}_k(W_n)$ the set of k-permutations of $W_n = \{w_1, w_2, \dots, w_n\}$. Note that, to each k-combination $\{a_1, a_2, \dots, a_k\}$ of W_n, there correspond $k!$ k-permutations $(a_{i_1}, a_{i_2}, \dots, a_{i_k})$ of W_n, which are formed by permuting its k elements in all $k!$ possible ways. Further, to each k-permutation, $(a_1, a_2, \dots, a_k)$, of W_n, there corresponds one and only one k-combination, $\{a_1, a_2, \dots, a_k\}$, of W_n. Therefore,

$$P(n, k) \equiv N(\mathcal{P}_k(W_n)) = k! \, N(\mathcal{C}_k(W_n)) \equiv k! C(n, k)$$

and, using (2.1), expression (2.6) is deduced.   ∎

REMARK 2.3   Note that every time a k-combination of the set $W_n = \{w_1, w_2, \dots, w_n\}$ is formed by selecting k elements, $n - k$ elements are left, forming an

$(n - k)$-combination of W_n. Hence to each k-combination, $\{a_1, a_2, \ldots, a_k\}$, of W_n, there uniquely corresponds an $(n - k)$-combination, $\{b_1, b_2, \ldots, b_{n-k}\} = \{w_1, w_2, \ldots, w_n\} - \{a_1, a_2, \ldots, a_k\}$, of W_n, and conversely. Consequently, $N(\mathcal{C}_k(W_n)) = N(\mathcal{C}_{n-k}(W_n))$ and

$$\binom{n}{k} = \binom{n}{n-k}.$$

This relation can also be shown algebraically, by using expression (2.6), as follows:

$$\binom{n}{k} = \frac{n!}{k!(n-k)!} = \frac{n!}{(n-k)!(n-(n-k))!} = \binom{n}{n-k}.$$

Note that the pair of sets $A = \{a_1, a_2, \ldots, a_k\}$ and $B = \{b_1, b_2, \ldots, b_{n-k}\}$ is a division of the set $W_n = \{w_1, w_2, \ldots, w_n\}$ (see Section 1.2.7). This point is examined in a more general set up in the next section.

It is worth noting that there exists a one-to-one correspondence (bijective mapping) of the set $\mathcal{M}_{k,n-k}(U_2)$, of permutations of two kinds of elements $U_2 = \{u_1, u_2\}$, with k and $n - k$ elements, respectively, onto the set $\mathcal{C}_k(W_n)$, of k-combinations of the set $W_n = \{w_1, w_2, \ldots, w_n\}$, of n elements. Specifically, to the permutation $(v_1, v_2, \ldots, v_n)$ in which the element u_1 occupies the k positions $\{i_1, i_2, \ldots, i_k\} \subseteq \{1, 2, \ldots, n\}$, the k-combination $\{w_{i_1}, w_{i_2}, \ldots, w_{i_k}\}$ of the set W_n, of n elements, is associated and conversely. Hence

$$M(k, n - k) = C(n, k),$$

in agreement with expressions (2.5) and (2.6). ∎

Example 2.15

Consider an organization with seven vacant positions, among which two require mathematics degrees, three require computer science degrees and the remaining two positions are open to candidates with either a mathematics or a computer science degree. After a screening, the selection committee prepares a shortlist of candidates of whom seven hold mathematics degrees and six computer science degrees. In how many ways can the vacant positions be filled?

From the seven candidates with mathematics degrees two can be selected (without regard to any order) in

$$\binom{7}{2} = \frac{7 \cdot 6}{2} = 21$$

ways and from the six candidates with computer science degrees three can be selected in

$$\binom{6}{3} = \frac{6 \cdot 5 \cdot 4}{2 \cdot 3} = 20$$

ways. Finally, from the remaining eight candidates, two can be selected in

$$\binom{8}{2} = \frac{8 \cdot 7}{2} = 28$$

ways. Thus, according to the multiplication principle, the vacant positions can be filled in

$$21 \cdot 20 \cdot 28 = 11{,}760$$

ways.   ☐

Example 2.16 Binary number system revisited

In the binary number system, calculate the number of five-digit binary sequences that contain exactly two zeros (see Example 1.4).

The first digit of a five-digit binary sequence is necessarily 1. A five-digit binary sequence $(1, a_1, a_2, a_3, a_4)$ that contains exactly two zeros uniquely corresponds to the set $\{i_1, i_2\}$ of the two positions, out of the four positions $\{1, 2, 3, 4\}$ that the two zeros occupy in the binary sequence. Thus the number of five-digit binary sequences $(1, a_1, a_2, a_3, a_4)$ that contain exactly two zeros equals

$$\binom{4}{2} = \frac{4 \cdot 3}{2} = 6,$$

the number of 2-combinations of 4.   ☐

Example 2.17 Distributions of indistinguishable balls into distinguishable urns of limited capacity

Suppose that k indistinguishable balls are successively distributed into n distinguishable urns $\{w_1, w_2, \ldots, w_m\}$, each with capacity limited to one ball. Find the number of different distributions.

Any distribution of k indistinguishable balls into n distinguishable urns may be represented by the subset $\{w_{i_1}, w_{i_2}, \ldots, w_{i_k}\}$ of urns in each of which a ball is placed. Thus, each distribution of k indistinguishable balls into n distinguishable urns corresponds to a selection of k urns from the set of the n urns without regard to order and conversely. Therefore, the number of ways of placing k indistinguishable balls into n distinguishable urns, each with capacity limited to one ball, equals

$$\binom{n}{k},$$

the number of k-combinations of the set $\{w_1, w_2, \ldots, w_n\}$ of n urns.   ☐

THEOREM 2.6 Pascal's triangle
The number $C(n, k) \equiv \binom{n}{k}$, of k-combinations of n, satisfies the "triangular" recurrence relation

$$\binom{n}{k} = \binom{n-1}{k} + \binom{n-1}{k-1}, \; k = 1, 2, \ldots, n, \; n = 1, 2, \ldots, \qquad (2.7)$$

with initial conditions

$$\binom{n}{0} = 1, \; n = 0, 1, \ldots, \quad \binom{n}{k} = 0, \; k > n.$$

PROOF   Let $\mathcal{C}_k(W_n)$ be the set of k-combinations of the set $W_n = \{w_1, w_2, \ldots, w_n\}$. If $\mathcal{A}$ is the set of the k-combinations of W_n that do not include the element w_n, and $\mathcal{B}$ the set of the k-combinations of W_n that include the element w_n, then $\mathcal{A} \cap \mathcal{B} = \emptyset$ and $\mathcal{C}_k(W_n) = \mathcal{A} + \mathcal{B}$. Hence, according to the addition principle,

$$N(\mathcal{C}_k(W_n)) = N(\mathcal{A}) + N(\mathcal{B}).$$

Apparently $\mathcal{A} = \mathcal{C}_k(W_{n-1})$ and $N(\mathcal{A}) = N(\mathcal{C}_k(W_{n-1}))$. Further, one of the k elements of a k-combination $\{a_1, a_2, \ldots, a_{k-1}, a_k\}$, belonging to $\mathcal{B}$, is the element w_n and, since the order in which the elements are written does matter, it may be assumed that $a_k = w_n$ and $a_r \in W_{n-1}, r = 1, 2, \ldots, k-1$. Thus, each k-combination $\{a_1, a_2, \ldots, a_{k-1}, w_n\}$ belonging to $\mathcal{B}$ corresponds to one and only one k-combination $\{a_1, a_2, \ldots, a_{k-1}\}$ belonging to $\mathcal{C}_{k-1}(W_{n-1})$, and conversely. Therefore $N(\mathcal{B}) = N(\mathcal{C}_{k-1}(W_{n-1}))$ and

$$N(\mathcal{C}_k(W_n)) = N(\mathcal{C}_k(W_{n-1})) + N(\mathcal{C}_{k-1}(W_{n-1})).$$

The last relation implies (2.7). The initial conditions $\binom{n}{0} = 1, n = 0, 1, \ldots,$ which are preferred to the conditions $\binom{n}{1} = n, n = 1, 2, \ldots,$ are compatible with (2.7) since

$$\binom{n}{0} = \binom{n+1}{1} - \binom{n}{1} = n + 1 - n = 1.$$

The recurrence relation (2.7) may also be shown algebraically, by using expression (2.6).   ∎

The numbers $C(n, k) = \binom{n}{k}$, known as *binomial coefficients*, can be tabulated by using Pascal's triangle and its initial conditions. Table 2.1 gives the binomial coefficients for $k = 0, 1, \ldots, n, n = 0, 1, \ldots, 12$.

COROLLARY 2.3
The number $C(n, k) \equiv \binom{n}{k}$, of k-combinations of n, satisfies the "vertical" recurrence relation

$$\binom{n}{k} = \sum_{r=k}^{n} \binom{r-1}{k-1}, \; k = 1, 2, \ldots, n, \; n = 1, 2, \ldots \qquad (2.8)$$

Table 2.1 *Binomial Coefficients* $C(n,k) \equiv \binom{n}{k}$

k \ n	0	1	2	3	4	5	6	7	8	9	10	11	12
0	1												
1	1	1											
2	1	2	1										
3	1	3	3	1									
4	1	4	6	4	1								
5	1	5	10	10	5	1							
6	1	6	15	20	15	6	1						
7	1	7	21	35	35	21	7	1					
8	1	8	28	56	70	56	28	8	1				
9	1	9	36	84	126	126	84	36	9	1			
10	1	10	45	120	210	252	210	120	45	10	1		
11	1	11	55	165	330	462	462	330	165	55	11	1	
12	1	12	66	220	495	792	924	792	495	220	66	12	1

and the "horizontal" recurrence relation

$$\binom{n}{k} = \sum_{r=0}^{k} (-1)^{k-r} \binom{n+1}{r}$$

$$= \sum_{j=k}^{n} (-1)^{k-j} \binom{n+1}{j+1}, \quad k = 1, 2, \dots, n, \; n = 1, 2, \dots. \quad (2.9)$$

PROOF   Summing both sides of the "triangular" recurrence relation

$$\binom{r-1}{k-1} = \binom{r}{k} - \binom{r-1}{k}$$

for $r = k, k+1, \dots, n$,

$$\sum_{r=k}^{n} \binom{r-1}{k-1} = \sum_{r=k}^{n} \binom{r}{k} - \sum_{r=k}^{n} \binom{r-1}{k}$$

and since $\binom{k-1}{k} = 0$, it follows that

$$\sum_{r=k}^{n} \binom{r-1}{k-1} = \sum_{r=k}^{n} \binom{r}{k} - \sum_{s=k}^{n-1} \binom{s}{k} = \binom{n}{k}.$$

Multiplying the "triangular" recurrence relation

$$\binom{n+1}{r} = \binom{n}{r} + \binom{n}{r-1}$$

by $(-1)^{k-r}$ and summing both sides for $r = 1, 2, \ldots, k$, we get

$$\sum_{r=1}^{k} (-1)^{k-r} \binom{n+1}{r} = \sum_{r=1}^{k} (-1)^{k-r} \binom{n}{r} + \sum_{r=1}^{k} (-1)^{k-r} \binom{n}{r-1}$$

and since $\binom{n+1}{0} = \binom{n}{0} = 1$, it follows that

$$\sum_{r=0}^{k} (-1)^{k-r} \binom{n+1}{r} = \sum_{r=0}^{k} (-1)^{k-r} \binom{n}{r} - \sum_{s=0}^{k-1} (-1)^{k-s} \binom{n}{s} = \binom{n}{k}.$$

The second part of (2.9) can be shown in a similar way. ■

REMARK 2.4 The "vertical" recurrence relation (2.8) may also be derived independently of the "triangular" recurrence relation (2.7). The following combinatorial derivation is of interest.

Let $\mathcal{C}_k(W_n)$ be the set of k-combinations of the set $W_n = \{w_1, w_2, \ldots, w_n\}$ and $\mathcal{A}_r$ the subset of $\mathcal{C}_k(W_n)$ containing the combinations that include w_r as the element with the largest subscript, $r = k, k+1, \ldots, n$. Then $\mathcal{A}_i \cap \mathcal{A}_j = \emptyset$, $i, j = k, k+1, \ldots, n, i \neq j$, $\mathcal{C}_k(W_n) = \mathcal{A}_k + \mathcal{A}_{k+1} + \cdots + \mathcal{A}_n$ and, according to the addition principle,

$$N(\mathcal{C}_k(W_n)) = N(\mathcal{A}_k) + N(\mathcal{A}_{k+1}) + \cdots + N(\mathcal{A}_n).$$

Note that, to each k-combination $\{a_1, a_2, \ldots, a_{k-1}, w_r\}$, with $a_j \in W_{r-1} = \{w_1, w_2, \ldots, w_{r-1}\}$, $j = 1, 2, \ldots, k-1$, that belongs to $\mathcal{A}_r$, there corresponds one and only one $(k-1)$-combination $\{a_1, a_2, \ldots, a_{k-1}\}$ that belongs to $\mathcal{C}_{k-1}(W_{r-1})$, $r = k, k+1, \ldots, n$. Therefore, $N(\mathcal{A}_r) = N(\mathcal{C}_{k-1}(W_{r-1}))$, $r = k, k+1, \ldots, n$, and

$$N(\mathcal{C}_k(W_n)) = N(\mathcal{C}_{k-1}(W_{k-1})) + N(\mathcal{C}_{k-1}(W_k)) + \cdots + N(\mathcal{C}_{k-1}(W_{n-1})).$$

The last relation implies (2.8). ■

Example 2.18
Consider a series of five tosses of a coin. Calculate the number of different outcomes of the first r tosses that include exactly two heads, for $r = 2, 3, 4$. Compare the sum of these numbers to the number of different outcomes of the five tosses that include exactly three heads.

An outcome of a series of r tosses of a coin $(a_1, a_2, \ldots, a_r)$ that includes two heads corresponds to the set $\{i_1, i_2\}$ of the two positions, out of the r positions $\{1, 2, \ldots, r\}$ that the two letters h occupy. Thus the number different outcomes of the first r tosses that include exactly two heads equals

$$\binom{r}{2}, \quad r = 2, 3, 4,$$

the number of 2-combinations of the set $\{1, 2, \ldots, r\}$. Summing these numbers we get

$$\sum_{r=2}^{4} \binom{r}{2} = 1 + 3 + 6 = 10.$$

These ten outcomes are the following

$$(h, h), (h, h, t), (h, t, h), (t, h, h), (h, h, t, t),$$

$$(h, t, h, t), (h, t, t, h), (t, h, h, t), (t, h, t, h), (t, t, h, h).$$

The number of different outcomes of the five tosses that include exactly three heads is given by

$$\binom{5}{3} = \binom{5}{2} = \frac{5 \cdot 4}{2} = 10$$

and equals the sum $\sum_{r=2}^{4} \binom{r}{2}$, in agreement with the vertical recurrence relation (2.8). These ten outcomes are the following

$$(h, h, h, t, t), (h, h, t, h, t), (h, t, h, h, t), (t, h, h, h, t), (h, h, t, t, h),$$

$$(h, t, h, t, h), (h, t, t, h, h), (t, h, h, t, h), (t, h, t, h, h), (t, t, h, h, h).$$

Note that, a one-to-one correspondence exists between the two different sets of outcomes. Remark 2.4 will be helpful in establishing it.   □

In the next theorem the number of combinations with repetition is derived.

THEOREM 2.7
The number of k-combinations of n with repetition, denoted by $E(n, k)$, is given by

$$E(n, k) = \binom{n + k - 1}{k}. \tag{2.10}$$

PROOF   Let $\mathcal{E}_k(W_n)$ be the set of k-combinations with repetition of the set $W_n = \{w_1, w_2, \ldots, w_n\}$ of n elements and $\mathcal{C}_k(W_{n+k-1})$ be the set of k-combinations without repetition of the set $W_{n+k-1} = \{w_1, w_2, \ldots, w_{n+k-1}\}$ of

$n + k - 1$ elements. Consider a k-combination $\{w_{i_1}, w_{i_2}, \dots, w_{i_k}\}$ that belongs to $\mathcal{E}_k(W_n)$ and suppose that the k subscripts, $i_1, i_2, \dots, i_k$, are numbered in rising order. Then $1 \leq i_1 \leq i_2 \leq \cdots \leq i_k \leq n$ and, putting $j_r = i_r + (r - 1)$, $r = 1, 2, \dots, k$, it follows that $1 \leq j_1 < j_2 < \cdots < j_k \leq n + k - 1$ and the corresponding k-combination $\{w_{j_1}, w_{j_2}, \dots, w_{j_k}\}$ belongs to $\mathcal{C}_k(W_{n+k-1})$. Further, the relation $j_r = i_r + (r - 1), r = 1, 2, \dots, k$, corresponds to each k-combination from $\mathcal{E}_k(W_n)$ one and only one k-combination from $\mathcal{C}_k(W_{n+k-1})$ and vice versa. Thus, $N(\mathcal{E}_k(W_n)) = N(\mathcal{C}_k(W_{n+k-1}))$ and, by virtue of Theorem 2.5, (2.10) is deduced. ∎

REMARK 2.5   The double sequence of the numbers $E(n, k) = \binom{n+k-1}{k}$, $k = 1, 2, \dots, n = 1, 2, \dots$, of k-combinations of n with repetition, satisfies recurrence relations analogous to that of the double sequence of the numbers $C(n, k) = \binom{n}{k}$, $k = 1, 2, \dots, n, n = 1, 2, \dots$, of k-combinations of n. Specifically, there hold:

(a) The "triangular" recurrence relation

$$E(n, k) = E(n - 1, k) + E(n, k - 1), \ k = 1, 2, \dots, \ n = 1, 2, \dots.$$

(b) The "vertical" recurrence relation

$$E(n, k) = \sum_{r=1}^{n} E(r, k - 1), \ k = 1, 2, \dots, \ n = 1, 2, \dots.$$

(c) The "horizontal" recurrence relation

$$E(n, k) = \sum_{r=0}^{k} E(n - 1, r), \ k = 1, 2, \dots, \ n = 1, 2, \dots.$$

These recurrence relations may be shown by employing the corresponding technique of deriving (2.7), (2.8) and (2.9). The following brief hints of a combinatorial derivation of the first two recurrences are quoted here.

(a) The set $\mathcal{E}_k(W_n)$ of the k-combinations of n with repetition, by considering the element $w_n \in W_n$, can be divided into the disjoint subsets $\mathcal{A}$, of the k-combinations of n with repetition that include w_n, and $\mathcal{B}$, of the k-combinations of n with repetition that do not include w_n. Then $N(\mathcal{A}) = N(\mathcal{E}_k(W_{n-1}))$, $N(\mathcal{B}) = N(\mathcal{E}_{k-1}(W_n))$ and

$$N(\mathcal{E}_k(W_n)) = N(\mathcal{E}_k(W_{n-1})) + N(\mathcal{E}_{k-1}(W_n)),$$

yielding the triangular recurrence relation.

(b) The set $\mathcal{E}_k(W_n)$ can be divided into n pairwise disjoint subsets $\mathcal{A}_1, \mathcal{A}_2, \dots, \mathcal{A}_n$, where $\mathcal{A}_r$ is the subset of $\mathcal{E}_k(W_n)$ containing the combinations that include w_r as the element with the largest subscript, $r = 1, 2, \dots, n$. Then $N(\mathcal{A}_r) = N(\mathcal{E}_{k-1}(W_r)), r = 1, 2, \dots, n$ and

$$N(\mathcal{E}_k(W_n)) = N(\mathcal{E}_{k-1}(W_1)) + N(\mathcal{E}_{k-1}(W_2)) + \cdots + N(\mathcal{E}_{k-1}(W_n)),$$

which implies the vertical recurrence relation.   ∎

Example 2.19

The number of distinguishable outcomes of a throw of k like (indistinguishable) dice equals $E(6, k) = \binom{k+5}{k}$, the number of k-combinations of the set $\{1, 2, 3, 4, 5, 6\}$, of the six faces of a die, with repetition (see Example 2.9b). For $k = 2$ the possible outcomes are the following 21:

$$\{1,1\}, \{1,2\}, \{1,3\}, \{1,4\}, \{1,5\}, \{1,6\}, \{2,2\}, \{2,3\}, \{2,4\}, \{2,5\}, \{2,6\},$$

$$\{3,3\}, \{3,4\}, \{3,5\}, \{3,6\}, \{4,4\}, \{4,5\}, \{4,6\}, \{5,5\}, \{5,6\}, \{6,6\}.$$

To these 2-combinations with repetition of the set $\{1, 2, 3, 4, 5, 6\}$ of six numbers, there correspond the following, in order, 2-combinations (without repetition) of the set $\{1, 2, 3, 4, 5, 6, 7\}$ of seven numbers:

$$\{1,2\}, \{1,3\}, \{1,4\}, \{1,5\}, \{1,6\}, \{1,7\}, \{2,3\}, \{2,4\}, \{2,5\}, \{2,6\}, \{2,7\},$$

$$\{3,4\}, \{3,5\}, \{3,6\}, \{3,7\}, \{4,5\}, \{4,6\}, \{4,7\}, \{5,6\}, \{5,7\}, \{6,7\}. \quad \square$$

Example 2.20 Distinguishable balls into distinguishable urns of unlimited capacity

Suppose that k indistinguishable balls are successively distributed (allocated) into n distinguishable urns $\{w_1, w_2, \ldots, w_n\}$ of unlimited capacity. Find the number of different distributions.

Any distribution of k indistinguishable balls into n distinguishable urns of unlimited capacity corresponds to a collection of k urns $\{w_{i_1}, w_{i_2}, \ldots, w_{i_k}\}$ from the set of the n urns, with repetition, where the selection of an urn r times corresponds to the placement of r balls into it, $r = 0, 1, \ldots, k$. Thus, the number of ways of placing k indistinguishable balls into n distinguishable urns equals

$$\binom{n+k-1}{k},$$

the number of k-combinations of the set $\{w_1, w_2, \ldots, w_n\}$, of the n urns, with repetition. $\square$

Example 2.21 Display of flags on poles

Consider the following "primitive" way of signaling. A display of flags of the same or different colors on poles in a row forms a signal. The absolute position of flags on a pole does not count and each pole has space for all the available flags. Assume that k flags and n poles are available. The number of different signals that can be transmitted is of interest.

Consider first the case of flags of the same color and let $s_{n,k}$ be the number of different signals. If $a_{n,r}$ is the number of different signals when r flags are used, then, by the addition principle,

$$s_{n,k} = \sum_{r=0}^{k} a_{n,r}.$$

Note that, to each signal with r flags of the same color displayed on n distinguishable poles, there corresponds a collection of r poles $\{p_{i_1}, p_{i_2}, \dots, p_{i_r}\}$ with repetition from the set $\{p_1, p_2, \dots, p_n\}$, of the n poles, where the selection of a pole j times corresponds to the display of j flags on it, $j = 0, 1, \dots, r$. Thus the number $a_{n,r}$ equals

$$a_{n,r} = \binom{n+r-1}{r},$$

the number of r-combinations of n with repetition. Therefore,

$$s_{n,k} = \sum_{r=0}^{k} \binom{n+r-1}{r}$$

and, by the vertical recurrence of r-combinations of n with repetition, this sum is given by (see Remark 2.5):

$$s_{n,k} = \binom{n+k}{k}.$$

Further note that, to each signal with r flags of the same color displayed on n poles, there correspond $r!$ signals with r flags of different colors displayed on n poles, which are obtained by permuting the r distinguishable flags in all positive ways (without altering the number of flags on each pole). Thus the number $b_{n,r}$ of signals with r flags of different colors displayed on n distinguishable poles equals

$$b_{n,r} = r! a_{n,r} = r! \binom{n+r-1}{r}.$$

Note also that r flags can be chosen from the k flags of different colors in

$$\binom{k}{r} = \frac{(k)_r}{r!}$$

ways. Thus, according to the addition principle, the number $c_{n,k}$ of signals with at most k flags of different colors displayed on n distinguishable poles is given by the sum

$$c_{n,k} = \sum_{r=0}^{k} \binom{k}{r} b_{n,r} = \sum_{r=0}^{k} (k)_r \binom{n+r-1}{r}. \qquad \square$$

2.4   DIVISIONS AND PARTITIONS OF A FINITE SET

As has been noted in Remark 2.3, to each k-combination $\{a_1, a_2, \ldots, a_k\}$ of the set $W_n = \{w_1, w_2, \ldots, w_n\}$ of n elements, there corresponds a division of W_n in two subsets $A = \{a_1, a_2, \ldots, a_k\}$ and $B = \{b_1, b_2, \ldots, b_{n-k}\} = W_n - \{a_1, a_2, \ldots, a_k\}$, with k and $n - k$ elements, respectively. Thus, the number of divisions (A, B) of a finite set W_n, with $N(W_n) = n$, in two (ordered) subsets with $N(A) = k \geq 0$ and $N(B) = n - k \geq 0$ equals

$$C(n, k) \equiv \binom{n}{k} = \frac{(n)_k}{k!} = \frac{n!}{k!(n-k)!},$$

the number of k-combinations of n. An extension of this result is presented in the following theorem.

THEOREM 2.8
The number of divisions $(A_1, A_2, \ldots, A_r)$ of a finite set W_n, with $N(W_n) = n$, in r (ordered) subsets, with $N(A_j) = k_j \geq 0$, $j = 1, 2, \ldots, r$, denoted by $C(n, k_1, k_2, \ldots, k_{r-1})$ or $\binom{n}{k_1, k_2, \ldots, k_{r-1}}$, is given by

$$C(n, k_1, k_2, \ldots, k_{r-1}) \equiv \binom{n}{k_1, k_2, \ldots, k_{r-1}} = \frac{n!}{k_1! k_2! \cdots k_{r-1}! k_r!}, \quad (2.11)$$

where $k_r = n - k_1 - k_2 - \cdots - k_{r-1}$.

PROOF   In a division $(A_1, A_2, \ldots, A_r)$ of a finite set W_n, with $N(W_n) = n$, the k_1 elements of the first set A_1 can be chosen from the set W_n, of n elements, in

$$\binom{n}{k_1}$$

ways. After the selection of the k_1 elements of the first set A_1, the k_2 elements of the second set A_2 can be chosen from the set $W_n - A_1$, of $n - k_1$ elements, in

$$\binom{n - k_1}{k_2}$$

ways. Continuing in this manner, after the selection of the $k_1, k_2, \ldots, k_{r-2}$ elements of the sets $A_1, A_2, \ldots, A_{r-2}$, respectively, the k_{r-1} elements of the set A_{r-1} can be chosen from the set $W_n - A_1 - A_2 - \cdots - A_{r-2}$, of $n - k_1 - k_2 - \cdots - k_{r-2}$ elements, in

$$\binom{n - k_1 - k_2 - \cdots - k_{r-2}}{k_{r-1}}$$

ways. Finally, the selection of the $k_1, k_2, \ldots, k_{r-1}$ elements of the sets $A_1, A_2,$ $\ldots, A_{r-1}$, respectively, determines the k_r elements of the last set A_r, since $W_n - A_1 - A_2 - \cdots - A_{r-1}$ is a set of $n - k_1 - k_2 - \cdots - k_{r-1} = k_r$ elements. Thus, by the multiplication principle, the number $C(n, k_1, k_2, \ldots, k_{r-1})$ is given by

$$C(n, k_1, k_2, \ldots, k_{r-1}) = \binom{n}{k_1} \binom{n - k_1}{k_2} \cdots \binom{n - k_1 - k_2 - \cdots - k_{r-2}}{k_{r-1}}$$

$$= \frac{n!}{k_1!(n - k_1)!} \frac{(n - k_1)!}{k_2!(n - k_1 - k_2)!} \cdots \frac{(n - k_1 - k_2 - \cdots - k_{r-2})!}{k_{r-1}!(n - k_1 - k_2 - \cdots - k_{r-1})!}$$

$$= \frac{n!}{k_1! k_2! \cdots k_{r-1}! k_r!}$$

and so (2.11) is established. ∎

REMARK 2.6 A comparison of (2.11) with (2.5) reveals that

$$C(n, k_1, k_2, \ldots, k_{r-1}) = M(k_1, k_2, \ldots, k_r),$$

where $k_r = n - k_1 - k_2 - \cdots - k_{r-1}$. This relation is not a simple coincidence. There exists a one-to-one correspondence of the set of divisions $(A_1, A_2, \ldots, A_r)$ of a finite set $W_n = \{w_1, w_2, \ldots, w_n\}$, of n elements, in r (ordered) subsets, with $N(A_j) = k_j \geq 0$, $j = 1, 2, \ldots, r$, $k_1 + k_2 + \cdots + k_r = n$ and the set of permutations $(a_1, a_2, \ldots, a_n)$ of r kinds of elements $U_r = \{u_1, u_2, \ldots, u_r\}$, with k_j elements equal to u_j, $j = 1, 2, \ldots, r$, $k_1 + k_2 + \cdots + k_r = n$. Specifically, to the division $(A_1, A_2, \ldots, A_r)$, in which w_i belongs to the subset A_j, there corresponds the permutation $(a_1, a_2, \ldots, a_n)$, in which the element $a_i = u_j$, where $i = 1, 2, \ldots, n$ and $j = 1, 2, \ldots, r$ and vice versa. For example, for $n = 7$ and $r = 3$, to the division (A_1, A_2, A_3) of the set $W_7 = \{w_1, w_2, \ldots, w_7\}$, with $A_1 = \{w_1, w_4, w_6, w_7\}$, $A_2 = \{w_3\}$ and $A_3 = \{w_2, w_5\}$, there corresponds the permutation $(u_1, u_3, u_2, u_1, u_3, u_1, u_1)$ of the three kinds of elements $U_3 = \{u_1, u_2, u_3\}$, while to the division (A_1, A_2, A_3), with $A_1 = W_7, A_2 = \emptyset, A_3 = \emptyset$, there corresponds the permutation $(u_1, u_1, u_1, u_1, u_1, u_1, u_1)$. ∎

REMARK 2.7 The notation adopted for the expression of the number (2.11) constitutes an extension of the notation for the expression of the number (2.6), to which it reduces for $r = 2$. On the contrary, the notation of the number (2.11) by

$$C(n, k_1, k_2, \ldots, k_r) \equiv \binom{n}{k_1, k_2, \ldots, k_r},$$

also used by several authors, includes in addition to the numbers $n, k_1, k_2, \ldots, k_{r-1}$ and the number k_r, which, when the previous numbers are known, is completely determined by the relation $k_r = n - (k_1 + k_2 + \cdots + k_{r-1})$. This notation for $r = 2$ leads to the notation

$$C(n, k, n - k) \equiv \binom{n}{k, n - k}$$

for the number (2.6) that is not coherent with the classical notation

$$C(n, k) \equiv \binom{n}{k}.$$

Further, note that from the relation $k_1 + k_2 + \cdots + k_r = n$ it follows that any k_i, $i = 1, 2, \ldots, r$, can be expressed as $k_i = n - (k_1 + \cdots + k_{i-1} + k_{i+1} + \cdots + k_r)$, a difference of the sum of the rest k_j, $j = 1, 2, \ldots, i - 1, i + 1, \ldots, r$ from n. Thus, the numbers in (2.11) possess the following symmetric property:

$$\binom{n}{k_1, k_2, \ldots, k_{r-2}, k_{r-1}} = \binom{n}{k_1, k_2, \ldots, k_{r-2}, k_r} = \cdots$$
$$= \binom{n}{k_2, k_3, \ldots, k_{r-1}, k_r},$$

which extends the property

$$\binom{n}{k} = \binom{n}{n - k},$$

presented in the first part of Remark 2.3.   ∎

A recurrence relation for the number of divisions of a finite set is derived in the next theorem.

THEOREM 2.9
The number $C(n, k_1, k_2, \ldots, k_{r-1}) = \binom{n}{k_1, k_2, \ldots, k_{r-1}}$, of divisions $(A_1, A_2, \ldots, A_r)$ of a finite set W_n, with $N(W_n) = n$, in r (ordered) subsets, with $N(A_j) = k_j \geq 0$, $j = 1, 2, \ldots, r$, satisfies the recurrence relation

$$\binom{n}{k_1, k_2, \ldots, k_{r-1}} = \binom{n-1}{k_1 - 1, k_2, \ldots, k_{r-1}} + \binom{n-1}{k_1, k_2 - 1, \ldots, k_{r-1}} +$$
$$\cdots + \binom{n-1}{k_1, k_2, \ldots, k_{r-1} - 1} + \binom{n-1}{k_1, k_2, \ldots, k_{r-1}}, \quad (2.12)$$

for $k_j = 1, 2, \ldots, n$, $j = 1, 2, \ldots, r - 1$, $k_1 + k_2 + \cdots + k_{r-1} \leq n$, $n = 1, 2, \ldots$, with initial conditions

$$\binom{n}{0, 0, \ldots, 0} = 1, \quad n = 0, 1, \ldots,$$

$$\binom{n}{k_1, k_2, \ldots, k_{r-1}} = 0, \quad k_1 + k_2 + \cdots + k_{r-1} > n$$

and

$$\binom{n}{k_1, k_2, \ldots, k_{r-2}, 0} = \binom{n}{k_1, k_2, \ldots, k_{r-2}}, \ldots, \binom{n}{k_1, 0} = \binom{n}{k_1}, \binom{n}{0} = 1.$$

PROOF   Let $\mathcal{C}_{k_1, k_2, \ldots, k_{r-1}}(W_n)$ be the set of divisions $(A_1, A_2, \ldots, A_r)$ of $W_n = \{w_1, w_2, \ldots, w_n\}$ in r (ordered) subsets with $N(A_j) = k_j \geq 1$, $j = 1, 2, \ldots, r$ (whence $k_r = n - k_1 - k_2 - \cdots - k_r$). If $\mathcal{A}_j$ is the set of divisions $(A_1, A_2, \ldots, A_r)$ of W_n in r (ordered) subsets, with $N(A_j) = k_j \geq 1$, $j = 1, 2, \ldots, r$, in which $w_n \in A_j$, $j = 1, 2, \ldots, r$, then,

$$\mathcal{A}_i \cap \mathcal{A}_j = \emptyset, \, i, j = 1, 2, \ldots, r, \, i \neq j,$$

$$\mathcal{C}_{k_1, k_2, \ldots, k_{r-1}}(W_n) = \mathcal{A}_1 + \mathcal{A}_2 + \cdots + \mathcal{A}_{r-1} + \mathcal{A}_r$$

and, by the addition principle,

$$N(\mathcal{C}_{k_1, k_2, \ldots, k_{r-1}}(W_n)) = N(\mathcal{A}_1) + N(\mathcal{A}_2) + \cdots + N(\mathcal{A}_{r-1}) + N(\mathcal{A}_r).$$

Since

$$N(\mathcal{A}_1) = N(\mathcal{C}_{k_1-1, k_2, \ldots, k_{r-1}}(W_{n-1})), \; N(\mathcal{A}_2) = N(\mathcal{C}_{k_1, k_2-1, \ldots, k_{r-1}}(W_{n-1})),$$

$$N(\mathcal{A}_{r-1}) = N(\mathcal{C}_{k_1, k_2, \ldots, k_{r-1}-1}(W_{n-1})), \; N(\mathcal{A}_r) = N(\mathcal{C}_{k_1, k_2, \ldots, k_{r-1}}(W_{n-1})),$$

it follows that

$$\begin{aligned}
N(\mathcal{C}_{k_1, k_2, \ldots, k_{r-1}}(W_n)) = \; & N(\mathcal{C}_{k_1-1, k_2, \ldots, k_{r-1}}(W_{n-1})) \\
& + N(\mathcal{C}_{k_1, k_2-1, \ldots, k_{r-1}}(W_{n-1})) \\
& + \cdots + N(\mathcal{C}_{k_1, k_2, \ldots, k_{r-1}-1}(W_{n-1})) \\
& + N(\mathcal{C}_{k_1, k_2, \ldots, k_{r-1}}(W_{n-1})).
\end{aligned}$$

This relation, by virtue of Theorem 2.8, implies (2.12).

Note that recurrence relation (2.12) can also be shown algebraically by using expression (2.11).   ∎

REMARK 2.8   Consider a finite set $W_n = \{w_1, w_2, \ldots, w_n\}$, with $n = kr$, and the particular case of the divisions $(A_1, A_2, \ldots, A_r)$ of W_n in r (ordered) subsets, with $N(A_j) = k$, $j = 1, 2, \ldots, r$. The number $C_{n,k,r}$ of these divisions of W_n, according to Theorem 2.8, equals

$$C_{n,k,r} = \frac{n!}{(k!)^r}.$$

Further, consider the particular case of the partitions $\{A_1, A_2, \ldots, A_r\}$ of W_n in r (unordered) subsets, with $N(A_j) = k$, $j = 1, 2, \ldots, r$. To each such partition there correspond $r!$ from the preceding divisions and specifically those that are formed by permuting the r subsets in all possible ways. Thus, if $B_{n,k,r}$ denotes the number of these partitions of W_n, then $C_{n,k,r} = r! B_{n,k,r}$, and so

$$B_{n,k,r} = \frac{n!}{r!(k!)^r}.$$

In order to exemplify this connection, consider the divisions (A, B) and the partitions $\{A, B\}$ of the set $W_4 = \{w_1, w_2, w_3, w_4\}$ in two subsets, with $N(A) = N(B) = 2$. The partitions are

$$\{\{w_1, w_2\}, \{w_3, w_4\}\}, \ \{\{w_1, w_3\}, \{w_2, w_4\}\}, \ \{\{w_1, w_4\}, \{w_2, w_3\}\}$$

and the corresponding divisions are

$$(\{w_1, w_2\}, \{w_3, w_4\}), (\{w_1, w_3\}, \{w_2, w_4\}), (\{w_1, w_4\}, \{w_2, w_3\}),$$

$$(\{w_3, w_4\}, \{w_1, w_2\}), (\{w_2, w_4\}, \{w_1, w_3\}), (\{w_2, w_3\}, \{w_1, w_4\}),$$

in agreement with $B_{4,2,2} = 3$ and $C_{4,2,2} = 6$. ∎

The next theorem is concerned with the number of partitions of a finite set and constitutes a generalization of the conclusion of the preceding remark.

THEOREM 2.10
The number of partitions $\{A_1, A_2, \ldots, A_r\}$ of a finite set W_n, with $N(W_n) = n$, in r (unordered) subsets, among which $r_j \geq 0$ include j elements, $j = 1, 2, \ldots, n$, denoted by $B(n, r; r_1, r_2, \ldots, r_n)$, equals

$$B(n, r; r_1, r_2, \ldots, r_n) = \frac{n!}{r_1!(1!)^{r_1} r_2!(2!)^{r_2} \cdots r_n!(n!)^{r_n}}, \qquad (2.13)$$

where $r_1 + r_2 + \cdots + r_n = r$ and $r_1 + 2r_2 + \cdots + nr_n = n$.

PROOF   Consider a partition $\{A_1, A_2, \ldots, A_r\}$ of W_n, in $r_j \geq 0$ subsets including j elements each, $j = 1, 2, \ldots, n$, with $r_1 + r_2 + \cdots + r_n = r$ and $r_1 + 2r_2 + \cdots + nr_n = n$. To it there correspond $r_1! r_2! \cdots r_n!$ divisions of W_n in r subsets, with $r_j \geq 0$ subsets including j elements each, $j = 1, 2, \ldots, n$, and, specifically, those that are formed by internal permutation of the $r_1 \geq 0$ subsets with one element, the $r_2 \geq 0$ subsets with two elements, and, finally, the $r_n \geq 0$ subsets with n elements in all possible ways. Thus,

$$C(n, 1, \ldots, 1, 2, \ldots, 2, \ldots) = r_1! r_2! \cdots r_n! B(n, r; r_1, r_2, \ldots, r_n)$$

and also by (2.11),

$$C(n, 1, \dots, 1, 2, \dots, 2, \dots) = \frac{n!}{(1!)^{r_1}(2!)^{r_2}\cdots(n!)^{r_n}}.$$

From these two relations, (2.13) is deduced. ∎

Example 2.22 Elevator passengers' discharge

Consider an elevator of a building of five floors that starts from the basement with eight passengers. Find the number of ways of discharging $k_1 = 0$, $k_2 = 3$, $k_3 = 1$, $k_4 = 2$ and $k_5 = 2$ passengers at the first, second, third, fourth and fifth floors, respectively.

Note that to each way of discharging $k_1 = 0$, $k_2 = 3$, $k_3 = 1$, $k_4 = 2$ and $k_5 = 2$ passengers at the first, second, third, fourth and fifth floors, respectively, there corresponds a division of the set $W_8 = \{w_1, w_2, \dots, w_8\}$, of eight passengers, in five (ordered) subsets $(A_1, A_2, \dots, A_5)$, with $N(A_1) = 0$, $N(A_2) = 3$, $N(A_3) = 1$, $N(A_4) = 2$ and $N(A_5) = 2$, where the discharge of the passenger w_i at the j-th floor corresponds to $w_i \in A_j$, $i = 1, 2, \dots, 8$, $j = 1, 2, \dots, 5$. Thus, the required number, according to Theorem 2.8, equals

$$\binom{8}{0, 3, 1, 2} = \frac{8!}{0!3!1!2!(8-3-1-2)!} = 1680. \quad \square$$

Example 2.23

Consider a series of 21 throws of a die. Find the number of different outcomes in which number i appears i times, $i = 1, 2, \dots, 6$, so that $1 + 2 + \cdots + 6 = 21$.

An outcome of a series of 21 tosses of a die $(a_1, a_2, \dots, a_{21})$ that includes i times the number i, $i = 1, 2, \dots, 6$, corresponds uniquely to a division $(A_1, A_2, \dots, A_6)$ of the set $\{1, 2, \dots, 21\}$, of the 21 throws of the die, in six ordered subsets, where A_i is the set of throws in which number i appears, with $N(A_i) = i$, $i = 1, 2, \dots, 6$. Thus, by Theorem 2.8, the required number equals

$$\binom{21}{1, 2, 3, 4, 5} = \frac{21!}{1!2!3!4!5!6!}. \quad \square$$

Example 2.24 Distributions of students in sections

Suppose that $n = kr$ students of the School of Sciences have chosen to attend a specific course from the department of mathematics. In order to provide better attendance conditions, these students are distributed in r sections, each with k students. Determine the number of such distributions.

(a) If the r sections are considered distinguishable (since the lectures take place either in different days of the week or in different hours of a day), then the required number, according to Theorem 2.8, equals

$$C_{n,k,r} = \frac{n!}{(k!)^r},$$

the number of divisions $(A_1, A_2, \ldots, A_r)$ of the set $W_n = \{w_1, w_2, \ldots, w_n\}$, of the n students, in r subsets (sections) with $N(A_j) = k, j = 1, 2, \ldots, r$.

(b) If the r sections are considered as indistinguishable (since they cover the same amount of knowledge), then the required number, according to Theorem 2.10, equals

$$B_{n,k,r} = \frac{n!}{r!(k!)^r},$$

the number of partitions $\{A_1, A_2, \ldots, A_r\}$ of the set $W_n = \{w_1, w_2, \ldots, w_n\}$, of the n students, in r subsets (sections) with $N(A_j) = k, j = 1, 2, \ldots, r$.   ☐

Example 2.25 *Distributions of distinguishable balls into indistinguishable urns*

Consider the distributions of n distinguishable balls into r indistinguishable urns, in which $r_j \geq 0$ urns include j balls, $j = 0, 1, 2, \ldots, n$, with $r_0 + r_1 + r_2 + \cdots + r_n = r$ and $r_1 + 2r_2 + \cdots + nr_n = n$. Find the number of such distributions.

Such a distribution corresponds to a partition of the set $B_n = \{b_1, b_2, \ldots, b_n\}$, of the n balls, in $r - r_0$ (unordered) subsets, among which $r_j \geq 0$ include j elements (balls), $j = 1, 2, \ldots, n$, with $r_1 + r_2 + \cdots + r_n = r - r_0$ and $r_1 + 2r_2 + \cdots + nr_n = n$. Thus, according to Theorem 2.10, the required number equals

$$B(n, r - r_0; r_1, r_2, \ldots, r_n) = \frac{n!}{r_1!(1!)^{r_1} r_2!(2!)^{r_2} \cdots r_n!(n!)^{r_n}}.$$

In particular, for $n = 10, r = 5$ and $r_0 = 0, r_1 = 2, r_2 = 1, r_3 = 2$, the number of distributions equals

$$\frac{10!}{2!(1!)^2 1!(2!)^1 2!(3!)^2} = 12{,}600. \quad ☐$$

2.5 INTEGER SOLUTIONS OF A LINEAR EQUATION

Consider the linear equation

$$x_1 + x_2 + \cdots + x_n = k, \tag{2.14}$$

where k is an integer. The calculation of the number of integer solutions $(r_1, r_2, \ldots, r_n)$ of (2.14) satisfying a given set of restrictions is reduced to the calculation of the number of combinations (with or without repetition) of the n indices $\{1, 2, \dots , n\}$ satisfying analogous restrictions.

In the following theorems and corollaries, the number of integer solutions of the linear equation (2.14) under several sets of restrictions is derived.

THEOREM 2.11

The number of solutions $(r_1, r_2, \dots , r_n)$ of the linear equation (2.14), with $x_i = 0$ or 1, $i = 1, 2, \dots , n$, and k a nonnegative integer, equals

$$\binom{n}{k},$$

the number of k-combinations of n.

PROOF   Note that, to each solution $(r_1, r_2, \dots , r_n)$ of (2.14), with $r_i = 0$ or 1, $i = 1, 2, \dots , n$, there corresponds an assignment of the k units of the right-hand side to the n distinguishable summands of the left-hand side of (2.14), assigning r_i units to the i-th summand, $i = 1, 2, \dots , n$. Since the possible values of r_i are 0 and 1, such an assignment may be expressed as $\{a_1, a_2, \dots , a_k\}$, where $a_j \in \{1, 2, \dots , n\}$ and $a_i \neq a_j$, by listing only the indices of the summands to which a unit is assigned (where the order of listing does not matter). This correspondence is obviously one to one and thus the number of solutions $(r_1, r_2, \dots , r_n)$ of (2.14), with $r_i = 0$ or 1, $i = 1, 2, \dots , n$, equals $\binom{n}{k}$, the number of k-combinations of n.   ∎

THEOREM 2.12

The number of nonnegative integer solutions $(r_1, r_2, \dots , r_n)$ of the linear equation (2.14), with k a nonnegative integer, equals

$$\binom{n+k-1}{k},$$

the number of k-combinations of n with repetition.

PROOF   Note that, to each nonnegative integer solution $(r_1, r_2, \dots , r_n)$ of (2.14), there corresponds an assignment of the k units of the right-hand side to the n distinguishable summands of the left-hand side of (2.14), which assigns $r_i \geq 0$ units to the i-th summand, $i = 1, 2, \dots , n$. Such an assignment may be expressed as $\{a_1, a_2, \dots , a_k\}$, where r_1 elements are equal to 1, r_2 elements are equal to $2, \dots , r_n$ elements are equal to n, by listing $r_i \geq 0$ times the index i, for $i = 1, 2, \dots , n$ (with the order of listing not counting). Thus, to each nonnegative integer solution $(r_1, r_2, \dots , r_n)$ of (2.14) there corresponds a k-combination

$\{a_1, a_2, \ldots, a_k\}$ of the n indices $\{1, 2, \ldots, n\}$ with repetition, in which the element (index) i is included $r_i \geq 0$ times, for $i = 1, 2, \ldots, n$. This correspondence is obviously one to one and so the number of nonnegative integer solutions of (2.14) equals $\binom{n+k-1}{k}$, the number of k-combinations of n with repetition. ∎

COROLLARY 2.4

(a) *The number of integer solutions* $(r_1, r_2, \ldots, r_n)$ *of the linear equation* (2.14), *with the restrictions*

$$x_i \geq s_i, \quad i = 1, 2, \ldots, n \tag{2.15}$$

for given integers s_i, $i = 1, 2, \ldots, n$, *with* $s = s_1 + s_2 + \cdots + s_n \leq k$, *equals*

$$\binom{n+k-s-1}{k-s} = \binom{n+k-s-1}{n-1}.$$

(b) *The number of integer solutions* $(r_1, r_2, \ldots, r_n)$ *of the linear equation* (2.14), *with the restrictions*

$$x_i \leq m_i, \quad i = 1, 2, \ldots, n \tag{2.16}$$

for given integers m_i, $i = 1, 2, \ldots, n$, *with* $m = m_1 + m_2 + \cdots + m_n \geq k$, *equals*

$$\binom{n+m-k-1}{m-k} = \binom{n+m-k-1}{n-1}.$$

PROOF   (a) On using the transformation $y_i = x_i - s_i$, $i = 1, 2, \ldots, n$, the linear equation (2.14) and the restrictions (2.15) reduce to the linear equation

$$y_1 + y_2 + \cdots + y_n = k - s, \tag{2.17}$$

with $y_i \geq 0$, $i = 1, 2, \ldots, n$ and $k - s$ a nonnegative integer. Since this transformation is one-to-one, to each integer solution $(r_1, r_2, \ldots, r_n)$ of the linear equation (2.14), with the restrictions (2.15), there corresponds one and only one nonnegative integer solution $(r_1 - s_1, r_2 - s_2, \ldots, r_n - s_n)$ of the linear equation (2.17) and vice versa. Thus, the required number equals the number of nonnegative integer solutions of (2.17), which, by Theorem 2.12, is given by

$$\binom{n+k-s-1}{k-s} = \binom{n+k-s-1}{n-1}.$$

(b) Similarly, upon using the transformation $z_i = m_i - x_i$, $i = 1, 2, \ldots, n$, the linear equation (2.14) and the restrictions (2.16) reduce to the linear equation

$$z_1 + z_1 + \cdots + z_n = m - k, \tag{2.18}$$

with $z_i \geq 0$, $i = 1, 2, \ldots, n$ and $m - k$ a nonnegative integer. Since this transformation (4.8) is also one-to-one, arguing as before, it follows that the required number equals the number of nonnegative integer solutions of (2.18), which, by Theorem 2.12, is given by

$$\binom{n + m - k - 1}{m - k} = \binom{n + m - k - 1}{n - 1}.$$

The proof of the corollary is thus completed. ∎

The particular case with $s_i = 1$, $i = 1, 2, \ldots, n$ in the restrictions (2.15), which corresponds to the calculation of the number of positive integer (integral) solutions of the linear equation (2.14), is worth a separate presentation.

COROLLARY 2.5
The number of positive integer solutions $(r_1, r_2, \ldots, r_n)$ *of the linear equation (2.14), with $k \geq n$, equals*

$$\binom{k - 1}{k - n} = \binom{k - 1}{n - 1}.$$

REMARK 2.9   The number of integer solutions $(r_1, r_2, \ldots, r_n)$ of the linear equation (2.14), with the restrictions

$$s_i \leq x_i \leq m_i, \quad i = 1, 2, \ldots, n,$$

for given integers s_i, m_i, $i = 1, 2, \ldots, n$, with

$$s \leq k \leq m, \ s = s_1 + s_2 + \cdots + s_n, \ m = m_1 + m_2 + \cdots + m_n,$$

is calculated in Chapter 4 (Theorem 4.3) by applying the inclusion and exclusion principle. ∎

REMARK 2.10   *Integer solutions of a linear equation and distributions of indistinguishable balls into distinguishable urns.* The existence of a one-to-one correspondence between the integer solutions of the linear equation (2.14) and the distributions of k indistinguishable balls into n distinguishable urns $\{w_1, w_2, \ldots, w_n\}$ is already indicated in the proofs of Theorems 2.11 and 2.12. Specifically, to an integer solution $(r_1, r_2, \ldots, r_n)$ of the linear equation (2.14) that satisfies a given set of restrictions, there uniquely corresponds the distribution of k indistinguishable balls into n distinguishable urns, $\{w_1, w_2, \ldots, w_n\}$, that assigns (allocates) r_i balls into urn w_i for all $i = 1, 2, \ldots, n$, and vice versa. ∎

Example 2.26

Consider a collection of 32 cards of four different colors (blue, green, red and yellow), with eight cards from each color. Find the number of different sub-collections of (a) five cards and (b) at most five cards.

(a) Let x_i be the number of cards of the i-th color, $i = 1, 2, 3, 4$. Then a sub-collection of five cards is a nonnegative integer solution of the linear equation

$$x_1 + x_2 + x_3 + x_4 = 5.$$

Consequently, by Theorem 2.12, the number of different sub-collections of five cards equals

$$\binom{8}{5} = \binom{8}{3} = \frac{8 \cdot 7 \cdot 6}{2 \cdot 3} = 56.$$

(b) A sub-collection of k cards is a nonnegative integer solution of the linear equation

$$x_1 + x_2 + x_3 + x_4 = k$$

and, according to Theorem 2.12, the number of different sub-collections of k cards equals

$$\binom{4 + k - 1}{k}, \quad k = 0, 1, \dots, 5.$$

Clearly, the number of sub-collections with at most five cards equals the sum of these numbers,

$$\sum_{k=0}^{5} \binom{4 + k - 1}{k} = 1 + 4 + 10 + 20 + 35 + 56 = 126.$$

Note that, by the vertical recurrence relation of the number of k-combinations of 4 with repetition, this sum is given by (see Remark 2.5)

$$\binom{4 + 5}{5} = \binom{9}{5} = \binom{9}{4} = \frac{9 \cdot 8 \cdot 7 \cdot 6}{2 \cdot 3 \cdot 4} = 126.$$

The number of sub-collections with at most five cards may also be derived by noting that a sub-collection of at most five cards is a nonnegative integer solution of the linear inequality

$$x_1 + x_2 + x_3 + x_4 \le 5.$$

The number of nonnegative integer solutions (r_1, r_2, r_3, r_4) of this linear inequality, upon introducing the surplus variable $x_5 = 5 - (x_1 + x_2 + x_3 + x_4)$, equals the number of nonnegative integer solutions $(r_1, r_2, r_3, r_4, r_5)$ of the linear equation

$$x_1 + x_2 + x_3 + x_4 + x_5 = 5,$$

which, according to Theorem 2.12, is given by

$$\binom{9}{5} = 126. \quad \square$$

Example 2.27 Transportation of products of different kinds

Let us consider a storehouse that has n kinds of a merchandized product, packed in boxes of equal size, with a proper identity label on each, in stocks of at least s boxes from each kind at any moment. Calculate the number of loads of different composition that a truck with capacity of k boxes may carry for the execution of orders requiring at least s boxes from each kind of the product .

Let $x_i \geq s$ be the number of boxes of the i-th kind that the truck may carry, $i = 1, 2, \ldots, n$. Then $x_1 + x_2 + \cdots + x_n$ is the total load of the truck, which, of course, does not exceed its capacity. Therefore, the required number equals the number of integer solutions $(r_1, r_2, \ldots, r_n)$ of the linear inequality

$$x_1 + x_2 + \cdots + x_n \leq k,$$

with the restrictions

$$x_i \geq s, \quad i = 1, 2, \ldots, n, \quad ns \leq k,$$

which, upon introducing the surplus variable $x_{n+1} = k - (x_1 + x_2 + \cdots + x_n)$, equals the number of integer solutions $(r_1, r_2, \ldots, r_n, r_{n+1})$, with $r_{n+1} = k - (r_1 + r_2 + \cdots + r_n)$, of the linear equation

$$x_1 + x_2 + \cdots + x_n + x_{n+1} = k,$$

with the restrictions

$$x_i \geq s, \quad i = 1, 2, \ldots, n, \quad x_{n+1} \geq 0, \quad ns \leq k.$$

Consequently, according to Corollary 2.4, this number is

$$\binom{k + n - ns}{k - ns} = \binom{k + n - ns}{n}. \qquad \Box$$

Example 2.28 Fibonacci numbers

Find the number $C_1(n, k)$ of k-combinations of the set $\{1, 2, \ldots, n\}$ that include no pair of consecutive integers and conclude the (total) number C_n of combinations of $\{1, 2, \ldots, n\}$ that include no pair of consecutive integers.

Let $\{i_1, i_2, \ldots, i_k\}$ be any k-combination of the set $\{1, 2, \ldots, n\}$. Since the order of the elements included in any combination does not count, it may be assumed that

$$1 \leq i_1 < i_2 < \cdots < i_k \leq n.$$

Consider the distance between the successive elements of this k-combination:

$$j_2 = i_2 - i_1, \quad j_3 = i_3 - i_2, \ldots, \quad j_k = i_k - i_{k-1}$$

and also

$$j_1 = i_1, \quad j_{k+1} = n - i_k.$$

Adding these equations by parts, the following linear equation

$$j_1 + j_2 + \cdots + j_k + j_{k+1} = n$$

is deduced. Thus, each k-combination $\{i_1, i_2, \ldots, i_k\}$ of the set $\{1, 2, \ldots, n\}$ that includes no pair of consecutive integers corresponds to an integer solution of this linear equation that satisfies the restrictions

$$j_1 \geq 1, \ j_2 \geq 2, \ j_3 \geq 2, \ldots, \ j_k \geq 2, \ j_{k+1} \geq 0,$$

with $s = s_1 + s_2 + \cdots + s_k + s_{k+1} = 1 + 2(k-1) = 2k - 1 \leq n$, and vice versa. Consequently, according to Corollary 2.4, the number $C_1(n, k)$ of k-combinations of the set $\{1, 2, \ldots, n\}$ that include no pair of consecutive integers is given by

$$C_1(n, k) = \binom{n - k + 1}{k}, \ k = 0, 1, \ldots, [(n+1)/2],$$

where $[x]$ denotes the integer part of x. Then, the (total) number C_n of combinations of $\{1, 2, \ldots, n\}$ that include no pair of consecutive integers is given by the sum

$$C_n = \sum_{k=0}^{[(n+1)/2]} \binom{n - k + 1}{k}, \ n = 0, 1, \ldots.$$

The numbers $f_0 = 1$, $f_n = C_{n-1}$, $n = 1, 2, \ldots$, are called *Fibonacci numbers*. The number f_n first appeared as the number of pairs of rabbits born from one pair at the end of the $(n-1)$-st month. This problem was posed in 1202 by Leonardo Fibonacci in his book on abacus (*Liber Abaci*). ▯

Example 2.29 Lucas numbers

Assume that the n numbers of the set $\{1, 2, \ldots, n\}$ are displayed on a circle, whence $(n, 1)$ is a pair of consecutive points. Find the number $B_1(n, k)$ of k-combinations of the set $\{1, 2, \ldots, n\}$ displayed on a circle that include no pair of consecutive points and conclude the (total) number K_n of combinations of the set $\{1, 2, \ldots, n\}$ displayed on a circle that include no pair of consecutive points.

The set of k-combinations of the set $\{1, 2, \ldots, n\}$ displayed on a circle that include no pair of consecutive points can be divided into two disjoint subsets: the subset of these combinations that include n and the subset of these combinations that do not include n. Further, (a) the number of these combinations that do not include n equals

$$C_1(n - 1, k) = \binom{n - k}{k},$$

the number of k combinations of the $n - 1$ consecutive integers $\{1, 2, \ldots, n-1\}$ displayed on a straight line, including no pair of consecutive integers. (b) The number of these combinations that include n, and thus do not include neither 1 nor $n-1$, equals

$$C_1(n-3, k-1) = \binom{n-k-1}{k-1},$$

the number of $(k-1)$-combinations of the $n-3$ consecutive positive integers $\{2, 3, \ldots, n-2\}$ displayed on a straight line, including no pair of consecutive integers. Hence, by the addition principle, the required number is given by the sum

$$B_1(n, k) = \binom{n-k}{k} + \binom{n-k-1}{k-1}$$

and since

$$k\binom{n-k}{k} = (n-k)\binom{n-k-1}{k-1},$$

it follows that

$$B_1(n, k) = \frac{n}{n-k}\binom{n-k}{k} = \frac{n}{k}\binom{n-k-1}{k-1}, \ k = 0, 1, \ldots, [n/2].$$

The (total) number K_n of combinations of the set $\{1, 2, \ldots, n\}$ displayed on a circle that include no pair of consecutive points is given by the sum

$$K_n = \sum_{k=0}^{[n/2]} \frac{n}{n-k}\binom{n-k}{k}, \ n = 1, 2, \ldots .$$

The numbers $g_0 = 2$, $g_n = K_n$, $n = 1, 2, \ldots$, which are connected with the Fibonacci numbers (see Example 2.28) by

$$g_n = f_n + f_{n-2}, \ n = 2, 3, \ldots ,$$

are called *Lucas numbers*. $\Box$

2.6 LATTICE PATHS

Consider an orthogonal system of axes on the plane xy and the *lattice* defined by the straight lines $x = i$, $i = 0, \pm1, \pm2, \ldots$, and $y = j$, $j = 0, \pm1, \pm2, \ldots$, that are orthogonal to the (horizontal) axis x and the (vertical) axis y, respectively.

A directed polygonal line on the lattice that leads from the point (r, s) to the point (n, k), $r \leq n$, $s \leq k$, through horizontal and vertical straight sections, positively directed, is called *lattice path* (or, more precisely, *minimal lattice path*) from the point (r, s) to the point (n, k).

A horizontal section of a lattice path between two successive points (i, j) and $(i + 1, j)$ (of the horizontal line $y = j$) is called *horizontal (unit) step*, while a vertical section of a lattice path between two successive points (i, j) and $(i, j + 1)$ (of the vertical line $x = i$) is called *vertical (unit) step*.

According to these definitions, a lattice path from the point (r, s) to the point (n, k), $r \leq n$, $s \leq k$, is completely determined in any of the following ways:

(a) by the set $\{(x_0, y_0), (x_1, y_1), \ldots, (x_{n+k-r-s}, y_{n+k-r-s})\}$ of the $n + k - r - s + 1$ points through which it passes, with $(x_0, y_0) = (r, s)$ and $(x_{n+k-r-s}, y_{n+k-r-s}) = (n, k)$,

(b) by the permutation $(s_1, s_2, \ldots, s_{n+k-r-s})$ of the $n - r$ symbols h and $k - s$ symbols v, where $s_i = h$ if the i-th step is a horizontal one and $s_i = v$ if the i-th step is a vertical one, $i = 1, 2, \ldots, n + k - r - s$,

(c) either by the set $\{i_1, i_2, \ldots, i_{n-r}\}$ of the positions of the horizontal steps, a subset of the set $\{1, 2, \ldots, n + k - r - s\}$, or by the set $\{j_1, j_2, \ldots, j_{k-s}\} = \{1, 2, \ldots, n + k - r - s\} - \{i_1, i_2, \ldots, i_{n-r}\}$ of the positions of the vertical steps.

The calculation of the lattice paths from the point (r, s) to the point (n, k) may be reduced, by a parallel transfer of the orthogonal system of axes, $z = x - r$, $w = y - s$, to the calculation of the lattice paths from the origin $(0, 0)$ to the point $(n - r, k - s)$. The following theorem and corollary are concerned with the enumeration of such paths.

THEOREM 2.13

The number of lattice paths from the origin $(0, 0)$ to the point (n, k), with $n \geq 0$, $k \geq 0$, is given by

$$L(n, k) = \frac{(n + k)!}{n! k!} = \binom{n + k}{k}. \tag{2.19}$$

PROOF   To each lattice path from the origin $(0, 0)$ to the point (n, k), which requires n horizontal (h) and k vertical (v) successive (unit) steps (positively directed), there corresponds a permutation $(s_1, s_2, \ldots, s_{n+k})$ of the n symbols h and the k symbols v and vice versa. Thus, the number of lattice paths from the origin $(0, 0)$ to the point (n, k) equals the number of permutations of $n + k$ symbols, among which n are h and k are v, and according to Theorem 2.4, is given by (2.19).   ∎

COROLLARY 2.6

The number of lattice paths from the point (r, s) to the point (n, k), with $n \geq r$,

$k \geq s$, *is given by*

$$L(r, s; n, k) = \binom{n + k - r - s}{k - s}. \tag{2.20}$$

In Figure 2.1, the number of lattice paths from the origin $(0,0)$ to the lattice point (n, k), for $n = 1, 2, \dots, 5$ and $k = 1, 2, \dots, 5$, is quoted.

FIGURE 2.1
Lattice paths

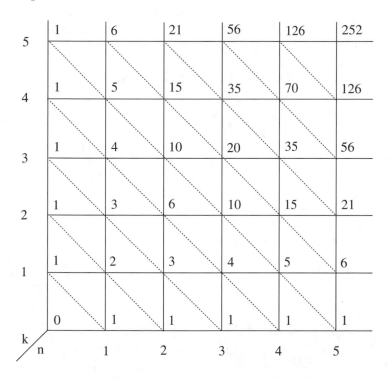

REMARK 2.11 (a) Each lattice path from the point (r, s) to the point (n, k) corresponds to an $(n - r)$-combination $\{i_1, i_2, \dots, i_{n-r}\}$, which determines the positions of the horizontal steps among the $n + k - r - s$ possible positions $\{1, 2, \dots, n + k - r - s\}$ and vice versa. Thus, the number of lattice paths from the point (r, s) to the point (n, k) equals the number of $(n - r)$-combinations of $n + k - r - s$, which, according to Theorem 2.5, is given by

$$\binom{n + k - r - s}{n - r},$$

which is equal to (2.20).

(b) Let w_i be the section of the vertical line $x = i$ between the points (i, s) and (i, k), $i = r, r + 1, \ldots, n$ and $W_{n-r+1} = \{w_r, w_{r+1}, \ldots, w_n\}$. Each of $k - s - 1$ horizontal lines $y = j$, $j = s + 1, s + 2, \ldots, k - 1$, divides every vertical section w_i into $k - s$ unit sections that are possible vertical steps of the path. So, the lattice path from the point (r, s) to the point (n, k) that has m_i vertical steps on the section w_i, with $0 \leq m_i \leq k - s$, for $i = r, r + 1, \ldots, n$ and $m_r + m_{r+1} + \cdots + m_n = k - s$, corresponds to the $(k - s)$-combination $\{w_{i_1}, w_{i_2}, \ldots, w_{i_{k-s}}\}$ of the set $W_{n-r+1} = \{w_r, w_{r+1}, \ldots, w_n\}$, of $n - r + 1$ sections, with repetition, in which the section w_i is included m_i times and vice versa. Thus, the number of lattice paths from the point (r, s) to the point (n, k) equals the number of $(k - s)$-combinations of $n - r + 1$ with repetition, which, according to Theorem 2.7, is given by (2.20). ∎

Let us consider the straight line $x = y + m$, with m an integer and two lattice points (r, s) and (n, k), with $r \leq n$, $s \leq k$, that lie to the same side of the line that is either (a) $r > s + m$ and $n > k + m$ or (b) $r < s + m$ and $n < k + m$. In the case $r > s + m$ and $n > k + m$, if $r \leq k + m$, as well as in the case $r < s + m$ and $n < k + m$, if $s + m \leq n$, there exist lattice paths from the point (r, s) to the point (n, k) that do not touch or intersect the line $x = y + m$. The calculation of these lattice paths may be reduced to the calculation of the lattice paths from the point that lies symmetric to the point (r, s), with respect to the line $x = y + m$, to the point (n, k), without any restriction. This reduction is achieved by applying the *reflection principle*. This principle, attributed to the French mathematician Désiré André (1840-1918) who essentially used it to solve the famous *ballot problem*, is formulated and proved in the next lemma.

LEMMA 2.1

Let $\mathcal{R}(r, s; n, k; m)$ be the set of lattice paths from the point (r, s) to the point (n, k), with $s + m < r \leq k + m < n$ or $r < s + m \leq n < k + m$, that do not touch or intersect the straight line $x = y + m$. Further, let $\mathcal{L}(s + m, r - m; n, k)$ be the set of lattice paths from the point $(s + m, r - m)$ to the point (n, k), without any restriction. Then a one-to-one correspondence exists between the sets $\mathcal{R}(r, s; n, k; m)$ and $\mathcal{L}(s + m, r - m; n, k)$ and hence,

$$N(\mathcal{R}(r, s; n, k; m)) = \binom{n + k - r - s}{k - r + m} = \binom{n + k - r - s}{n - s - m}. \quad (2.21)$$

PROOF   Note that each lattice path from the set $\mathcal{R}(r, s; n, k; m)$ meets the straight line $x = y + m$, for the first time, at one of the points

$$(i, i - m), \, i = r, r + 1, \ldots, k + m \text{ if } s + m < r \leq k + m < n$$

FIGURE 2.2
Reflection principle

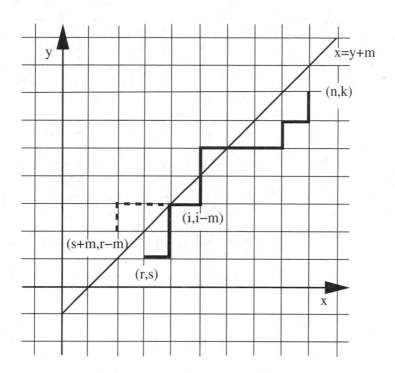

or at one of the points

$$(j + m, j), \; j = s, s + 1, \ldots , n - m \; \text{ if } \; r < s + m < n \leq k + m.$$

Assume that $s + m < r \leq k + m \leq n$ and consider a lattice path from the set $\mathcal{R}(r, s; n, k; m)$ that meets the straight line $x = y + m$, for the first time, at a specific point $(i, i - m)$. This point separates the lattice path in two sections: (a) from the point (r, s) to the point $(i, i - m)$ and (b) from the point $(i, i - m)$ to the point (n, k) (see Figure 2.2). To this lattice path there corresponds a lattice path from the set $\mathcal{L}(s + m, r - m; n, k)$ that passes from the point $(i, i - m)$; its section from $(s + m, r - m)$ to $(i, i - m)$ is the reflection of (a) with respect to the straight line $x = y + m$, that is, the point (u, v) is reflected to the point $(v + m, u - m)$, $u = r, r + 1, \ldots , i$, $v = s, s + 1, \ldots , i - m$, while its section from $(i, i - m)$ to (n, k) coincides with (b). This correspondence is one-to-one and thus $N(\mathcal{R}(r, s; n, k; m)) = N(\mathcal{L}(s + m, r - m; n, k))$, whence on using (2.20), the expression (2.21) is deduced. The other case $r < s + m \leq n < k + m$ is similarly treated. ∎

Example 2.30 The ballot problem

Consider a ballot between two candidates $\mathcal{N}$ and $\mathcal{K}$, in which $\mathcal{N}$ receives n votes and $\mathcal{K}$ receives k votes. (a) If $n > k$, calculate the number $\Psi_{n,k}$ of ways of counting the votes such that, at any step, the votes for $\mathcal{N}$ are more than the votes for $\mathcal{K}$. (b) If $n \geq k$, calculate the number $\psi_{n,k}$ of ways of counting the votes such that, at any step, the votes for $\mathcal{K}$ are not more than the votes for $\mathcal{N}$.

Let x_m and y_m be the numbers of votes for $\mathcal{N}$ and $\mathcal{K}$, respectively, after counting m votes, $m = 1, 2, \ldots, n + k$. Then, the ordered pairs (x_m, y_m), $m = 1, 2, \ldots, n + k$, are points of the lattice $x = i, y = j, i, j = 0, \pm 1, \pm 2, \ldots$ and

(a) $\Psi_{n,k}$ is the number of lattice paths from the origin $(0, 0)$ to the point (n, k) such that $x_m > y_m$, $m = 1, 2, \ldots, n + k$. In order to calculate this number, consider the division of the lattice paths from the origin $(0, 0)$ to the point (n, k) according to whether the first step is horizontal or vertical. The lattice paths from $(0, 0)$ to (n, k) with the first step vertical and thus passing from the point $(0, 1)$ (with the first vote for $\mathcal{K}$) are not of interest because $x_1 = 0 < 1 = y_1$. Further, the set of lattice paths with the first step horizontal, and thus passing from the point $(1, 0)$ (with the first vote for $\mathcal{N}$), and satisfying the conditions $x_m > y_m$, $m = 2, 3, \ldots, n + k$, may be expressed as a difference of two sets. Specifically, consider the set A of lattice paths from the point $(1, 0)$ to the point (n, k) and the set B of lattice paths from the point $(1, 0)$ to the point (n, k) that touch or intersect the straight line $x = y$. Then

$$\Psi_{n,k} = N(A - B), \quad B \subseteq A$$

and, using part (c) of Theorem 1.4,

$$\Psi_{n,k} = N(A - B) = N(A) - N(B).$$

According to Corollary 2.6 and Lemma 2.1,

$$N(A) = L(n - 1, k) = \binom{n + k - 1}{k}, \quad N(B) = L(n, k - 1) = \binom{n + k - 1}{k - 1}$$

and hence

$$\Psi_{n,k} = \binom{n + k - 1}{k} - \binom{n + k - 1}{k - 1} = \frac{n}{n + k}\binom{n + k}{k} - \frac{k}{n + k}\binom{n + k}{k}$$

$$= \frac{n - k}{n + k}\binom{n + k}{k}.$$

(b) Also, $\psi_{n,k}$ is the number of lattice paths from the origin $(0, 0)$ to the point (n, k) such that $x_m \geq y_m$, $m = 1, 2, \ldots, n + k$ or, equivalently, $x_m > y_m - 1$, $m = 1, 2, \ldots, n + k$. In order to calculate this number, let us consider the set C of lattice paths from the origin $(0, 0)$ to the point (n, k) and the set D of the lattice paths from $(0,0)$ to (n,k) that touch or intersect the straight line $x = y - 1$. Then $D \subseteq C$ and

$$\psi_{n,k} = N(C - D) = N(C) - N(D).$$

Therefore

$$\psi_{n,k} = \binom{n+k}{k} - \binom{n+k}{k-1} = \frac{n-k+1}{n+k+1}\binom{n+k+1}{k} = \frac{n-k+1}{n+1}\binom{n+k}{k}.$$

In the particular case of $n = k$, it follows that

$$C_{n+1} \equiv \psi_{n,n} = \frac{1}{n+1}\binom{2n}{n}.$$

The numbers $\psi_{n,k}$, $k = 0, 1, \dots, n$, $n = 0, 1, 2, \dots$, are called *ballot numbers*, while the numbers C_n, $n = 0, 1, 2, \dots$, are called *Catalan numbers*. $\quad\square$

Example 2.31 The cashier problem

Consider a cashier of a theater with an entrance fee of $10, in front of which n persons, each with a $10 bill, and k persons, each with a $20 bill, are waiting. If each person buys only one ticket and the cashier has initially m $10 bills, calculate the number $c_{n,k,m}$ of ways the $n + k$ persons can form a queue in front of the cashier so that no one will have to wait for change.

Let x_r be the number of persons each with a $10 bill and y_r the number of persons each with a $20 bill, among the first r persons of the queue, $r = 1, 2, \dots, n+k$. Then, the ordered pairs (x_r, y_r) are points of the lattice $x = i$, $y = j$, $i, j = 0, \pm 1, \pm 2, \dots$ and $c_{n,k,m}$ is the number of lattice paths from the origin $(0,0)$ to the point (n,k), with $x_r + m \geq y_r$, $r = 1, 2, \dots, n+k$ or, equivalently, with $x_r - y_r > -(m+1)$, $r = 1, 2, \dots, n+k$. In order to calculate this number, let us consider the set A of lattice paths from the origin $(0,0)$ to the point (n,k) and the set B of lattice paths from the origin $(0,0)$ to the point (n,k) that touch or intersect the straight line $x = y - (m+1)$. Then

$$c_{n,k,m} = N(A - B), \quad B \subseteq A$$

and using part (c) of Theorem 1.4, it follows that

$$c_{n,k,m} = N(A - B) = N(A) - N(B),$$

where, according to Theorem 2.13,

$$N(A) = L(n,k) = \binom{n+k}{n}.$$

As regards the calculation of the number $N(B)$, note that (a) If $k < m + 1$, the points $(0,0)$ and (n,k) lie below the straight line $x = y - (m+1)$ and there is no lattice path from $(0,0)$ to (n,k) touching or intersecting this line, whence $B = \emptyset$ and so $N(B) = 0$. (b) If $m + 1 \leq k \leq n + m$, then according to Lemma 2.1,

$$N(B) = \binom{n+k}{k-m+1} = \binom{n+k}{n+m-1}.$$

Further, (c) if $k > n + m$, the point (n,k) lies either on the straight line $x = y - (m+1)$ or above it, while the origin $(0,0)$ lies below it; thus all the lattice paths from the origin $(0,0)$ to the point (n,k) touch or intersect this line, whence $B = A$, and so $N(A) = N(B)$. Therefore,

$$c_{n,k,m} = \begin{cases} \binom{n+k}{k}, & k < m+1, \\ \binom{n+k}{k} - \binom{n+k}{k-m-1}, & m+1 \leq k \leq n+m \end{cases}$$

and $c_{n,k,m} = 0$, for $k > n + m$. $\quad\square$

2.7 PROBABILISTIC APPLICATIONS

Some interesting probabilistic applications of permutations and combinations are presented in this section. Specifically, four classical problems in discrete probability, problems of ordered and unordered samples in statistics and three classical probabilistic models of statistical mechanics are examined.

2.7.1 Classical problems in discrete probability

(a) *The most beneficial bet.* Chevalier de Meré, a professional player of games of chance, in 1654 proposed the following problem to the famous French mathematician Blaise Pascal (1623-1662) for solution: which is more beneficial for a player, to bet on the appearance of at least one 6 in four throws of a die or of at least one double 6 in 24 throws of a pair of distinguishable dice?

Note that in a throw of a die there are six equiprobable possible results, $\{1,2,3,4,5,6\}$, and thus the probability of the appearance of 6, according to classical definition of probability (see Section 1.4), is 1/6. Also, in a throw of a pair of distinguishable dice, according to Example 2.8, there are $6^2 = 36$ equiprobable possible results, $\{(i,j) : i = 1,2,\ldots,6, \ j = 1,2,\ldots,6\}$, and thus the probability of the appearance of $(6,6)$ is 1/36. Since the probability of the appearance of 6 in a throw of a die is six times the probability of the appearance of $(6,6)$ in a throw of a pair of distinguishable dice and the total number of throws of the pair of distinguishable dice is six times the total number of throws of a die, Chevalier de Meré concluded that the two required probabilities are equal. But, in practice, he noted a difference between the two probabilities and complained to Pascal that mathematics led to wrong conclusions.

Pascal corresponded this problem to the famous French mathematician Pierre Fermat (1608 - 1665). Both suggested different solutions to this problem. Pascal's solution, is the following.

Let A be the event of the appearance of at least one 6 in four throws of a die and B the event of the appearance of at least one $(6,6)$ in 24 throws of a pair of distinguishable dice. Then

$$P(A) = 1 - P(A'), \quad P(B) = 1 - P(B'),$$

where

$$P(A') = \frac{5^4}{6^4}, \quad P(B') = \frac{35^{24}}{36^{24}},$$

since $N(A') = 5^4$, which is the number of the 4-permutations with repetition of the five numbers (excluding the number 6), $N(\Omega_1) = 6^4$ and $N(B') = 35^{24}$, which is the number of the 24-permutations with repetition of the 35 pairs (excluding the pair $(6,6)$), $N(\Omega_2) = 36^{24}$. Thus

$$P(A) = 1 - \left(\frac{5}{6}\right)^4 \cong 0,518, \quad P(B) = 1 - \left(\frac{35}{36}\right)^{24} \cong 0,491.$$

(b) *Distribution of shares.* Consider two players $\mathcal{K}$ and $\mathcal{R}$ contending in a series of games in which the winner is the one who first wins n games. Assume that the probability for each player to win a game is $1/2$. Further, suppose that for some reason the series of games is interrupted when $\mathcal{K}$ has won $n - k$ games and $\mathcal{R}$, $n - r$ games, $k < n, r < n$. In this case, in what shares might the total stake of s dollars be divided?

Note that this is the oldest known probabilistic problem. It was initially published in 1494 by Lucas de Burgo Pacioli; however, the solution given by Pacioli was wrong. Later, Nicolo Tartaglia in 1556 and Francesco Peverone in 1581 engaged in this problem, but the solutions they proposed were also wrong. This is not surprising since then the notion of probability was not yet known. Fermat and Pascal, each independent of the other, provided the first correct solution of this problem. One of the solutions proposed by Fermat is the following.

Assume that the series of games is continued and let A be the event in which player $\mathcal{K}$ is finally the winner and A_j the event in which player $\mathcal{K}$ completes the series of the rest of k wins at the j-th game (after the resumption of the series of games), $j = k, k + 1, \ldots, k + r - 1$. Then

$$A_i \cap A_j = \emptyset, \ i \neq j, \ A = A_k + A_{k+1} + \cdots + A_{k+r-1}$$

and so

$$P(A) = \sum_{j=k}^{k+r-1} P(A_j).$$

Further, $N(\Omega_j) = 2^j$, which is the number of j-permutations with (unrestricted) repetition of the $n = 2$ symbols w (win for $\mathcal{K}$) and d (defeat for $\mathcal{K}$) and

$$N(A_j) = \binom{j-1}{k-1},$$

which is the number of different selections of the $k - 1$ positions for the symbol w from a total of $j - 1$ positions (the j-th position being occupied by the k-th w). Thus, by the classical definition of probability,

$$P(A_j) = \binom{j-1}{k-1}\frac{1}{2^j}$$

and so

$$P(A) = \sum_{j=k}^{k+r-1} \binom{j-1}{k-1}\frac{1}{2^j}.$$

This expression can be reduced as follows: the probability $P(A')$ that player $\mathcal{K}$ does not finally win and thus player $\mathcal{R}$ is finally the winner, which can be similarly evaluated, equals

$$P(A') = \sum_{j=r}^{r+k-1} \binom{j-1}{r-1}\frac{1}{2^j}$$

and also, according to $P(A') = 1 - P(A)$, is given by

$$P(A') = 1 - \sum_{j=k}^{k+r-1} \binom{j-1}{k-1}\frac{1}{2^j}.$$

Equating the two expressions of $P(A')$ and setting $r = k$, we conclude that

$$\sum_{j=k}^{2k-1} \binom{j-1}{k-1}\frac{1}{2^j} = \frac{1}{2}.$$

Therefore

$$P(A) = \begin{cases} \dfrac{1}{2} + \displaystyle\sum_{j=2k}^{r+k-1} \binom{j-1}{k-1}\dfrac{1}{2^j}, & k < r \\[4mm] \dfrac{1}{2}, & k = r \\[4mm] \dfrac{1}{2} - \displaystyle\sum_{j=r+k}^{2k-1} \binom{j-1}{k-1}\dfrac{1}{2^j}, & k > r \end{cases}$$

and, from the total stake of s dollars, player $\mathcal{K}$ might get a share of $sP(A)$ dollars and player $\mathcal{R}$ a share of $s[1 - P(A)]$ dollars.

(c) *The ballot problem (continuation).* In a ballot between two candidates $\mathcal{N}$ and $\mathcal{K}$, $\mathcal{N}$ receives n votes and $\mathcal{K}$ receives k votes, with $n > k$. Calculate the probability that the votes for $\mathcal{N}$ are more than the votes for $\mathcal{K}$ at any step of a random counting of the votes.

This problem was formulated by Bertrand in 1887. An elegant solution given in the same year by the French mathematician Desiré André (1840 - 1917) is based on the reflection principle (see Section 2.6).

Let E be the event in which the votes for candidate $\mathcal{N}$ are constantly more than the votes for candidate $\mathcal{K}$ during the counting of the votes. Since the votes are randomly taken out of the urn, the elementary events (sample points) are equiprobable and Laplace's formula of classical probability,

$$P(E) = N(E)/N,$$

can be applied. Note that $N(E)$ is the number of ways of counting the votes so that, at any step, the votes for $\mathcal{N}$ are more than the votes for $\mathcal{K}$ and N is the number of ways of counting the votes without any restriction. According to Example 2.30,

$$N(E) = \Psi_{n,k} = \frac{n - k}{n + k} \binom{n + k}{k}.$$

Further,

$$N = \binom{n + k}{k}$$

and so

$$P(E) = \frac{n - k}{n + k}.$$

(d) *The Banach matchbox problem.* This problem, which was inspired by the smoking habits of the great Polish mathematician Banach, was presented by H. Steinhaus during a ceremony in honor of Banach.

A mathematician always carries one matchbox in his right pocket and another in his left pocket. Whenever he needs a match to light his cigarette, he randomly chooses one of the two matchboxes. Suppose that each box initially contains n matches and consider the moment when, for the first time, the mathematician discovers that a box is empty. Calculate the probability $p_{n,k}$ that, at this moment, the other box contains k matches.

Let us denote by A the event in which the box in the right pocket contains k matches at the moment the box in the left pocket is discovered to be empty. Also let us denote by w_1 the selection of the box in the left pocket and by w_2 the selection of the box in the right pocket. Then

$$A = \{(a_1, a_2, \ldots, a_{2n-k}, w_1) : n \text{ of } a's \text{ are } w_1 \text{ and } n - k \text{ are } w_2\}.$$

Note that the number of sample points that belong to A equals

$$N(A) = \frac{(2n-k)!}{n!(n-k)!} = \binom{2n-k}{n},$$

the number of permutations of the total of $2n - k$ symbols among which n symbols are w_1 and $n - k$ symbols are w_2. Further, each elementary event (sample point)

$$B = \{(a_1, a_2, \dots, a_{2n-k}, w_1) : n \text{ specific } a's \text{ are } w_1$$
$$\text{and the other } n - k \text{ are } w_2\}$$

can be written as

$$B = B_1 \times B_2 \times \cdots \times B_{2n-k} \times B_{2n-k+1},$$

where $B_{2n-k+1} = \{w_1\}$ and n specific B_j are equal to $\{w_1\}$ and the other $n - k$ B_j are equal to $\{w_2\}$. Thus $B_j \subseteq \{w_1, w_2\}$, $P(B_j) = 1/2$ and so (see Section 1.4),

$$P(B) = P(B_1)P(B_2) \cdots P(B_{2n-k})P(B_{2n-k+1}) = \frac{1}{2^{2n-k+1}}.$$

Therefore,

$$P(A) = N(A)P(B) = \binom{2n-k}{n} \frac{1}{2^{2n-k+1}}.$$

The same probability is derived for the event in which the box in the left pocket contains k matches at the moment that the box in the right pocket is discovered to be empty; so the required probability $p_{n,k}$ is given by

$$p_{n,k} = 2P(A) = \binom{2n-k}{n} \frac{1}{2^{2n-k}}.$$

2.7.2 Ordered and unordered samples

In statistical inference the set of elements under consideration, in the study of a problem, is called *population*. Thus, a population constitutes the inhabitants of a city as well as the revenues of the public servants. Since the examination of all the elements (units) of a population may be expensive and time consuming, in statistics, we restrict ourselves to the study of a subset (part of) the population called *sample*. The number k of the elements of a sample is called the *size* of it. Among the various ways of selecting a sample, only a few are mentioned in this section.

Sampling without replacement. Once selected, an element is not returned to the population. According to whether the order of selection of the elements of a sample counts or does not count, we refer to *ordered* and *unordered samples.* Then, in a population of n elements, *the number of ordered samples of size k is* $(n)_k$, while *the number of unordered samples (sub-populations) of size k is* $\binom{n}{k}$.

Sampling with replacement. An element selected is returned to the population before the selection of the next element and so it can be chosen again. In this case, *the number of ordered samples of size k is* n^k, while *the number of unordered samples of size k is* $\binom{n+k-1}{k}$.

Usually equal probabilities are assigned to all the different samples that may be obtained from a population; such a sample is called *random.* So, a *random ordered sample* of size k from a population of n elements has probability to be selected $1/n^k$ in the case of sampling with replacement, and $1/(n)_k$ in the case of sampling without replacement. For a *random unordered sample,* the respective probabilities are $1/\binom{n+k-1}{k}$ and $1/\binom{n}{k}$. The next probabilistic model, which leads to the classical hypergeometric and binomial distributions, constitutes a simple application of these notions and results.

(a) *Hypergeometric distributions.* A random ordered or unordered sample of k balls is selected, without or with replacement, from an urn containing s white and $n - s$ black balls. Let A_r be the event in which the sample contains exactly r white balls, $r = 0, 1, \ldots, k$. Calculate the probability $P(A_r)$, $r = 0, 1, \ldots, k$.

Let us first consider the case of sampling without replacement and assume that the order of selection of the balls is not of interest. The number of unordered samples of size k is $N(\Omega_1) = \binom{n}{k}$, while the number of such samples that belong to A_r is $N(A_r) = \binom{s}{r}\binom{n-s}{k-r}$, since the r white balls can be selected in $\binom{s}{r}$ different ways and the $k - r$ black balls in $\binom{n-s}{k-r}$ different ways. Thus, by Laplace's definition of probability,

$$P(A_r) = \binom{s}{r}\binom{n-s}{k-r} \bigg/ \binom{n}{k}, \quad r = 0, 1, \ldots, k.$$

The probability distribution defined by this sequence of probabilities is called *hypergeometric distribution.*

In the case of sampling with replacement, the number of unordered samples of size k is $N(\Omega_2) = \binom{n+k-1}{k}$, while the number of such samples that belong to A_r is $N(A_r) = \binom{s+r-1}{r}\binom{n-s+k-r-1}{k-r}$ and thus,

$$P(A_r) = \binom{s+r-1}{r}\binom{n-s+k-r-1}{k-r} \bigg/ \binom{n+k-1}{k}, \quad r = 0, 1, \ldots, k.$$

The probability distribution defined by this sequence of probabilities is called *negative hypergeometric distribution.*

Let us now suppose that the s white balls are numbered from 1 to s and the $n - s$ black balls are numbered from $s + 1$ to n and assume that the order of selection of the balls is of interest. In the case of sampling without replacement, the number of ordered samples of size k is $N(\Omega_3) = (n)_k$. The number of such samples that belong to A_r is $N(A_r) = \binom{k}{r}(s)_r(n-s)_{k-r}$. Indeed, there are $\binom{k}{r}$ ways of selecting the r positions for the white balls (the $k - r$ positions for the black balls being thus specified) and, further, the r positions for the white balls can be occupied in $(s)_r$ ways, while the remaining $k - r$ positions for the black balls can be occupied in $(n-s)_{k-r}$ ways. Hence

$$P(A_r) = \binom{k}{r}\frac{(s)_r(n-s)_{k-r}}{(n)_k}, \quad r = 0, 1, \ldots, k.$$

Since

$$\binom{k}{r}\frac{(s)_r(n-s)_{k-r}}{(n)_k} = \frac{(s)_r}{r!}\frac{(n-s)_{k-r}}{(k-r)!}\bigg/\frac{(n)_k}{k!} = \binom{s}{r}\binom{n-s}{k-r}\bigg/\binom{n}{k},$$

it follows that the probability $P(A_r)$ is the same for ordered as well as for unordered samples.

(b) *Binomial distribution.* For the last model and in the case of sampling with replacement, the number of ordered samples of size k is $N(\Omega_4) = n^k$. The number of such samples that belong to A_r is $N(A_r) = \binom{k}{r}s^r(n-s)^{k-r}$, since there are $\binom{k}{r}$ ways of selecting the r positions for the white balls and these positions can be occupied in s^r ways, while the remaining $k - r$ positions for the black balls can be occupied in $(n-s)^{k-r}$ ways. Hence

$$P(A_r) = \binom{k}{r}\left(\frac{s}{n}\right)^r\left(1 - \frac{s}{n}\right)^{k-r}, \quad r = 0, 1, \ldots, k.$$

This sequence of probabilities constitutes a particular case of the probability function of the *binomial distribution.*

(c) *The birthday problem.* A random ordered sample of size k is chosen with replacement from a population of n elements. Calculate the probability that the elements of the sample are all different.

Consider the set Ω of ordered samples of size k that can be selected, with replacement, from a population of n elements. Let A be the event in which the elements of such a sample are all different. Then $N(\Omega) = n^k$, $N(A) = (n)_k$ and so

$$P(A) = \frac{(n)_k}{n^k}.$$

An interesting application of this probability is given in the following birthday problem. The birthdays of k people form an ordered sample of size k from the population of all days in the year. With $n = 365$, the probability that all k birthdays are different is

$$P(A) = \frac{(365)_k}{365^k},$$

while the probability that at least two persons have a common birthday is

$$P(A') = 1 - P(A) = 1 - \frac{(365)_k}{365^k}.$$

For $k = 23$, it follows that $P(A') > 1/2$.

2.7.3 Probability models in statistical mechanics

Consider a mechanical system consisting of k particles. To each particle of mass m there correspond its position (x, y, z) and its momentum (mu, mv, mw), where u, v and w are the coordinates of its velocity. In statistical mechanics the 6-dimensional space, in which the vectors (x, y, z, mu, mv, mw) belong, is subdivided into a large number of n small regions or cells so that each particle, in accordance with its coordinates, is assigned to one cell. According to the hypotheses imposed on the particles (distinguishable or indistinguishable) and on the cells (distinguishable or indistinguishable, of limited or unlimited capacity), several models may be introduced.

(a) *Maxwell-Boltzman model.* Assume that the k particles as well as the n cells are distinguishable. Thus, there are n^k distributions (allocations) of the particles into the cells. Further, assume that these distributions are equiprobable and so each has probability $1/n^k$. Note that this assumption may be achieved by an appropriate definition of the cells. The number of distributions resulting in r_j particles in j-th cell, $j = 1, 2, \ldots, n$, equals

$$\binom{k}{r_1}\binom{k - r_1}{r_2}\cdots\binom{k - r_1 - \cdots - r_{n-2}}{r_{n-1}} = \frac{k!}{r_1!r_2!\cdots r_n!},$$

with $r_1 + r_2 + \cdots + r_n = k$. Therefore the probability that r_j particles are in the j-th cell, $j = 1, 2, \ldots, n$, with $r_1 + r_2 + \cdots + r_n = k$, is

$$p(r_1, r_2, \ldots, r_n; k) = \frac{k!}{r_1!r_2!\cdots r_n!} \cdot \frac{1}{n^k}.$$

It should be noted that in modern theory it was shown beyond doubt that this model does not describe the behavior of any known particles.

(b) *Bose-Einstein model.* Assume that the k particles are indistinguishable and the n cells are distinguishable. Thus, there are $\binom{n+k-1}{k}$ distributions of the particles into the cells. Further, suppose that these distributions are equiprobable, each with probability $1/\binom{n+k-1}{k}$. This assumption holds true for photons, nuclei and atoms containing an even number of elementary particles. Thus, the probability that a given cell contains r particles is

$$p(r; n, k) = \binom{n + k - r - 2}{k - r}\Big/\binom{n + k - 1}{k},$$

since the remaining $k - r$ particles can be distributed into the remaining $n - 1$ cells in $\binom{n+k-r-2}{k-r}$ ways. The probability that s cells remain empty is

$$q(s; n, k) = \binom{n}{s} \binom{k-1}{n-s-1} \Big/ \binom{n+k-1}{k},$$

since there are $\binom{n}{s}$ ways of selecting the empty cells and, for each of these selections, $\binom{k-1}{n-s-1}$ ways of distributing the k indistinguishable particles into the remaining $n - s$ cells so that no one of these remains empty.

(c) *Fermi-Dirac model.* It is assumed that the k particles are indistinguishable and the n cells distinguishable, with capacity limited to one particle. The second assumption requires $k \leq n$. Thus, there are $\binom{n}{k}$ distributions of the particles into the cells. Further, it is assumed that these distributions are equiprobable, each with probability $1/\binom{n}{k}$. Therefore, the probability that a group of s given cells contains r particles equals

$$q(r, s; n, k) = \binom{s}{r} \binom{n-s}{k-r} \Big/ \binom{n}{k}.$$

Note that in statistical mechanics it was shown that electrons, neutrons and protons satisfy the assumptions of this model.

2.8   BIBLIOGRAPHIC NOTES

The French mathematicians Blaise Pascal (1623-1662) and Pierre Fermat (1608-1665), motivated by the probabilistic problems of the *most beneficial bet* and *the distribution of shares*, discussed in Section 2.7, initiated in 1654 the unification of the long existing methods and results on the enumeration of permutations and combinations of a finite set. Pascal's *Triangle Arithmétique*, which is now referred to as *Pascal's triangle*, was published after his death in 1665. The first textbook, *Ars Conjectandi*, dealing with problems of combinatorial and probabilistic nature was written by James Bernoulli (1654-1705). It was published posthumously in 1713 by Nicolas and John Bernoulli. The first comprehensive textbook on counting permutations and combinations was written by W. A. Whitworth (1867) and is available in reprint form.

Illustrating the theory on enumeration of integer solutions of a linear equation, presented in Section 2.5, we examined the enumeration of linear and circular combinations that include no pair of consecutive integers; the Fibonacci and Lucas numbers emerged. Generating functions and other properties of these numbers will be discussed in subsequent chapters. Even more can be learned from the book by V. E. Hoggatt (1969).

As we have noted in Section 2.7, the famous *ballot problem* was formulated by J. Bertrand (1887). An elegant solution by using the reflection principle was given by D. André (1887). In recent years a lot of papers dealing with extensions and generalizations of this problem were published and led to the development of lattice path combinatorics. We have only touched upon this subject. The interested reader is referred to the books by W. Feller (1968, Chapter 3), F. Spitzer (1964), L. Takacs (1967a), S. G. Mohanty (1979) and T. V. Narayana (1979).

2.9 EXERCISES

1. Calculate the number of different seatings in a row of four boys and three girls (a) so that no two girls are seated next to each other and (b) without any restriction.

2. Suppose that 16 teams take part in a soccer premier league. For the championship of a season, each team contests with every other team in two matches, one home and the other away. Calculate the total number of matches.

3. In how many ways can five mathematics books, three physics books and two books of two other different subjects be placed on a shelf so that the books of the same subject are not interrupted?

4. An elevator of a four-floor building starts from the basement with three passengers. Find the number of ways of discharging these passengers. In how many of these ways does one passenger get off at the fourth floor?

5. Find the number of different ways the president, vice-president, secretary and treasurer of a seven-member committee can be chosen.

6. Consider a series of three throws of a die. Find the number of possible outcomes that include three consecutive numbers.

7. Calculate the number of four-digit integers that (a) have no digits other than 0, 1 or 2, (b) are even (numbers) and (c) are odd (numbers).

8. Suppose that ten letters $\{l_1, l_2, \ldots, l_{10}\}$ from an alphabet are given. Find the number of four-letter words that can be formed (a) without repeated letters and (b) without any restriction.

9. Find the number of permutations of five kinds of elements $\{w_1, w_2, w_3, w_4, w_5\}$, with two like elements of each kind, in which the two like elements of each of the two kinds $\{w_1, w_2\}$ are consecutive.

10. Find the number of permutations of the ten digits $\{0, 1, \ldots, 9\}$ in which 0 precedes 1 and 2 precedes 3.

11. *A problem of Galilei.* Suppose that three distinguishable dice (one white, one black and one red) are rolled and let $S_3 = \{(w, b, r) : w = 1, 2, \ldots, 6, \ b = 1, 2, \ldots, 6, \ r = 1, 2, \ldots, 6\}$ be the set of possible results. Also, let A be the subset of possible results (w, b, r) with sum $w + b + r = 9$ and B the subset of possible results (w, b, r) with sum $w + b + r = 10$. Calculate the numbers $N(A)$ and $N(B)$, and conclude that $P(A) < P(B)$.

12. *Circular permutations.* An arrangement $(i_1, i_2, \ldots, i_n)$ of the elements of a set $\{1, 2, \ldots, n\}$ on a circle, so that i_1 succeeds i_n, is called circular permutation. Show that the number of circular permutations of the set $\{1, 2, \ldots, n\}$ equals $(n - 1)!$

13. Suppose that seven persons $\{w_1, w_2, \ldots, w_7\}$ are arranged (a) in a row and (b) on a circle. In each case, find the number of different arrangements in which three persons stand between w_1 and w_2.

14. Calculate the number of different seating of seven married couples (a) around a straight table, and (b) around a circular table, so that the husband and wife of three specified couples are seated next to each other.

15. Suppose that two distinguishable dice are rolled. Find the number of outcomes (pairs) in which both numbers are less than or equal (a) to 4 and (b) to 3. Combine the two results to find (c) the number of outcomes (pairs) in which 4 is the maximum of the two numbers.

16. (*Continuation*). Find the number of outcomes (pairs) in which 4 is the minimum of the two numbers.

17. A domino piece is marked by an unordered pair of two sets of dots. Each set contains 0 to 6 dots. Find the number of different domino pieces.

18. In how many ways can a group of six boys and six girls be divided into two equals groups so that each group contains odd numbers of boys and girls?

19. Consider five points on the circumference of a circle. Find (a) the number of cords that can be constructed by connecting these points in all possible ways, (b) the number of triangles that can be constructed with these points as vertices. Note that there is a one-to-one correspondence between the set of cords and the set of triangles that can be constructed using these five points.

20. In how many ways can a five-member council of a club of n couples be formed (a) if two women should be included, (b) if both husband and wife can not be included and (c) if no restriction exists.

21. Consider a batch of ten new books that are to be placed on five available shelves of a library and assume that each shelf has room for up to ten books. Calculate the number of different arrangements of the books on the shelves in the cases where the books are (a) distinguishable and (b) indistinguishable in their external appearance.

22. Find the number of eight-digit ternary sequences (of 0s, 1s or 2s) which include two 0s and three 1s.

23. Find the number of permutations $(i_1, i_2, \dots, i_{10})$ of the ten digits $\{0, 1, \dots, 9\}$ in which (a) $i_1 < i_2$, (b) $i_1 < i_2 < i_3$ and (c) $i_1 < i_2$ and $i_3 < i_4$.

24. Find the number of ways of distributing ten different cards into four distinguishable boxes so that j cards are placed into the j-th box, $j = 1, 2, 3, 4$.

25. Find the number of ways of equally distributing six different cards (a) to three children and (b) into three indistinguishable boxes.

26. Suppose that ten cards numbered from 0 to 9 are equally distributed (a) to five children and (b) into five indistinguishable boxes. In each case, find the number of different distributions and the number of these distributions in which each child (box) receives one card with an even number and the other card with an odd number.

27. Assume that three qualified candidates contest for a position of assistant professor in the department of mathematics. The 11-member selection committee is about to decide by voting whom to hire for the position. Calculate the number of different election outcomes (a) without any restriction and (b) with the restriction that a candidate receives a majority.

28. Every morning a man walks eight blocks to his office, which is located five blocks east and three blocks north of his house. For simplicity, assume that all the blocks are rectangular and there are no blind streets. Find the number of different paths he may follow if he walks only to the east and to the north.

29*. Consider the following procedure of partitioning a finite set of n elements into n one-element subsets. Initially this set is partitioned into two subsets. If one of the two subsets is an one-element set and the other contains at least two elements, then the subset with at least two elements is partitioned into two subsets; if both the subsets contain at least two elements, then any of these is partitioned into two subsets. Generally, at the k-th stage any of the subsets obtained at the preceding stages and containing at least two elements is partitioned into two subsets, $k = 1, 2, \dots, n-1$. So, after $n - 1$ stages the set of n elements is partitioned into n one-element subsets. Show that the number of ways this procedure can be executed is given by

$$\binom{n}{2}\binom{n-1}{2}\cdots\binom{3}{2}\binom{2}{2} = \frac{n!(n-1)!}{2^{n-1}}.$$

30*. Suppose that n generals place secret documents in a safe that can be opened only when the majority of them is present. For this reason, the safe is locked with a number of lockers and each general has keys for some of them. Calculate (a) the least number a_n of lockers needed and (b) the number b_n of keys a general should have.

31. Let r, s and n be positive integers. Prove combinatorially the relations

$$\sum_{k=0}^{n}\binom{r}{k}\binom{s}{n-k} = \binom{r+s}{n}$$

and

$$\sum_{k=0}^{n}\binom{r+k-1}{k}\binom{s+n-k-1}{n-k} = \binom{r+s+n-1}{n}.$$

32. Show, (a) algebraically and (b) combinatorially, that

$$\binom{n}{s}\binom{s}{k} = \binom{n}{k}\binom{n-k}{s-k} = \binom{n}{s-k}\binom{n-s+k}{k}$$

and

$$\binom{n}{r}\binom{n-r}{k} = \binom{n}{k}\binom{n-k}{r} = \binom{n}{r+k}\binom{r+k}{r}.$$

33. Show that

$$n\binom{n}{k} = (k+1)\binom{n}{k+1} + k\binom{n}{k}$$

and

$$\binom{n}{s}\binom{n}{k} = \sum_{j=0}^{m}\binom{s}{j}\binom{k+s-j}{s}\binom{n}{k+s-j}, \quad m = \min\{k,s\}.$$

34. Show that

$$n\binom{n}{k} = k\binom{n+1}{k+1} + \binom{n}{k+1}$$

and

$$\binom{n}{s}\binom{n}{k} = \sum_{j=0}^{m}\binom{s}{j}\binom{k}{j}\binom{n+j}{s+k}, \quad m = \min\{k,s\}.$$

35. Prove that

$$\sum_{r=k}^{n}(-1)^{r-k}\binom{n}{r} = \binom{n-1}{k-1}$$

and

$$\sum_{s=r}^{n}(-1)^{s-r}\binom{s}{k}\binom{n}{s} = \binom{n}{k}\binom{n-k-1}{r-k-1}.$$

36. Using Pascal's triangle, show that

$$\sum_{r=s}^{k}(-1)^{k-r}\binom{n}{r} = (-1)^{k-s}\binom{n-1}{s-1} + \binom{n-1}{k}.$$

37. Show that the number of injective maps f of the set $X = \{x_1, x_2, \ldots, x_k\}$ into the set $Y = \{y_1, y_2, \ldots, y_n\}$ is $(n)_k$.

38. Show that the number of maps f of the set $\{1, 2, \ldots, k\}$ into the set $\{1, 2, \ldots, n\}$ that are (a) strictly increasing is $\binom{n}{k}$ and (b) increasing is $\binom{n+k-1}{k}$.

39. Let $B(n, k, s)$ be the number of k-combinations of the set $W_{n+1} = \{w_0, w_1, \ldots, w_n\}$, of $n+1$ elements, with repetition and the restriction that the element w_0 is allowed to appear at most s times and each of the other n elements at most once. Show that

$$B(n, k, s) = \sum_{j=0}^{m}\binom{n}{k-j}, \quad m = \min\{k, s\}.$$

40. Let $E_2(n, k)$ be the number of k-combinations of n with repetition and the restriction that each element is allowed to appear at most twice. Show that

$$E_2(n, k) = \sum_{j=0}^{m}\binom{n}{j}\binom{n-j}{r}, \quad m = \min\{n, [k/2]\}.$$

41. (*Continuation*). Let $A(n, r, k)$ be the number of k-combinations of $n + r$ with repetition and the restriction that each of n specified elements may appear at most twice, while each of the other r elements may appear at most once. Show that

$$A(n, r, k) = \sum_{j=0}^{m}\binom{n}{j}\binom{n+r-j}{k-2j}, \quad m = \min\{n, [k/2]\}.$$

Also derive the expression

$$A(n, r, k) = \sum_{i=0}^{s} \binom{r}{i} E_2(n, k - i), \quad s = \min\{r, k\}.$$

42. Consider n elements belonging in r different kinds, with $k_1, k_2, \ldots, k_r$ elements, respectively. Let $C(k_1, k_2, \ldots, k_r; k)$ be the number of k-combinations of the $k_1 + k_2 + \cdots + k_r = n$ elements and $C(k_2, k_3, \ldots, k_r; k - j)$ be the number of $(k - j)$-combinations of the $k_2 + k_3 + \cdots + k_r = n - k_1$ elements. Show that

$$C(k_1, k_2, \ldots, k_r; k) = \sum_{j=0}^{k_1} C(k_2, k_3, \ldots, k_r; k - j).$$

43*. Let $S(n, k)$ be the number of partitions of a set of n elements in k subsets. (a) Prove that

$$S(n, k) = \sum \frac{n!}{r_1!(1!)^{r_1} r_2!(2!)^{r_2} \cdots r_n!(n!)^{r_n}},$$

where the summation is extended over all $r_j \geq 0$, $j = 1, 2, \ldots, n$, with $r_1 + r_2 + \cdots + r_n = k$ and $r_1 + 2r_2 + \cdots + nr_n = n$. Further, derive (b) the triangular recurrence relation

$$S(n, k) = S(n - 1, k - 1) + kS(n - 1, k), \quad k = 1, 2, \ldots, \ n = 1, 2, \ldots,$$

with initial conditions $S(0, 0) = 1$, $S(n, k) = 0$, $k > n$, and (c) the vertical recurrence relation

$$S(n, k) = \sum_{r=0}^{n-1} \binom{n - 1}{r} S(r, k - 1).$$

44*. (*Continuation*). If $T(n, r, k)$ is the number of maps f of $X = \{x_1, x_2, \ldots, x_n\}$ into $Y = \{y_1, y_2, \ldots, y_r\}$, with range space $f(X)$ containing k elements, (a) show that

$$T(n, r, k) = S(n, k)(r)_k$$

and (b) conclude that

$$\sum_{k=1}^{n} S(n, k)(r)_k = r^n.$$

(c) Multiplying both sides of the last relation by $(-1)^{j-r} \binom{j}{r}$, $r \leq j \leq n$ and summing for $r = 1, 2, \ldots, j$, deduce the expression

$$S(n, j) = \frac{1}{j!} \sum_{r=1}^{j} (-1)^{j-r} \binom{j}{r} r^n, \quad j = 1, 2, \ldots, n, \ n = 1, 2, \ldots.$$

The numbers $S(n, j)$, $j = 1, 2, \ldots, n$, $n = 1, 2, \ldots$, are called *Stirling numbers of the second kind*.

45*. (*Continuation*). Let $W(n, k)$ be the number of divisions of a set of n elements in k nonempty subsets. Show that

$$W(n, k) = k! S(n, k) = \sum_{r=1}^{k} (-1)^{k-r} \binom{k}{r} r^n.$$

46*. (*Continuation*). Let B_n be the (total) number of partitions of a set of n elements. (a) Prove that

$$B_n = \sum \frac{n!}{r_1!(1!)^{r_1} r_2!(2!)^{r_2} \cdots r_n!(n!)^{r_n}},$$

where the summation is extended over all $r_i \geq 0$, $i = 1, 2, \ldots, n$, with $r_1 + 2r_2 + \cdots + nr_n = n$. Further, (b) derive the recurrence relation

$$B_n = \sum_{k=0}^{n-1} \binom{n-1}{k} B_k, \quad k = 1, 2, \ldots, \quad B_0 = 1.$$

The numbers B_n, $n = 0, 1, 2, \ldots$, are called *Bell numbers*.

47*. (*Continuation*). Prove that

$$B_n = \sum_{k=1}^{n} S(n, k)$$

and

$$B_n = e^{-1} \sum_{r=0}^{\infty} \frac{r^n}{r!}.$$

48. Let k indistinguishable balls be distributed into n distinguishable urns of unlimited capacity. Find the number of distributions of the balls into the urns in which no urn remains empty.

49*. *Runs of like elements in permutations.* Let $(a_1, a_2, \ldots, a_n)$ be a permutation of k zeros and $n - k$ ones. If

$$a_i = 1, \; a_j = 0, \; j = i + 1, i + 2, \ldots, i + r, \; a_{i+r+1} = 1,$$

then $(a_{i+1}, a_{i+2}, \ldots, a_{i+r})$ is called *run of zeros of length* r of the permutation $(a_1, a_2, \ldots, a_n)$. Show that the number of permutations of k zeros and $n - k$ ones that include (a) s runs of zeros is

$$\binom{n-k+1}{s} \binom{k-1}{s-1}$$

and (b) s runs of zeros, among which $s_r \geq 0$ are of length r, $r = 1, 2, \ldots, k$, so that $s_1 + s_2 + \cdots + s_k = s$, is

$$\binom{n-k+1}{s} \frac{s!}{s_1! s_2! \cdots s_k!}.$$

50*. (*Continuation*). *Runs of consecutive elements in combinations.* Let $C_k = \{i_1, i_2, \ldots, i_k\}$ be a k-combination of the set $\{1, 2, \ldots, n\}$. If,

$$S_r = \{i+1, i+2, \ldots, i+r\} \subseteq C_k, \ i \notin C_k, \ i+r+1 \notin C_k,$$

then the r-combination S_r of the set $\{1, 2, \ldots, n\}$ is called *run of consecutive elements of length* r of the k-combination C_k. Show that, to each k-combination C_k of $\{1, 2, \ldots, n\}$ there uniquely corresponds a permutation $U_{k,n-k}$ of k zeros and $n-k$ ones and, to each run of consecutive elements of length r of C_k there uniquely corresponds a run of zeros of length r of $U_{k,n-k}$. Thus, conclude that the number of k-combinations of $\{1, 2, \ldots, n\}$ that include (a) s runs of consecutive elements is

$$\binom{n-k+1}{s} \binom{k-1}{s-1}$$

and (b) s runs of consecutive elements, among which $s_r \geq 0$ are of length r, $r = 1, 2, \ldots, k$, so that $s_1 + s_2 + \cdots + s_k = s$, is

$$\binom{n-k+1}{s} \frac{s!}{s_1! s_2! \cdots s_k!}.$$

51*. (*Continuation*). *Runs of consecutive elements in combinations of circularly ordered elements.* Consider the first n integral numbers $\{1, 2, \ldots, n\}$ displayed on a circle so that n and 1 are consecutive. Show that the number of (circular) k-combinations of these n elements that include (a) s runs of consecutive elements is

$$\frac{n}{n-k} \binom{n-k}{s} \binom{k-1}{s-1}$$

and (b) s runs of consecutive elements, among which $s_r \geq 0$ are of length r, $r = 1, 2, \ldots, k$, so that $s_1 + s_2 + \cdots + s_k = s$, is

$$\frac{n}{n-k} \binom{n-k}{s} \frac{s!}{s_1! s_2! \cdots s_k!}.$$

52. Find the number of integer solutions of the linear equation

$$x_1 + x_2 + \cdots + x_n = k, \quad k \text{ integer}$$

with the restrictions (a) $x_i \geq s$, $i = 1, 2, \dots, n$, s integer such that $sn \leq k$ and (b) $x_i \leq m$, $i = 1, 2, \dots, n$, m integer such that $mn \geq k$.

53. *Partial derivatives of a function.* Find the number of partial derivatives of order k of an analytic function of n variables $f(x_1, x_2, \dots, x_n)$.

54. Let $a_{n,k}$ be the number of monomials in the most general polynomial in n variables $x_1, x_2, \dots, x_n$ of degree k. Show that

$$a_{n,k} = \binom{n+k}{k}.$$

55. Let $C_s(n, k)$ be the number of k-combinations of the n numbers $\{1, 2, \dots, n\}$ possessing the property: between any two points belonging to such a combination there are at least s points that do not belong to it. Prove that

$$C_s(n, k) = \binom{n - (k-1)s}{k}.$$

56. Let $B_s(n, k)$ be the number of k-combinations of the n numbers $\{1, 2, \dots, n\}$, displayed on a circle, with the property: between any two points belonging to such a combination there are at least s points that do not belong to it. Prove that

$$B_s(n, k) = \binom{n - ks}{k} + s\binom{n - ks - 1}{k - 1} = \frac{n}{n - ks}\binom{n - ks}{k}.$$

57*. Let $C(n, k; s_{k-1})$ be the number of k-combinations $\{i_1, i_2, \dots, i_k\}$, $i_1 < i_2 < \cdots < i_k$, of $\{1, 2, \dots, n\}$, with differences $d_m = i_{m+1} - i_m > a_m$, $m = 1, 2, \dots, k - 1$, where a_m, $m = 1, 2, \dots, k - 1$, are given positive integers and $s_{k-1} = a_1 + a_2 + \cdots + a_{k-1}$. Further, let $B(n, k; s_{k-1}, r)$ be the number of the preceding k-combinations which, in addition, have span $d = i_k - i_1 < n - r$, where r is a given positive integer. Prove that

$$C(n, k; s_{k-1}) = \binom{n - s_{k-1}}{k}$$

and

$$B(n, k; s_{k-1}, r) = \binom{n - s_{k-1} - r}{k} + r\binom{n - s_{k-1} - r - 1}{k - 1}$$
$$= \frac{n - s_{k-1} + (k-1)r}{n - s_{k-1} - r}\binom{n - s_{k-1} - r}{k}.$$

58. *Fibonacci numbers.* Let Q_n be the number of n-permutations of the two elements 0 and 1, with repetition and the restriction that no two zeros are consecutive. Showing that the number $Q_{n,k}$, of n-permutations of the two elements 0 and 1, with repetition, which include k zeros, no two of which are consecutive, is

$$Q_{n,k} = \binom{n-k+1}{k}, \quad k = 0, 1, \ldots, [(n+1)/2],$$

conclude that

$$Q_n = \sum_{k=0}^{[(n+1)/2]} \binom{n-k+1}{k}, \quad n = 0, 1, \ldots$$

and

$$Q_n = Q_{n-1} + Q_{n-2}, \; n = 2, 3, \ldots, \; Q_0 = 1, \; Q_1 = 2.$$

The numbers $f_0 = 1$, $f_n = Q_{n-1}$, $n = 1, 2, \ldots$, are the *Fibonacci numbers*.

59*. Prove that the number of permutations $(a_1, a_2, \ldots, a_{2n})$ of n zeros and n ones, among the first k elements of which there are at least as many zeros as ones, for every $k = 1, 2, \ldots, 2n$, equals

$$C_{n+1} = \frac{1}{n+1}\binom{2n}{n},$$

the *Catalan number*.

60*. *Lattice paths with diagonal steps.* Consider paths from one point of a lattice to another in which, besides a horizontal step from (i, j) to $(i+1, j)$ and a vertical step from (i, j) to $(i, j+1)$, a diagonal step from (i, j) to $(i+1, j+1)$ is allowed. Prove that (a) the number of lattice paths from the origin $(0, 0)$ to the point (n, k), with r diagonal steps, equals

$$L(n, k, r) = \frac{(n+k-r)!}{r!(n-r)!(k-r)!} = \binom{n+k-r}{r, k-r}$$

and (b) the number of lattice paths from the origin $(0, 0)$ to the point (n, k), with r diagonal steps, that do not touch (with the exception of the origin) or intersect the straight line $x = y$ equals

$$\Psi(n, k, r) = \frac{n-k}{n+k-r}\binom{n+k-r}{r, k-r}.$$

61*. (*Continuation*). Consider a ballot district, where each voter is allowed to vote for up to two candidates of the same party. Assume that two candidates $\mathcal{N}$ and $\mathcal{K}$, of the same party, receive n and k votes, with $n > k$, respectively. Calculate the probability that, at any stage of the counting procedure, the votes for $\mathcal{N}$ are more than the votes for $\mathcal{K}$, given that r voters vote for both $\mathcal{N}$ and $\mathcal{K}$.

62. From an urn containing n tickets numbered $1, 2, \ldots, n$, tickets are successively drawn one after the other without replacement. Show that the probability $p_{n,k}$ that the number r is drawn at the k-th drawing is independent of k and equals $p_{n,k} = 1/n$.

63. From an urn containing n distinguishable balls among which r are black and $n - r$ are white, balls are successively drawn one after the other without replacement, until for the first time a white ball is drawn. Show that the probability $p_{n,r,k}$ that k drawings are required is given by

$$p_{n,r,k} = \frac{(n-r)(r)_{k-1}}{(n)_k}$$

and conclude that

$$\sum_{j=1}^{r} \frac{(r)_j}{(n-1)_j} = \frac{r}{n-r}.$$

64. *Spread of rumors.* In a town of $n + 1$ inhabitants, a person tells a rumor to a second person, who in turn repeats it to a third person, etc. At each step, the recipient of the rumor is chosen at random from the n people available. Calculate the probability that the rumor will be told k times without (a) returning to the originator and (b) being repeated to any person.

65. Consider a sequence of tosses of a fair coin and let p_n be the probability that n tosses are required before the appearance of a run of heads of length 2. Show that

$$p_n = f_n/2^{n+2}, \quad n = 0, 1, \ldots,$$

where

$$f_n = \sum_{k=0}^{[n/2]} \binom{n-k}{k}, \quad n = 0, 1, \ldots,$$

are the *Fibonacci numbers*.

Chapter 3

FACTORIALS, BINOMIAL AND MULTINOMIAL COEFFICIENTS

3.1 INTRODUCTION

The number $(n)_k$, of k-permutations of n, and the number $\binom{n}{k}$, of k-combinations of n, constitute particular cases of the factorial and the binomial of a real number x of order k, respectively. The concept of a factorial is as important as the concept of a power. In the calculus of finite differences, a major subject of discrete mathematics, factorials occupy the same central position as the powers in the infinitesimal calculus. The finite differences of factorials possess many of the nice properties of the derivatives of powers. The binomial (coefficient), closely related to factorial, is of equal importance. In discrete probability theory and in theoretical computer science, which concern finite differences and sums of functions, the expansion into factorial series is more advantageous than the expansion into power series. The multinomial (coefficient) is a multivariate extension of the binomial (coefficient).

In this chapter we only touch upon the broad subject of factorials, binomial and multinomial coefficients. Specifically, after introducing the notion of a factorial of a real number, Vandermonde's factorial convolution formula is presented. Also, an invaluable formula in the approximation of probability distributions, known as Stirling's approximation formula of $n!$, is obtained. Further, Newton's binomial and the negative binomial formulae are combinatorially derived. Then, the monotonicity of the binomial coefficients is examined. Cauchy's binomial convolution formula is presented as a corollary of Vandermonde's factorial convolution formula. Finally, applications of these formulae in the evaluation of sums involving binomial coefficients and in the derivation of combinatorial identities are exemplified. In the last section of this chapter, the multinomial formula is combinatorially deduced.

3.2 FACTORIALS

The factorial of order k, as well as the power of order k, is initially defined for k a positive integer and successively extended to $k = 0$ and to k a negative integer. Thus, let x be a real number and k a positive integer. The *factorial of x of order k*, denoted by $(x)_k$, is defined by

$$(x)_k = x(x - 1) \cdots (x - k + 1).$$

If r is also a positive integer, then

$$(x)_{k+r} = x(x - 1) \cdots (x - k - r + 1)$$
$$= [x(x - 1) \cdots (x - k + 1)][(x - k)(x - k - 1) \cdots (x - k - r + 1)]$$

and since

$$(x - k)(x - k - 1) \cdots (x - k - r - 1) = (x - k)_r,$$

we deduce the following *fundamental property* of the factorials:

$$(x)_{k+r} = (x)_k (x - k)_r. \tag{3.1}$$

Requiring the validity of this fundamental property to be preserved, the definition of the factorial may be extended to zero or negative order. Specifically, it is required that formula (3.1) is valid for any integer values of k and r. Then, putting into it $k = 0$, it follows that

$$(x)_r = (x)_0 (x)_r,$$

for any integer r. This equation, if $x \neq 0, 1, \ldots, r - 1$, whence $(x)_r \neq 0$, implies

$$(x)_0 = 1,$$

while, if $x = 0, 1, \ldots, r - 1$, reduces to an identity for any value $(x)_0$ is required to represent. Further, from (3.1) with r a positive integer and $k = -r$, it follows that

$$(x)_{-r} (x + r)_r = 1$$

and for $x \neq -1, -2, \ldots, -r$,

$$(x)_{-r} = \frac{1}{(x + r)_r} = \frac{1}{(x + r)(x + r - 1) \cdots (x + 1)}, \quad r = 1, 2, \ldots.$$

Summarizing, the next general definition of the factorial is adopted.

DEFINITION 3.1   *The factorial of x of order k, denoted by* $(x)_k$*, is defined for every real number x and every integer number k by*

$$. \ (x)_k = x(x-1)\cdots(x-k+1), \quad k = 1, 2, \ldots, \quad (x)_0 = 1 \qquad (3.2)$$

and

$$(x)_{-k} = \frac{1}{(x+k)_k} = \frac{1}{(x+k)(x+k-1)\cdots(k+1)}, \quad k = 1, 2, \ldots, \quad (3.3)$$

for $x \neq -1, -2, \ldots, -k.$

REMARK 3.1   In addition to the notation $(x)_k$ of the factorial of x of order k, which is almost established in combinatorics, the notation $x^{(k)}$, which indicates the relation of the factorials to the powers, is also used.

The factorial of x of order k, which is defined by (3.2) and (3.3), is called *falling* in distinction to the *rising* factorial of x of order k, which, for $k = 1, 2, \ldots$ is defined by the product

$$x(x+1)\cdots(x+k-1).$$

This distinction is needless since both factorials can be expressed by the same notation, only the argument being different. Specifically, this product, with the notation we have adopted, equals $(x+k-1)_k$. Also,

$$(x+k-1)_k = (-1)^k(-x)(-x-1)\cdots(-x-k+1) = (-1)^k(-x)_k.$$

Finally, it is worth noting that (3.3), for $x = 0$, yields

$$(0)_{-r} = 1/r!, \quad r = 1, 2, \ldots,$$

an interesting particular case.   ∎

A factorial convolution formula, due to Vandermonde, is derived in the following theorem.

THEOREM 3.1 Vandermonde's formula
Let x and y be real numbers and n a positive integer. Then

$$(x+y)_n = \sum_{k=0}^{n} \binom{n}{k} (x)_k (y)_{n-k}. \qquad (3.4)$$

PROOF   Note first that (3.4) holds true for $n = 1$. Assume that (3.4) holds for $n - 1$, that is,

$$(x+y)_{n-1} = \sum_{k=0}^{n-1} \binom{n-1}{k} (x)_k (y)_{n-k-1}.$$

It will be shown that (3.4) holds also for n. From (3.2), with $x + y$ instead of x and the induction hypothesis, we have

$$(x + y)_n = (x + y - n + 1)(x + y)_{n-1}$$

$$= (x + y - n + 1) \sum_{k=0}^{n-1} \binom{n-1}{k} (x)_k (y)_{n-k-1}.$$

Since

$$(x + y - n + 1)(x)_k (y)_{n-k-1} = (x)_k (y)_{n-k} + (x)_{k+1} (y)_{n-k-1},$$

we successively get

$$(x + y)_n = \sum_{k=0}^{n-1} \binom{n-1}{k} (x)_k (y)_{n-k} + \sum_{k=0}^{n-1} \binom{n-1}{k} (x)_{k+1} (y)_{n-k-1}$$

$$= \sum_{k=0}^{n-1} \binom{n-1}{k} (x)_k (y)_{n-k} + \sum_{k=1}^{n} \binom{n-1}{k-1} (x)_k (y)_{n-k}$$

$$= (y)_n + \sum_{k=1}^{n-1} \left\{ \binom{n-1}{k} + \binom{n-1}{k-1} \right\} (x)_k (y)_{n-k} + (x)_n.$$

Therefore, on using Pascal's triangle (2.7), we deduce

$$(x + y)_n = \sum_{k=0}^{n} \binom{n}{k} (x)_k (y)_{n-k}.$$

Thus, (3.4) holds true for n and, according to the principle of mathematical induction, it holds true for every positive integer n. ∎

Example 3.1 Mean of hypergeometric probabilities

Suppose that n balls are randomly drawn one after the other, without replacement, from an urn containing r white and s black balls. Clearly, the probability of drawings k white balls in the n drawings is given by (see Section 2.7.2)

$$p_{n,k} = \binom{n}{k} \frac{(r)_k (s)_{n-k}}{(r + s)_n}, \quad k = 0, 1, \dots, n.$$

Note that $p_{n,k} \geq 0$, $k = 0, 1, \dots, n$ and, according to Vandermonde's formula,

$$\sum_{k=0}^{n} p_{n,k} = \frac{1}{(r + s)_n} \sum_{k=0}^{n} \binom{n}{k} (r)_k (s)_{n-k} = 1,$$

implying that this is a legitimate probability mass function. The *expected value* (or the *mean*) of the number of white balls drawn in n drawings is given by the sum

$$\mu = \sum_{k=1}^{n} k p_{n,k} = \frac{1}{(r+s)_n} \sum_{k=1}^{n} k \binom{n}{k} (r)_k (s)_{n-k}.$$

This sum, upon using the expressions

$$k \binom{n}{k} = n \binom{n-1}{k-1}, \quad (r)_k = r(r-1)_{k-1},$$

reduces to

$$\mu = \frac{nr}{(r+s)_n} \sum_{k=1}^{n} \binom{n-1}{k-1} (r-1)_{k-1} (s)_{n-k}.$$

Thus, by virtue of Vandermonde's formula and since $(r+s)_n = (r+s) \cdot (r+s-1)_{n-1}$, we get

$$\mu = \frac{nr}{r+s}. \qquad \square$$

As has already been noted in Remark 2.1, $n! = 1 \cdot 2 \cdot 3 \cdots n$ increases rapidly with increasing n. A very useful approximation is given by Stirling's formula, the proof of which requires the *Wallis formula*, stated in the following lemma.

LEMMA 3.1
The sequence

$$w_n = \frac{2^{2n} (n!)^2}{(2n)! \sqrt{n}}, \quad n = 1, 2, \ldots$$

is convergent and

$$\lim_{n \to \infty} \frac{2^{2n} (n!)^2}{(2n)! \sqrt{n}} = \sqrt{\pi}. \tag{3.5}$$

PROOF   Let us consider the sequence of integrals

$$I_r = \int_0^{\pi/2} \sin^r t \, dt, \quad r = 0, 1, \ldots.$$

Clearly, $I_0 = \pi/2$, $I_1 = 1$ and, integrating the identity

$$\frac{d}{dt} (\sin^{r-1} t \cos t) = (r-1) \sin^{r-2} t - r \sin^r t, \quad r = 2, 3, \ldots,$$

in the interval $[0, \pi/2]$, we deduce the recurrence relation

$$I_r = \frac{r-1}{r} I_{r-2}, \quad r = 2, 3, \ldots .$$

Putting $r = 2n$ and applying successively the resulting recurrence relation, we conclude that

$$I_{2n} = \frac{(2n)!}{2^{2n}(n!)^2} \cdot \frac{\pi}{2}.$$

Similarly, from the recurrence relation with $r = 2n + 1$, it follows that

$$I_{2n+1} = \frac{2^{2n}(n!)^2}{(2n+1)!} = \frac{1}{(2n+1)I_{2n}} \cdot \frac{\pi}{2}.$$

Note that, for $0 \le t \le \pi/2$, we have $0 \le \sin t \le 1$ and so $\sin^{2n+1} t \le \sin^{2n} t \le \sin^{2n-1} t$. Therefore

$$I_{2n+1} \le I_{2n} \le I_{2n-1}$$

and

$$\frac{1}{I_{2n}I_{2n-1}} \le \frac{1}{I_{2n}^2} \le \frac{1}{I_{2n}I_{2n+1}}. \tag{3.6}$$

Further,

$$I_{2n}I_{2n+1} = \frac{\pi}{2(2n+1)}$$

and since $I_{2n+1} = 2nI_{2n-1}/(2n+1)$,

$$I_{2n}I_{2n-1} = \frac{\pi}{4n}.$$

Introducing the above expressions of I_{2n}, $I_{2n}I_{2n-1}$ and $I_{2n}I_{2n+1}$ in (3.6), we deduce for the sequence

$$w_n = \frac{2^{2n}(n!)^2}{(2n)!\sqrt{n}} > 0, \quad n = 1, 2, \ldots ,$$

the double inequality

$$\pi \le w_n^2 \le \pi \left(1 + \frac{1}{2n} \right).$$

Taking the limit for $n \to \infty$, we conclude the relation

$$\lim_{n \to \infty} w_n^2 = \pi,$$

which implies formula (3.5) of Wallis, since $w_n > 0$. ∎

We are now in position to derive Stirling's approximation formula of n factorial.

THEOREM 3.2 Stirling's formula
If n is a positive integer, then

$$n! \cong e^{-n} n^{n+1/2} \sqrt{2\pi}, \tag{3.7}$$

where the symbol $\cong$ is used in the sense that

$$\lim_{n \to \infty} \frac{n!}{e^{-n} n^{n+1/2}} = \sqrt{2\pi}. \tag{3.8}$$

PROOF   Note that

$$\log n! = \sum_{k=1}^{n} \log k$$

and, since $\log x$ is a monotone function of x, it follows that

$$\int_{k-1}^{k} \log x\, dx < \log k < \int_{k}^{k+1} \log x\, dx, \quad k = 1, 2, \dots, n,$$

whence, summing for $k = 1, 2, \dots, n$, we get the double inequality

$$\int_{0}^{n} \log x\, dx < \log n! < \int_{1}^{n+1} \log x\, dx.$$

Therefore

$$n \log n - n < \log n! < (n+1) \log(n+1) - n.$$

This double inequality suggests comparing $\log n!$ with some quantity close to the mean of the extreme members. The simplest such quantity is $(n + 1/2) \log n - n$. Thus, consider the sequence

$$a_n = \log n! - (n + 1/2) \log n + n, \quad n = 1, 2, \dots \tag{3.9}$$

and note that

$$a_n - a_{n+1} = \left(n + \frac{1}{2}\right) \log \frac{n+1}{n} - 1$$

and

$$\frac{n+1}{n} = \frac{1 + 1/(2n+1)}{1 - 1/(2n+1)}.$$

Then, using the expansion

$$\frac{1}{2} \log \frac{1+t}{1-t} = \sum_{r=0}^{\infty} \frac{t^{2r+1}}{2r+1}, \quad -1 < t < 1,$$

with $t = 1/(2n + 1)$, the difference $a_n - a_{n+1}$ may be expressed as

$$a_n - a_{n+1} = \sum_{r=1}^{\infty} \frac{1}{(2r + 1)} \left(\frac{1}{2n + 1} \right)^{2r} > 0.$$

Therefore, the sequence a_n, $n = 1, 2, \ldots$ is decreasing. Further, since the right-hand side of the last expression is less than the geometric series

$$\frac{1}{3} \sum_{r=1}^{\infty} \left(\frac{1}{2n + 1} \right)^{2r} = \frac{1}{3[(2n + 1)^2 - 1]} = \frac{1}{12n} - \frac{1}{12(n + 1)},$$

whence

$$\left(a_n - \frac{1}{12n} \right) - \left(a_{n+1} - \frac{1}{12(n + 1)} \right) < 0,$$

it follows that the sequence $d_n = a_n - \frac{1}{12n}$, $n = 1, 2, \ldots$, is increasing. Consequently, the limit of the sequence a_n, $n = 1, 2, \ldots$ exists and, according to (3.9), so does the limit of the sequence

$$b_n = e^{a_n} = \frac{n!}{e^{-n} n^{n+1/2}}, \quad n = 1, 2, \ldots .$$

Thus, on using Wallis formula (3.5), the limit b of the last sequence is obtained as

$$b = \lim_{n \to \infty} \frac{b_n^2}{b_{2n}} = \sqrt{2} \lim_{n \to \infty} \frac{2^{2n} (n!)^2}{(2n)! \sqrt{n}} = \sqrt{2\pi},$$

establishing (3.8).   ∎

REMARK 3.2   The error of the approximation (3.7) increases indefinitely with increasing n, but the relative error

$$(n! - e^{-n} n^{n+1/2} \sqrt{2\pi})/n!$$

decreases steadily and vanishes at the limit. The approximation (3.7) is accurate even for small n. For example, for $n = 1, 2$ and 5, the exact values are $1! = 1$, $2! = 2$ and $5! = 120$, while the corresponding approximate values of the right-hand side of (3.7) are $0,922$, $1,199$ and $118,019$.   ∎

3.3   BINOMIAL COEFFICIENTS

The number $\binom{n}{k}$ of k-combinations of n is called *binomial coefficient* since it is the coefficient of the general term of Newton's binomial formula. This formula is combinatorially derived in the next theorem.

THEOREM 3.3 Newton's binomial formula
Let x and y be real numbers and n a positive integer. Then

$$(x+y)^n = \sum_{k=0}^{n} \binom{n}{k} x^k y^{n-k}. \tag{3.10}$$

PROOF   Note that

$$(x+y)^n = p_1(x,y)p_2(x,y)\cdots p_n(x,y),$$

where $p_i(x,y) = x+y$, $i = 1, 2, \ldots, n$. Executing the operations (multiplications and additions) by selecting each time k out of the n factors $\{p_1(x,y), p_2(x,y), \ldots, p_n(x,y)\}$ and multiplying the terms x of these together with the terms y of the remaining $n-k$ factors, a summand of the form

$$a_{n,k} x^k y^{n-k}, \quad k = 0, 1, \ldots, n$$

is produced, where the coefficient $a_{n,k}$ is the number of ways of selecting k out of the n factors. Since the order of selection does not count, because it does not affect the result of the multiplications, the number $a_{n,k}$ equals $\binom{n}{k}$, the number of k-combinations of n. Consequently, according to the summation principle, Newton's binomial formula (3.10) is established.   ∎

*Example 3.2 **Mean of binomial probabilities***
Suppose that n balls are randomly drawn one after the other, with replacement, from an urn containing r white and s black balls. The probability of drawing k white balls in the n drawings is given by (see Section 2.7.2)

$$p_{n,k} = \binom{n}{k} p^k q^{n-k}, \quad k = 0, 1, \ldots, n$$

where $p = r/(r+s)$ and $q = s/(r+s)$. Notice that $p_{n,k} > 0$, $k = 0, 1, \ldots, n$ and, according to Newton's binomial formula,

$$\sum_{k=0}^{n} p_{n,k} = \sum_{k=0}^{n} \binom{n}{k} p^k q^{n-k} = (p+q)^n = 1,$$

implying that this is a legitimate probability mass function. The *mean* of the number of white balls drawn in n drawings is given by the sum

$$\mu = \sum_{k=1}^{n} k p_{n,k} = \sum_{k=1}^{n} k \binom{n}{k} p^k q^{n-k}.$$

This sum, upon using the expression

$$k \binom{n}{k} = n \binom{n-1}{k-1},$$

reduces to

$$\mu = np \sum_{k=1}^{n} \binom{n-1}{k-1} p^{k-1} q^{n-k}.$$

Thus, by virtue of Newton's binomial formula, we get

$$\mu = np. \quad \Box$$

Definition 3.1 of the factorial of x of degree k allows the following useful extension of the notion of the binomial coefficient.

DEFINITION 3.2   *The binomial coefficient of order k is defined for every real number x and a nonnegative integer k by*

$$\binom{x}{k} = \frac{(x)_k}{k!}, \quad k = 0, 1, \dots . \tag{3.11}$$

Note that, in addition to the case $x = n$ is a positive integer, for which the binomial coefficient $\binom{n}{k}$ is a positive integer and gives the number of k-combinations of n, in the case $x = -n$, with n a positive integer, the binomial coefficient $\binom{-n}{k}$ is also an integer and has the sign of $(-1)^k$. The positive integer $(-1)^k \binom{-n}{k}$ gives the number of k-combinations of n, with repetition, as it follows from the relation (see Remark 3.1)

$$\binom{n+k-1}{k} = (-1)^k \binom{-n}{k}. \tag{3.12}$$

An extension of Newton's binomial formula (3.10) to the case where the exponent of the left-hand side is a negative integer is given in the next theorem.

THEOREM 3.4 Newton's negative binomial formula
Let x and y be real numbers, with $y \neq 0$, $-1 < x/y < 1$, and n a positive integer. Then

$$(x + y)^{-n} = \sum_{k=0}^{\infty} \binom{-n}{k} x^k y^{-n-k}. \tag{3.13}$$

PROOF   Note first that (3.13) is equivalent to

$$(1 - t)^{-n} = \sum_{k=0}^{\infty} \binom{n+k-1}{k} t^k, \quad -1 < t < 1, \tag{3.14}$$

as it follows directly by putting $t = -x/y$, $y \neq 0$ and using (3.12). Further,

$$(1 - t)^{-n} = f_1(t) f_2(t) \cdots f_n(t),$$

where $f_i(t) = (1 - t)^{-1}$, $i = 1, 2, \dots , n$, and, using the expression of the sum of an infinite number of terms of a geometric progression with ratio t,

$$f_i(t) = 1 + t + t^2 + \cdots = \sum_{r_i=0}^{\infty} t^{r_i}, \quad i = 1, 2, \dots , n.$$

If the operations (multiplications and additions) are executed by selecting the term t^{r_i}, $r_i \geq 0$, from the factor $f_i(t)$, $i = 1, 2, \dots , n$ and forming the product $t^{r_1} t^{r_2} \cdots t^{r_n} = t^{r_1 + r_2 + \cdots + r_n}$, with $r_1 + r_2 + \cdots + r_n = k$, a summand of the form

$$b_{n,k} t^k, \quad k = 0, 1, \dots$$

is produced, where the coefficient $b_{n,k}$ is the number of ways of selecting the terms from the n factors. Since, to each such selection there corresponds a nonnegative integer solution $(r_i \geq 0, i = 1, 2, \dots , n)$ of the linear equation

$$r_1 + r_2 + \cdots + r_n = k,$$

and vice versa, the number $b_{n,k}$, according to Theorem 2.12, equals $\binom{n+k-1}{k}$. Consequently, by the summation principle, expression (3.14) is deduced and the proof of the theorem is completed. $\blacksquare$

The next theorem is concerned with the monotonicity of the sequence of the terms of the binomial expansion.

THEOREM 3.5
Let

$$a_{n,k}(x, y) = \binom{n}{k} x^k y^{n-k}, \quad k = 0, 1, \dots , n,$$

with n a fixed positive integer and x and y given positive real numbers. Also let $m = (n + 1)/(1 + y/x)$.
    (a) *If m is an integer, then*

$$a_{n,0}(x, y) < a_{n,1}(x, y) < \cdots < a_{n,m-1}(x, y),$$

$$a_{n,m-1}(x, y) = a_{n,m}(x, y),$$

$$a_{n,m}(x, y) > a_{n,m+1}(x, y) > \cdots > a_{n,n}(x, y).$$

    (b) *If m is not an integer and $r = [m]$, the integer part of m, then*

$$a_{n,0}(x, y) < a_{n,1}(x, y) < \cdots < a_{n,r}(x, y),$$

$$a_{n,r}(x,y) > a_{n,r+1}(x,y) > \cdots > a_{n,n}(x,y).$$

PROOF   On using the relation

$$\binom{n}{k} = \frac{n(n-1)\cdots(n-k+2)}{(k-1)!} \cdot \frac{n-k+1}{k} = \frac{n-k+1}{k}\binom{n}{k-1},$$

the following recurrence relation for the sequence $a_{n,k}(x,y)$, $k = 0, 1, \ldots, n$, is obtained:

$$a_{n,k}(x,y) = \left(\frac{n-k+1}{k} \cdot \frac{x}{y}\right)\binom{n}{k-1}x^{k-1}y^{n-(k-1)}$$

$$= \frac{n-k+1}{(y/x)k}a_{n,k-1}(x,y), \quad k = 1, 2, \ldots, n.$$

Setting $m = (n+1)/(1+y/x)$ and since

$$\frac{n-k+1}{(y/x)k} = 1 + \frac{(n+1)-(1+y/x)k}{(y/x)k} = 1 + \frac{1+y/x}{y/x} \cdot \frac{m-k}{k},$$

it follows that

$$a_{n,k}(x,y) = \left(1 + \frac{x+y}{y} \cdot \frac{m-k}{k}\right)a_{n,k-1}(x,y), \quad k = 1, 2, \ldots, n.$$

Note that $1 < m < n+1$ and, by the last recurrence relation, we conclude that:
(a) If m is an integer, then

$$a_{n,k-1}(x,y) < a_{n,k}(x,y), \quad k = 1, 2, \ldots, m-1,$$

$$a_{n,m-1}(x,y) = a_{n,m}(x,y),$$

$$a_{n,k-1}(x,y) > a_{n,k}(x,y), \quad k = m+1, m+2, \ldots, n.$$

(b) If m is not an integer, then $m - k \neq 0$, $k = 1, 2, \ldots, n$ and

$$a_{n,k-1}(x,y) < a_{n,k}(x,y), \quad k = 1, 2, \ldots, r, \quad r = [m]$$

$$a_{n,k-1}(x,y) > a_{n,k}(x,y), \quad k = r+1, r+2, \ldots, n.$$

The proof of the theorem is thus completed.   ∎

The next corollary is concerned with a particular case, $x = y = 1$, of special interest.

COROLLARY 3.1
The sequence of binomial coefficients

$$C(n,k) = \binom{n}{k}, \quad k = 0, 1, \ldots, n,$$

(a) *for a given odd positive integer n, strictly increases for* $k = 0, 1, \ldots ,$ $m - 1$, $m = (n + 1)/2$, *assumes its maximum value at the points* $m - 1$ *and* m *and strictly decreases for* $k = m, m + 1, \ldots , n$, *while*

(b) *for a given even positive integer n, strictly increases for* $k = 0, 1, \ldots , r$, $r = n/2$, *assumes its maximum value at the point* r *and strictly decreases for* $k = r, r + 1, \ldots , n$.

Isaac Newton obtained in 1676 an extension of the binomial formula (3.10), as well as the negative binomial formula (3.13), to the case where the exponent of the left-hand side is a real number. This general binomial formula is stated in the following theorem, without proof. It is worth noting that there does not exist a combinatorial derivation of this general formula, since the binomial coefficient $\binom{x}{k}$ emerging in it has not, for every real number x, a combinatorial meaning. An analytic proof can be found in most advanced books on infinitesimal calculus. Newton's general binomial formula will be revisited in Chapter 6 on generating functions.

THEOREM 3.6 Newton's general binomial formula
Let t and x be real numbers, with $-1 < t < 1$. *Then*

$$(1 + t)^x = \sum_{k=0}^{\infty} \binom{x}{k} t^k. \tag{3.15}$$

Cauchy's binomial convolution formula, which by virtue of (3.11) may be considered a reformulation of Vandermonde's factorial convolution formula, is stated in the following corollary of Theorem 3.1. This corollary may also be deduced from Newton's general binomial formula. Two particular cases of such a derivation are worked out in Examples 3.6 and 3.7 that follow.

COROLLARY 3.2
Let x and y be real numbers and n a positive integer. Then

$$\binom{x + y}{n} = \sum_{k=0}^{n} \binom{x}{k}\binom{y}{n - k}. \tag{3.16}$$

Some interesting applications of these theorems in the evaluation of sums involving binomial coefficients and in the derivation of combinatorial identities are worked out in the following examples.

Example 3.3 The number of subsets of a finite set
Let us consider a finite set $W_n = \{w_1, w_2, \ldots , w_n\}$, of n elements, and let $\mathcal{A}_k$ be the set of all subsets A_k of W_n, with $N(A_k) = k$. Then, according to the definition

of a k-combination of n (without repetition), $\mathcal{A}_k = C_k(W_n)$ and $N(\mathcal{A}_k) = \binom{n}{k}$.
Evaluate the sums:

$$S_1 = \sum_{k=0}^{n} \binom{n}{k}, \quad S_2 = \sum_{j=0}^{[n/2]} \binom{n}{2j}, \quad S_3 = \sum_{j=0}^{[(n-1)/2]} \binom{n}{2j+1}.$$

Newton's binomial formula (3.10), with $x = y = 1$, yields

$$S_1 = \sum_{k=0}^{n} \binom{n}{k} = 2^n.$$

Therefore, the (total) number of subsets of W_n, with $N(W_n) = n$, equals 2^n (in agreement with the result of Example 1.7).

For the evaluation of the other two sums, note that

$$S_2 + S_3 = \sum_{j=0}^{[n/2]} \binom{n}{2j} + \sum_{j=0}^{[(n-1)/2]} \binom{n}{2j+1} = \sum_{k=0}^{n} \binom{n}{k} = 2^n$$

and

$$S_2 - S_3 = \sum_{j=0}^{[n/2]} \binom{n}{2j} - \sum_{j=0}^{[(n-1)/2]} \binom{n}{2j+1} = \sum_{k=0}^{n} (-1)^k \binom{n}{k}.$$

Further, Newton's binomial formula (3.10), with $x = -1$ and $y = 1$, yields

$$S_2 - S_3 = \sum_{k=0}^{n} (-1)^k \binom{n}{k} = 0.$$

Thus, the number S_2 of subsets with an even number of elements equals the number S_3 of subsets with an odd number of elements for any number n of elements of W_n. Consequently,

$$S_2 = S_3 = \frac{S_1}{2} = 2^{n-1}. \quad \square$$

Example 3.4
Evaluate the sums

$$S_4 = \sum_{k=1}^{n} k \binom{n}{k}, \quad S_5 = \sum_{k=0}^{n} (r + sk) \binom{n}{k}.$$

The first sum, on using the relation

$$k \binom{n}{k} = n \binom{n-1}{k-1},$$

may be written as

$$S_4 = n \sum_{k=1}^{n} \binom{n-1}{k-1} = n \sum_{j=0}^{n-1} \binom{n-1}{j}.$$

The last sum, by Newton's binomial formula with $n-1$ instead of n, equals

$$\sum_{j=0}^{n-1} \binom{n-1}{j} = 2^{n-1}$$

and so

$$S_4 = n2^{n-1}.$$

The second sum may be split up into two sums as

$$S_5 = r \sum_{k=0}^{n} \binom{n}{k} + s \sum_{k=0}^{n} k \binom{n}{k} = rS_1 + sS_4$$

and so

$$S_5 = r2^n + sn2^{n-1} = (2r + sn)2^{n-1}.$$

Note that the sum S_4 can be evaluated by taking the derivative of (3.10) with respect to x at a suitable point as follows: putting $y = 1$ in (3.10) and then taking the derivative of the resulting expression with respect to x,

$$\sum_{k=1}^{n} k \binom{n}{k} x^{k-1} = n(1+x)^{n-1},$$

at $x = 1$, we get

$$S_4 = \sum_{k=1}^{n} k \binom{n}{k} = n2^{n-1}. \qquad \square$$

Example 3.5
Evaluate the sums

$$S_6 = \sum_{k=0}^{n} \frac{1}{k+1} \binom{n}{k}, \quad S_7 = \sum_{k=0}^{n} \frac{(-1)^{k+1}}{k+1} \binom{n}{k}, \quad S_8 = \sum_{r=0}^{[n/2]} \frac{1}{2r+1} \binom{n}{2r}.$$

The first sum, on using the relation

$$\frac{1}{k+1} \binom{n}{k} = \frac{1}{n+1} \binom{n+1}{k+1},$$

may be written as

$$S_6 = \frac{1}{n+1} \sum_{k=0}^{n} \binom{n+1}{k+1} = \frac{1}{n+1} \sum_{j=1}^{n+1} \binom{n+1}{j}$$

and since, by Newton's binomial formula,

$$\sum_{j=1}^{n+1} \binom{n+1}{j} = \sum_{j=0}^{n+1} \binom{n+1}{j} - \binom{n+1}{0} = 2^{n+1} - 1,$$

it follows that

$$S_6 = \frac{2^{n+1} - 1}{n+1}.$$

In the same way

$$S_7 = \frac{1}{n+1} \sum_{k=0}^{n} (-1)^{k+1} \binom{n+1}{k+1} = \frac{1}{n+1} \sum_{j=1}^{n+1} (-1)^j \binom{n+1}{j}$$

and since

$$\sum_{j=1}^{n+1} (-1)^j \binom{n+1}{j} = \sum_{j=0}^{n+1} (-1)^j \binom{n+1}{j} - \binom{n+1}{0} = -1,$$

it follows that

$$S_7 = -\frac{1}{n+1}.$$

In order to evaluate the third sum, note first that

$$S_6 - S_7 = \sum_{k=0}^{n} \frac{1}{k+1} \binom{n}{k} - \sum_{k=0}^{n} \frac{(-1)^{k+1}}{k+1} \binom{n}{k} = \sum_{k=0}^{n} \frac{1 - (-1)^{k+1}}{k+1} \binom{n}{k}.$$

Since

$$1 - (-1)^{k+1} = \begin{cases} 2, & k = 2r, \quad r = 0, 1, \ldots \\ 0, & k = 2r + 1, \quad r = 0, 1, \ldots \end{cases}$$

it follows that

$$S_6 - S_7 = 2 \sum_{r=0}^{[n/2]} \frac{1}{2r+1} \binom{n}{2r} = 2S_8.$$

Therefore

$$S_8 = \frac{2^n}{n+1}.$$

Note that these sums can be evaluated by an integration of (3.10) with respect to x in a suitable interval as follows: putting $y = 1$ in (3.10) and integrating the resulting expression with respect to x in the interval $[x_1, x_2]$,

$$\sum_{k=0}^{n} \binom{n}{k} \int_{x_1}^{x_2} x^k \, dx = \int_{x_1}^{x_2} (x+1)^n \, dx,$$

and since

$$\int_{x_1}^{x_2} x^k \, dx = \left[\frac{x^{k+1}}{k+1} \right]_{x_1}^{x_2} = \frac{x_2^{k+1} - x_1^{k+1}}{k+1},$$

$$\int_{x_1}^{x_2} (x+1)^n \, dx = \left[\frac{(x+1)^{n+1}}{n+1} \right]_{x_1}^{x_2} = \frac{(x_2+1)^{n+1} - (x_1+1)^{n+1}}{n+1},$$

the following expression is deduced:

$$\sum_{k=0}^{n} \frac{x_2^{k+1} - x_1^{k+1}}{k+1} \binom{n}{k} = \frac{(x_2+1)^{n+1} - (x_1+1)^{n+1}}{n+1}.$$

By suitably choosing the limits x_1 and x_2 of the integration interval, it follows in particular that:

(a) For $x_1 = 0$ and $x_2 = 1$,

$$S_6 = \sum_{k=0}^{n} \frac{1}{k+1} \binom{n}{k} = \frac{2^{n+1} - 1}{n+1}.$$

(b) For $x_1 = -1$ and $x_2 = 0$,

$$-S_7 = \sum_{k=0}^{n} \frac{(-1)^k}{k+1} \binom{n}{k} = \frac{1}{n+1}.$$

(c) For $x_1 = -1$ and $x_2 = 1$,

$$S_8 = \sum_{r=0}^{[n/2]} \frac{1}{2r+1} \binom{n}{2r} = \frac{1}{2} \sum_{k=0}^{n} \frac{1 - (-1)^{k+1}}{k+1} \binom{n}{k} = \frac{2^n}{n+1}. \qquad \square$$

Example 3.6

For r, s and n positive integers, (a) derive analytically the combinatorial identity

$$\sum_{k=0}^{n} \binom{r}{k} \binom{s}{n-k} = \binom{r+s}{n}$$

and (b) conclude the identities

$$\sum_{k=0}^{r-j} \binom{r}{k} \binom{r}{j+k} = \binom{2r}{r+j}, \quad \sum_{k=0}^{r} \binom{r}{k}^2 = \binom{2r}{r},$$

$$\left[\sum_{k=0}^{r}\binom{r}{k}\right]^{2} = \sum_{k=0}^{2r}\binom{2r}{k}.$$

The expansion of the identity

$$(x+1)^{r}(x+1)^{s} = (x+1)^{r+s},$$

using Newton's binomial formula, yields

$$\left[\sum_{k=0}^{r}\binom{r}{k}x^{k}\right]\cdot\left[\sum_{j=0}^{s}\binom{s}{j}x^{j}\right] = \sum_{n=0}^{r+s}\binom{r+s}{n}x^{n}.$$

Executing the multiplication in the left-hand side, the general term $a_n x^n$, $n = 0, 1, \ldots, r+s$, is formed by multiplying the term $\binom{r}{k}x^k$ of the first factor by the term $\binom{s}{n-k}x^{n-k}$ of the second factor for all values $k = 0, 1, \ldots, n$. Thus, by the addition principle,

$$a_n = \sum_{k=0}^{n}\binom{r}{k}\binom{s}{n-k}$$

and

$$\sum_{n=0}^{r+s}\left\{\sum_{k=0}^{n}\binom{r}{k}\binom{s}{n-k}\right\}x^n = \sum_{n=0}^{r+s}\binom{r+s}{n}x^n$$

and, equating the coefficients of x^n of both sides the combinatorial identity,

$$\sum_{k=0}^{n}\binom{r}{k}\binom{s}{n-k} = \binom{r+s}{n}$$

is deduced. Note that this relation constitutes a particular case, with $x = r$ and $y = s$ positive integers, of Cauchy's binomial convolution formula (3.16).

Putting in (a) $s = r$, $n = r - j$ and since

$$\binom{r}{r-j-k} = \binom{r}{j+k}, \quad \binom{2r}{r-j} = \binom{2r}{r+j},$$

we conclude that

$$\sum_{k=0}^{j}\binom{r}{k}\binom{r}{j+k} = \binom{2r}{r+j}.$$

Also, setting in (a) $r = s = n$, and since

$$\binom{r}{r-k} = \binom{r}{k},$$

we find

$$\sum_{k=0}^{r}\binom{r}{k}^2 = \binom{2r}{r}.$$

Finally, executing the multiplication in the right-hand side of the expression

$$\left[\sum_{k=0}^{r} \binom{r}{k}\right]^2 = \left[\sum_{k=0}^{r} \binom{r}{k}\right] \cdot \left[\sum_{j=0}^{r} \binom{r}{j}\right],$$

we get

$$\left[\sum_{k=0}^{r} \binom{r}{k}\right]^2 = \sum_{n=0}^{2r} \left\{\sum_{k=0}^{n} \binom{r}{k}\binom{r}{n-k}\right\},$$

where the inner sum, by virtue of (a) with $s = r$, equals

$$\sum_{k=0}^{n} \binom{r}{k}\binom{r}{n-k} = \binom{2r}{n}.$$

Therefore

$$\left[\sum_{k=0}^{r} \binom{r}{k}\right]^2 = \sum_{n=0}^{2r} \binom{2r}{n}. \qquad \square$$

Example 3.7

For r, s and n positive integers, derive analytically the combinatorial identity

$$\sum_{k=0}^{n} \binom{r+k-1}{k}\binom{s+n-k-1}{n-k} = \binom{r+s+n-1}{n}.$$

The expansion of the identity

$$(1-t)^{-r}(1-t)^{-s} = (1-t)^{-(r+s)},$$

using Newton's negative binomial formula (3.14), yields

$$\left[\sum_{k=0}^{\infty} \binom{r+k-1}{k}t^k\right] \cdot \left[\sum_{j=0}^{\infty} \binom{s+j-1}{j}t^j\right] = \sum_{n=0}^{\infty} \binom{r+s+n-1}{n}t^n$$

and

$$\sum_{n=0}^{\infty} \left\{\sum_{k=0}^{n} \binom{r+k-1}{k}\binom{s+n-k-1}{n-k}\right\}t^n = \sum_{n=0}^{\infty} \binom{r+s+n-1}{n}t^n,$$

for every real number $t \in (-1, 1)$. Therefore

$$\sum_{k=0}^{n} \binom{r+k-1}{k}\binom{s+n-k-1}{n-k} = \binom{r+s+n-1}{n}.$$

Note that this relation also constitutes a particular case, with $x = -r$ and $y = -s$, of Cauchy's binomial convolution formula (3.16). □

Example 3.8

For s, k and n positive integers, with $s, k \leq n$, derive analytically the combinatorial identity

$$\sum_{r=k}^{n+k-s} \binom{r-1}{k-1}\binom{n-r}{s-k} = \binom{n}{s}.$$

Multiplying by t^s both members of the equivalent to Newton's negative binomial formula,

$$\sum_{j=0}^{\infty} \binom{s+j}{s} t^j = (1-t)^{-s-1},$$

and putting $n = s + j$, we get the expansion

$$\sum_{n=s}^{\infty} \binom{n}{s} t^n = t^s(1-t)^{-s-1}.$$

In the same way we deduce that:

$$\sum_{r=k}^{\infty} \binom{r-1}{k-1} t^r = t^k(1-t)^{-k}, \quad \sum_{j=s-k}^{\infty} \binom{j}{s-k} t^j = t^{s-k}(1-t)^{-s+k-1}.$$

Expanding the identity

$$[t^k(1-t)^{-k}][t^{s-k}(1-t)^{-s+k-1}] = t^s(1-t)^{-s-1},$$

by using the preceding three formulas, we get

$$\left[\sum_{r=k}^{\infty} \binom{r-1}{k-1} t^r\right] \cdot \left[\sum_{j=s-k}^{\infty} \binom{j}{s-k} t^j\right] = \sum_{n=s}^{\infty} \binom{n}{s} t^n$$

and

$$\sum_{n=s}^{\infty} \left\{ \sum_{r=k}^{n+k-s} \binom{r-1}{k-1}\binom{n-r}{s-k} \right\} t^n = \sum_{n=s}^{\infty} \binom{n}{s} t^n,$$

for every real number $t \in (-1, 1)$. Therefore

$$\sum_{r=k}^{n+k-s} \binom{r-1}{k-1}\binom{n-r}{s-k} = \binom{n}{s}. □$$

3.4 MULTINOMIAL COEFFICIENTS

The number

$$\binom{n}{k_1, k_2, \ldots, k_{r-1}} = \frac{n!}{k_1! k_2! \cdots k_{r-1}! k_r!},$$

where $k_r = n - (k_1 + k_2 + \cdots + k_{r-1})$, $k_i = 0, 1, \ldots, n$, $i = 1, 2, \ldots, r$, $n = 1, 2, \ldots$, is called *multinomial coefficient*, since it is the coefficient of the general term of the multinomial expansion. This expansion is combinatorially derived in the next theorem.

THEOREM 3.7 Multinomial formula
Let x_i, $i = 1, 2, \ldots, r$ be real numbers and n a positive integer. Then

$$(x_1 + x_2 + \cdots + x_r)^n = \sum \binom{n}{k_1, k_2, \ldots, k_{r-1}} x_1^{k_1} x_2^{k_2} \cdots x_r^{k_r}, \quad (3.17)$$

where the summation is extended over all $k_i = 0, 1, \ldots, n$, $i = 1, 2, \ldots, r$, such that $k_1 + k_2 + \cdots + k_{r-1} + k_r = n$.

PROOF   Note that each term in the expansion of $(x_1 + x_2 + \cdots + x_r)^n$ is of degree n with respect to the variables $x_1, x_2, \ldots, x_r$ and can be formed by executing the multiplication of the n factors $p_j = p_j(x_1, x_2, \ldots, x_r) = x_1 + x_2 + \cdots + x_r$, $j = 1, 2, \ldots, n$. The general term of this expansion, $x_1^{k_1} x_2^{k_2} \cdots x_r^{k_r}$, is produced by selecting x_i from k_i different factors, $i = 1, 2, \ldots, r$. To each such selection, there corresponds a division of the set $\{p_1, p_2, \ldots, p_n\}$, of the n factors, into r subsets that include $k_1, k_2, \ldots, k_r$ factors, respectively, with $k_i \geq 0$, $i = 1, 2, \ldots, r$ and $k_1 + k_2 + \cdots + k_r = n$. Consequently, according to Theorem 2.8, the general term $x_1^{k_1} x_2^{k_2} \cdots x_r^{k_r}$ appears

$$\binom{n}{k_1, k_2, \ldots, k_{r-1}}$$

times in the expansion. Thus, by the addition principle, (3.17) is deduced.   ∎

REMARK 3.3   The number of terms in expansion (3.17) equals the number of nonnegative integer solutions of the linear equation $k_1 + k_2 + \cdots + k_r = n$ and, according to Theorem 2.12, is equal to $\binom{r+n-1}{n}$.   ∎

Putting $x_i = 1$, $i = 1, 2, \ldots, n$ in (3.17) and taking into account Theorem 2.8, the next corollary is deduced.

COROLLARY 3.3

The number of divisions $(A_1, A_2, \ldots, A_r)$ *of a finite set* W_n, *with* $N(W_n) = n$, *in* r *(ordered) subsets, equals*

$$\sum \binom{n}{k_1, k_2, \ldots, k_{r-1}} = r^n,$$

where the summation is extended over all $k_i = 0, 1, \ldots, n$, $i = 1, 2, \ldots, r$, *such that* $k_1 + k_2 + \cdots + k_r = n$.

3.5   BIBLIOGRAPHIC NOTES

Although James Stirling (1676‑1770) was first to recognize the importance of the factorial in his work *Methodus Differentiallis* (1730), he did not use any special notation for it. The approximate formula for n factorial to which his name was attached was actually developed by Abraham De Moivre (1667‑1754). The first notation of the factorial was introduced by N. Vandermonde (1772), who also derived the factorial convolution formula. Further, he extended the definition of the factorial to negative order. The binomial formula appeared in various forms in early manuscripts, but it was James Bernoulli in his *Ars Conjectandi* (1713) who stated and proved it. The negative and general binomial formulae were worked out by Isaac Newton in 1676. It seems that the first notation of the binomial coefficient was introduced by Leonhard Euler (1707‑1783) in 1781. The notation, most in use now, is attributed to J. L. Raabe (1851). The combinatorial identities in Exercise 29 are due to E. S. Andersen (1953).

3.6   EXERCISES

1. Let a and b be real numbers and n a positive integer. Evaluate the sum

$$s_{n,r} = \sum_{k=r}^{n} (k)_r \binom{n}{k} (a)_k (b)_{n-k},$$

2*. Let x and y be real numbers and n a positive integer. Using Vandermonde's formula, show that

$$\sum_{k=0}^{n} \binom{n}{k} \frac{(x)_k}{(x+y)_k} = \frac{(x+y+n)_n}{(y+n)_n}, \quad y \neq -1, -2, \ldots, -n$$

and conclude that

$$\sum_{k=0}^{n}(-1)^{n-k}\binom{n}{k}\frac{x}{x-k}=\binom{x-1}{n}^{-1},$$

and

$$\sum_{k=0}^{n}(-1)^{k}\binom{n}{k}\frac{y}{y+k}=\binom{y+n}{n}^{-1}.$$

3*. Let x and y be real numbers and n a positive integer. Using Vandermonde's formula, show that

$$\sum_{k=0}^{n}(-1)^{k}\binom{n}{k}\frac{(x)_k}{(x+y)_k}=\frac{(y)_n}{(x+y)_n},\quad x+y\neq 0,1,2,\ldots,n-1$$

and conclude that

$$\sum_{k=0}^{n}(-1)^{k}\binom{n}{k}\frac{(x)_k}{(x+n)_k}=\sum_{k=0}^{n}(-1)^{k}\binom{n}{k}\binom{x}{k}\binom{x+n}{k}^{-1}=\binom{x+n}{n}^{-1}.$$

4. For r, s and n positive integers, with $n\leq r$, show that

$$\sum_{k=0}^{s}\binom{n+k-1}{k}\frac{(s)_k}{(r+s)_{n+k}}=\frac{1}{(r)_n}$$

and conclude that

$$\sum_{k=0}^{s}\frac{(s)_k}{(r+s)_k}=\frac{r+s+1}{r+1}.$$

5. (*Continuation*) Evaluate the sum

$$s_n=\sum_{k=1}^{s}k\binom{n+k-1}{k}\frac{(s)_k}{(r+s)_{n+k}}$$

and, more generally, the sum

$$s_{n,m}=\sum_{k=m}^{s}(k)_m\binom{n+k-1}{k}\frac{(s)_k}{(r+s)_{n+k}}.$$

6. Let $0<p<1$, $q=1-p$ and n a positive integer. Evaluate the sum

$$s_{n,r}=\sum_{k=r}^{n}(k)_r\binom{n}{k}p^k q^{n-k},\quad r=1,2,\ldots,n.$$

7. For $0 < p < 1$, $q = 1 - p$ and n a positive integer, evaluate the sum

$$s_{n,r} = \sum_{k=r}^{\infty} (k)_r \binom{n+k-1}{k} p^n q^k, \quad r = 1, 2, \ldots .$$

8. Evaluate the sums

$$S_1 = \sum_{k=1}^{n} (-1)^{k-1} k^2 \binom{n}{k}, \quad S_2 = \sum_{k=1}^{n} (-1)^{k-1} k^4 \binom{n}{k},$$

$$S_3 = \sum_{k=0}^{n} (2k+1) \binom{n}{k}, \quad S_4 = \sum_{k=1}^{n} 2k(k-2) \binom{n}{k}.$$

9. Evaluate the sums

$$S_1 = \sum_{k=0}^{n} \frac{k-1}{k+1} \binom{n}{k}, \quad S_2 = \sum_{k=0}^{n} \frac{k(k+4)+2}{(k+1)(k+2)} \binom{n}{k}.$$

10. For n and k positive integers, show that

$$\sum_{s=k}^{n} \binom{s}{k} \binom{n}{s} = 2^{n-k} \binom{n}{k}, \quad n \geq k,$$

and

$$\sum_{s=k}^{n} (-1)^{s-k} \binom{s}{k} \binom{n}{s} = \delta_{k,n}, \quad n \geq k,$$

where $\delta_{n,n} = 1$ and $\delta_{k,n} = 0$, $k \neq n$.

11. Consider the sum

$$S_n = \sum_{k=1}^{n} (-1)^{k-1} \frac{1}{k} \binom{n}{k}, \quad n = 1, 2, \ldots .$$

Show that

$$S_n = S_{n-1} + \frac{1}{n}, \quad n = 2, 3, \ldots$$

and since $S_1 = 1$, conclude that

$$S_n = \sum_{k=1}^{n} \frac{1}{k}.$$

12. Expanding the function $f_n(t) = [1 - (1-t)^n]/t$, where t is a real number different from 0 and n a positive integer, derive the expression

$$\sum_{k=1}^{n} (-1)^{k-1} \binom{n}{k} t^{k-1} = \sum_{k=1}^{n} (1-t)^{k-1};$$

integrating it in the interval $[0, 1]$, show that

$$\sum_{k=1}^{n}(-1)^{k-1}\frac{1}{k}\binom{n}{k} = \sum_{k=1}^{n}\frac{1}{k}.$$

13. (*Continuation*). Show that

$$\sum_{k=1}^{n}(-1)^{k-1}\frac{1}{k^2}\binom{n}{k} = \sum_{k=1}^{n}\sum_{r=1}^{k}\frac{1}{kr}.$$

14. Let

$$S_n(z) = \sum_{k=0}^{n}(-1)^k\frac{z}{z+k}\binom{n}{k}, \quad z \neq -1, -2, \ldots, -n, \; n = 0, 1, \ldots.$$

Using Pascal's triangle, show that

$$S_n(z) = \frac{n}{z+n}S_{n-1}(z), \quad n = 1, 2, \ldots, \quad S_0(z) = 1$$

and thus conclude that

$$S_n(z) = \binom{z+n}{n}^{-1}.$$

15. (*Continuation*). Show that

$$S_1 = \sum_{k=0}^{n}(-1)^k\frac{1}{1+2k}\binom{n}{k} = \frac{2^{2n}}{2n+1}\binom{2n}{n}^{-1}$$

and

$$S_2 = \sum_{k=0}^{n}(-1)^k\frac{1}{1-2k}\binom{n}{k} = 2^{2n}\binom{2n}{n}^{-1}.$$

16. *Leibnitz numbers.* Let

$$L(r,n) = \sum_{k=0}^{n}(-1)^{n-k}\frac{1}{r-k+1}\binom{n}{k}, \quad n = 0, 1, \ldots, r, \; r = 0, 1, \ldots.$$

Show that

$$L(r,n) = \frac{n}{r+1}L(r-1, n-1), \quad n = 1, 2, \ldots, r, \; r = 1, 2, \ldots,$$

with $L(r, 0) = 1/(r+1)$ and conclude that

$$L(r,n) = \frac{n!}{(r+1)_{n+1}} = \left[(r+1)\binom{r}{n}\right]^{-1}.$$

The numbers $L(r, n)$, $n = 0, 1, \ldots, r$, $r = 0, 1, \ldots$ are called *Leibnitz numbers*.

17*. Let n and r be positive integers and x a real number. Show that

$$\sum_{k=0}^{n} \binom{n}{k}\binom{x+k}{n} = \sum_{k=0}^{n} \binom{n}{k}\binom{x}{k} 2^k$$

and conclude that

$$\sum_{k=0}^{n} \binom{n}{k}\binom{r+k}{n} = \sum_{k=0}^{n} \binom{n}{k}\binom{r}{k} 2^k = \sum_{k=0}^{n} (-1)^{n-k}\binom{n}{k}\binom{r+k}{k} 2^k.$$

18*. For r a positive integer, show that

$$\sum_{k=0}^{r} \binom{2r}{k}^2 = \frac{1}{2}\binom{4r}{2r} + \frac{1}{2}\binom{2r}{r}^2$$

and

$$\sum_{k=0}^{r} \binom{2r+1}{k}^2 = \frac{1}{2}\binom{4r+2}{2r+1}.$$

19. Let n and k be positive integers with $n \geq k$. Show that

$$\sum_{r=k}^{n} (-1)^{n-r}\binom{n}{r}\binom{n+r}{n+k} = \binom{n}{k}.$$

20. Show that

$$\sum_{k=1}^{n} k\binom{r}{k}\binom{s}{n-k} = r\binom{r+s-1}{n-1}$$

and conclude that

$$\sum_{k=1}^{n} k\binom{n}{k}^2 = (2n-1)\binom{2n-2}{n-1}.$$

21. Show that

$$\sum_{k=0}^{n} (-1)^{n-k}\binom{r}{n-k}\binom{s+k}{k} = \binom{s-r+n}{n}, \quad s \geq r$$

and conclude that

$$\sum_{k=0}^{n} (-1)^{n-k}\binom{n}{k}\binom{r+k}{n} = 1.$$

22. (*Continuation*). Show that

$$\sum_{k=0}^{n}(-1)^k \binom{n}{k}\binom{r}{k}\binom{n+r}{k}^{-1} = \binom{n+r}{r}^{-1}$$

and conclude that

$$\sum_{k=0}^{n}(-1)^k \binom{n}{k}^2 \binom{2n}{k}^{-1} = \binom{2n}{n}^{-1}.$$

23. Let r, s and n be positive integers with $s \geq n$. Show that

$$\sum_{k=0}^{n}\binom{r+k}{k}\binom{s-k}{n-k} = \binom{r+s+1}{n}.$$

24*. For r, s and n positive integers with $s \leq r \leq n$, show that

$$\sum_{k=0}^{r-s}(-1)^{r-s-k}\binom{r-s}{k}\binom{n+k}{s+k} = \binom{n}{r}.$$

25*. Let r, s, j and n be nonnegative integers with $j \leq r \leq j+n$. Show that

$$\sum_{k=0}^{n}(-1)^k \binom{r}{j+k}\binom{s+k-1}{k} = \binom{r-s}{r-j}$$

and conclude that

$$\sum_{k=0}^{n}(-1)^k \binom{r}{s+k}\binom{s+k-1}{k} = 1.$$

26. *Recurrence relations for the binomial coefficients.* Prove that the binomial coefficient $\binom{x}{k}$, where x is a real number and k a nonnegative integer, satisfies:

(a) The "triangular" recurrence relation

$$\binom{x}{k} = \binom{x-1}{k} + \binom{x-1}{k-1}$$

(b) The "vertical" recurrence relation

$$\binom{x}{k} = \binom{x-s-1}{k} + \sum_{r=0}^{s}\binom{x-r-1}{k-1}$$

(c) The "horizontal" recurrence relation

$$\binom{x}{k} = \sum_{r=0}^{k}(-1)^{k-r}\binom{x+1}{r}.$$

27. Expanding the identity

$$(1+t+u+tu)^x = (1+t)^x(1+u)^x$$

into powers of t and u, show that

$$\binom{x}{n}\binom{x}{k} = \sum_{j=0}^{r}\binom{n}{j}\binom{n+k-j}{n}\binom{x}{n+k-j},\ r = \min\{n,k\},$$

for every real number x and positive integers n and k and conclude that

$$(x)_n(x)_k = \sum_{j=0}^{r}\binom{n}{k}\binom{k}{j}j!(x)_{n+k-j}.$$

28. Show that

$$(-1)^k 2^{2k}\binom{-1/2}{k} = \binom{2k}{k}$$

and

$$\sum_{k=0}^{n}\binom{2n-2k}{n-k}\binom{2k}{k} = 2^{2n}.$$

29*. Let n and r be positive integers with $r \le n$ and x a real number. Prove that

$$\sum_{k=0}^{r}\binom{x}{k}\binom{-x}{n-k} = -\sum_{k=r+1}^{n}\binom{x}{k}\binom{-x}{n-k} = \frac{n-r}{n}\binom{x-1}{r}\binom{-x}{n-r}$$

and

$$\sum_{k=0}^{r}\binom{x}{k}\binom{1-x}{n-k+1} = \frac{n(1-x)-r}{n(n+1)}\binom{x-1}{r}\binom{-x}{n-r}.$$

30. For x_i, $i = 1, 2, \ldots, r$, real numbers and n a positive integer, show that

$$\sum\binom{x_1}{k_1}\binom{x_2}{k_2}\cdots\binom{x_2}{k_r} = \binom{x_1+x_2+\cdots+x_r}{n},$$

where the summation is extended over all $k_i \ge 0$, $i = 1, 2, \ldots, r$, with $k_1 + k_2 + \cdots + k_r = n$; conclude that

$$\sum\frac{n!}{k_1!k_2!\cdots k_r!}(x_1)_{k_1}(x_2)_{k_2}\cdots(x_r)_{k_r} = (x_1+x_2+\cdots+x_r)_n,$$

where the summation is taken as in the preceding sum.

Chapter 4

THE PRINCIPLE OF INCLUSION AND EXCLUSION

4.1 INTRODUCTION

The basic principles of counting the elements of a union of n finite and pairwise disjoint sets (principle of addition) and of a subset of the Cartesian product of n finite sets (multiplication principle) were presented in Chapter 1. The principle of inclusion and exclusion, or the sieve formula of Eratosthenes, is also a powerful counting tool and this chapter is devoted to its presentation. The number of elements in a union (and in the intersection of the complements) of any n subsets of a finite set is computed. More generally, the numbers of elements that are contained in k and in at least k among n subsets of a finite set are evaluated. Further, a series of alternating inequalities for these numbers, due to Bonferroni, is derived. In the case of ordered subsets of a set, the notion of the rank of an element is introduced. Then, the elements of this set of a given rank are enumerated. The important case of exchangeable sets, in which the number of elements that are contained in the intersection of any r of them depends only on r, is separately examined in all the preceding enumeration problems. The Möbius inversion formula, which is also a useful counting tool, is presented as a complement in the Exercises.

Note that the principle of inclusion and exclusion and, more generally, the problem of counting the elements of a finite set are presented by several authors, equivalently, according to the definition of a set: $A = \{a : a$ has property $P\}$, by using the corresponding characteristic properties of the sets. So, in this presentation, the number of elements that have at least one among n given properties expresses the cardinal of the union of the corresponding n sets, while the number of elements with none of n properties expresses the cardinal of the intersection of the complements of the n sets.

4.2 NUMBER OF ELEMENTS IN A UNION OF SETS

The number of elements (cardinal) in a union of two sets is derived in the next theorem.

THEOREM 4.1

If A and B are any finite sets, then

$$N(A \cup B) = N(A) + N(B) - N(AB). \tag{4.1}$$

PROOF   In order to calculate the number of elements of the union $A \cup B$, the elements that belong either to A or B should be counted, but each element should be included only once. Since in the sum $N(A) + N(B)$ the elements that belong to both sets A and B are counted twice, the number $N(AB)$ should be subtracted from it and so expression (4.1) is established.   ∎

REMARK 4.1   Expression (4.1) constitutes a generalization of the addition principle, $N(A + B) = N(A) + N(B)$, to which it reduces if the finite sets A and B are supposed to be disjoint, whence $N(AB) = 0$. Note also that (4.1) may be derived by using the addition principle as follows: if A and B are any finite sets, then the sets A and $B - A = A' \cap B$ are disjoint, $A \cap (A' \cap B) = (A \cap A') \cap B = \emptyset \cap B = \emptyset$ and $A \cup (A' \cap B) = (A \cup A') \cap (A \cup B) = \Omega \cap (A \cup B) = A \cup B$. Consequently, by the addition principle,

$$N(A \cup B) = N(A + (B - A)) = N(A) + N(B - A)$$

and, since $N(B - A) = N(B) - N(AB)$, (see Theorem 1.4), expression (4.1) is deduced.   ∎

COROLLARY 4.1

If A and B are any subsets of a finite set Ω, with $N(\Omega) = N$, then

$$N(A'B') = N - N(A) - N(B) + N(AB). \tag{4.2}$$

PROOF   Since by De Morgan's formula $A'B' = (A \cup B)'$, expressing the number of elements of $(A \cup B)'$ in terms of the number of elements of $A \cup B$,

$$N(A'B') = N((A \cup B)') = N - N(A \cup B)$$

and then, using (4.1), expression (4.2) is established.   ∎

Example 4.1

Calculate the number of positive integers less than or equal to 100 that are divisible by 3 or by 5.

Let A and B be the sets of positive integers less than or equal to 100 that are divisible by 3 and 5, respectively. Then $N(A \cup B)$ is the number of positive integers less than or equal to 100 that are not divisible by 3 or 5 and, by (4.1),

$$N(A \cup B) = N(A) + N(B) - N(AB).$$

Denoting by $[x]$ the integer part of x, we have

$$N(A) = [100/3] = 33, \ N(B) = [100/5] = 20, \ N(AB) = [100/15] = 6.$$

Therefore
$$N(A \cup B) = 33 + 20 - 6 = 47. \quad \Box$$

Example 4.2

Suppose that the elevator of a four-floor building starts from the basement with three passengers. Calculate the number of different ways of discharging at least one passenger at each of the first two floors.

Each way that the three passengers may get off the elevator may be represented by an ordered selection of three floors $(f_{i_1}, f_{i_2}, f_{i_3})$, with repetition, from the set $\{f_1, f_2, f_3, f_4\}$ of the four floors, where f_{i_r} is the floor at which the r-th passenger gets off the elevator, $r = 1, 2, 3$.

One may argue that there are $\binom{3}{2} = 3$ two-element subsets of the set $\{p_1, p_2, p_3\}$ of the three passengers and, for each of these subsets, the two passengers get off the elevator, one passenger at the first and the other at the second floor, in $2! = 2$ different ways. Since it is required that at least one passenger get off the elevator at each of the first two floors, there are four different choices for the third passenger to get off the elevator. Thus, by this argument and according to the multiplication principle, the number of different ways of discharging at least one passenger at each of the first two floors equals

$$3 \cdot 2 \cdot 4 = 24.$$

Although all the steps of this derivation seem to be justified, a well-disguised error allows certain ways of discharging the passengers to be counted twice. A closer inspection reveals that the following 3-permutations of the set $\{f_1, f_2, f_3, f_4\}$ are counted twice:

$$(f_1, f_2, f_2), \ (f_2, f_1, f_2), \ (f_2, f_2, f_1), \ (f_2, f_1, f_1), \ (f_1, f_2, f_1), \ (f_1, f_1, f_2).$$

For example the first of these permutations is counted as one of the four permutations

$$(f_1, f_2, f_r), \ r = 1, 2, 3, 4$$

and also as one of the four permutations

$$(f_1, f_r, f_2), \quad r = 1, 2, 3, 4.$$

Consequently, the correct number of different ways of discharging at least one passenger at each of the first two floors equals

$$24 - 6 = 18.$$

This number can be effectively evaluated by the aid of Corollary 4.1 as follows. Let Ω be the set of different ways of discharging the three passengers. Also, let A and B be the sets of different ways of discharging the three passengers without stopping at the first and second floor, respectively. Then $N(A'B')$ is the number of different ways of discharging at least one passenger at each of the first two floors, and so, by (4.2),

$$N(A'B') = N - N(A) - N(B) + N(AB).$$

Clearly, $N = N(\Omega) = 4^3 = 64$, which is the number of the 3-permutations, with repetition, of the set $\{f_1, f_2, f_3, f_4\}$, of the four floors. Also, $N(A) = 3^3 = 27$, which is the number of the 3-permutations, with repetition, of the set $\{f_2, f_3, f_4\}$, of three floors, and similarly, $N(B) = 3^3 = 27$. Further, $N(AB) = 2^3 = 8$, which is the number of the 3-permutations, with repetition, of the set $\{f_3, f_4\}$, of two floors. Hence, the required number equals

$$N(A'B') = 64 - 2 \cdot 27 + 8 = 18. \quad \square$$

The next theorem constitutes a generalization of Theorem 4.1 to n sets.

THEOREM 4.2 Inclusion and exclusion principle
The number $L_{n,1} = N(A_1 \cup A_2 \cup \cdots \cup A_n)$, of elements that are contained in at least one among n finite sets $A_1, A_2, \ldots, A_n$, is given by

$$L_{n,1} = S_{n,1} - S_{n,2} + \cdots + (-1)^{n-1} S_{n,n}, \tag{4.3}$$

with

$$S_{n,1} = \sum_{i=1}^{n} N(A_i), \quad S_{n,2} = \sum_{i=1}^{n-1} \sum_{j=i+1}^{n} N(A_i A_j)$$

and, in general,

$$S_{n,r} = \sum N(A_{i_1} A_{i_2} \cdots A_{i_r}), \quad r = 1, 2, \ldots, n \tag{4.4}$$

where the summation is extended over all r-combinations $\{i_1, i_2, \ldots, i_r\}$ of the n indices $\{1, 2, \ldots, n\}$.

PROOF   The number $L_{n,1} = N(A_1 \cup A_2 \cup \cdots \cup A_n)$ is deduced by counting the elements included either in A_1 or in $A_2, \ldots,$ or in A_n, but each element should be counted only once. This deduction may be achieved initially by taking the sum

$$S_{n,1} = \sum_{i=1}^{n} N(A_i).$$

Note that, in this sum, the elements included in any one set A_i only are counted once, in the term $N(A_i)$, $i = 1, 2, \ldots, n$, while the elements common to any two sets A_i and A_j, with $\{i, j\} \subseteq \{1, 2, \ldots, n\}$, are counted twice, in each of the terms $N(A_i)$ and $N(A_j)$. Thus the sum

$$S_{n,2} = \sum_{i=1}^{n-1} \sum_{j=i+1}^{n} N(A_i A_j)$$

should be subtracted from $S_{n,1}$. Further, the elements common to any three sets A_i, A_j and A_s, with $\{i, j, s\} \subseteq \{1, 2, \ldots, n\}$, are counted in the sum $S_{n,1}$ three times, in each of the terms $N(A_i)$, $N(A_j)$ and $N(A_s)$, and in the sum $S_{n,2}$ again three times, in each of the terms $N(A_i A_j)$, $N(A_i A_s)$ and $N(A_j A_s)$. Thus, these elements are not counted, in total, in the difference $S_{n,2} - S_{n,1}$ and, consequently, the sum

$$S_{n,3} = \sum N(A_i A_j A_s),$$

should be added to it. Continuing this process, having considered the elements common to any $n-1$ sets $A_{i_1}, A_{i_2}, \ldots, A_{i_{n-1}}, \{i_1, i_2, \ldots, i_{n-1}\} \subseteq \{1, 2, \ldots, n\}$, and after algebraically adding the sum

$$S_{n,n-1} = \sum N(A_{i_1}, A_{i_2} \cdots A_{i_{n-1}}),$$

we end up with the algebraic sum $S_{n,1} - S_{n,2} + \cdots + (-1)^{n-2} S_{n,n-1}$, in which these elements are counted, in total, only once. Finally, the elements common to the n sets $A_1, A_2, \ldots, A_n$ are counted $\binom{n}{k}$ times in the sum $S_{n,k}$, in each of the terms $N(A_{i_1} A_{i_2} \cdots A_{i_k})$, $\{i_1, i_2, \ldots, i_k\} \subseteq \{1, 2, \ldots, n\}$, of it, $k = 1, 2, \ldots, n$. According to Newton's binomial formula (3.10),

$$\sum_{k=1}^{n-1} (-1)^{k-1} \binom{n}{k} = 1 + (-1)^n - \sum_{k=0}^{n} (-1)^k \binom{n}{k} = 1 + (-1)^n$$

and so these elements are counted, in total, zero times if n is odd and two times if n is even. Consequently, the term

$$(-1)^{n-1} S_{n,n} = (-1)^{n-1} N(A_1 A_2 \cdots A_n)$$

should be added to the algebraic sum $S_{n,1} - S_{n,2} + \cdots + (-1)^{n-2} S_{n,n-1}$, and so (4.3) is established.   ∎

REMARK 4.2   The term *inclusion and exclusion* refers to the process in which, initially, all the elements under consideration are indiscriminately included; and those elements that should not be included are excluded, then those elements that should not be excluded are included, and so on, alternately including and excluding.   ∎

REMARK 4.3   The following proof of (4.3) by mathematical induction is of interest. Note first that, according to (4.1), (4.3) holds for $n = 2$. Then, supposing that (4.3) holds for $n - 1$, it should be shown that it also holds for n. From (4.1), with $A = A_1 \cup A_2 \cup \cdots \cup A_{n-1}$, $B = A_n$ and since $AB = (A_1 A_n) \cup (A_2 A_n) \cup \cdots \cup (A_{n-1} A_n)$, it follows that

$$N(A_1 \cup A_2 \cup \cdots \cup A_n) = N(A_1 \cup A_2 \cup \cdots \cup A_{n-1}) + N(A_n)$$
$$-N(B_1 \cup B_2 \cup \cdots \cup B_{n-1}),$$

where $B_1 = A_1 A_n$, $B_2 = A_2 A_n, \ldots, B_{n-1} = A_{n-1} A_n$. Thus, by the induction hypothesis,

$$N(A_1 \cup A_2 \cup \cdots \cup A_n) = \sum_{r=1}^{n-1}(-1)^{r-1} S_{n-1,r} + N(A_n) - \sum_{r=1}^{n-1}(-1)^{r-1} Q_{n-1,r},$$

with

$$S_{n-1,r} = \sum N(A_{i_1} A_{i_2} \cdots A_{i_r}),$$

$$Q_{n-1,r} = \sum N(B_{i_1} B_{i_2} \cdots B_{i_r}) = \sum N(A_{i_1} A_{i_2} \cdots A_{i_r} A_n),$$

where, in both sums, the summation is extended over all r-combinations $\{i_1, i_2, \ldots, i_r\}$ of the $n - 1$ indices $\{1, 2, \ldots, n - 1\}$. Note that

$$S_{n-1,1} + N(A_n) = \sum_{i=1}^{n-1} N(A_i) + N(A_n) = \sum_{i=1}^{n} N(A_i) = S_{n,1}.$$

Further, for $r = 2, 3, \ldots, n - 1$,

$$S_{n-1,r} + Q_{n-1,r-1} = \sum N(A_{i_1} A_{i_2} \cdots A_{i_r}) + \sum N(A_{i_1} A_{i_2} \cdots A_{i_{r-1}} A_n)$$

is the sum of the terms $N(A_{i_1} A_{i_2} \cdots A_{i_r})$, with the summation extended over the

$$\binom{n-1}{r}$$

r-combinations $\{i_1, i_2, \ldots, i_r\}$ of the n indices $\{1, 2, \ldots, n - 1, n\}$ that do not include the index n and over the

$$\binom{n-1}{r-1}$$

r-combinations $\{i_1, i_2, \ldots, i_{r-1}, n\}$ of the n indices $\{1, 2, \ldots, n-1, n\}$ that include the index n. Thus,

$$S_{n-1,r} + Q_{n-1,r-1} = \sum N(A_{i_1} A_{i_2} \cdots A_{i_r}) = S_{n,r}, \quad r = 2, 3, \ldots, n-1,$$

where the summation is extended over the

$$\binom{n-1}{r} + \binom{n-1}{r-1} = \binom{n}{r}$$

r-combinations $\{i_1, i_2, \ldots, i_r\}$ of the n indices $\{1, 2, \ldots, n\}$. Also

$$Q_{n-1,n-1} = N(B_1 B_2 \cdots B_{n-1}) = N(A_1 A_2 \cdots A_{n-1} A_n) = S_{n,n}.$$

Consequently,

$$N(A_1 \cup A_2 \cup \cdots \cup A_n) = S_{n,1} - S_{n,2} + \cdots + (-1)^{n-1} S_{n,n}$$

and so it is shown that (4.3) also holds for n. Therefore, according to the principle of mathematical induction, (4.3) holds true for every integer $n \geq 2$.   ∎

COROLLARY 4.2

If $A_1, A_2, \ldots, A_n$ are subsets of a finite set Ω, then the number $N_{n,0} = N(A_1' A_2' \cdots A_n')$ of elements of Ω that are not contained in any of these n sets is given by

$$N_{n,0} = S_{n,0} - S_{n,1} + \cdots + (-1)^n S_{n,n}, \tag{4.5}$$

where $S_{n,0} = N(\Omega)$ and $S_{n,r}$, $r = 1, 2, \ldots, n$, is given by (4.4).

PROOF   Since by De Morgan's formula $(A_1 \cup A_2 \cup \cdots \cup A_n)' = A_1' A_2' \cdots A_n'$, upon using the expression $N(A') = N(\Omega) - N(A)$, with $A = A_1 \cup A_2 \cup \cdots \cup A_n$, it follows that

$$N(A_1' A_2' \cdots A_n') = N(\Omega) - N(A_1 \cup A_2 \cup \cdots \cup A_n).$$

Introducing expression (4.3) into it, we conclude (4.5).   ∎

The inclusion and exclusion principle enables completion of the discussion, started in Section 2.5, on the number of integer solutions of the linear equation

$$x_1 + x_2 + \cdots + x_n = k, \tag{4.6}$$

which satisfy a given set of restrictions, with k an integer. The next theorem is concerned with the enumeration of the integer solutions of (4.6) that are bounded both from above and below.

THEOREM 4.3
The number of integer solutions $(r_1, r_2, \ldots, r_n)$ of the linear equation (4.6), with the restrictions

$$s_i \leq x_i \leq m_i, \quad i = 1, 2, \ldots, n, \tag{4.7}$$

for given integers s_i, m_i, $i = 1, 2, \ldots, n$, with $s \leq k \leq m$, $s = s_1 + s_2 + \cdots + s_n$, $m = m_1 + m_2 + \cdots + m_n$, and for $u_i = m_i - s_i \geq 0$, $i = 1, 2, \ldots, n$, is given by

$$A_{n,k}(u_1, u_2, \ldots, u_n) = \binom{n + k - s - 1}{n - 1}$$

$$+ \sum_{r=1}^{n} (-1)^r \sum \binom{n + k - s - u_{i_1} - u_{i_2} - \cdots - u_{i_r} - r - 1}{n - 1}, \tag{4.8}$$

where, in the inner sum, the summation is extended over all r-combinations $\{i_1, i_2, \ldots, i_r\}$ of the n indices $\{1, 2, \ldots, n\}$.

PROOF   On using the transformation $y_i = x_i - s_i$, $i = 1, 2, \ldots, n$, the linear equation (4.6) and the restrictions (4.7) reduce to

$$y_1 + y_2 + \cdots + y_n = k - s, \tag{4.9}$$

with $k - s$ a nonnegative integer, and

$$0 \leq y_i \leq u_i, \; i = 1, 2, \ldots, n, \; u = u_1 + u_2 + \cdots + u_n \geq k - s. \tag{4.10}$$

Since the transformation is one-to-one, the number $A_{n,k}(u_1, u_2, \ldots, u_n)$ of the integer solutions of (4.6), with the restrictions (4.7), equals the number of integer solutions of (4.9), with the restrictions (4.10). In order to evaluate this number, consider the set Ω of nonnegative integer solutions of (4.9), without any restriction, and let A_i be the subset of nonnegative integer solutions of (4.9), with the restriction $y_i \geq u_i + 1$, $i = 1, 2, \ldots, n$. Then

$$A_{n,k}(u_1, u_2, \ldots, u_n) = N(A_1' A_2' \cdots A_n')$$

and, for its evaluation, Corollary 4.2 can be applied. Note first that $S_{n,0} = N(\Omega)$ is the number of nonnegative integer solutions of (4.9) which, according to Theorem 2.12, equals

$$S_{n,0} = \binom{n + k - s - 1}{k - s} = \binom{n + k - s - 1}{n - 1}.$$

Further, $N(A_{i_1} A_{i_2} \cdots A_{i_r})$ is the number of nonnegative integer solutions of (4.9) with the restrictions

$$y_{i_j} \geq u_{i_j} + 1, \quad j = 1, 2, \ldots, r,$$

which, according to Corollary 2.4, is given by

$$N(A_{i_1} A_{i_2} \cdots A_{i_r}) = \binom{n + k - s - u_{i_1} - u_{i_2} - \cdots - u_{i_r} - r - 1}{n - 1}$$

if $u_{i_1} + u_{i_2} + \cdots + u_{i_r} + r \leq k - s$ and

$$N(A_{i_1} A_{i_2} \cdots A_{i_r}) = 0$$

if $u_{i_1} + u_{i_2} + \cdots + u_{i_r} + r > k - s$. Introducing these expressions into (4.5), the required expression (4.8) is deduced. ∎

REMARK 4.4 *Combinations with restricted repetition.* As has been noted in Section 2.5, to each nonnegative integer solution $(r_1, r_2, \ldots, r_n)$ of (4.6), there uniquely corresponds a k-combination $\{a_1, a_2, \ldots, a_k\}$ of the set $W_n = \{w_1, w_2, \ldots, w_n\}$, with repetition, in which the element w_i is included $r_i \geq 0$ times, for $i = 1, 2, \ldots, n$. Thus, the number of k-combinations of n in which the element w_i is included no less than s_i and no more than m_i times, for $i = 1, 2, \ldots, n$, is given by (4.8). ∎

Example 4.3

Consider a collection of seven nickels, four dimes and two quarters. Assume that coins of any denomination are identical. Find the number of different sub-collections of ten coins.

Consider a sub-collection of ten coins and let x_1 be the number of nickels, x_2 the number of dimes and x_3 the number of quarters contained in it. Then the number of different sub-collections of ten coins equals the number of integer solutions of the linear equation

$$x_1 + x_2 + x_3 = 10,$$

with the restrictions $0 \leq x_1 \leq 7, 0 \leq x_2 \leq 5$ and $0 \leq x_3 \leq 2$. Applying (4.8), with $n = 3, k = 10$ and $u_1 = m_1 = 7, u_2 = m_2 = 5, u_3 = m_3 = 2$, we conclude that the required number equals

$$\binom{12}{2} - \left\{ \binom{4}{2} + \binom{6}{2} + \binom{9}{2} \right\} + \binom{3}{2} = 66 - (6 + 15 + 36) + 3 = 12.$$

Clearly, these 12 solutions are the following:

$$(7, 3, 0), \ (7, 2, 1), \ (7, 1, 2), \ (6, 4, 0), \ (6, 3, 1), \ (6, 2, 2),$$

$$(5, 5, 0), \ (5, 4, 1), \ (5, 3, 2), \ (4, 5, 1), \ (4, 4, 2), \ (3, 5, 2). \quad \square$$

Example 4.4 **Euler's function**

Let n be a positive integer and let us denote by $\phi(n)$ the number of positive integers less than or equal to n that are relatively prime to n. The function $\phi(n)$

is called the Euler's function of n. In order to compute this function, consider the decomposition of n into prime factors:

$$n = s_1^{k_1} s_2^{k_2} \cdots s_r^{k_r},$$

where s_i are prime numbers and k_i positive integers, $i = 1, 2, \ldots, r$ and let Ω be the set of positive integers that are less than or equal to n. If A_i is the set of positive integers less than or equal to n that are divided by s_i, $i = 1, 2, \ldots, r$, then $\phi(n) = N(A_1' A_2' \cdots A_r')$ is the number of positive integers less than or equal n that are relatively prime to n. Thus, according to Corollary 4.2,

$$\phi(n) = N(A_1' A_2' \cdots A_r') = \sum_{k=0}^{r} (-1)^k S_{r,k},$$

with

$$S_{r,0} = N(\Omega), \quad S_{r,k} = \sum N(A_{i_1} A_{i_2} \cdots A_{i_k}), \quad k = 1, 2, \ldots, r,$$

where the summation is extended over all k-combinations $\{i_1, i_2, \ldots, i_k\}$ of the r indices $\{1, 2, \ldots, r\}$. Clearly,

$$S_{r,0} = n, \quad N(A_{i_1} A_{i_2} \cdots A_{i_k}) = \frac{n}{s_{i_1} s_{i_2} \cdots s_{i_k}}$$

and so the Euler's function of n is given by the expression

$$\phi(n) = n + \sum_{k=1}^{r} (-1)^k \sum \frac{n}{s_{i_1} s_{i_2} \cdots s_{i_k}}.$$

This expression may be written in the form

$$\phi(n) = n \left(1 - \frac{1}{s_1} \right) \left(1 - \frac{1}{s_2} \right) \cdots \left(1 - \frac{1}{s_r} \right).$$

Table 4.1 gives Euler's function $\phi(n)$ for $n = 1, 2, \ldots, 15$.   ☐

Table 4.1 *Euler's Function*

n	1	2	3	4	5	6	7	8	9	10	11	12	13	14	15
$\phi(n)$	1	1	2	2	4	2	6	4	6	4	10	4	12	6	8

Example 4.5 The ménages problem
  A classical enumeration problem formulated and solved by E. Lucas in 1891, known as the ménages problem, asks the following. What is the number of different

seatings of n married couples (ménages) around a circular table so that men and women alternate and no man is next to his wife?

Assume that n of the $2n$ seats are green and the rest n seats are red and that the green and red seats are arranged around the table in alternate positions. Suppose (without any loss of generality) that the wives are seated first. Since men and women ought to be in alternate positions, the n women may sit either on the green seats, which may be done in $n!$ ways, or on the red seats, which also may be done in $n!$ ways. Consequently, according to the addition principle, the n women can be placed around the table in $2n!$ ways. Let us now number (a) the n women from 1 to n in the ordinary direction (counterclockwise) starting from any one of them, (b) the n empty seats from 1 to n in the ordinary direction starting from the seat that is to the left of the woman with the number 1 and (c) the n men by assigning to every man the number of his wife. In this way the enumeration of the different ways of placing the n men on the n empty seats so that no man is seated next to his wife is reduced to the enumeration of the number M_n of permutations $(j_1, j_2, \ldots, j_n)$ of $\{1, 2, \ldots, n\}$ that satisfy the restrictions:

$$j_r \neq r, \ j_r \neq r + 1, \ r = 1, 2, \ldots, n - 1, \ j_n \neq n, \ j_n \neq 1.$$

In order to enumerate the number M_n, let us consider the set Ω of the $n!$ permutations $(j_1, j_2, \ldots, j_n)$ of $\{1, 2, \ldots, n\}$ and its subsets $A_1, A_2, \ldots, A_n$, which are defined as follows: A_{2r-1} is the set of permutations with $j_r = r, r = 1, 2, \ldots, n$, A_{2r} the set of permutations with $j_r = r + 1, \ r = 1, 2, \ldots, n - 1$ and A_{2n} the set of permutations with $j_n = 1$. Then $M_n = N(A'_1 A'_2 \cdots A'_{2n})$ and, according to Corollary 4.2,

$$M_n = \sum_{k=0}^{2n} (-1)^k S_{2n,k},$$

with

$$S_{2n,0} = N(\Omega) = n!, \ S_{2n,k} = \sum N(A_{i_1} A_{i_2} \cdots A_{i_k}), \ k = 1, 2, \ldots, 2n,$$

where the summation is extended over all k-combinations $\{i_1, i_2, \ldots, i_k\}$ of the $2n$ indices $\{1, 2, \ldots, 2n\}$. Note that

$$N(A_{i_1} A_{i_2} \cdots A_{i_k}) = 0$$

in the case where, between the indicated sets, there are at least two with consecutive indices: A_{i_s} and A_{i_s+1}, since if $i_s = 2r - 1$, an odd number, then $j_r = r$ and $j_r = r + 1$, while if $i_s = 2r$, an even number, then $j_r = r + 1$ and $j_{r+1} = r + 1$, which cannot happen in the same permutation $(j_1, j_2, \ldots, j_n)$ of $\{1, 2, \ldots, n\}$. Therefore

$$S_{2n,k} = 0 \ \text{for} \ k = n + 1, n + 2, \ldots, 2n,$$

since all the terms of this sum include at least two sets with consecutive indices. Further, the number of nonzero terms of $S_{2n,k}, k = 0, 1, \ldots, n$, equals

$$\frac{2n}{2n - k} \binom{2n - k}{k},$$

the number of k-combinations $\{i_1, i_2, \ldots, i_k\}$ of the $2n$ indices $\{1, 2, \ldots, 2n\}$ displayed on a circle that include no pair of consecutive integers (see Example 2.29). In addition, in the case where, between the indicated sets, there is no pair with consecutive indices, the general term of $S_{2n,k}$, $k = 0, 1, \ldots, n$, equals

$$N(A_{i_1} A_{i_2} \cdots A_{i_k}) = (n - k)!,$$

the number of permutations of the remaining $n - k$ elements, the positions of which are not determined by the indicated sets. Hence

$$M_n = \sum_{k=0}^{n} (-1)^k \frac{2n}{2n - k} \binom{2n - k}{k} (n - k)!.$$

The number M_n is called the *reduced ménages number*. The total number of different seatings of n married couples around a circular table so that men and women alternate and no man is next to his wife equals $2n! M_n$. Table 2.2 gives reduced ménages numbers M_n for $n = 2, 3, \ldots, 12$.   □

Table 4.2   Reduced Ménages Numbers

n	2	3	4	5	6	7	8	9	10	11	12
M_n	0	1	2	13	80	579	4738	43387	439792	4890741	59216642

Some interesting corollaries may be deduced from Theorem 4.2 and Corollary 4.2 in the case of the existence of some kind of symmetry with respect to the sets. Such symmetry, appearing in several applications, constitutes the case where the sets are exchangeable in the following sense.

DEFINITION 4.1   *The sets $A_1, A_2, \ldots, A_n$ are called exchangeable if, for every collection of r indices $\{i_1, i_2, \ldots, i_r\}$ from the n indices $\{1, 2, \ldots, n\}$, the number*

$$N(A_{i_1} A_{i_2} \cdots A_{i_r}) = \nu_r, \quad r = 1, 2, \ldots, n, \tag{4.11}$$

depends only on r and not on the specific collection of indices.

COROLLARY 4.3
Let $A_1, A_2, \ldots, A_n$ be exchangeable subsets of a finite set Ω. Then

$$N(A_1 \cup A_2 \cup \cdots \cup A_n) = \sum_{r=1}^{n} (-1)^{r-1} \binom{n}{r} \nu_r \tag{4.12}$$

and

$$N(A'_1 A'_2 \cdots A'_n) = \sum_{r=0}^{n} (-1)^r \binom{n}{r} \nu_r, \tag{4.13}$$

where $\nu_0 = N(\Omega)$ *and* $\nu_r = N(A_{i_1} A_{i_2} \cdots A_{i_r})$, $r = 1, 2, \ldots, n$.

PROOF The assumption of the exchangeability of the n sets $A_1, A_2, \ldots,$ A_n entails that each of the $\binom{n}{r}$ terms of the sum in the right-hand side of (4.4) equals $N(A_{i_1} A_{i_2} \cdots A_{i_r}) = \nu_r$ and so

$$S_{n,r} = \binom{n}{r} \nu_r, \quad r = 1, 2, \ldots, n.$$

Expressions (4.12) and (4.13) are then immediately deduced from (4.3) and (4.5), respectively. ∎

Example 4.6

A secretary with mathematical interests is about to place four letters, numbered $\{1, 2, 3, 4\}$, into four envelopes, also numbered $\{1, 2, 3, 4\}$. She is wondering in how many ways this can be done so that no letter is placed in the correct envelope.

Note first that each placement of the four letters into the four envelopes may be represented by a permutation of the four letters in which the positions represent the four envelopes. Consider the set Ω of permutations of the four letters and let $A_i \subseteq \Omega$ be the subset of permutations of the four letters in which the i-th letter is placed in the i-th position, $i = 1, 2, 3, 4$. Then $N(A'_1 A'_2 A'_3 A'_4)$ is the number of different placements in which no letter is placed in the correct envelope. Further, for any collection of r indices $\{i_1, i_2, \ldots, i_r\}$ from the set $\{1, 2, 3, 4\}$,

$$\nu_r = N(A_{i_1} A_{i_2} \cdots A_{i_r}) = (4 - r)!, \quad r = 1, 2, 3, 4,$$

which is the number of permutations of the other $4 - r$ numbers. Thus, the sets A_1, A_2, A_3, A_4 are exchangeable and, according (4.13), the required number is given by

$$N(A'_1 A'_2 A'_3 A'_4) = 4! \left(1 - 1 + \frac{1}{2!} - \frac{1}{3!} + \frac{1}{4!} \right) = 9.$$

Clearly, these nine placements are the following:

$$(2, 1, 4, 3), \ (2, 4, 1, 3), \ (2, 3, 4, 1), \ (3, 1, 4, 2), \ (3, 4, 1, 2),$$

$$(3, 4, 2, 1), \ (4, 1, 2, 3), \ (4, 3, 1, 2), \ (4, 3, 2, 1).$$

This problem constitutes a particular case of the more general problem of evaluation of permutations with a given number of fixed (unchanged) points, which is presented at length in Chapter 5. ☐

Example 4.7 Combinations with restricted repetition

Find the number of k-combinations of the set $W_n = \{w_1, w_2, \dots, w_n\}$ with repetition and the restriction that each element is allowed to appear at most s times.

Consider the set of k-combinations of W_n with (unrestricted) repetition and let A_i be its subset that includes the k-combinations, in which the element w_i appears at least $s + 1$ times, $i = 1, 2, \dots, n$. Then $E_s(n, k) = N(A_1' A_2' \cdots A_n')$ is the number of k-combinations of W_n with repetition and the restriction that each element is allowed to appear at most s times. Further, for any collection of r indices $\{i_1, i_2, \dots, i_r\}$ from the n indices $\{1, 2, \dots, n\}$, we have

$$\nu_r = N(A_{i_1} A_{i_2} \cdots A_{i_r}) = \binom{n + k - r(s+1) - 1}{k - r(s+1)}$$
$$= \binom{n + k - r(s+1) - 1}{n - 1}, \quad r = 1, 2, \dots, n,$$

since the $r(s + 1)$ elements of each k-combination are determined, while the remaining $k - r(s+1)$ elements are chosen with (unrestricted) repetition from the n. Therefore the sets $A_1, A_2, \dots, A_n$ are exchangeable and, according to (4.13),

$$E_s(n, k) = \sum_{r=0}^{n} (-1)^r \binom{n}{r} \binom{n + k - r(s+1) - 1}{n - 1}.$$

Note that, for $s \geq k$ all the terms of this sum, except the first ($r = 0$), vanish and this number reduces to

$$E_s(n, k) = \binom{n + k - 1}{n - 1} = \binom{n + k - 1}{k},$$

which is the number of k-combinations of W_n with unrestricted repetition. ⬜

4.3 NUMBER OF ELEMENTS IN A GIVEN NUMBER OF SETS

The number $N_{n,k}$ of elements that are contained in k among n subsets $A_1, A_2, \dots, A_n$ of a finite set Ω, in the particular case $k = 0$ has been expressed in the form of an algebraic sum of the number $S_{n,r} = \sum N(A_{i_1} A_{i_2} \cdots A_{i_r})$ (see Corollary 4.2). An analogous expression in the general case with $0 \leq k \leq n$ is given in the next theorem.

THEOREM 4.4 General inclusion and exclusion principle
*The number $N_{n,k}$ of elements that are contained in k among n subsets A_1, A_2,
$\dots , A_n$, of a finite set Ω, is given by*

$$N_{n,k} = \sum_{r=k}^{n} (-1)^{r-k} \binom{r}{k} S_{n,r}, \quad k = 0, 1, \dots , n, \qquad (4.14)$$

where $S_{n,0} = N(\Omega)$ and $S_{n,r}$, $r = 1, 2, \dots , n$, is given by (4.4).

PROOF   Clearly,

$$N_{n,k} = \sum_{i_1, \dots , i_k} N(A_{i_1} A_{i_2} \cdots A_{i_k} A'_{j_1} A'_{j_2} \cdots A'_{j_{n-k}}), \qquad (4.15)$$

where the summation is extended over all k-combinations $\{i_1, i_2, \dots , i_k\}$ of the n
indices $\{1, 2, \dots , n\}$ and $\{j_1, j_2, \dots , j_{n-k}\} = \{1, 2, \dots , n\} - \{i_1, i_2, \dots , i_k\}$.

Let us first calculate the number $N(A_{i_1} A_{i_2} \cdots A_{i_k} A'_{j_1} A'_{j_2} \cdots A'_{j_{n-k}})$, of elements of Ω that are contained in k specified subsets among the n subsets. Applying
the expression $N(AB') = N(A - B) = N(A) - N(AB)$ (see Theorem 1.4) to
the sets $A = A_{i_1} A_{i_2} \cdots A_{i_k}$ and $B = A_{j_1} \cup A_{j_2} \cup \cdots \cup A_{j_{n-k}}$, whence $B' = A'_{j_1} A'_{j_2} \cdots A'_{j_{n-k}}$, and putting $C_{j_s} = A_{i_1} A_{i_2} \cdots A_{i_k} A_{j_s}$, $s = 1, 2, \dots , n - k$,
we deduce the relation

$$N(A_{i_1} A_{i_2} \cdots A_{i_k} A'_{j_1} A'_{j_2} \cdots A'_{j_{n-k}})$$
$$= N(A_{i_1} A_{i_2} \cdots A_{i_k}) - N(C_{j_1} \cup C_{j_2} \cup \cdots \cup C_{j_{n-k}}),$$

from which, using (4.3), we conclude the expression

$$N(A_{i_1} A_{i_2} \cdots A_{i_k} A'_{j_1} A'_{j_2} \cdots A'_{j_{n-k}}) = \sum_{r=k}^{n} (-1)^{r-k} Q_{n,k,r}(i_1, i_2, \dots , i_k),$$

with

$$Q_{n,k,r}(i_1, i_2, \dots , i_k) = \sum_{h_1, \dots , h_{r-k}} N(A_{i_1} A_{i_2} \cdots A_{i_k} A_{h_1} A_{h_2} \cdots A_{h_{r-k}}),$$

for $r = k, k + 1, \dots , n$, where the summation is extended over all $(r - k)$-
combinations $\{h_1, h_2, \dots , h_{r-k}\}$ of the $n - k$ indices $\{j_1, j_2, \dots , j_{n-k}\}$. Introducing this expression into expression (4.15) of the number $N_{n,k}$, we get

$$N_{n,k} = \sum_{i_1, \dots , i_k} \sum_{r=k}^{n} (-1)^{r-k} Q_{n,k,r}(i_1, i_2, \dots , i_k)$$
$$= \sum_{r=k}^{n} (-1)^{r-k} (-1)^{r-k} S_{n,k,r}, \qquad (4.16)$$

where

$$S_{n,k,r} = \sum_{i_1,\dots,i_k} Q_{n,k,r}(i_1, i_2, \dots, i_k)$$

$$= \sum_{i_1,\dots,i_k} \sum_{h_1,\dots,h_{r-k}} N(A_{i_1} A_{i_2} \cdots A_{i_k} A_{h_1} A_{h_2} \cdots A_{h_{r-k}}).$$

Note that this sum includes

$$\binom{n}{k}\binom{n-k}{r-k}$$

terms of the form: $N(A_{m_1} A_{m_2} \cdots A_{m_r})$, $\{m_1, m_2, \dots, m_r\} \subseteq \{1, 2, \dots, n\}$. Further, among these terms, the distinct ones are equal to

$$\binom{n}{r},$$

the number of r-combinations $\{m_1, m_2 \dots, m_r\}$ of the n indices $\{1, 2, \dots, n\}$ and each such term is included in the sum as many times as the number

$$\binom{r}{k}$$

of ways of selecting k indices $\{i_1, i_2, \dots, i_k\}$ from the set $\{m_1, m_2, \dots, m_r\}$, so that

$$\binom{n}{r}\binom{r}{k} = \binom{n}{k}\binom{n-k}{r-k},$$

which is a simple combinatorial identity. Consequently,

$$S_{n,k,r} = \binom{r}{k} S_{n,r}, \quad S_{n,r} = \sum_{m_1,\dots,m_r} N(A_{m_1} A_{m_2} \cdots A_{m_r}),$$

where the summation is extended over all r-combinations $\{m_1, m_2, \dots, m_r\}$ of the n indices $\{1, 2, \dots, n\}$. Introducing it into expression (4.16), we deduce (4.14). ∎

REMARK 4.5   The following proof of (4.14) is of broader interest. Its technique can be used in the proof of (4.3) and other analogous expressions.

Note that, in general, an arbitrary element ω in Ω is contained in s subsets $A_{i_1}, A_{i_2}, \dots, A_{i_s}$ among the n subsets $A_1, A_2, \dots, A_n$, with $0 \le s \le n$. If $0 \le s < k$, then this element is not counted in expression (4.14), while if $s = k$, it is counted only once in the sum

$$S_{n,k} = \sum N(A_{m_1} A_{m_2} \cdots A_{m_r}),$$

specifically in the term $N(A_{i_1} A_{i_2} \cdots A_{i_k})$, and is not counted in the sum

$$S_{n,r} = \sum N(A_{m_1} A_{m_2} \cdots A_{m_r}), \quad r = k+1, k+2, \ldots, n.$$

If $k < s \le n$, then the element ω is counted in the sum $S_{n,r}$, $r = k, k+1$, $\ldots, s$, as many times as is the number $\binom{s}{r}$ of terms $N(A_{m_1} A_{m_2} \cdots A_{m_r})$, with $\{m_1, m_2, \ldots, m_r\} \subseteq \{i_1, i_2, \ldots, i_s\}$, and is not counted in the sum $S_{n,r}$, $r = s+1, s+2, \ldots, n$. Consequently, this element is counted in (4.14) as many times as is the number

$$\sum_{r=k}^{s} (-1)^{r-k} \binom{r}{k} \binom{s}{r} = \binom{s}{k} \sum_{r=k}^{s} (-1)^{r-k} \binom{s-k}{r-k} = 0.$$

Therefore, in (4.14) only the elements of Ω that are contained in k among the n subsets $A_1, A_2, \ldots, A_n$ are counted. ∎

COROLLARY 4.4

Let $A_1, A_2, \ldots, A_n$ be exchangeable subsets of a finite set Ω. Then the number $N_{n,k}$ of elements of Ω that are contained in k among these n subsets is given by

$$N_{n,k} = \binom{n}{k} \sum_{j=0}^{n-k} (-1)^j \binom{n-k}{j} \nu_{k+j}, \quad k = 0, 1, \ldots, n, \qquad (4.17)$$

where $\nu_0 = N(\Omega)$ and $\nu_r = N(A_{i_1} A_{i_2} \cdots A_{i_r})$, $r = 1, 2, \ldots, n$.

PROOF The sum

$$S_{n,r} = \sum N(A_{i_1} A_{i_2} \cdots A_{i_r}), \quad r = k, k+1, \ldots, n,$$

where the summation is extended over all r-combinations $\{i_1, i_2, \ldots, i_r\}$ of the n indices $\{1, 2, \ldots, n\}$, under the assumption $N(A_{i_1} A_{i_2} \cdots A_{i_r}) = \nu_r$, of the exchangeability of the n sets $A_1, A_2, \ldots, A_n$, is given by

$$S_{n,r} = \binom{n}{r} \nu_r, \quad r = k, k+1, \ldots, n.$$

Introducing it into expression (4.14) of the number $N_{n,k}$, we get

$$N_{n,k} = \sum_{r=k}^{n} (-1)^{r-k} \binom{r}{k} \binom{n}{r} \nu_r = \binom{n}{k} \sum_{r=k}^{n} (-1)^{r-k} \binom{n-k}{r-k} \nu_r$$

$$= \binom{n}{k} \sum_{j=0}^{n-k} (-1)^j \binom{n-k}{j} \nu_{k+j}, \quad k = 0, 1, \ldots, n,$$

which is the required expression. ∎

As regards the number $L_{n,k}$ of elements that are contained in at least k among n subsets of a finite set Ω, which is connected with the number $N_{n,s}$, we deduce the next corollary of Theorem 4.4 .

COROLLARY 4.5

The number $L_{n,k}$ of elements that are contained in at least k among n subsets $A_1, A_2, \ldots, A_n$, of a finite set Ω, is given by

$$L_{n,k} = \sum_{r=k}^{n} (-1)^{r-k} \binom{r-1}{k-1} S_{n,r}, \quad k = 1, 2, \ldots, n, \qquad (4.18)$$

where the sum $S_{n,r}$ is given by (4.4). If, in addition, the sets $A_1, A_2, \ldots, A_n$ are exchangeable, then

$$L_{n,k} = n \binom{n-1}{k-1} \sum_{j=0}^{n-k} (-1)^j \binom{n-k}{j} \frac{\nu_{k+j}}{k+j}, \quad k = 1, 2, \ldots, n, \qquad (4.19)$$

where $\nu_r = N(A_{i_1} A_{i_2} \cdots A_{i_r})$, $r = 1, 2, \ldots, n$.

PROOF   Since an arbitrary element ω is contained in at least k among the n subsets of Ω if and only if it is contained either in k or in $k + 1, \ldots$, or in all the n subsets, applying the addition principle and using (4.14), it follows that

$$L_{n,k} = \sum_{s=k}^{n} N_{n,s} = \sum_{s=k}^{n} \sum_{r=s}^{n} (-1)^{r-s} \binom{r}{s} S_{n,r}$$

$$= \sum_{r=k}^{n} (-1)^{r-k} \left\{ \sum_{s=k}^{r} (-1)^{k-s} \binom{r}{s} \right\} S_{n,r},$$

where, according to (2.8),

$$\sum_{s=k}^{r} (-1)^{s-k} \binom{r}{s} = \binom{r-1}{k-1}.$$

If the sets $A_1, A_2, \ldots, A_n$ are exchangeable, then

$$S_{n,r} = \binom{n}{r} \nu_r$$

and

$$L_{n,k} = \sum_{r=k}^{n} (-1)^{r-k} \binom{r-1}{k-1} \binom{n}{r} \nu_r$$

$$= n \binom{n-1}{k-1} \sum_{r=k}^{n} (-1)^{r-k} \binom{n-k}{r-k} \frac{\nu_r}{r}$$

$$= n \binom{n-1}{k-1} \sum_{j=0}^{n-k} (-1)^{j} \binom{n-k}{j} \frac{\nu_{k+j}}{k+j},$$

which completes the proof. ∎

Example 4.8

Consider an ordered series of eight lottery-urns, each containing ten balls numbered from 0 to 9. In a drawing, one ball is drawn from each of the eight urns and an eight-digit number is formed. Calculate the different six-digit numbers in which only half of the ten digits appear.

Let A_i be the set of eight-digit numbers, in which the digit $i - 1$ does not appear, $i = 1, 2, \ldots, 10$. Then the different eight-digit numbers in which (exactly) five of the ten digits appear equals the number of elements that are contained in (exactly) five of the ten sets $A_1, A_2, \ldots, A_{10}$. These sets are exchangeable. Indeed, for any collection of r indices $\{i_1, i_2, \ldots, i_r\}$ from the ten indices $\{1, 2, \ldots, 10\}$, we have

$$\nu_r = N(A_{i_1} A_{i_2} \cdots A_{i_r}) = (10 - r)^8, \quad r = 1, 2, \ldots, 10,$$

which is the number of the 8-permutations with repetition of the $10 - r$ non-excluded numbers. Thus, by Corollary 4.4, the required number equals

$$N_{10,5} = \binom{10}{5} \sum_{j=0}^{5} (-1)^j \binom{5}{j} (5 - j)^8 = 31{,}752{,}000.$$

Note that the total number of 8-digit numbers is 100,000,000. □

Example 4.9 Divisions with a given number of empty sets

A division of a finite set W in r subsets is an ordered r-tuple of sets $(A_1, A_2, \ldots, A_r)$ that are pairwise disjoint subsets of W and their union is W. In a division of a set, the inclusion of one or more empty sets is not excluded (see Section 1.2.7). Let Ω be the set of divisions of W, with $N(W) = n$, in r subsets. The number of elements of this set equals $N(\Omega) = r^n$, the number of n-permutations of r with repetition (see Corollary 3.3). Calculate the number of divisions of the set W, of n elements, in r subsets with (a) k empty subsets and (b) at least k empty subsets.

Let $D_i \subseteq \Omega$ be the set of divisions $(A_1, A_2, \ldots, A_r)$ of the set W of n elements in r subsets, in which $A_i = \emptyset$, $i = 1, 2, \ldots, r$. Note that, for any collection of s indices $\{i_1, i_2, \ldots, i_s\}$ from the r indices $\{1, 2, \ldots, r\}$, we have

$$\nu_0 = N(\Omega) = r^n, \quad \nu_s = N(D_{i_1} D_{i_2} \cdots D_{i_s}) = (r - s)^n, \quad s = 1, 2, \ldots, n,$$

since the number of divisions of the set W of n elements in r subsets, in which s specified sets are empty, equals the number of divisions of the set W of n elements in $r - s$ subsets. Consequently the sets $D_1, D_2, \ldots, D_r$ are exchangeable.

(a) The number of divisions of the set W of n elements in r subsets among which k are empty, according to Corollary 4.4, equals

$$N_{r,k} = \binom{r}{k} \sum_{j=0}^{r-k} (-1)^j \binom{r-k}{j} (r - k - j)^n$$

$$= \binom{r}{k} \sum_{i=0}^{r-k} (-1)^{r-k-i} \binom{r-k}{i} i^n.$$

(b) The number of divisions of the set W of n elements in r subsets among which at least k are empty, according to Corollary 4.5, equals

$$L_{r,k} = r \binom{r-1}{k-1} \sum_{j=0}^{r-k} (-1)^j \binom{r-k}{j} \frac{(r-k-j)^n}{k+j}$$

$$= r \binom{r-1}{k-1} \sum_{i=0}^{r-k} (-1)^{r-k-i} \binom{r-k}{i} \frac{i^n}{r-i}.$$

Note that

$$N_{r,k} = (r)_{r-k} S(n, r - k),$$

where

$$S(n, r) = \frac{1}{r!} \sum_{i=0}^{r} (-1)^{r-i} \binom{r}{i} i^n, \quad r = 0, 1, \ldots, n, \quad n = 0, 1, \ldots$$

are the *Stirling numbers of the second kind* . These numbers appear in several interesting problems in combinatorics and for this reason are thoroughly examined in Chapter 8.   ☐

Example 4.10 Sum of multinomial coefficients
Evaluate combinatorially the sum

$$C_{n,r} = \sum \frac{n!}{k_1! k_2! \cdots k_{r-1}! k_r!},$$

where the summation is extended over all $k_j \geq 1, j = 1, 2, \ldots, r$, with $k_1 + k_2 + \cdots + k_{r-1} + k_r = n$.

The combinatorial evaluation of a sum requires, at first, a suitable combinatorial interpretation. The multinomial coefficients and the sum $C_{n,r}$ admit of several combinatorial interpretations. Let us consider the following combinatorial interpretation relative to the divisions of a finite set with a given number of empty sets, which thus allows the use of the conclusion of the previous example. The multinomial coefficient

$$\frac{n!}{k_1! k_2! \cdots k_{r-1}! k_r!},$$

according to Theorem 2.8, equals the number of divisions $(A_1, A_2, \ldots, A_r)$ of a finite set W, of n elements, in r subsets with $N(A_j) = k_j$, $j = 1, 2, \ldots, r$. Consequently, the sum $C_{n,r}$ expresses the number of divisions of the set W of n elements in r non-empty subsets, since $N(A_j) \geq 1$, $j = 1, 2, \ldots, r$, and so, according to conclusion (a) of the previous example with $k = 0$, equals

$$C_{n,r} = \sum_{i=0}^{r} (-1)^{r-i} \binom{r}{i} i^n = r! S(n, r),$$

where $S(n, r)$ is the *Stirling number of the second kind.* □

Example 4.11 Coupon collector's problem

For the promotion of sales of an industrial product, the producing company has printed s series of coupons numbered from 1 to s. A coupon is placed in each package box of the product. A customer who collects k different coupons gets k units of the product free, while in the exceptional case of collecting a complete series of coupons, the prize is a special gift from the company. Calculate the number of collections of n coupons that include (a) k different coupons and (b) a complete series of coupons.

Consider the set Ω of collections of n coupons and let $A_i \subseteq \Omega$ be the set of collections of n coupons that do not include any coupon bearing the number i, $i = 1, 2, \ldots, r$. Then for any collection of m indices $\{i_1, i_2, \ldots, i_m\}$ from the set of r indices $\{1, 2, \ldots, r\}$, we have

$$\nu_m = N(A_{i_1} A_{i_2} \cdots A_{i_m}) = \binom{s(r-m)}{n}, \quad m = 0, 1, \ldots, r,$$

which is the number of n-combinations of the remaining $s(r - m)$ coupons, after excluding the sm coupons bearing the m specified number. Consequently, the sets $A_1, A_2, \ldots, A_r$ are exchangeable.

(a) The number of collections of n coupons that include k different coupons equals the number of elements of Ω that are contained in $r - k$ among the r

subsets $A_1, A_2, \ldots, A_r$ and according to Corollary 4.4,

$$N_{r,r-k} = \binom{r}{r-k} \sum_{j=0}^{k} (-1)^j \binom{k}{j} \nu_{r-k+j},$$

is given by

$$N_{r,r-k} = \binom{r}{k} \sum_{j=0}^{k} (-1)^j \binom{k}{j} \binom{s(k-j)}{n}$$

$$= \binom{r}{k} \sum_{i=0}^{k} (-1)^{k-i} \binom{k}{i} \binom{si}{n}.$$

(b) The number of collections of n coupons that include a complete series of coupons equals

$$N_{r,0} = \sum_{i=0}^{r} (-1)^{r-i} \binom{r}{i} \binom{si}{n}.$$

Note that

$$N_{r,r-k} = \frac{(r)_k}{n!} C(n, k; s),$$

where the numbers

$$C(n, k; s) = \frac{1}{k!} \sum_{i=0}^{k} (-1)^{k-i} \binom{k}{i} (si)_n, \quad r = 0, 1, \ldots, n, \ n = 0, 1, \ldots,$$

which are called *generalized factorial coefficients*, are thoroughly examined, together with the Stirling numbers with which they are connected, in Chapter 8. ⬜

4.4   BONFERRONI INEQUALITIES

A series of alternating inequalities for the number $N_{n,k}$ of elements that are contained in k and the number $L_{n,k}$ of elements that are contained in at least k among n subsets of a finite set may be deduced from relations (4.14) and (4.18) and their inverses, respectively. The next lemma is concerned with the inversion of (4.14) and (4.18).

LEMMA 4.1
The inverse relations of (4.14) and (4.18) are

$$S_{n,r} = \sum_{k=r}^{n} \binom{k}{r} N_{n,k} \tag{4.20}$$

and

$$S_{n,r} = \sum_{k=r}^{n} \binom{k-1}{r-1} L_{n,k}, \qquad (4.21)$$

respectively.

PROOF Multiplying (4.14) by $\binom{k}{j}$ and summing for $k = j, j+1, \ldots, n$, we get

$$\sum_{k=j}^{n} \binom{k}{j} N_{n,k} = \sum_{k=j}^{n} \sum_{r=k}^{n} (-1)^{r-k} \binom{k}{j} \binom{r}{k} S_{n,r}$$

$$= \sum_{r=j}^{n} \left\{ \sum_{k=j}^{r} (-1)^{r-k} \binom{k}{j} \binom{r}{k} \right\} S_{n,r},$$

where the inner sum

$$s_{r,j} = \sum_{k=j}^{r} (-1)^{r-k} \binom{k}{j} \binom{r}{k},$$

is evaluated by the aid of the relation

$$\binom{k}{j} \binom{r}{k} = \binom{r}{j} \binom{r-j}{k-j}$$

and is equal to

$$s_{r,j} = \binom{r}{j} \sum_{k=j}^{r} (-1)^{r-k} \binom{r-j}{k-j} = \binom{r}{j} (1-1)^{r-j} = \delta_{r,j},$$

the Kronecker delta: $\delta_{j,j} = 1, \delta_{r,j} = 0, r \neq j$. Consequently,

$$\sum_{k=j}^{n} \binom{k}{j} N_{n,k} = \sum_{r=j}^{n} \delta_{r,j} S_{n,r} = S_{n,j}.$$

Multiplying (4.18) by $\binom{k-1}{j-1}$ and summing for $k = j, j+1, \ldots, n$, we deduce in the same way (4.21). ∎

THEOREM 4.5 Bonferroni inequalities
Let $N_{n,k}$ be the number of elements that are contained in k and $L_{n,k}$ the number of elements that are contained in at least k among n subsets $A_1, A_2, \ldots, A_n$ of a

finite set Ω. *Then, for* $s = k, k+1, \ldots, n$, *the following alternating inequalities hold true :*

$$(-1)^{s-k+1} \left\{ N_{n,k} - \sum_{r=k}^{s} (-1)^{r-k} \binom{r}{k} S_{n,r} \right\} \geq 0, \qquad (4.22)$$

$k = 0, 1, \ldots, n$, *and*

$$(-1)^{s-k+1} \left\{ L_{n,k} - \sum_{r=k}^{s} (-1)^{r-k} \binom{r-1}{k-1} S_{n,r} \right\} \geq 0, \qquad (4.23)$$

$k = 1, 2, \ldots, n$, *where the sum* $S_{n,r}$, $r = 0, 1, \ldots, n$, *is given by (4.4).*

PROOF   Consider, for fixed n and $s = k, k+1, \ldots, n$, $k = 0, 1, \ldots, n$, the sequence

$$M_{n,k,s} = (-1)^{s-k+1} \left\{ N_{n,k} - \sum_{r=k}^{s} (-1)^{r-k} \binom{r}{k} S_{n,r} \right\},$$

which, by virtue of (4.20), vanishes for $s = n$ and any $k = 0, 1, \ldots, n$, while, for $s = k, k+1, \ldots, n-1$ is given by

$$M_{n,k,s} = \sum_{r=s+1}^{n} (-1)^{r-s-1} \binom{r}{k} S_{n,r}, \quad k = 0, 1, \ldots, n-1.$$

Upon using (4.20), this sequence can be written as a linear combination of $N_{n,j}$, $j = s+1, s+2, \ldots, n$:

$$M_{n,k,s} = \sum_{r=s+1}^{n} \sum_{j=r}^{n} (-1)^{r-s-1} \binom{r}{k} \binom{j}{r} N_{n,j}$$

$$= \sum_{j=s+1}^{n} \left\{ \sum_{r=s+1}^{j} (-1)^{r-s-1} \binom{r}{k} \binom{j}{r} \right\} N_{n,j}.$$

Note that the coefficient of the general term of this expression,

$$c_{j,s} = \sum_{r=s+1}^{j} (-1)^{r-s-1} \binom{r}{k} \binom{j}{r} = \binom{j}{k} \sum_{r=s+1}^{j} (-1)^{r-s-1} \binom{j-k}{r-k},$$

according to (2.8), is equal to

$$c_{j,s} = \binom{j}{k} \sum_{i=0}^{j-s-1} (-1)^{j-s-i-1} \binom{j-k}{i} = \binom{j}{k} \binom{j-k-1}{j-s-1},$$

that is, is positive, which implies (4.22). The inequalities (4.23) are deduced in the same way from (4.18) and (4.21). ∎

REMARK 4.6 The inequalities (4.22) in the particular case $k = 0$ reduce to

$$(-1)^{s+1} \left\{ N(A'_1 A'_2 \cdots A'_n) - \sum_{r=0}^{s} (-1)^r S_{n,r} \right\} \geq 0, \ s = 0, 1, \ldots, n,$$

while the inequalities (4.23) in the particular case $k = 1$ reduce to

$$(-1)^s \left\{ N(A_1 \cup A_2 \cup \cdots \cup A_n) - \sum_{r=1}^{s} (-1)^{r-1} S_{n,r} \right\} \geq 0, \ s = 1, 2, \ldots, n,$$

which, for $s = 1$, yields the *Boole inequality*

$$N(A_1 \cup A_2 \cup \cdots \cup A_n) \leq \sum_{i=1}^{n} N(A_i).$$

Further, from (4.22) and (4.23), with $s = k$ and $s = k + 1$, we deduce the double inequalities,

$$S_{n,k} - (k+1)S_{n,k+1} \leq N_{n,k} \leq S_{n,k}$$

and

$$S_{n,k} - kS_{n,k+1} \leq L_{n,k} \leq S_{n,k},$$

respectively. ∎

4.5 NUMBER OF ELEMENTS OF A GIVEN RANK

In the enumeration of the elements that are contained in a given number of subsets among n subsets of a finite set, there does not enter any ordering of these subsets. When the considered subsets are ordered, some interesting problems of enumeration of elements, according to the rank (order) of the first set in which they are contained, emerges. This is clarified in the next combinatorial model.

Let $W_1, W_2, \ldots, W_n$ be finite sets and let Ω be their Cartesian product,

$$\Omega = W_1 \times W_2 \times \cdots \times W_n = \{(w_1, w_2, \ldots, w_n) : w_j \in W_j, \ j = 1, 2, \ldots, n\}.$$

Let us consider, for a given element $w_j \in W_j$, the subset A_j of Ω that includes the elements $\omega = (w_1, w_2, \ldots, w_n) \in \Omega$ for which the j-th coordinate is w_j, $j = 1, 2, \ldots, n$. The ordering $(w_1, w_2, \ldots, w_n)$ of the elements w_j, $j = 1, 2, \ldots, n$ induced the ordering $(A_1, A_2, \ldots, A_n)$ of the corresponding subsets A_j, $j = 1, 2, \ldots, n$. The enumeration of the elements of Ω that, for a given index k, are not contained in any of the sets $A_1, A_2, \ldots, A_{k-1}$ and are contained in A_k (the first in-order set in which they are contained) is one of the interesting problems that emerged.

Note that, in the corresponding probabilistic model, Ω is the sample space of a sequence of Bernoulli experiments and A_j is the event of a success at the j-th experiment, $j = 1, 2, \ldots$. The calculation of the probability that the first success is realized at the j-th experiment requires the enumeration of the sample points (elements of Ω) that are not contained in the events (subsets of Ω) $A_1, A_2, \ldots, A_{k-1}$ and are contained in A_k.

As regards the general problem, the notion of the rank of an element is formally introduced in the next definition.

DEFINITION 4.2　　Let $(A_1, A_2, \ldots, A_n, \ldots)$ be ordered subsets of a set Ω. An element $\omega \in \Omega$ that is contained in the set $A_1' A_2' \cdots A_{k-1}' A_k$ is said to be of rank k. In the case of a finite number of n ordered subsets $(A_1, A_2, \ldots, A_n)$, an element $\omega \in \Omega$ that is not contained in any of the n subsets is said, by convention, to be of rank $n + 1$.

The next theorem is concerned with the number of elements of a given rank.

THEOREM 4.6

Let $(A_1, A_2, \ldots, A_n, \ldots)$ be ordered and finite subsets of a set Ω. Then the number $R_k = N(A_1' A_2' \cdots A_{k-1}' A_k)$, of elements of Ω of rank k, is given by,

$$R_k = Q_{k-1,0} - Q_{k-1,1} + \cdots + (-1)^{k-1} Q_{k-1,k-1}, \quad k = 1, 2, \ldots, \quad (4.24)$$

with

$$Q_{k-1,0} = N(A_k), \quad Q_{k-1,k-1} = N(A_1 A_2 \cdots A_{k-1} A_k),$$

and

$$Q_{k-1,r} = \sum N(A_{i_1} A_{i_2} \cdots A_{i_r} A_k), \quad r = 1, 2, \ldots, k - 2, \quad (4.25)$$

where the summation is extended over all r-combinations $\{i_1, i_2, \ldots, i_r\}$ of the $k - 1$ indices $\{1, 2, \ldots, k - 1\}$.

PROOF　　Applying the expression $N(AB') = N(A) - N(AB)$ (see Theorem 1.4) to the sets $A = A_k$ and $B = A_1 \cup A_2 \cup \cdots \cup A_{k-1}$, whence $B' =$

$A'_1 A'_2 \cdots A'_{k-1}$, and setting $C_j = A_j A_k$, $j = 1, 2, \ldots, k-1$, we get the expression

$$R_k = N(A'_1 A'_2 \cdots A'_{k-1} A_k) = N(A_k) - N(C_1 \cup C_2 \cup \cdots \cup C_{k-1}).$$

Further, using (4.3) and since

$$C_{i_2} C_{i_3} \cdots C_{i_r} = A_{i_1} A_{i_2} \cdots A_{i_r} A_k, \quad \{i_1, i_2, \ldots, i_r\} \subseteq \{1, 2, \ldots, k-1\},$$

we deduce (4.24). ∎

REMARK 4.7 An alternative expression of the number of elements of a given rank may be derived by an application of the expression $N(AB') = N(A) - N(AB)$ to the sets $A = A'_1 A'_2 \cdots A'_{k-1}$ and $B = A'_k$, whence $B' = A_k$. Then

$$N(A'_1 A'_2 \cdots A'_{k-1} A_k) = N(A'_1 A'_2 \cdots A'_{k-1}) - N(A'_1 A'_2 \cdots A'_k A'_k)$$

and so

$$R_k = N_{k-1,0} - N_{k,0},$$

where $N_{n,0}$ is given by (4.5). ∎

COROLLARY 4.6

Let $(A_1, A_2, \ldots, A_n, \ldots)$ be ordered and finite subsets of a set Ω. If, in addition, for a given k the sets $A_1, A_2, \ldots, A_k$ are exchangeable, then the number of elements of Ω of rank k is given by

$$R_k = \binom{k-1}{0} \nu_1 - \binom{k-1}{1} \nu_2 + \cdots + (-1)^{k-1} \binom{k-1}{k-1} \nu_k, \quad (4.26)$$

where $\nu_r = N(A_{i_1} A_{i_2} \cdots A_{i_r})$, $r = 1, 2, \ldots, k$.

PROOF The assumption of exchangeability of the sets $A_1, A_2, \ldots, A_k$ implies that each of the $\binom{k-1}{r}$ terms of the sum $Q_{k-1,r}$ in (4.25) is given by

$$N(A_{i_1} A_{i_2} \cdots A_{i_r} A_k) = N(A_1 A_2 \cdots A_{r+1}) = \nu_{r+1}, \quad r = 0, 1, \ldots, k-1.$$

Therefore

$$Q_{k-1,r} = \binom{k-1}{r} \nu_{r+1}, \quad r = 0, 1, \ldots, k-1$$

and, by (4.24), expression (4.26) is deduced. ∎

Example 4.12
Consider a lottery-urn containing ten balls numbered from 0 to 9 and assume that balls are drawn one after the other without replacement. This procedure is

terminated at the k-th trial if number k is drawn, $k = 1, 2, \ldots, 9$; otherwise it is terminated at the tenth trial whether or not number 0 is drawn. Calculate the number T_k of different series of drawings that are terminated at the k-th trial, $k = 1, 2, \ldots, 10$.

Let Ω be the set of the 10! different series of drawings and consider the ordered subsets $(A_1, A_2, \ldots, A_{10})$ of Ω that are defined as follows: the set A_j contains those series of drawings in which number j is drawn at the j-th trial, $j = 1, 2, \ldots, 9$. Especially the set A_{10} contains those series of drawings in which number 0 is drawn at the tenth trial. These sets are exchangeable. Indeed,

$$\nu_r = N(A_{i_1} A_{i_2} \cdots A_{i_r}) = N(A_1 A_2 \cdots A_r) = (10 - r)!, \; r = 1, 2, \ldots, 10,$$

which is the number of different series of drawings of the other $10 - r$ numbers. The number R_k of different series of drawings of rank k, using Corollary 4.6, is obtained as

$$R_k = \sum_{r=0}^{k-1} (-1)^r \binom{k-1}{r} (9 - r)!, \;\; k = 1, 2, \ldots, 10.$$

Especially, the number $R_{11} = N(A_1' A_2' \cdots A_{10}')$ of different series of drawings of rank 11, using Corollary 4.3, is obtained as

$$R_{11} = \sum_{r=0}^{10} (-1)^r \binom{10}{r} (10 - r)!.$$

Clearly, the number T_k of different series of drawings that are terminated at the k-th trial is given by $T_k = R_k$, $k = 1, 2, \ldots, 9$ and $T_{10} = R_{10} + R_{11}$.

Note that, from a probabilistic point of view, the quotient

$$p_k = T_k / 10!, \;\; k = 1, 2, \ldots, 10,$$

gives the probability that the drawing is terminated at the k-th trial. □

4.6 BIBLIOGRAPHIC NOTES

This chapter was devoted to a thorough presentation of the counting principle of inclusion and exclusion. In probability theory books, Theorem 4.2, expressed in terms of probabilities, is attributed to H. Poincaré (1896) and is referred to as Poincaré formula. However, it seems that this formula was also known to J. Sylvester (1883) and to Euler (1760) when he evaluated $\phi(n)$, now known as Euler's function. Abraham De Moivre (1718) used this formula in the particular case of exchangeable sets (events); he was able to generalize results previously discussed by P. R. Montmort (1708). Also, De Moivre (1730) derived Theorem 4.3 for $s_i = 1$, and $m_i = m$, $i = 1, 2, \ldots, n$. After De Moivre, C. Jordan (1867) derived expressions analogous to those of Theorem 4.4 and its Corollary 4.4. Ch. Jordan (1926, 1927a,b, 1934, 1939a,b) obtained Theorems 4.4 and 4.6 and their Corollaries 4.4 and 4.6 and presented several applications. More general formulations of the inclusion and exclusion principle were given by M. Frèchet (1940, 1943), K. L. Chung (1941, 1943a,b,c) and L. Takacs (1967b), who furnished a very extensive reference list. Further, L. Takacs (1991) derived a limiting form of Theorem 4.4 as the number n of sets (events) increases to infinity. The work of C. E. Bonferroni (1936) is considered as the foundation of the inequalities in Theorem 4.5. The mènages problem was presented and solved by E. Lucas (1891); before that under a different formulation this problem had been examined by A. Cayley (1878a,b) and T. Muir (1878). The expression for the reduced mènages number given in Example 4.5 is due to J. Touchard (1934, 1953). A general method of enumeration of permutations with restricted positions was developed by I. Kaplansky and J. Riordan in a series of papers. For details, see Chapters 7 and 8 of the book on combinatorial analysis by J. Riordan (1958).

4.7 EXERCISES

1. Calculate the number of four-digit university campus telephone numbers in which each of the digits 1, 2 and 3 appear at least once.

2. Calculate the number of positive integers less than or equal to 1000 that are neither perfect squares nor perfect cubes.

3. Suppose five pairs of gloves are to be distributed to five people by a child. Calculate the number of different distributions of two gloves to each of the five people with no one getting a matching pair.

4. Suppose that four gloves are randomly chosen from a drawer containing five different pairs of gloves. Calculate the number of selections that include at least one pair of gloves.

5. Calculate the number of 6-combinations of the set $W_4 = \{w_1, w_2, w_3, w_4\}$ with repetition and the restriction that the element w_i is allowed to appear, at most, i times, $i = 1, 2, 3, 4$.

6. In the base 5 number system each number is represented by an ordered sequence of the digits from the set $\{0, 1, 2, 3, 4\}$ with repetition. For example, the numbers 625, 859 and 3125, which are expressed in terms of powers of 5 as $625 = 1 \cdot 5^4 + 0 \cdot 5^3 + 0 \cdot 5^2 + 0 \cdot 5^1 + 0 \cdot 5^0$, $859 = 1 \cdot 5^4 + 1 \cdot 5^3 + 4 \cdot 5^2 + 1 \cdot 5^1 + 4 \cdot 5^0$ and $3125 = 1 \cdot 5^5 + 0 \cdot 5^4 + 0 \cdot 5^3 + 0 \cdot 5^2 + 0 \cdot 5^1 + 0 \cdot 5^0$, are represented by the ordered sequences $(1, 0, 0, 0, 0)$, $(1, 1, 4, 1, 4)$ and $(1, 0, 0, 0, 0, 0)$. Note that each of 2500 integers that are greater than or equal to $5^4 = 625$ and less that $5^5 = 3125$ may be represented by a 5-permutation of the set $\{0, 1, 2, 3, 4\}$ with repetition, in which the first position is occupied by digit 1. Calculate the number of integers greater than or equal to 625 and less than 3125 for which only three of the five digits $\{0, 1, 2, 3, 4\}$ appear in the last four positions of their 5-nary sequence representation.

7. Show that the number of positive integers, less than or equal to $70n$, that are not divisible by any of the prime numbers 2, 5 and 7 equals $24n$.

8. Show that the number of positive integers, less than or equal to 100, that have not repeated prime factors equals 61.

9. Let $s_1, s_2, \ldots, s_r$, be positive integers, relatively prime, and n a positive integer. Show that the number $N(n; s_1, s_2, \ldots, s_r)$ of positive integers, less than or equal to n, that are not divisible by any of the numbers $s_1, s_2, \ldots, s_r$ is given by

$$N(n; s_1, s_2, \ldots, s_r) = n + \sum_{k=1}^{r} (-1)^k \sum \left[\frac{n}{s_{i_1} s_{i_2} \cdots s_{i_k}} \right],$$

where, in the inner sum, the summation is extended over all k-combinations $\{i_1, i_2, \ldots, i_k\}$ of the r indices $\{1, 2, \ldots, r\}$ and $[x]$ denotes the integral part of x.

10. *The sieve of Eratosthenes.* Let $a_1 = 2$, $a_2 = 3$, $a_3 = 5, \ldots$, be the sequence of primes (in increasing order) and let $E(s)$, $s > 0$, be the number of prime numbers that are less than or equal to s. Show that

$$E(r) - E(\sqrt{r}) = S_{n,0} - S_{n,1} + S_{n,2} - \cdots + (-1)^n S_{n,n},$$

with $n = E(\sqrt{r})$ and

$$S_{n,0} = r - 1, \quad S_{n,k} = \sum \left[\frac{r}{a_{i_1} a_{i_2} \cdots a_{i_k}} \right], \quad k = 1, 2, \ldots, n,$$

where the summation is extended over all k-combinations $\{i_1, i_2, \ldots, i_k\}$ of the n indices $\{1, 2, \ldots, n\}$.

11. (*Continuation*). Show that the number $E(100)$ of prime numbers less than or equal to 100 equals 25.

12. Show that

$$\binom{n}{k} = \sum_{r=0}^{[k/2]} (-1)^r \binom{n}{r} \binom{n + k - 2r - 1}{n - 1},$$

by a suitable combinatorial interpretation of the binomial coefficients and application of the inclusion and exclusion principle.

13. Prove the identity

$$\binom{k-1}{n-1} = \sum_{r=0}^{n-1}(-1)^r\binom{n}{r}\binom{n+k-r-1}{k}$$

by a suitable combinatorial interpretation of the binomial coefficients and application of the inclusion and exclusion principle.

14. *The Galilei problem.* Let $G(n,k)$ be the number of different outcomes of a throw of n distinguishable dice, in each of which the sum of the numbers on the upmost faces of the dice equals k. Show that

$$G(n,k) = \sum_{r=0}^{s}(-1)^r\binom{n}{r}\binom{k-6r-1}{n-1}, \quad s = [(k-n)/6].$$

15. Let $U(n,k,r)$ be the number of k-permutations $(j_1, j_2, \ldots, j_k)$ of the set $\{1, 2, \ldots, n\}$, with repetition, in which $j_1 + j_2 + \cdots + j_k \leq r$. Show that

$$U(n,k,r) = \sum_{s=0}^{m}(-1)^s\binom{k}{s}\binom{r-ns}{k}, \quad m = [(r-k)/n]$$

and conclude that

$$n^k = \sum_{j=1}^{k}(-1)^{k-j}\binom{k}{j}\binom{nj}{k}.$$

16*. *Runs of consecutive elements in combinations.* Let $C(n,k,s)$ be the number of k-combinations of the set $\{1, 2, \ldots, n\}$ which include no run of s consecutive integers. Show that

$$C(n,k,s) = \sum_{r=0}^{[k/2]}(-1)^r\binom{n-k+1}{r}\binom{n-rs}{n-k}.$$

17*. *Runs of consecutive elements in combinations of circularly ordered elements.* Let $B(n,k,s)$ be the number of (circular) k-combinations of the set $\{1, 2, \ldots, n\}$ of n integers, displayed on a circle (whence n and 1 are consecutive), which include no run of s consecutive points. Show that

$$B(n,k,s) = \sum_{r=0}^{[k/s]}(-1)^r\binom{n-k}{r}\frac{n}{n-rs}\binom{n-rs}{n-k}.$$

18*. Let $C(n, k; a_1, b_1, \ldots, a_{k-1}, b_{k-1})$ be the number of k-combinations $\{i_1, i_2, \ldots, i_k\}$, $i_1 < i_2 < \cdots < i_k$, of $\{1, 2, \ldots, n\}$, with differences $d_m = i_{m+1} - i_m$, $m = 1, 2, \ldots, k - 1$, satisfying the inequalities $a_m < d_m \leq b_m$, $m = 1, 2, \ldots, k - 1$, where $a_m < b_m$, $m = 1, 2, \ldots, k - 1$, are given positive integers. Show that

$$C(n, k; a_1, b_1, \ldots, a_{k-1}, b_{k-1})$$
$$= \sum_{j=0}^{k-1} (-1)^j \sum \binom{n - r_{k-1} - (c_{m_1} + c_{m_2} + \cdots + c_{m_j})}{k},$$

where $r_{k-1} = a_1 + a_2 + \cdots + a_{k-1}$, $c_m = b_m - a_m$, $m = 1, 2, \ldots, k - 1$, and in the inner sum the summation is extended over all j-combinations $\{m_1, m_2, \ldots, m_j\}$ of the $k - 1$ indices $\{1, 2, \ldots, k - 1\}$. In the particular case $a_m = r$, $b_m = s$, $m = 1, 2, \ldots, k - 1$, setting $C(n, k; r, s, \ldots, r, s) \equiv C(n, k; r, s)$, conclude that

$$C(n, k; r, s) = \sum_{j=0}^{k-1} (-1)^j \binom{k-1}{j} \binom{n - (k-1)r - j(s-r)}{k}.$$

19*. Let $B(n, k; a_1, b_1, \ldots, a_k, b_k)$ be the number of k-combinations $\{i_1, i_2, \ldots, i_k\}$, $i_1 < i_2 < \cdots < i_k$, of $\{1, 2, \ldots, n\}$, with differences $d_m = i_{m+1} - i_m$, $m = 1, 2, \ldots, k - 1$, satisfying the inequalities $a_m < d_m \leq b_m$, $m = 1, 2, \ldots, k - 1$, and span $d = i_k - i_1$ satisfying the inequality $a_k < n - d \leq b_k$, where $a_m < b_m$, $m = 1, 2, \ldots, k$, are given positive integers. Show that

$$B(n, k; a_1, b_1, \ldots, a_k, b_k)$$
$$= \sum_{j=0}^{k-1} (-1)^j \sum \frac{n - r_{k-1} + (k-1)a_k - (c_{m_1} + \cdots + c_{m_j})}{n - r_{k-1} - a_k - (c_{m_1} + \cdots + c_{m_j})}$$
$$\times \binom{n - r_{k-1} - a_k - (c_{m_1} + \cdots + c_{m_j})}{k}$$
$$- \sum_{j=0}^{k-1} (-1)^j \sum \frac{n - r_{k-1} + (k-1)b_k - (c_{m_1} + \cdots + c_{m_j})}{n - r_{k-1} - b_k - (c_{m_1} + \cdots + c_{m_j})}$$
$$\times \binom{n - r_{k-1} - b_k - (c_{m_1} + \cdots + c_{m_j})}{k},$$

where $r_{k-1} = a_1 + a_2 + \cdots + a_{k-1}$, $c_m = b_m - a_m$, $m = 1, 2, \ldots, k - 1$, and in the inner sums the summation is extended over all j-combinations $\{m_1, m_2, \ldots, m_j\}$ of the $k - 1$ indices $\{1, 2, \ldots, k - 1\}$. In the particular case $a_m = r$, $b_m = s$, $m = 1, 2, \ldots, k - 1$, and $a_k = u$, $b_k = w$ setting

$B(n, k; ; r, s, \ldots, r, s, u, w) \equiv B(n, k; r, s, u, w)$, conclude that

$$
\begin{aligned}
B(n, k; r, s, u, w) = & \sum_{j=0}^{k-1} (-1)^j \binom{k-1}{j} \frac{n - (k-1)(r-u) - j(s-r)}{n - (k-1)r - u - j(s-r)} \\
& \times \binom{n - (k-1)r - u - j(s-r)}{k} \\
& - \sum_{j=0}^{k-1} (-1)^j \binom{k-1}{j} \frac{n - (k-1)(r-w) - j(s-r)}{n - (k-1)r - w - j(s-r)} \\
& \times \binom{n - (k-1)r - w - j(s-r)}{k}.
\end{aligned}
$$

20. *The ménages problem at a straight table.* Let $2n!L_n$ be the number of ways of seating n married couples at a straight table so that men and women alternate and no man is next to his wife. Show that

$$
L_n = \sum_{k=0}^{n} (-1)^k \binom{2n-k}{k} (n-k)!.
$$

21. (*Continuation*). Let M_n and L_n be the reduced ménages number for a circular and a straight table, respectively. Show that

$$
M_n = L_n - L_{n-1},
$$

$$
(n-2)M_n = n(n-2)M_{n-1} + nM_{n-2} + 4(-1)^{n+1},
$$

$$
(n-1)L_n = (n^2 - n - 1)L_{n-1} + nL_{n-2} + 2(-1)^{n+1}.
$$

22. *A modified ménages problem.* Let H_n be the number of ways of seating n married couples around a circular table so that no man is next to his wife. Show that

$$
H_n = \sum_{k=0}^{n} (-1)^k \binom{n}{k} 2^k (2n - k - 1)!
$$

23. (*Continuation*). Let W_n be the number of ways of seating n married couples at a straight table so that no woman is next to her husband. Show that

$$
W_n = \sum_{k=0}^{n} (-1)^k \binom{n}{k} 2^k (2n - k)!.
$$

24. Show that the number $D_{n,k}$ of the terms in the development of a determinant of order n that contain k diagonal elements is given by

$$D_{n,k} = \frac{n!}{k!} \sum_{j=0}^{n-k} \frac{(-1)^j}{j!}, \quad k = 0, 1, \dots, n.$$

25. Consider a series of k throws of a die and let $Q(k,r)$ be the number of different outcomes, in each of which $r \le 6$ specified faces appear. Show that

$$Q(k,r) = \sum_{s=0}^{r} (-1)^s \binom{r}{s} (6-s)^k.$$

26. Let $\Omega_k = \{(\omega_1, \omega_2, \dots, \omega_k) : \omega_i = (u_i, v_i), u_i, v_i = 1, 2, \dots, 6, i = 1, 2, \dots, k\}$ be the set of different outcomes of a series of k throws of a pair of distinguishable dice and let A_k be the subset of Ω_k containing the outcomes, in each of which all six pairs $(u, u), u = 1, 2, \dots, 6$ appear. Show that

$$N(A_k) = \sum_{s=0}^{6} (-1)^s \binom{6}{s} (36-s)^k.$$

27. *Permutations with restricted repetition.* Let $U_1(n,k)$ be the number of k-permutations of n with repetition and the restriction that each element appears at least once. Show that

$$U_1(n,k) = n!S(k,n) = \sum_{j=0}^{n} (-1)^{n-j} \binom{n}{j} j^k,$$

where $S(k,n)$ is the Stirling number of the second kind.

28. *Combinations with restricted repetition.* Let $E_{r,s}(n,k)$ be the number of k-combinations of n with repetition and the restriction that each element appears at least r and at most s times. Show that

$$E_{r,s}(n,k) = \sum_{j=0}^{k} (-1)^j \binom{n}{j} \binom{n+k-rn-j(s-r+1)-1}{n-1}.$$

29. Consider $2n$ elements that belong to n different kinds, with two like elements of each kind, and let Q_n be the number of permutations of these elements in which no two like elements are consecutive. Show that

$$Q_n = 2^{-n} \sum_{k=0}^{n} (-1)^k \binom{n}{k} 2^k (2n-k)!.$$

30. Consider an urn containing n balls numbered from 1 to n. Assume that, in a drawing, r numbers are simultaneously drawn from the urn and returned to it before the next drawing. An s-tuple drawing consists of s consecutive drawings. If $L(n, r, s)$ denotes the number of s-tuple drawings, in each of which all the n numbers are drawn (each of them at least once), show that

$$L(n, r, s) = \sum_{k=0}^{n} (-1)^{n-k} \binom{n}{k} \binom{k}{r}^s.$$

31. Let $D(n, r, s)$ be the number of divisions $(A_1, A_2, \ldots, A_r)$ of a finite set W, with $N(W) = n$, in r subsets with $N(A_j) \geq s$, $i = 1, 2, \ldots, r$. Show that

$$D(n, r, s+1) = \sum_{j=0}^{r} (-1)^j \binom{r}{j} \frac{(n)_{sj}}{(s!)^j} D(n - sj, r - j, s),$$

for $s = 0, 1, \ldots, [n/r] - 1$ and conclude that

$$D(n, r, 1) = r! S(n, r) = \sum_{j=0}^{r} (-1)^j \binom{r}{j} (r - j)^n,$$

where $S(n, r)$ is the *Stirling number of the second kind*.

32. (*Continuation*). Consider the sum of multinomial coefficients

$$S_2(n, r) = \frac{1}{r!} \sum \binom{n}{k_1, k_2, \ldots, k_{r-1}},$$

where the summation is extended over all $k_i \geq 2$, $i = 1, 2, \ldots, r$, with $k_1 + k_2 + \cdots + k_{r-1} + k_r = n$. Show that

$$S_2(n, r) = \sum_{j=0}^{r} (-1)^j \binom{n}{j} S(n - j, r - j).$$

The number $S_2(n, r)$ is called *associated Stirling number of the second kind*.

33. Let

$$C(n, r; s) = \frac{n!}{r!} \sum \binom{s}{k_1} \binom{s}{k_2} \cdots \binom{s}{k_r},$$

where the summation is extended over all $k_i \geq 1$, $i = 1, 2, \ldots, r$, with $k_1 + k_2 + \cdots + k_r = n$. Show that

$$C(n, r; s) = \frac{1}{r!} \sum_{j=0}^{r} (-1)^{r-j} \binom{r}{j} (sj)_n.$$

The number $C(n, r; s)$ is called *generalized factorial coefficient*.

34. (*Continuation*). Let

$$C_2(n,r;s) = \frac{n!}{r!} \sum \binom{s}{k_1}\binom{s}{k_2}\cdots\binom{s}{k_r},$$

where the summation is extended over all $k_i \geq 2$, $i = 1, 2, \ldots, r$, with $k_1 + k_2 + \cdots + k_r = n$. Show that

$$C_2(n,r;s) = \sum_{j=0}^{r}(-1)^j\binom{n}{j}s^j C(n-j, r-j; s).$$

The number $C_2(n,r;s)$ is called *associated generalized factorial coefficient*.

35*. *Fréchet inequality* . Let $A_1, A_2, \ldots, A_n$ be subsets of a finite set Ω and

$$S_{n,0} = N(\Omega), \; S_{n,r} = \sum N(A_{i_1} A_{i_2} \cdots A_{i_r}), \; r = 1, 2, \ldots, n,$$

where the summation is extended over all r-combinations $\{i_1, i_2, \ldots, i_r\}$ of the n indices $\{1, 2, \ldots, n\}$. Show that

$$S_{n,r}\Big/\binom{n}{r} \leq S_{n,r-1}\Big/\binom{n}{r-1}, \quad r = 1, 2, \ldots, n.$$

36*. (*Continuation*). *Gumbel inequality.* Show that

$$\left\{\binom{n}{r}S_{n,0} - S_{n,r}\right\}\Big/\binom{n-1}{r-1} \leq \left\{\binom{n}{r-1}S_{n,0} - S_{n,r-1}\right\}\Big/\binom{n-1}{r-2},$$

for $r = 2, 3, \ldots, n$.

37*. *Möbius function on partially ordered and locally finite sets.* A set $P = \{\ldots, x, y, z, \ldots\}$ supplied with an order relation, $x \leq y$ (x is less than or is included in or is ahead of y), which is valid for some pairs of elements (x, y) and with an equality relation, $x = y$, so that

    (1) $x \leq x$ for every element x in P
    (2) if $x \leq y$ and $y \leq z$, then $x \leq z$
    (3) if $x \leq y$ and $y \leq x$, then $x = y$

is called *partially ordered*. Note that the notation $y \geq x$ may be used as an alternative form of $x \leq y$. Also $x < y$ (or $y > x$) if $x \leq y$ (or $y \geq x$) and $x \neq y$. A partially ordered set P is called *locally finite* if the number of elements in every interval $[x, y] = \{z : x \leq z \leq y\}$ is finite. Let F be the set of real functions $f(x, y)$ for x, y elements in P such that $f(x, y) = 0$ if $x \not\leq y$ (that is $x > y$ or x and y are not comparable). The *Kronecker delta function* defined by

$$\delta(x, y) = 1 \; \text{if} \; x = y \; \text{and} \; \delta(x, y) = 0 \; \text{if} \; x \neq y$$

and the *zeta function* defined by

$$\zeta(x,y) = 1 \text{ if } x \leq y \text{ and } \zeta(x,y) = 0 \text{ otherwise,}$$

belong to the set F.

(a) Let f be a function in F. Prove that there exist functions g and h in F such that

$$\delta(x,y) = \sum_{x \leq z \leq y} f(x,z)g(z,y), \ \delta(x,y) = \sum_{x \leq z \leq y} h(x,z)f(z,y),$$

if and only if $f(x,x) \neq 0$ for every element x in P.

(b) Let $g(x,y)$ and $h(x,y)$ be solutions of the equations in (a). Prove that $h(x,y) = g(x,y)$ for all elements x,y in P. This common solution g is called the *inverse* of f.

(c) The inverse of the zeta function, denoted by $\mu(x,y)$ for x,y elements in P is called the *Möbius function* of P. Therefore

$$\mu(x,x) = 1$$

for every element x in P and

$$\mu(x,y) = - \sum_{x \leq z \leq y} \mu(x,z), \ \ \mu(x,y) = - \sum_{x \leq z \leq y} \mu(z,y)$$

for $x < y$ fixed points in P.

38*. (*Continuation*). *Möbius inversion formula.* Let P be a partially ordered and locally finite set with a zero element 0. If $f(x)$ and $g(x)$ are functions defined for every element x in P and $\mu(x,y)$ is the Möbius function of P, prove that the relation

$$g(x) = \sum_{y \leq x} f(y)$$

implies the relation

$$f(x) = \sum_{y \leq x} g(y)\mu(y,x)$$

and vice versa.

39*. (*Continuation*). Let $\mathcal{P}$ be the set of the subsets of a finite set S with respect to the order relation $\subseteq$ (is included, is a subset of). Prove that the Möbius function of $\mathcal{P}$ is given by

$$\mu(X,Y) = (-1)^{N(Y)-N(X)}, \ \ X \subseteq Y,$$

where $N(X)$ and $N(Y)$ are the numbers of elements of S that belong to X and Y, respectively.

40*. (*Continuation*). Consider n subsets $A_1, A_2, \ldots, A_n$ of a finite set Ω and let $S = \{1, 2, \ldots, n\}$. To each element $\omega \in \Omega$, there corresponds a set of indices $J \subseteq S$ such that $\omega \in A_j$ for every $j \in J$. Let $f(K)$ be the number of elements of Ω that belong only to the intersection of the sets A_j, $j \notin K$ for K a subset of S. Then $g(I) = \sum_{K \subseteq I} f(K)$ is the number of elements of Ω that belong (not necessarily exclusively) to the intersection of the sets A_j, $j \notin I$. Prove that

$$f(K) = \sum_{I \subseteq K} (-1)^{N(K) - N(I)} g(I),$$

where $N(K)$ and $N(I)$ are the numbers of elements of S that belong to K and I, respectively. Deduce the number $N_{n,0}$ of elements of Ω that do not belong to any of the n sets.

Chapter 5

PERMUTATIONS WITH FIXED POINTS AND SUCCESSIONS

5.1 INTRODUCTION

One of the classical combinatorial problems in the theory of probability is the famous *problem of coincidences (probleme des recontres)*, which constitutes in the computation of the ways of putting n cards, numbered from 1 to n, in a series so that, in a given number of cards their position coincides with their number. This problem was initially formulated in the particular case $n = 13$ by Montmort (1678 - 1719) and then in the general case by De Moivre (1667 - 1754), whose solution essentially constitutes an application of the inclusion and exclusion principle. Later, several reformulations and generalizations of this problem were followed.

In this chapter targeting to the problem of coincidences, the number of permutations of the first n positive integers $\{1, 2, \ldots, n\}$ (a) with a given number of fixed points (with respect to their natural order $(1, 2, \ldots, n)$) and (b) of a given rank are computed by applying the inclusion and exclusion principle. In the same way, the computation of the permutations of the first n positive integers with a given number of pairs of successive points is carried out.

5.2 PERMUTATIONS WITH FIXED POINTS

DEFINITION 5.1   Let $(j_1, j_2, \ldots, j_n)$ be a permutation of the set $\{1, 2, \ldots, n\}$. The point j_r is called fixed point of this permutation if $j_r = r$, $r = 1, 2, \ldots, n$.

Note that, according to this definition, the problem of coincidences constitutes in the computation of the permutations of the set $\{1, 2, \ldots, n\}$, with a given number of fixed points. The permutations without a fixed point are particularly interesting and deserve special attention. A particular name is given to such a permutation.

DEFINITION 5.2   *A permutation $(j_1, j_2, \ldots, j_n)$ of the set $\{1, 2, \ldots, n\}$ is called a derangement if it does not have a fixed point, that is, if $j_r \neq r$ for all $r = 1, 2, \ldots, n$.*

These notions are illustrated in the following simple example.

Example 5.1
Consider the permutations (j_1, j_2, j_3, j_4) of the set $\{1, 2, 3, 4\}$. Clearly, these 24 permutations can be classified according to the fixed points they have as follows.
(a) The derangements are the following 9:

$$(2, 1, 4, 3), \ (2, 4, 1, 3), \ (2, 3, 4, 1), \ (3, 1, 4, 2), \ (3, 4, 1, 2),$$

$$(3, 4, 2, 1), \ (4, 1, 2, 3), \ (4, 3, 1, 2), \ (4, 3, 2, 1).$$

(b) The permutations with one fixed point are the following 8:

$$(1, 3, 4, 2), \ (1, 4, 2, 3), \ (3, 2, 4, 1), \ (4, 2, 1, 3),$$

$$(2, 4, 3, 1), \ (4, 1, 3, 2), \ (3, 1, 2, 4), \ (2, 3, 1, 4).$$

(c) The permutations with two fixed points are the following 6:

$$(1, 2, 4, 3), \ (1, 4, 3, 2), \ (1, 3, 2, 4), \ (4, 2, 3, 1), \ (3, 2, 1, 4), \ (2, 1, 3, 4).$$

(d) There are no permutations with three fixed points, while

$$(1, 2, 3, 4)$$

is the only permutation with four fixed points. $\Box$

The following two theorems are concerned with the number of derangements.

THEOREM 5.1
The number D_n of derangements of the set $\{1, 2, \ldots, n\}$ equals

$$D_n = n! \sum_{k=0}^{n} \frac{(-1)^k}{k!} \tag{5.1}$$

and satisfies the recurrence relation

$$D_n = nD_{n-1} + (-1)^n, \quad n = 1, 2, \ldots, \quad D_0 = 1. \tag{5.2}$$

PROOF Consider the set Ω of the permutations $(j_1, j_2, \ldots, j_n)$ of the set $\{1, 2, \ldots, n\}$ and let A_r be the subset of permutations $(j_1, j_2, \ldots, j_n)$ for which the point j_r is a fixed point, $r = 1, 2, \ldots, n$. Then, the number D_n of derangements of the set $\{1, 2, \ldots, n\}$ is given by

$$D_n = N(A'_1 A'_2 \cdots A'_n).$$

Clearly, for any selection of k indices $\{i_1, i_2, \ldots, i_k\}$ out of the n indices $\{1, 2, \ldots, n\}$,

$$\nu_k = N(A_{i_1} A_{i_2} \cdots A_{i_k}) = (n-k)!, \quad k = 1, 2, \ldots, n,$$

which is the number of permutations of the $n - k$ non-fixed points. This expression implies the exchangeability of the sets $A_1, A_2, \ldots, A_n$ and so, according to Corollary 4.3,

$$D_n = n! \sum_{k=0}^{n} (-1)^k \binom{n}{k} (n-k)! = n! \sum_{k=0}^{n} \frac{(-1)^k}{k!}.$$

Using this expression, recurrence relation (5.2) is readily deduced:

$$D_n = n(n-1)! \sum_{k=0}^{n-1} \frac{(-1)^k}{k!} + (-1)^n = nD_{n-1} + (-1)^n.$$

As regards the initial value of this recurrence relation, it is clear that $D_1 = 0$, which is equivalent to $D_0 = 1$. $\blacksquare$

THEOREM 5.2 Euler's recurrence relation
The number D_n of derangements of the set $\{1, 2, \ldots, n\}$ satisfies the recurrence relation

$$D_n = (n-1)\{D_{n-1} + D_{n-2}\}, \ n = 2, 3, \ldots, \ D_0 = 1, \ D_1 = 0. \quad (5.3)$$

PROOF Note that the last position of a derangement $(j_1, j_2, \ldots, j_n)$ of the set $\{1, 2, \ldots, n\}$ can be occupied by any of the $n - 1$ numbers $\{1, 2, \ldots, n - 1\}$. Assume that $j_n = k$, $k \neq n$. Then, the derangements $(j_1, j_2, \ldots, j_{n-1})$ of the set $\{1, 2, \ldots, k - 1, k + 1, \ldots, n\}$, can be distinguished according to whether n does or does not occupy the k-th position. If $j_k = n$, then the number of these derangements is equal to the number D_{n-2} of the derangements $(j_1, j_2, \ldots, j_{k-1}, j_{k+1}, \ldots, j_{n-1})$ of the $n - 2$ numbers $\{1, 2, \ldots, k - 1, k + 1, \ldots, n - 1\}$. If $j_k \neq n$, then the number of these derangements equals the number of permutations $(j_1, j_2, \ldots, j_{n-1})$ of the $n - 1$ numbers $\{1, 2, \ldots, k - 1, k + 1, \ldots, n\}$, with $j_r \neq r, r = 1, 2, \ldots, k - 1, k + 1, \ldots, n - 1, j_k \neq n$.

The latter number equals the number D_{n-1} of the derangements $(j_1, j_2, \ldots, j_{n-1})$ of the $n - 1$ numbers $\{1, 2, \ldots, n - 1\}$. Consequently

$$D_n = (n - 1)\{D_{n-1} + D_{n-2}\}, \; n = 2, 3, \ldots, \; D_0 = 1, \; D_1 = 0,$$

which is the required recurrence relation.   ∎

REMARK 5.1   The similarity of recurrence relations (5.2) and (5.3) to the corresponding recurrence relations for the factorials,

$$n! = n(n - 1)!, \; n = 1, 2, \ldots, \; 0! = 1,$$

$$n! = (n - 1)\{(n - 1)! + (n - 2)!\}, \; n = 2, 3, \ldots, \; 0! = 1, \; 1! = 1,$$

led Whitworth to call the number D_n n-subfactorial. Using recurrence relation (5.2) or (5.3), a table of subfactorials can be constructed. Table 5.1 gives the numbers D_n for $n = 0, 1, \ldots, 10$.   ∎

Table 5.1   Subfactorials D_n

n	0	1	2	3	4	5	6	7	8	9	10
D_n	1	0	1	2	0	44	265	1854	148331	133496	1334961

The next theorem is concerned with the number of permutations with a given number of fixed points.

THEOREM 5.3
The number $D_{n,k}$ of permutations of the set $\{1, 2, \ldots, n\}$ with k fixed points equals

$$D_{n,k} = \frac{n!}{k!} \sum_{j=0}^{n-k} \frac{(-1)^j}{j!} = \binom{n}{k} D_{n-k}. \tag{5.4}$$

PROOF   Consider the set Ω of the permutations $(j_1, j_2, \ldots, j_n)$ of the set $\{1, 2, \ldots, n\}$ and let A_r be the subset of permutations $(j_1, j_2, \ldots, j_n)$ for which the point j_r is a fixed point, $r = 1, 2, \ldots, n$. Then, the number $D_{n,k}$ equals the number of elements of the set Ω that are contained in k among the n exchangeable sets $A_1, A_2, \ldots, A_n$. This number, according to Corollary 4.4 and since $v_r = N(A_{i_1} A_{i_2} \cdots A_{i_r}) = (n - r)!$, is given by

$$D_{n,k} = \binom{n}{k} \sum_{j=0}^{n-k} (-1)^j \binom{n-k}{j} (n - k - j)! = \frac{n!}{k!} \sum_{j=0}^{n-k} \frac{(-1)^j}{j!}.$$

Further

$$D_{n,k} = \binom{n}{k}(n-k)! \sum_{j=0}^{n-k} \frac{(-1)^j}{j!}$$

and so, using (5.1), the second part of (5.4) is deduced. ∎

Example 5.2 The classical problem of coincidences
Consider an urn containing n balls numbered from 1 to n and assume that all the balls are successively drawn one after the other without replacement. The draw of the j-th ball at the j-th trial, $j = 1, 2, \ldots, n$, is called a *coincidence*. Calculate the number of series of drawings of the n balls that result in k coincidences.

Clearly, to each series of drawings of the n balls, there corresponds a permutation $(j_1, j_2, \ldots, j_n)$ of the set $\{1, 2, \ldots, n\}$ and to each coincidence there corresponds a fixed point of the permutation. Therefore the number of series of drawings of the n balls that result in k coincidences equals

$$D_{n,k} = \frac{n!}{k!} \sum_{j=0}^{n-k} \frac{(-1)^j}{j!},$$

the number of permutations of the set $\{1, 2, \ldots, n\}$ with k fixed points. □

REMARK 5.2   The number $D_{n,k}$ is known in the bibliography as the *co-incidence number* , a name due to the preceding example. Using the relation $D_{n,k} = \binom{n}{k}D_{n-k}$, Table 5.1 of subfactorials and Table 2.1 of binomial coefficients, a table of the numbers $D_{n,k}$ can be constructed. Table 5.2 gives the coincidence numbers $D_{n,k}$ for $k = 1, 2, \ldots, n, n = 1, 2, \ldots, 9$. ∎

Table 5.2   Coincidence Numbers $D_{n,k}$

n \ k	1	2	3	4	5	6	7	8	9
1	1								
2	0	1							
3	3	0	1						
4	8	6	0	1					
5	45	20	10	0	1				
6	264	135	40	15	0	1			
7	1855	924	315	70	21	0	1		
8	14832	7420	2464	630	112	28	0	1	
9	133497	66744	22260	5544	1134	168	36	0	1

5.3   RANKS OF PERMUTATIONS

DEFINITION 5.3   *Let $(j_1, j_2, \ldots, j_n)$ be a permutation of the set $\{1, 2, \ldots, n\}$.
If*

- $$j_r \neq r, \ r = 1, 2, \ldots, k-1, \ \ j_k = k, \ 2 \leq k \leq n,$$

*the permutation is said to be of rank k. In particular, if $j_1 = 1$, the permutation is
said to be of rank 1, while if $j_r \neq r, r = 1, 2, \ldots, n$, it is said, by convention, to
be of rank $n + 1$.*

Note that several problems in combinatorics and in probability theory,
known as *problems of duration of trials (drawings)*, reduce to the enumeration of the permutations of the set $\{1, 2, \ldots, n\}$ of a given rank.

Example 5.3
The 24 permutations (j_1, j_2, j_3, j_4) of the set $\{1, 2, 3, 4\}$ can be classified according
to their rank as follows.
(a) The permutations of rank 1 are the following 6:

$$(1, 2, 3, 4), \ (1, 2, 4, 3), \ (1, 3, 2, 4), \ (1, 3, 4, 2), \ (1, 4, 2, 3), \ (1, 4, 3, 2).$$

(b) The permutations of rank 2 are the following 4:

$$(3, 2, 1, 4), \ (3, 2, 4, 1), \ (4, 2, 1, 3), \ (4, 2, 3, 1).$$

(c) The permutations of rank 3 are the following 3:

$$(2, 1, 3, 4), \ (2, 4, 3, 1), \ (4, 1, 3, 2).$$

(d) The permutations of rank 4 are the following 2:

$$(2, 3, 1, 4), \ (3, 1, 2, 4).$$

(e) The permutations of rank 5 are the following 9:

$$(2, 1, 4, 3), \ (2, 4, 1, 3), \ (2, 3, 4, 1), \ (3, 1, 4, 2), \ (3, 4, 1, 2),$$

$$(3, 4, 2, 1), \ (4, 1, 2, 3), \ (4, 3, 1, 2), \ (4, 3, 2, 1).$$

Note that the permutations of rank 5 are, by convention, the derangements of the
set $\{1, 2, 3, 4\}$.   ☐

The next theorem is concerned with the number of permutations of a
given rank.

THEOREM 5.4
The number $R_{n,k}$ of permutations of the set $\{1, 2, \dots, n\}$ of rank k equals

$$R_{n,k} = \sum_{s=0}^{k-1}(-1)^s\binom{k-1}{s}(n-s-1)!, \tag{5.5}$$

for $k = 1, 2, \dots, n$ and $R_{n,n+1} = D_n$. Further, it satisfies the recurrence relation

$$R_{n,k} = R_{n,k-1} - R_{n-1,k-1}, \; k = 2, 3, \dots, n, \; n = 2, 3, \dots, \tag{5.6}$$

with $R_{n,1} = (n-1)!$, $R_{n,k} = 0$, $k > n + 1$.

PROOF   Consider the set Ω of the permutations $(j_1, j_2, \dots, j_n)$ of the set $\{1, 2, \dots, n\}$ and let A_r be the subset of permutations $(j_1, j_2, \dots, j_n)$ for which the point j_r is a fixed point, $r = 1, 2, \dots, n$. The number $R_{n,k}$ equals the number of elements of Ω of rank k with respect to the n ordered and exchangeable sets $A_1, A_2, \dots, A_n$. Since $\nu_s = N(A_{i_1}A_{i_2}\cdots A_{i_s}) = (n-s)!$, $s = 1, 2, \dots, n$, this number, according to Corollary 4.6, is

$$R_{n,k} = \sum_{s=0}^{k-1}(-1)^s\binom{k-1}{s}(n-s-1)!,$$

for $k = 1, 2, \dots, n$. Clearly $R_{n,n+1} = N(A'_1 A'_2 \cdots A'_n) = D_n$. Further, expression (5.5), on using Pascal's triangle, can be written as

$$R_{n,k} = (n-1)! + \sum_{s=1}^{k-1}(-1)^s\left\{\binom{k-2}{s} + \binom{k-2}{s-1}\right\}(n-s-1)!$$

and so

$$R_{n,k} = \sum_{s=0}^{k-2}(-1)^s\binom{k-2}{s}(n-s-1)! + \sum_{s=1}^{k-1}(-1)^s\binom{k-2}{s-1}(n-s-1)!$$

$$= \sum_{s=0}^{k-2}(-1)^s\binom{k-2}{s}(n-s-1)! - \sum_{i=0}^{k-2}(-1)^i\binom{k-2}{i}(n-i-2)!,$$

which, by virtue of (5.5), implies recurrence relation (5.6).   ∎

Example 5.4 Duration of a series of drawings

Consider a lottery-urn containing n balls numbered from 1 to n and assume that balls are drawn one after the other without replacement. This series of drawings is terminated with occurrence of the first coincidence. Calculate the number of the series of drawings with a given duration of k drawings.

The required number, according to the Definition 3.1 of the rank of a permutation, is equal to the number of permutations of $\{1, 2, \ldots, n\}$ of rank k which, according to Theorem 3.1, is

$$R_{n,k} = \sum_{s=0}^{k-1} (-1)^s \binom{k-1}{s} (n-s-1)!, \quad k = 1, 2, \ldots, n. \quad \square$$

REMARK 5.3   The number $R_{n,k}$ has been studied by M. Fréchet (1943); J. Riordan (1958) called it *rank number*. Using recurrence relation (5.6), Table 5.3 of the numbers $R_{n,k}$, for $k = 1, 2, \ldots, n+1$, $n = 1, 2, \ldots, 8$, is constructed. ∎

Table 5.3   **Rank Numbers $R_{n,k}$**

k \ n	1	2	3	4	5	6	7	8	9
1	1	0							
2	1	0	1						
3	2	1	1	2					
4	6	4	3	2	9				
5	24	18	14	11	9	44			
6	120	96	78	64	53	44	265		
7	720	600	504	426	362	309	265	1854	
8	5040	4320	3720	3216	2790	2428	2119	1854	14833

5.4   PERMUTATIONS WITH SUCCESSIONS

DEFINITION 5.4   *Let $(j_1, j_2, \ldots, j_j)$ be a permutation of the set $\{1, 2, \ldots, n\}$. If*

$$j_{r+1} = j_r + 1, \quad 1 \leq r \leq n-1,$$

then the pair (j_r, j_{r+1}) is called a succession of the permutation.

Example 5.5

Consider the permutations (j_1, j_2, j_3, j_4) of the set $\{1, 2, 3, 4\}$. These 24 permutations can be classified according to their successions as follows.

(a) The permutations with the no succession are the following 11:

$$(1,3,2,4), \ (1,4,3,2), \ (2,1,4,3), \ (2,4,1,3), \ (2,4,3,1), \ (3,1,4,2),$$

$$(3,2,1,4), \ (3,2,4,1), \ (4,1,3,2), \ (4,2,1,3), \ (4,3,2,1).$$

(b) The permutations with one succession are the following 9:

$$(1,2,4,3), \ (3,1,2,4), \ (4,3,1,2), \ (2,3,4,1), \ (4,2,3,1),$$

$$(1,4,2,3), \ (3,4,2,1), \ (1,3,4,2), \ (2,1,3,4).$$

(c) The permutations with two successions are the following 3:

$$(2,3,4,1), \ (3,4,1,2), \ (4,1,2,3).$$

(d) The permutation

$$(1,2,3,4)$$

is the only permutation with three successions.   ⬜

The next theorem is concerned with the number of permutations without a succession.

THEOREM 5.5
The number S_n of permutations of the set $\{1,2,\dots,n\}$ without a succession is equal to

$$S_n = (n-1)! \sum_{k=0}^{n-1} (-1)^k \frac{n-k}{k!} \tag{5.7}$$

and satisfies the recurrence relation

$$nS_n = (n^2-1)S_{n-1} - (-1)^n, \ n = 2,3,\dots, \ S_1 = 1. \tag{5.8}$$

PROOF   Consider the set Ω of the permutations $(j_1,j_2,\dots,j_n)$ of the set $\{1,2,\dots,n\}$ and let A_r be the set of these permutations for which the pair (j_r, j_{r+1}) is a succession, $r = 1,2,\dots,n-1$. Then the number S_n of permutations of the set $\{1,2,\dots,n\}$ without a succession is given by $S_n = N(A_1' A_2' \cdots A_{n-1}')$. Further, the sets $A_1, A_2, \dots, A_{n-1}$ are exchangeable and

$$\nu_0 = N(\Omega) = n!, \ \nu_k = N(A_{i_1} A_{i_2} \cdots A_{i_k}) = (n-k)!, \ k = 1,2,\dots,n-1.$$

Indeed, in the case where the set of indices $\{i_1, i_2, \dots, i_k\}$, with $1 \leq i_1 < i_2 < \cdots < i_k \leq n$, consists of consecutive integers, $i_r = i + r - 1$, $r = 1,2,\dots,k$, $1 \leq i \leq n-k$, the k successions constitute a run $\rho = (j_i, j_{i+1}, \dots, j_{i+k})$. Thus, the remaining $n - k + 1$ integers $\{j_1, j_2, \ldots, j_{i-1}, j_{i+k+1}, \ldots, j_n\}$, along with the run ρ, constitute a set of $n - k$ elements and the number $N(A_{i_1} A_{i_2} \cdots A_{i_k})$ equals $(n - k)!$, the number of permutations of this set. If the set of indices $I = \{i_1, i_2, \ldots, i_k\}$, with $1 \le i_1 < i_2 < \cdots < i_k \le n$, does not consist of consecutive integers, then it is partitioned in s subsets $I_1, I_2, \ldots, I_s$, where I_c, with $N(I_c) = m_c, c = 1, 2, \ldots, s$, is either a one-point set or a set of consecutive integers and $m_1 + m_2 + \cdots + m_s = k$. In this case, the k successions constitute s runs, $\rho_1, \rho_2, \ldots, \rho_s$, of consecutive integers, which are formed by $(m_1 + 1) + (m_2 + 1) + \cdots + (m_s + 1) = k + s$ integers. Thus the remaining $n - k - s$ integers along with the s runs constitute a set of $n - k$ elements and the number $N(A_{i_1} A_{i_2} \cdots A_{i_k})$ is equal to $(n - k)!$, the number of permutations of this set. Therefore, according to Corollary 4.3, the number S_n of permutations of the set $\{1, 2, \ldots, n\}$ without a succession is

$$S_n = \sum_{k=0}^{n-1} (-1)^k \binom{n-1}{k} (n - k)! = (n - 1)! \sum_{k=0}^{n-1} (-1)^k \frac{n - k}{k!}.$$

Further, multiplying both members of the identity

$$\frac{n(n - k)}{k!} = \frac{(n + 1)(n - k - 1)}{k!} + \frac{1}{k!} + \frac{1}{(k - 1)!}$$

by $(-1)^k (n - 1)!$ and summing for $k = 0, 1, \ldots, n - 1$, we get the expression

$$n(n - 1)! \sum_{k=0}^{n-1} (-1)^k \frac{n - k}{k!} = (n^2 - 1)(n - 2)! \sum_{k=0}^{n-2} (-1)^k \frac{n - k - 1}{k!} + (-1)^{n-1},$$

which, by virtue of (5.7), implies (5.8).   ∎

REMARK 5.4   The number S_n can be expressed in terms of the number D_r, of r-subfactorial. Indeed, rewriting expression (5.7) as

$$S_n = n(n - 1)! \sum_{k=0}^{n-1} \frac{(-1)^k}{k!} - (n - 1)! \sum_{k=1}^{n-1} \frac{(-1)^k}{(k - 1)!}$$

$$= n! \sum_{k=0}^{n} \frac{(-1)^k}{k!} + (n - 1)! \sum_{j=0}^{n-1} \frac{(-1)^j}{j!}$$

and, using (5.1), we get the relation $S_n = D_n + D_{n-1}$, $n = 1, 2, \ldots$, which, combined with the recurrence relation (5.3), yields for S_n the recurrence relation

$$S_n = (n - 1)S_{n-1} + (n - 2)S_{n-2}, \; n = 3, 4, \ldots, \; S_1 = S_2 = 1. \quad (5.9)$$

Using either of the recurrence relations (5.8) or (5.9), a table of the numbers S_n can be constructed. Table 5.4 gives the numbers S_n for $n = 1, 2, \ldots, 10$.   ∎

Table 5.4 *The Numbers S_n*

n	1	2	3	4	5	6	7	8	9	10
S_n	1	1	3	11	53	509	2119	16687	148329	1468457

The next theorem is concerned with the number of permutations with a given number of successions.

THEOREM 5.6

The number $S_{n,k}$ of permutations of the set $\{1, 2, \dots, n\}$ with k successions is equal to

$$S_{n,k} = \frac{(n-1)!}{k!} \sum_{j=0}^{n-k-1} (-1)^j \frac{n-k-j}{j!} = \binom{n-1}{k} S_{n-k}. \qquad (5.10)$$

PROOF   Consider the set Ω of the permutations $(j_1, j_2, \dots, j_n)$ of the set $\{1, 2, \dots, n\}$ and let A_r be the set of these permutations for which the pair (j_r, j_{r+1}) is a succession, $r = 1, 2, \dots, n-1$. Then the number $S_{n,k}$ equals the number of elements of the set Ω, which are contained in k among the $n-1$ sets $A_1, A_2, \dots, A_{n-1}$. As it was shown in the proof of Theorem 5.5, the sets $A_1, A_2, \dots, A_{n-1}$ are exchangeable and

$$\nu_0 = N(\Omega) = n!, \quad \nu_r = N(A_{i_1} A_{i_2} \cdots A_{i_r}) = (n-r)!, \quad r = 1, 2, \dots, n-1.$$

Thus, according to Corollary 4.4, the number $S_{n,k}$ is given by

$$S_{n,k} = \binom{n-1}{k} \sum_{j=0}^{n-k-1} (-1)^j \binom{n-k-1}{j} (n-k-j)!$$

$$= \frac{(n-1)!}{k!} \sum_{j=0}^{n-k-1} (-1)^j \frac{n-k-j}{j!} = \binom{n-1}{k} S_{n-k},$$

with the last equality implied by virtue of (5.7).   ∎

Using (5.10), Table 5.4 of the numbers S_n and Table 2.1 of binomial coefficients, a table of *succession numbers* $S_{n,k}$ can be constructed. Table 5.5 gives the numbers $S_{n,k}$ for $k = 1, 2, \dots, n-1$, $n = 2, 3, \dots, 10$.

Table 5.5 Succession Numbers $S_{n,k}$

A123513 ED)

k / n	1	2	3	4	5	6	7	8	9
2	1								
3	2	1							
4	9	3	1						
5	44	18	4	1					
6	265	110	30	5	1				
7	1854	795	220	45	6	1			
8	14833	6489	1855	385	63	7	1		
9	133496	59332	17304	3710	616	84	8	1	
10	1334961	600732	177996	38934	6678	924	108	9	1

Example 5.6 Caravan in the desert

A caravan of n camels travels across a desert in a journey that lasts for several days. The camel-drivers find it boring to see the same camel in front of them every day. The calculation of the number of permutations of the camels so that the next day each camel follows a different camel is then of interest. For this purpose, let us consider the initial order of the camels and number them from 1 to n. Then the required number is equal to the number of permutations of the set $\{1, 2, \ldots, n\}$ that do not include any of the $n - 1$ pairs: $(1, 2), (2, 3), \ldots, (n - 1, n)$. According to Theorem 5.5, this number equals

$$S_n = (n - 1)! \sum_{k=0}^{n-1} (-1)^k \frac{(n - k)!}{k!}.$$

If, more generally, the interest is in the number of permutations of the n camels so that the next day each of k camels follows the same camel, then, according to Theorem 5.6, this number equals

$$S_{n,k} = \frac{(n - 1)!}{k!} \sum_{j=0}^{n-k-1} (-1)^j \frac{n - k - j}{j!}. \qquad \square$$

5.5 CIRCULAR PERMUTATIONS WITH SUCCESSIONS

Consider the positive integers $1, 2, \ldots, n$ displayed on a circle $(j_1, j_2, \ldots, j_n)$, where j_1 follows j_n. Thus a circularly ordered permutation, called a *cir-*

cular permutation of the set $\{1, 2, \dots, n\}$, is formed. Note that, to each circular permutation $(j_1, j_2, \dots, j_n)$, there correspond n linear permutations $(j_1, j_2, \dots, j_n)$, $(j_2, j_3, \dots, j_n, j_1)$, $\dots$, $(j_n, j_1, \dots, j_{n-1})$ and so the number of circular permutations of the set $\{1, 2, \dots, n\}$ is equal to $n!/n = (n-1)!$. In the case of the circular permutations of the set $\{1, 2, \dots, n\}$, in addition to the pairs $(1, 2)$, $(2, 3)$, $\dots$, $(n-1, n)$, the pair $(n, 1)$ is also a succession.

The next theorem is concerned with the number of circular permutations without a succession.

THEOREM 5.7
The number C_n of circular permutations of the set $\{1, 2, \dots, n\}$ without a succession equals

$$C_n = n! \sum_{k=0}^{n-1} \frac{(-1)^k}{(n-k)k!} + (-1)^n. \tag{5.11}$$

PROOF   Consider the set Ω of the circular permutations $(j_1, j_2, \dots, j_n)$ of the set $\{1, 2, \dots, n\}$ and let A_r be the subset, of Ω, of circular permutations in which the pair (j_r, j_{r+1}) constitutes a succession, $r = 1, 2, \dots, n-1$. In addition let A_n be the subset, of Ω, of circular permutations in which the pair (j_n, j_1) constitutes a succession. Then the number C_n of circular permutations of the set $\{1, 2, \dots, n\}$ without a succession is given by $C_n = N(A_1' A_2' \cdots A_n')$. Further, the sets $A_1, A_2, \dots, A_n$ are exchangeable and

$$\nu_0 = N(\Omega) = (n-1)!,$$

$$\nu_k = N(A_{i_1} A_{i_2} \cdots A_{i_k}) = (n-k-1)!, \quad k = 1, 2, \dots, n-1.$$

Indeed, as in the proof of Theorem 5.5, the cases in which the indices $i_1, i_2, \dots, i_k$ are or are not consecutive integers are separately examined. In both cases, the number $N(A_{i_1} A_{i_2} \cdots A_{i_k})$ is reduced to the number $(n-k-1)!$, of circular permutation of $n - k$ distinct elements. Also

$$\nu_n = N(A_1 A_2 \cdots A_n) = 1,$$

which is the only circular permutation, $(1, 2, \dots, n)$, belonging in all n sets. Thus, by virtue of Corollary 4.3, the number C_n is given by

$$C_n = \sum_{k=0}^{n-1} (-1)^k \binom{n}{k} (n-k-1)! + (-1)^n$$

$$= n! \sum_{k=0}^{n-1} \frac{(-1)^k}{(n-k)k!} + (-1)^n,$$

which is the required formula.   ∎

REMARK 5.5   The number C_n can be expressed in terms of the number D_r, of r-subfactorial. Indeed, rewriting expression (5.11) as

$$C_n = (n-1)! \sum_{k=0}^{n-1} (-1)^k \frac{(n-k)+k}{(n-k)k!} + (-1)^n$$

$$= (n-1)! \sum_{k=0}^{n-1} \frac{(-1)^k}{k!} + (n-1)! \sum_{k=1}^{n-1} \frac{(-1)^k}{(n-k)(k-1)!} + (-1)^n$$

$$= (n-1)! \sum_{k=0}^{n-1} \frac{(-1)^k}{k!} - \left\{ (n-1)! \sum_{j=0}^{n-2} \frac{(-1)^j}{(n-1-j)j!} + (-1)^{n-1} \right\}$$

and, using (5.1), we get

$$C_n + C_{n-1} = D_{n-1}, \quad n = 3, 4, \ldots . \tag{5.12}$$

Writing this relation in the form

$$(-1)^{n-k} C_k + (-1)^{n-k} C_{k-1} = (-1)^{n-k} D_{k-1}, \quad k = 3, 4, \ldots$$

and summing for $k = 3, 4, \ldots, n$, it follows that

$$\sum_{k=3}^{n} (-1)^{n-k} C_k + \sum_{k=3}^{n} (-1)^{n-k} C_{k-1} = \sum_{k=3}^{n} (-1)^{n-k} D_{k-1}.$$

The left-hand side of this relation is equal to $C_n + (-1)^{n-3} C_2$ and, since $C_2 = 0$,

$$C_n = \sum_{k=3}^{n} (-1)^{n-k} D_{k-1}, \quad n = 3, 4, \ldots .$$

Furthermore, note that (5.12), by using recurrence relations (5.2) and (5.3), yields the recurrence relations:

$$C_n = (n-2)C_{n-1} + (n-1)C_{n-2} + (-1)^{n-1}, \quad n = 4, 5, \ldots, \tag{5.13}$$

with $C_2 = 0$, $C_3 = 1$, and

$$C_n = (n-3)C_{n-1} + 2(n-2)C_{n-2} + (n-2)C_{n-3}, \quad n = 5, 6, \ldots, \tag{5.14}$$

with $C_2 = 0$, $C_3 = 1$, $C_4 = 1$. Using either of these recurrence relations, a table of the numbers C_n can be constructed. Table 5.6 gives the numbers C_n for $n = 2, 3, \ldots, 12$.   ∎

Table 5.6   The Numbers C_n

n	2	3	4	5	6	7	8	9	10	11	12
C_n	0	1	1	8	36	229	1625	13208	120288	1214673	13469897

A000757

THEOREM 5.8
The number $C_{n,k}$ of circular permutations of the set $\{1, 2, \dots, n\}$ with k successions equals

$$C_{n,k} = \frac{n!}{k!}\left\{\sum_{j=0}^{n-k-1}\frac{(-1)^j}{(n-k-j)j!} + \frac{(-1)^{n-k}}{(n-k)!}\right\} = \binom{n}{k}C_{n-k}. \quad (5.15)$$

PROOF   Consider the set Ω of the circular permutations of the set $\{1, 2, \dots, n\}$ and the n exchangeable sets $A_1, A_2, \dots, A_n$, subsets of Ω, defined in the proof of Theorem 5.7. Then, the number $C_{n,k}$ equals the number of elements of Ω that are contained in k among the n subsets. Since

$$\nu_0 = N(\Omega) = (n-1)!,$$

$$\nu_r = N(A_{i_1}A_{i_2}\cdots A_{i_r}) = (n-r-1)!, \quad k = 1, 2, \dots, n-1,$$

$$\nu_n = N(A_1 A_2 \cdots A_n) = 1,$$

it follows from Corollary 4.4 that

$$C_{n,k} = \binom{n}{k}\left\{\sum_{j=0}^{n-k-1}(-1)^j\binom{n-k}{j}(n-k-j-1)! + (-1)^{n-k}\right\}$$

$$= \frac{n!}{k!}\left\{\sum_{j=0}^{n-k-1}\frac{(-1)^j}{(n-k-j)j!} + \frac{(-1)^{n-k}}{(n-k)!}\right\} = \binom{n}{k}C_{n-k},$$

with the last equality implied by virtue of (5.11).   ∎

Example 5.7   A round-table conference
  Consider n representatives participating in a round-table conference scheduled to continue for several days. It is decided that every day they sit around the table so that each has to his right a different person. At their first seating, let us number the n representatives from 1 to n in the ordinary direction (counterclockwise), starting from any one of them. The problem of their second seating is reduced to the

calculation of the circular permutations of the set $\{1, 2, \ldots, n\}$ that do not include any of the n pairs: $(1, 2), (2, 3), \ldots, (n - 1, n), (n, 1)$. According to Theorem 5.7, this number is

$$C_n = n! \sum_{k=0}^{n-1} \frac{(-1)^k}{(n-k)k!} + (-1)^n.$$

More generally, if the interest is on the number of ways of creating a second seating of the n representatives so that k of them have to their right the same person, then this number, according to Theorem 5.8, equals

$$C_{n,k} = \frac{n!}{k!} \left\{ \sum_{j=0}^{n-k-1} \frac{(-1)^j}{(n-k-j)j!} + \frac{(-1)^{n-k}}{(n-k)!} \right\}. \qquad \square$$

5.6   BIBLIOGRAPHIC NOTES

The famous problem of coincidences (matches, recontres) was initially treated in the particular case of 13 cards by P. R. Montmort (1708) and J. Bernoulli (1714). Abraham De Moivre (1718) examined the general case of n cards by using the inclusion and exclusion principle. L. Euler (1751) derived a recurrence relation for the number of derangements. W. A. Whitworth (1867), in his book on the combinatorics of the games of chance, studied the matching problem and contributed to its popularization. P. MacMahon (1902, 1915, 1916) provided a generating function of the number of arrangements of the n cards with k matches. The number of permutations of a given rank has been examined by M. Fréchet (1945) and J. Riordan (1958). Generalizations and applications of the matching problem have been discussed by several authors; the interested reader is referred to the bibliography furnished in the review article of D. E. Barton (1958). The two sections on the enumeration of the linear and circular permutations by successions were inspired by the problem of the caravan in the desert posed and solved by N. Y. Vilenkin (1971).

5.7   EXERCISES

1. Let D_n be the n-subfactorial. Show that

$$|D_n - n!e^{-1}| < 1/(n+1), \quad n = 1, 2, \ldots$$

and thus conclude, for large n, the asymptotic expression

$$D_n \cong n!e^{-1}.$$

2. Let $D_{n,k}$ be the number of permutations of the set $\{1, 2, \ldots, n\}$ with k fixed points. Show that (a)

$$D_{n,k} = nD_{n-1,k} + (-1)^{n-k}\binom{n}{k}$$

and (b)

$$D_{n,k} = D_{n-1,k-1} + (n-1)\{D_{n-1,k} + D_{n-2,k} - D_{n-2,k-1}\}.$$

3. Let $R_{n,k}$ be the rank number and D_r r-subfactorial. Show that

$$R_{n,k} = \sum_{r=0}^{n-k}\binom{n-k}{r}D_{k+r-1}, \quad k = 1, 2, \ldots, n.$$

4. Let $R_{n,k}$ be the rank number and

$$R_n = \sum_{k=1}^{n+1} R_{n,k}, \quad E_n = \sum_{k=1}^{n+1} kR_{n,k}.$$

Using the recurrence relation

$$R_{n,k} = R_{n,k-1} - R_{n-1,k-1}, \quad k = 1, 2, \ldots, n,$$

with $R_{n,1} = (n-1)!$, $R_{n,n} = D_{n-1}$, $R_{n,n+1} = D_n$, show that

$$R_{n-1} = (n-1)!, \quad E_{n-1} = n! - D_n, \quad n = 1, 2, \ldots.$$

5. Show that the rank number $R_{n,k}$ satisfies the recurrence relation

$$R_{n,k} = (n-k)R_{n-1,k} + (k-1)R_{n-1,k-1},$$

for $k = 2, 3, \ldots, n-1$, $n = 3, 4, \ldots$.

6. Let $E_{n,r}$ be the number of permutations of the set $\{1, 2, \ldots, n\}$ without a fixed point in r predetermined positions. Show that

$$E_{n,r} = \sum_{j=0}^{r}(-1)^j\binom{r}{j}(n-j)!.$$

Derive the recurrence relations

$$E_{n,r} = E_{n,r-1} - E_{n-1,r-1}, \; r = 1, 2, \ldots, n, \; E_{n,0} = n!, \; E_{n,n} = D_n,$$

$$E_{n,r} = (n-r)E_{n-1,r} + rE_{n-1,r-1}, \; r = 1, 2, \ldots, n-1, \quad n = 2, 3, \ldots .$$

and, if $R_{n,k}$ is the rank number, conclude that

$$E_{n,r} = R_{n+1,r+1}.$$

7. (*Continuation*). Let $E_{n,r,k}$ be the number of permutations of the set $\{1, 2, \ldots, n\}$ with k fixed points in r predetermined positions. Show that

$$E_{n,r,k} = \binom{r}{k} \sum_{j=0}^{r-k} (-1)^j \binom{r-k}{j} (n-k-j)!$$

and conclude that

$$E_{n,r,1} = r R_{n,r}.$$

8. (*Continuation*). Show that

$$E_{n,r,k} = (n-k)E_{n-1,r,k} + (k+1)E_{n-1,k+1}.$$

9. (*Continuation*). Show that

$$E_{n,r,k} = \binom{r}{k} \sum_{j=0}^{n-r} \binom{n-r}{j} D_{n-k-j},$$

where D_n is the n-subfactorial and, using the recurrence relation

$$D_n = nD_{n-1} + (-1)^n,$$

conclude that

$$E_{n,r,k} = (n-r)E_{n-1,r,k} + rE_{n-1,r-1,k}.$$

10. *Double coincidences.* Let Q_n be the number of ways of placing n letters and n invoices in n envelopes, one letter and one invoice in each, so that no envelope contains its corresponding letter and invoice. Show that

$$Q_n = n! \sum_{k=0}^{n} (-1)^n \frac{(n-k)!}{k!}.$$

11. (*Continuation*). Show that

$$Q_n = n^2 Q_{n-1} + n(n-1)Q_{n-2} + (-1)^n, \; n = 2, 3, \ldots, \; Q_0 = 1, \; Q_1 = 0.$$

12. (*Continuation*). Let $Q_{n,k}$ be the number of ways of placing n letters and n invoices in n envelopes, one letter and one invoice in each, so that each of k envelopes contains its corresponding letter and invoice. Show that

$$Q_{n,k} = \frac{n!}{k!} \sum_{j=0}^{n-k} (-1)^j \frac{(n-k-j)!}{j!}.$$

13. *Multiple coincidences.* Consider r urns, each containing n balls numbered from 1 to n. Assume that a ball is drawn from each urn and, without replacement, this multiple drawing is repeated until all the balls are drawn. The drawing of the j-th ball from each of the urns at the j-th r-tuple trial constitutes an r-tuple coincidence.

(a) Let $B_{n,r}$ be the number of r-tuple drawings without an r-tuple coincidence. Show that

$$B_{n,r} = \sum_{j=0}^{n} (-1)^{n-j} \binom{n}{j} (j!)^r.$$

(b) If $B_{n,r,k}$ is the number of r-tuple drawings with k r-tuple coincidences, show that

$$B_{n,r,k} = \binom{n}{k} B_{n-k,r}.$$

14. (*Continuation*). Show that the number $B_{n,r}$ satisfies the relations

$$B_{n,r} = n \sum_{s=0}^{n-1} (-1)^{n-s-1} \binom{n-1}{s} (s!)^r (s+1)^{r-1} + (-1)^n,$$

$$B_{n,r} = \sum_{s=0}^{n-1} (-1)^{n-s-1} \binom{n-1}{s} (s!)^r (s+1)^r - B_{n-1,r}$$

and proving that

$$\sum_{j=0}^{n-s-1} (-1)^j \binom{n-s-1}{j} j! S(k,j;n-j) = (s+1)^k,$$

where

$$S(k,j;u) = \frac{1}{j!} \sum_{i=0}^{j} (-1)^{j-1} \binom{j}{i} (i+u)^k,$$

conclude the recurrence relations

$$B_{n,r} = \sum_{j=1}^{r} (n)_j S(r-1,j-1;n-j+1) B_{n-j,r} + (-1)^n,$$

$$B_{n,r} = (n^r - 1)B_{n-1,r} + \sum_{j=2}^{n+1}(n-1)_{j-1}S(r,j-1;n-j+1)B_{n-j,r}.$$

The number $S(k,j;u)$ is called *non-central Stirling number of the second kind*.

15. *A generalization of the problem of coincidences.* Consider an urn containing s like series of balls with n balls, numbered from 1 to n, in each series. Assume that all the sn balls are drawn one after the other without replacement.

(a) Let $G_{n,s}$ be the number of drawings without a coincidence. Show that

$$G_{n,s} = \sum_{j=0}^{n}(-1)^j \binom{n}{j}s^j(sn-j)!.$$

(b) If $G_{n,k,s}$ is the number of drawings with k coincidences, show that

$$G_{n,k,s} = \binom{n}{k}\sum_{j=0}^{n-k}(-1)^j\binom{n-k}{j}s^{k+j}(sn-k-j)!.$$

16. (*Continuation*). Assume that the successive drawing of balls is terminated with the appearance of the first coincidence. If $H_{n,k,s}$ is the number of drawings with a given duration of k trials, show that

$$H_{n,k,s} = \sum_{j=0}^{k-1}(-1)^j\binom{k-1}{j}s^{j+1}(sn-j-1)!.$$

17. Let $S_{n,k}$ be the number of permutations of the set $\{1,2,\ldots,n\}$ with k successions. Show that

$$S_{n+1,k} = (k+1)S_{n,k+1} + (n-k)S_{n,k} + S_{n,k-1},$$

for $k = 1,2,\ldots,n$, $n = 2,3,\ldots$, with $S_{1,0} = 1$, $S_{n,n-1} = 1$, $S_{n,k} = 0$, $k \geq n$.

18. Let $C_{n,k}$ be the number of circular permutations of the set $\{1,2,\ldots, n\}$ with k successions. Show that

$$C_{n,k} + C_{n-1,k} - C_{n-1,k-1} = D_{n-1,k},$$

where $D_{n,k}$ is the coincidence number, and conclude that

$$C_{n,k} = (n-2)C_{n-1,k} + C_{n-1,k-1} + (n-1)C_{n-2,k}$$
$$+(n-1)C_{n-2,k-1} + (-1)^{n-k-1}\binom{n-1}{k}.$$

19. *Permutations with a given number of transpositions.* A permutation can always be represented as a product of circular permutations. A circular permutation of length 2 is called a *transposition.* Let $T_{n,k}$ be the number of permutations of the set $\{1, 2, \ldots, n\}$ which can be written as a product of k transpositions. Show that

$$T_{n,k} = \frac{n!(-1)^k}{k!2^k} \sum_{j=0}^{m-k} \frac{(-1)^j}{j!2^j}, \quad m = [n/2].$$

20. (*Continuation*). Show that

$$T_{2r,k} = 2rT_{2r-1,k} + (-1)^r \frac{(2r)!}{k!(r-k)!2^k},$$

and

$$T_{2r+1,k} = (2r+1)T_{2r,k}.$$

Chapter 6

GENERATING FUNCTIONS

6.1 INTRODUCTION

The generating functions constitute an important means for a unified treatment of combinatorial and probabilistic problems. P. S. Laplace, their inventor, has first introduced them in the form of power series. Later, generating functions of a more general than the power series form have been used. Moreover, for their combinatorial uses, they are to be regarded, following E. T. Bell, as tools in the study of an algebra of sequences; thus, despite all appearances they belong to algebra and not to analysis.

A generating function expanded generates a sequence of numbers a_k, $k = 0, 1, \ldots$, the coefficients of the expansion, which in combinatorics express the number of elements of a certain finite set. In this respect its study is a natural complement to the study of these numbers. In the next section, a justification of the use of generating functions in the derivation of the number of elements of certain finite sets is provided by a brief presentation of two simple combinatorial problems preceding the formal definitions. After introducing the basic forms of generating functions, several examples illustrating their use are discussed. Also, the general form of a generating function is introduced and exemplified. In order to facilitate the derivation and expansion of generating functions, the Maclaurin, Newton and Lagrange series are briefly discussed.

A general method for constructing generating functions for combinations and permutations is presented and illustrated in a separate section. The factorial moments of a sequence a_k, $k = 0, 1, \ldots$, of combinatorial numbers (or probabilities or masses) are closely connected with the general term of the inclusion and exclusion formula, which is one of the basic counting expressions. The inclusion of a section on the moment generating functions is thus justified. In ending this chapter, bivariate and multivariate generating functions are examined.

6.2   UNIVARIATE GENERATING FUNCTIONS

6.2.1   Definitions and basic properties

Before proceeding to the formal definition and study of generating functions, let us examine two simple combinatorial problems that suggest the use of a certain form of generating functions. The first problem is concerned with the derivation of the number of k-combinations of n. Let us consider first three elements labelled x_1, x_2 and x_3 and form the algebraic product

$$(1 + x_1 t)(1 + x_2 t)(1 + x_3 t),$$

which, after executing the multiplication and arranging the terms in ascending powers of t, is equal to

$$1 + (x_1 + x_2 + x_3)t + (x_1 x_2 + x_1 x_3 + x_2 x_3)t^2 + (x_1 x_2 x_3)t^3.$$

Introducing the *elementary symmetric functions* with respect to the three variables x_1, x_2 and x_3,

$$a_0 = a_0(x_1, x_2, x_3) = 1, \quad a_1 = a_1(x_1, x_2, x_3) = x_1 + x_2 + x_3,$$

$$a_2 = a_2(x_1, x_2, x_3) = x_1 x_2 + x_1 x_3 + x_2 x_3, \quad a_3 = a_3(x_1, x_2, x_3) = x_1 x_2 x_3,$$

it follows that

$$\prod_{i=1}^{3}(1 + x_i t) = \sum_{k=0}^{3} a_k(x_1, x_2, x_3)t^k.$$

Note that the function $a_k = a_k(x_1, x_2, x_3)$, $k = 1, 2, 3$, which is the coefficient of t^k, contains one term for each k-combination of the three elements. Hence, the number $C(3, k)$ of the k-combinations of 3 is given by the value of $a_k = a_k(x_1, x_2, x_3)$ at the point $(1, 1, 1)$, that is,

$$C(3, k) = a_k(1, 1, 1), \quad k = 0, 1, 2, 3,$$

and so, setting $x_i = 1$, $i = 1, 2, 3$, we conclude that

$$(1 + t)^3 = \sum_{k=0}^{3} a_k(1, 1, 1)t^k = \sum_{k=0}^{3} C(3, k)t^k.$$

Similarly, in the case of n distinct elements labelled $x_1, x_2, \dots, x_n$, we obtain

$$\prod_{i=1}^{n}(1 + x_i t) = \sum_{k=0}^{n} a_k(x_1, x_2, \dots, x_n)t^k, \tag{6.1}$$

with $a_k = a_k(x_1, x_2, \ldots, x_n)$, $k = 0, 1, \ldots, n$, the *elementary symmetric functions*, with respect to the n variables $x_1, x_2, \ldots, x_n$, which are defined by the sum

$$a_k = a_k(x_1, x_2, \ldots, x_n) = \sum x_{i_1} x_{i_2} \cdots x_{i_k}, \quad k = 0, 1, \ldots, n, \qquad (6.2)$$

where the summation is extended over all k-element subsets $\{i_1, i_2, \ldots, i_k\}$ of the set of n indices $\{1, 2, \ldots, n\}$. It is clear that the symmetric function $a_k(x_1, x_2, \ldots, x_n)$ contains one term for each k-combination of n. Hence the number $C(n, k)$ of the k-combinations of n is given by the value of $a_k(x_1, x_2, \ldots, x_n)$ at the point $(1, 1, \ldots, 1)$,

$$C(n, k) = a_k(1, 1, \ldots, 1), \quad k = 0, 1, \ldots, n.$$

Consequently, setting in (6.1) $x_i = 1$, $i = 1, 2, \ldots, n$, we conclude that

$$(1 + t)^n = \sum_{k=0}^{n} a_k(1, 1, \ldots, 1) t^k = \sum_{k=0}^{n} C(n, k) t^k. \qquad (6.3)$$

The function $A(t) = (1 + t)^n$, which generates, in the sense of (6.3), the sequence of the numbers $C(n, k) = a_k(1, 1, \ldots, 1)$, $k = 0, 1, \ldots, n$, may be called *enumerating generating function of combinations* of n distinct elements.

Let us now examine, from the same viewpoint, the problem of the derivation of the number of the k-permutations of n. Since the products $x_1 x_2$ and $x_2 x_1$ are not distinguished in the usual algebraic product, it is not possible to define a function analogous to (6.2) containing one term for each k-permutation of n and, consequently, it is impossible to define a function similar to (6.1). Nevertheless, an enumerating generating function of the number $P(n, k)$ of the k-permutations of n is easy to find. Indeed, from (6.3), on using the relation $C(n, k) = P(n, k)/k!$, it follows that

$$(1 + t)^n = \sum_{k=0}^{n} P(n, k) \frac{t^k}{k!}. \qquad (6.4)$$

The function $E(t) = (1 + t)^n$, which generates, in the sense of (6.4), the sequence of the number $P(n, k)$, $k = 0, 1, \ldots, n$, may be called *enumerating generating function of permutations* of n distinct elements.

The preceding discussion of the enumerating generating functions for combinations and permutations suggests the study of the following generating functions.

DEFINITION 6.1   Let a_k, $k = 0, 1, \ldots$, be a sequence of real numbers. The *sum*

$$A(t) = \sum_{k=0}^{\infty} a_k t^k \qquad (6.5)$$

is called (ordinary) generating function, while the sum

$$E(t) = \sum_{k=0}^{\infty} a_k \frac{t^k}{k!} \tag{6.6}$$

is called exponential generating function of the sequence a_k, $k = 0, 1, \ldots$.

The existence of the generating functions $A(t)$ and $E(t)$ requires the absolute convergence of the corresponding series for $t_0 < t < t_1$, with t_0 and t_1 real numbers. The algebra of the corresponding sequences of the (ordinary) generating functions is known as Cauchy algebra and of the exponential generating functions is known as Blissard or symbolic calculus. It is not necessary to burden this presentation of the generating functions with formal definitions and study of the relative algebras. Note only that the equality of two generating functions entails the equality of the coefficients of the powers of t of the same degree.

The operations of summation and multiplication, as well as differentiation and integration with respect to t, of the generating functions can be used to derive relations between the corresponding coefficients. In order to clarify this, let us consider the generating functions

$$A(t) = \sum_{k=0}^{\infty} a_k t^k, \quad B(t) = \sum_{k=0}^{\infty} b_k t^k, \quad C(t) = \sum_{k=0}^{\infty} c_k t^k.$$

A direct consequence of the equality of generating functions in the following pairs of relations is that the one relation entails the other:

(a) *Sum*

$$C(t) = A(t) + B(t) \text{ iff } c_k = a_k + b_k, \quad k = 0, 1, \ldots. \tag{6.7}$$

(b) *Product*

$$C(t) = A(t)B(t) \text{ iff } c_k = \sum_{j=0}^{k} a_j b_{k-j}, \quad k = 0, 1, \ldots. \tag{6.8}$$

Note that relations (6.7) and (6.8) define the sum and the product of sequences in Cauchy algebra. The sequence c_k, $k = 0, 1, \ldots$, defined by (6.8) is called *convolution* of the sequences a_k, $k = 0, 1, \ldots$, and b_k, $k = 0, 1, \ldots$. Several applications of the product of generating functions (6.8) may be obtained by specifying one or two of the generating functions. Such an application is given in the following example.

Example 6.1 Sum of terms of an arithmetic progression
   The sum
$$s_k = 1 + 2 + \cdots + k, \quad k = 1, 2, \ldots$$

may be evaluated by using (6.8) as follows. Note first that this sum is of the form

$$c_k = a_0 b_k + a_1 b_{k-1} + \cdots + a_k b_0, \quad k = 0, 1, \ldots ,$$

where

$$c_0 = 0, \quad c_k \equiv s_k = 1 + 2 + \cdots + k, \quad k = 1, 2, \ldots ,$$

$$a_k = k, \quad b_k = 1, \quad k = 0, 1, \ldots .$$

Thus, according to (6.8),

$$C(t) = A(t) B(t)$$

and the evaluation of the sum $s_k \equiv c_k$ of an arithmetic progression reduces to the derivation of the generating functions $A(t)$ and $B(t)$. By the geometric series, it follows that

$$B(t) = \sum_{k=0}^{\infty} b_k t^k = \sum_{k=0}^{\infty} t^k = (1-t)^{-1}, \quad |t| < 1.$$

In order to evaluate the generating function

$$A(t) = \sum_{k=0}^{\infty} a_k t^k = \sum_{k=1}^{\infty} k t^k,$$

we differentiate the relation

$$\sum_{k=0}^{\infty} t^k = (1-t)^{-1}$$

and get

$$\sum_{k=1}^{\infty} k t^{k-1} = (1-t)^{-2}.$$

Therefore,

$$A(t) = t \sum_{k=1}^{\infty} k t^{k-1} = t(1-t)^{-2}$$

and

$$C(t) = A(t) B(t) = t(1-t)^{-3}.$$

The generating function $C(t)$, on using Newton's negative binomial formula (see Theorem 3.4), with $n = 3$, is expanded into powers of t as

$$C(t) = t \sum_{r=0}^{\infty} \binom{3+r-1}{r} t^r = \sum_{r=0}^{\infty} \binom{2+r}{2} t^{r+1} = \sum_{k=1}^{\infty} \binom{k+1}{2} t^k.$$

Thus

$$s_k \equiv c_k = \binom{k+1}{2} = \frac{k(k+1)}{2}, \quad k = 1, 2, \dots . \qquad \square$$

Example 6.2 The Catalan numbers

Consider a product of n numbers in a specified order, $y = x_1 x_2 \cdots x_n$, and let C_n be the number of ways of evaluating this product by successive multiplication of exactly two numbers at each step and without altering the initial order. Note that $C_2 = 1 : x_1 x_2$, $C_3 = 2 : x_1(x_2 x_3)$, $(x_1 x_2)x_3$, $C_4 = 5 :$ $x_1(x_2(x_3 x_4))$, $x_1((x_2 x_3)x_4)$, $(x_1 x_2)(x_3 x_4)$, $(x_1(x_2 x_3))x_4$, $((x_1 x_2)x_3)x_4$. In order to determine a recurrence relation for the numbers C_n, $n = 2, 3, \dots$, note that the last multiplication, with which the evaluation of the product is completed, is $y = y_k y_{n-k}$, where y_k is the result of the product $x_1 x_2 \cdots x_k$ and y_{n-k} is the result of the product $x_{k+1} x_{k+2} \cdots x_n$, with $k = 1, 2, \dots, n-1$. The number of ways of evaluating the product $y_k = x_1 x_2 \cdots x_k$ is C_k, while the number of ways of evaluating the product $y_{n-k} = x_{k+1} x_{k+2} \cdots x_n$ is C_{n-k}, $k = 1, 2, \dots, n-1$. According to the multiplication principle, the number of ways of evaluating the products y_k and y_{n-k} is $C_k C_{n-k}$, $k = 1, 2, \dots, n-1$. Thus, by the addition principle, the total number C_n of ways of evaluating the product y is equal to

$$C_n = C_1 C_{n-1} + C_2 C_{n-2} + \cdots + C_{n-1} C_1, \quad n = 2, 3, \dots .$$

In the particular case of $n = 2$, it holds $C_2 = C_1^2$, which, by virtue of $C_2 = 1$, entails $C_1 = 1$ and so the initial conditions are

$$C_1 = 1, \quad C_2 = 1.$$

Using the recurrence relation and the initial conditions, we construct Table 6.1 of Catalan numbers C_n for $n = 1, 2, \dots, 12$.

Table 6.1   Catalan Numbers

n	1	2	3	4	5	6	7	8	9	10	11	12
C_n	1	1	2	5	14	42	132	429	1430	4862	16796	58786

In order to determine the generating function

$$C(t) = \sum_{n=1}^{\infty} C_n t^n,$$

note that the right-hand side of the recurrence relation is the convolution of the sequence $a_n = C_{n+1}$, $n = 0, 1, \dots$ with itself $b_n = C_{n+1}$, $n = 0, 1, \dots$. Thus, multiplying the recurrence relation by t^n and summing for $n = 2, 3, \ldots$, by virtue of (6.8), we get

$$\sum_{n=2}^{\infty} C_n t^n = \sum_{n=2}^{\infty} \left(\sum_{k=1}^{n-1} C_k C_{n-k} \right) t^n = \left(\sum_{k=1}^{\infty} C_k t^k \right) \left(\sum_{j=1}^{\infty} C_j t^j \right)$$

and so

$$C(t) - t = [C(t)]^2.$$

Therefore

$$[C(t)]^2 - C(t) + t = 0$$

and

$$C_1(t) = \frac{1}{2} \left(1 - \sqrt{1 - 4t} \right), \quad C_2(t) = \frac{1}{2} \left(1 + \sqrt{1 - 4t} \right).$$

Since $C(0) = 0$, the solution $C_2(t)$ is rejected because $C_2(0) = 1$ and so

$$C(t) = \frac{1}{2} \left(1 - \sqrt{1 - 4t} \right).$$

This generating function, on using Newton's general binomial formula (see Theorem 3.6), is expanded into powers of t as

$$C(t) = \frac{1}{2} - \frac{1}{2}(1 - 4t)^{1/2} = \sum_{n=1}^{\infty} (-1)^{n-1} 2^{2n-1} \binom{1/2}{n} t^n.$$

Therefore

$$C_n = (-1)^{n-1} 2^{2n-1} \binom{1/2}{n} = (-1)^{n-1} 2^{2n-1} \frac{(1/2)_n}{n!}$$

$$= \frac{1 \cdot 3 \cdots (2n - 3)}{n!} \cdot 2^{n-1} \cdot \frac{2 \cdot 4 \cdots (2n - 2)}{2 \cdot 4 \cdot (2n - 2)} = \frac{(2n - 2)!}{n!(n - 1)!}$$

and

$$C_n = \frac{1}{n} \binom{2n - 2}{n - 1}, \quad n = 1, 2, \ldots . \qquad \Box$$

Let us now consider the exponential generating functions

$$E(t) = \sum_{k=0}^{\infty} a_k \frac{t^k}{k!}, \quad F(t) = \sum_{k=0}^{\infty} b_k \frac{t^k}{k!}, \quad G(t) = \sum_{k=0}^{\infty} c_k \frac{t^k}{k!}.$$

In this case we have
(a) *Sum*

$$G(t) = E(t) + F(t) \quad \text{iff} \quad c_k = a_k + b_k, \quad k = 0, 1, \ldots . \qquad (6.9)$$

(b) *Product*

$$G(t) = E(t)F(t) \text{ iff } c_k = \sum_{j=0}^{k} \binom{k}{j} a_j b_{k-j}, \quad k = 0, 1, \ldots . \quad (6.10)$$

Note that (6.9) is the same as its corresponding relation (6.7) for the ordinary generating functions, while (6.10) differs from its corresponding relation (6.8) through the presence of the binomial coefficients. This difference suggests a basic device of the *symbolic calculus*, which may informally be presented as follows. The subscripts of a sequence a_k, $k = 0, 1, \ldots$, are replaced by exponents transforming it into the sequence a^k, $k = 0, 1, \ldots$. Then a^k is treated as an ordinary power (with a^0 not necessarily equal to 1 and a^1 not necessarily equal to a) until symbolic formulae are transformed back into ordinary formulae by transforming a^k, $k = 0, 1, \ldots$, into a_k, $k = 0, 1, \ldots$. A reminder of such a transformation, of the form $a^k \equiv a_k$, should accompany a symbolic formula. Thus, with this device, the second relation of (6.10) is written as

$$c_k = (a + b)^k, \quad a^j \equiv a_j, \quad b^r \equiv b_r \quad (6.11)$$

and the exponential generating functions take the form

$$E(t) = e^{at}, \quad a^k \equiv a_k, \quad F(t) = e^{bt}, \quad b^k \equiv b_k, \quad G(t) = e^{ct}, \quad c^k \equiv c_k$$

and the first relation of (6.10) becomes

$$e^{ct} = e^{at} e^{bt} = e^{(a+b)t}, \quad c^k \equiv c_k, \quad a^k \equiv a_k, \quad b^k \equiv b_k.$$

Expanding its first and last members and equating the coefficients of t^k, (6.11) is deduced.

Note that the derivative of $E(t) = e^{at}$, $a^k \equiv a_k$, with respect to t, gives

$$\frac{d}{dt} E(t) = a e^{at}, \quad a^k \equiv a_k.$$

This expression, expanded into powers of t, takes the ordinary form

$$\frac{d}{dt} E(t) = a \sum_{k=0}^{\infty} a^k \frac{t^k}{k!} = \sum_{k=0}^{\infty} a^{k+1} \frac{t^k}{k!} = \sum_{k=0}^{\infty} a_{k+1} \frac{t^k}{k!} = \sum_{r=1}^{\infty} a_r \frac{t^{r-1}}{(r-1)!}.$$

Applications of the symbolic calculus are given in the following examples.

Example 6.3 Exponential Bell numbers
Let

$$E(t) = \sum_{k=0}^{\infty} a_k \frac{t^k}{k!}, \quad F(t) = \sum_{k=0}^{\infty} b_k \frac{t^k}{k!},$$

and

$$F(t) = \exp\{E(t) - E(0)\}. \tag{6.12}$$

The numbers b_k, $k = 0, 1, \ldots$, the exponential generating function of which is connected with the generating function of the numbers, a_k, $k = 0, 1, \ldots$, by the exponential relation (6.12) are called *exponential Bell numbers*.

The derivation of a recurrence relation for the numbers b_k, $k = 0, 1, \ldots$, can be efficiently carried out by using the symbolic calculus as follows. The defining relation (6.12) is written symbolically as

$$e^{bt} = \exp(e^{at} - a_0), \quad b^k \equiv b_k, \quad a^r \equiv a_r.$$

Differentiating it with respect to t, we get

$$be^{bt} = ae^{at}\exp(e^{at} - a_0) = ae^{at}e^{bt} = ae^{(a+b)t}, \quad b^k \equiv b_k, \quad a^r \equiv a_r.$$

Expanding the first and last member of this relation into powers of t,

$$\sum_{k=0}^{\infty} b^{k+1}\frac{t^k}{k!} = \sum_{k=0}^{\infty} a(a+b)^k\frac{t^k}{k!}, \quad b^k \equiv b_k, \quad a^r \equiv a_r,$$

and equating the coefficients of $t^k/k!$ in both sides, we conclude that

$$b^{k+1} = a(a+b)^k = \sum_{j=0}^{k}\binom{k}{j}a^{j+1}b^{k-j}, \quad b^k \equiv b_k, \quad a^r \equiv a_r$$

and

$$b_{k+1} = \sum_{j=0}^{k}\binom{k}{j}a_{j+1}b_{k-j}, \quad k = 0, 1, \ldots,$$

which is the required recurrence relation. In the particular case of $a_k = 1$, $k = 0, 1, \ldots$, $b_0 = 1$, we have

$$b_{k+1} = \sum_{j=0}^{k}\binom{k}{j}b_{k-j}, \quad k = 0, 1, \ldots, \quad b_0 = 1.$$

The *Bell numbers* b_k, $k = 0, 1, \ldots$, in this particular case, express the number of partitions of a finite set of k elements (see Exercise 2.42). ☐

Example 6.4 **Exponential generating function of subfactorials**

Let D_n be the number of permutations $(j_1, j_2, \ldots, j_n)$ of the set $\{1, 2, \ldots, n\}$ without a fixed point, that is, $j_r \neq r$, $r = 1, 2, \ldots, n$. D_n is *n-subfactorial* (see Section 5.2). Let us determine the exponential generating function

$$D(t) = \sum_{n=0}^{\infty} D_n\frac{t^n}{n!} = e^{Dt}, \quad D^n \equiv D_n.$$

Note that the n-subfactorial D_n is connected with the coincidence number $D_{n,k}$ by the relation (see Theorem 5.3)

$$D_{n,k} = \binom{n}{k} D_{n-k}, \quad k = 0, 1, \ldots, n, \quad n = 0, 1, \ldots.$$

Consequently,

$$P_n \equiv \sum_{k=0}^{n} D_{n,k} = \sum_{k=0}^{n} \binom{n}{k} D_{n-k} = (D+1)^n, \quad D^k \equiv D_k$$

and if

$$P(t) = \sum_{n=0}^{\infty} P_n \frac{t^n}{n!} = e^{Pt}, \quad P^n \equiv P_n,$$

then

$$e^{Pt} = \sum_{n=0}^{\infty} (D+1)^n \frac{t^n}{n!} = e^{(D+1)t} = e^{Dt} e^t.$$

Therefore

$$D(t) = e^{Dt} = e^{Pt} e^{-t} = P(t) e^{-t}.$$

But

$$P_n = \sum_{k=0}^{n} D_{n,k} = n!, \quad P(t) = \sum_{n=0}^{\infty} P_n \frac{t^n}{n!} = \sum_{n=0}^{\infty} t^n = (1-t)^{-1}$$

and so

$$D(t) = (1-t)^{-1} e^{-t}. \quad \Box$$

The ordinary and the exponential generating functions are particular classes of the following general form of a generating function.

DEFINITION 6.2   *Let a_k, $k = 0, 1, \ldots$, be a sequence of real numbers and $g_k(t)$, $k = 0, 1, \ldots$, be a sequence of linearly independent real functions. The sum*

$$G(t) = \sum_{k=0}^{\infty} a_k g_k(t) \tag{6.13}$$

is called generating function of the sequence a_k, $k = 0, 1, \ldots$, with respect to the sequence $g_k(t)$, $k = 0, 1, \ldots$.

The existence of the generating function $G(t)$ requires the absolute convergence of the series for $t_0 < t < t_1$, with t_0 and t_1 real numbers. The assumption of the linear independence of the sequence $g_k(t)$, $k = 0, 1, \ldots$ assures the uniqueness of the definition of a generating function. The general form (6.13) of a generating function directly refers to algebra where it belongs. The most frequently used sequences of linearly independent functions in combinatorics and probability theory are $g_k(t) = t^k$, $k = 0, 1, \ldots$ and $g_k(t) = t^k/k!$, $k = 0, 1, \ldots$, which correspond to the ordinary and the exponential generating functions, respectively. The sequences $g_k(t) = (t)_k$, $k = 0, 1, \ldots$ and $g_k(t) = \binom{t}{k}$, $k = 0, 1, \ldots$ are also widely used. Their use requires particular names for the corresponding generating functions. Specifically, the following definitions, which are particular cases of Definition 6.2, are introduced.

DEFINITION 6.3   *Let a_k, $k = 0, 1, \ldots$, be a sequence of real numbers. The sum*

$$F(t) = \sum_{k=0}^{\infty} a_k (t)_k \qquad (6.14)$$

is called factorial generating function, while the sum

$$B(t) = \sum_{k=0}^{\infty} a_k \binom{t}{k} \qquad (6.15)$$

is called binomial generating function of the sequence a_k, $k = 0, 1, \ldots$.

Example 6.5 Stirling numbers of the second kind
Let $S(n, k)$, $k = 0, 1, \ldots, n$, be the number of partitions of a finite set of n elements into k subsets. Determine the factorial generating function

$$g_n(t) = \sum_{k=0}^{n} S(n, k)(t)_k.$$

Note first that, to each partition $\{A_1, A_2, \ldots, A_k\}$ of a finite set W_n of n elements into k subsets, there correspond $k!$ divisions $(A_{j_1}, A_{j_2}, \ldots, A_{j_k})$ of W_n into k nonempty subsets, which are obtained from it by permuting the k sets in all possible ways. Hence $W(n, k) = k!S(n, k)$ is the number of divisions of a finite set of n elements into k nonempty subsets. Further, the number of divisions of a finite set of n elements into r subsets among which k are nonempty is given by (see Example 4.9)

$$\binom{r}{k} W(n, k) = (r)_k S(n, k), \quad k = 0, 1, \ldots, r, \quad r = 0, 1, \ldots, n,$$

while, by Corollary 3.3, the total number of divisions of a finite set of n elements into r subsets equals r^n. Consequently, by the addition principle,

$$\sum_{k=0}^{n} S(n,k)(r)_k = r^n.$$

Since this formula is an identity for all $r = 0, 1, \ldots, n$, it follows that

$$g_n(t) = \sum_{k=0}^{n} S(n,k)(t)_k = t^n$$

holds for all real numbers t. The numbers $S(n,k)$ are called *Stirling numbers of the second kind.*   ▢

6.2.2   Power, factorial and Lagrange series

The derivation of certain generating functions is facilitated by the use of some well-known expansions of functions. Every function expanded into a power or a factorial series can be considered as a generating function. The Maclaurin, Newton and Lagrange series are presented (without proof) in this section.

Maclaurin power series

$$f(t) = \sum_{k=0}^{\infty} \left[\frac{1}{k!} \frac{d^k}{dt^k} f(t) \right]_{t=0} \cdot t^k = \sum_{k=0}^{\infty} \left[\frac{d^k}{dt^k} f(t) \right]_{t=0} \cdot \frac{t^k}{k!}. \tag{6.16}$$

Note that, with

$$a_k = \left[\frac{1}{k!} \frac{d^k}{dt^k} f(t) \right]_{t=0}, \quad k = 0, 1, \ldots,$$

(6.16) is of the form (6.5),

$$f(t) = \sum_{k=0}^{\infty} a_k t^k, \tag{6.17}$$

while for

$$b_k = \left[\frac{d^k}{dt^k} f(t) \right]_{t=0}, \quad k = 0, 1, \ldots,$$

(6.16) is of the form (6.6),

$$f(t) = \sum_{k=0}^{\infty} b_k \frac{t^k}{k!}. \tag{6.18}$$

Example 6.6

Let us consider the function $f(t) = e^t$. Its derivatives are

$$\frac{d^k}{dt^k} e^t = e^t, \quad k = 0, 1, \ldots .$$

Thus

$$b_k = \left[\frac{d^k}{dt^k} e^t \right]_{t=0} = 1, \quad k = 0, 1, \ldots$$

and, by (6.18), we find the expansion of the exponential function:

$$e^t = \sum_{k=0}^{\infty} \frac{t^k}{k!}.$$

The derivatives of the function $f(t) = -\log(1 - t)$ are

$$\frac{d^k}{dt^k} \{ -\log(1 - t) \} = (k - 1)!(1 - t)^{-k}, \quad k = 1, 2, \ldots .$$

Thus

$$a_0 = f(0) = -\log 1 = 0,$$

$$a_k = \left[\frac{1}{k!} \frac{d^k}{dt^k} \{ -\log(1 - t) \} \right]_{t=0} = \frac{1}{k}, \quad k = 1, 2, \ldots$$

and, by (6.17), we get the expansion of the logarithm:

$$-\log(1 - t) = \sum_{k=1}^{\infty} \frac{t^k}{k}.$$

Setting $u = -t$, this expansion can also be written as

$$\log(1 + u) = \sum_{k=1}^{\infty} (-1)^{k-1} \frac{u^k}{k}.$$

The derivatives of the function $f(t) = (1 + t)^x$ are

$$\frac{d^k}{dt^k} (1 + t)^x = (x)_k (1 + t)^{k-x}, \quad k = 0, 1, \ldots .$$

Thus

$$a_k = \left[\frac{1}{k!} \frac{d^k}{dt^k} (1 + t)^x \right]_{t=0} = \frac{(x)_k}{k!} = \binom{x}{k}, \quad k = 0, 1, \ldots$$

and, by (6.17), we deduce the general binomial formula:

$$(1 + t)^x = \sum_{k=0}^{\infty} \binom{x}{k} t^k. \quad \square$$

Example 6.7 Stirling numbers of the first kind

Expanding the factorial of t of degree n, $(t)_n$, into a Maclaurin's series and putting

$$s(n,k) = \left[\frac{1}{k!} \frac{d^k}{dt^k} (t)_n \right]_{t=0}, \quad k = 0, 1, \ldots, n,$$

we get

$$(t)_n = \sum_{k=0}^{n} s(n,k) t^k.$$

The coefficients $s(n,k)$ are called *Stirling numbers of the first kind.* Note that the sequence $s(n,k)$, $k = 0, 1, \ldots, n$, for fixed n, has generating function $g_n(t) = (t)_n$. $\square$

Newton factorial series

$$f(t) = \sum_{k=0}^{\infty} \left[\frac{1}{k!} \Delta^k f(t) \right]_{t=0} \cdot (t)_k = \sum_{k=0}^{\infty} [\Delta^k f(t)]_{t=0} \cdot \binom{t}{k}, \qquad (6.19)$$

where $\Delta^k f(t)$ denotes the difference of order k of the function $f(t)$.

Note that, with

$$a_k = \left[\frac{1}{k!} \Delta^k f(t) \right]_{t=0}, \quad k = 0, 1, \ldots,$$

(6.19) is of the form (6.14),

$$f(t) = \sum_{k=0}^{\infty} a_k (t)_k. \qquad (6.20)$$

Setting

$$b_k = [\Delta^k f(t)]_{t=0}, \quad k = 0, 1, \ldots,$$

(6.19) is of the form (6.15),

$$f(t) = \sum_{k=0}^{\infty} b_k \binom{t}{k}. \qquad (6.21)$$

Example 6.8

Let us consider the function $f(t) = (u+t)_n$, where n is a positive integer and u is a fixed real number. Then

$$\Delta(u+t)_n = (u+t+1)_n - (u+t)_n = n(u+t)_{n-1}$$

and, in general,

$$\Delta^k (u+t)_n = (n)_k (u+t)_{n-k}, \quad k = 0, 1, \ldots .$$

Thus

$$a_k = \left[\frac{1}{k!} \Delta^k (u+t)_n \right]_{t=0} = \binom{n}{k} (u)_{n-k}, \quad k = 0, 1, \ldots$$

and so, from (6.20), we deduce Vandermonde's formula:

$$(u+t)_n = \sum_{k=0}^{n} \binom{n}{k} (u)_{n-k} (t)_k. \quad \Box$$

Lagrange series

The Lagrange series appears in two basic forms:

$$f(t) = f(0) + \sum_{k=1}^{\infty} \left[\frac{d^{k-1}}{dt^{k-1}} \left(g^k(t) \frac{d}{dt} f(t) \right) \right]_{t=0} \frac{u^k}{k!}, \quad u = \frac{t}{g(t)} \qquad (6.22)$$

and

$$f(t) \left(1 - u \frac{d}{dt} g(t) \right)^{-1} = \sum_{k=0}^{\infty} \left[\frac{d^k}{dt^k} \left(g^k(t) f(t) \right) \right]_{t=0} \frac{u^k}{k!}, \quad u = \frac{t}{g(t)}. \quad (6.23)$$

Note that these two series constitute a basic means for the solution of equations of implicit form and for the inversion of series. They are also used as generating functions with variables defined implicitly. More specifically, with

$$a_0 = f(0), \quad a_k = \left[\frac{d^{k-1}}{dt^{k-1}} \left(g^k(t) \frac{d}{dt} f(t) \right) \right]_{t=0}, \quad h_k(t) = \frac{t^k}{k! g^k(t)},$$

$$G(t) = f(t) \left(1 - \frac{t}{g(t)} \frac{d}{dt} g(t) \right)^{-1}, \quad b_k = \left[\frac{d^k}{dt^k} \left(g^k(t) f(t) \right) \right]_{t=0},$$

(6.22) and (6.23) take the forms

$$f(t) = \sum_{k=0}^{\infty} a_k h_k(t), \quad G(t) = \sum_{k=0}^{\infty} b_k h_k(t),$$

which are of the general form, (6.13), of a generating function. Some interesting applications of such generating functions are given in the following examples.

Example 6.9 Abel's formula
Let $g(t) = e^{zt}$ and $f(t) = e^{xt}$. Then from (6.22) and since

$$\left[\frac{d^{k-1}}{dt^{k-1}}\left(g^k(t)\frac{d}{dt}f(t)\right)\right]_{t=0} = \left[\frac{d^{k-1}}{dt^{k-1}}xe^{(x+kz)t}\right]_{t=0} = x(x+kz)^{k-1},$$

we deduce the expansion

$$e^{xt} = \sum_{k=0}^{\infty}x(x+kz)^{k-1}\frac{u^k}{k!}, \quad u = te^{-zt}. \tag{6.24}$$

Similarly, from (6.23), we get

$$\frac{e^{xt}}{1-zt} = \sum_{k=0}^{\infty}(x+kz)^k\frac{u^k}{k!}, \quad u = te^{-zt}. \tag{6.25}$$

The coefficients in (6.24),

$$A_k(x,z) = \frac{x(x+kz)^{k-1}}{k!}, \quad k = 0,1,\dots,$$

are called *Abel polynomials*. Expanding both members of the identity

$$e^{(x+y)t} = e^{xt}e^{yt}$$

into powers of $u = te^{-zt}$, using (6.24), we get

$$\sum_{n=0}^{\infty}A_n(x+y,z)u^n = \left(\sum_{k=0}^{\infty}A_k(x,z)u^k\right)\left(\sum_{j=0}^{\infty}A_j(y,z)u^j\right)$$

$$= \sum_{n=0}^{\infty}\left\{\sum_{k=0}^{n}A_k(x,z)A_{n-k}(y,z)\right\}u^n$$

and, consequently,

$$A_n(x+y,z) = \sum_{k=0}^{n}A_k(x,z)A_{n-k}(y,z).$$

Further, consider the identity

$$\frac{e^{(x+w)t}}{1-zt} = e^{xt}\frac{e^{wt}}{1-zt}$$

and expand both its members into powers of $u = te^{-zt}$, using (6.24) and (6.25). We get the relation

$$\sum_{n=0}^{\infty}(x+w+nz)^n\frac{u^n}{n!} = \left(\sum_{k=0}^{\infty}x(x+kz)^{k-1}\frac{u^k}{k!}\right)\left(\sum_{j=0}^{\infty}(w+jz)^j\frac{u^j}{j!}\right)$$

which, according to (6.10), implies *Abel's formula*:

$$(x + w + nz)^n = \sum_{k=0}^{n} \binom{n}{k} x(x + kz)^{k-1}(w + nz - kz)^{n-k}.$$

Putting $y = w + nz$, we find the expression

$$(x + y)^n = \sum_{k=0}^{n} \binom{n}{k} x(x + kz)^{k-1}(y - kz)^{n-k}.$$

Note that Newton's binomial formula is a particular case of Abel's formula, corresponding to $z = 0$. ☐

Example 6.10 Gould's formula

Let $f(t) = (1 + t)^x$ and $g(t) = (1 + t)^z$. Then from (6.22) and since

$$\left[\frac{d^{k-1}}{dt^{k-1}} \left(g^k(t) \frac{d}{dt} f(t) \right) \right]_{t=0} = \left[\frac{d^{k-1}}{dt^{k-1}} x(1 + t)^{x+kz-1} \right]_{t=0}$$

$$= \frac{x}{x + kz} \binom{x + kz}{k},$$

we deduce the expansion

$$(1 + t)^x = \sum_{k=0}^{\infty} \frac{x}{x + kz} \binom{x + kz}{k} u^k, \quad u = t(1 + t)^{-z}. \tag{6.26}$$

Similarly, from (6.23), we get

$$\frac{(1 + t)^{x+1}}{1 + t - zt} = \sum_{k=0}^{\infty} \binom{x + kz}{k} u^k, \quad u = t(1 + t)^{-z}. \tag{6.27}$$

The coefficients in (6.26),

$$G_k(x, z) = \frac{x}{x + kz} \binom{x + kz}{k} = \frac{x(x + kz - 1)_{k-1}}{k!}, \quad k = 0, 1, \dots,$$

are called *Gould polynomials*. Expanding both members of the identity

$$(1 + t)^{x+y} = (1 + t)^x (1 + t)^y$$

into powers of $u = t(1 + t)^{-z}$, using (6.26), we get the relation

$$\sum_{n=0}^{\infty} G_n(x + y, z) u^n = \left(\sum_{k=0}^{\infty} G_k(x, z) u^k \right) \left(\sum_{j=0}^{\infty} G_j(y, z) u^j \right)$$

which, according to (6.8), implies

$$G_n(x+y,z) = \sum_{k=0}^{n} G_k(x,z)G_{n-k}(y,z).$$

Expanding both members of the identity

$$\frac{(1+t)^{x+w+1}}{1+t-zt} = (1+t)^x \frac{(1+t)^{w+1}}{1+t-zt}$$

into powers of $u = t(1+t)^{-z}$, using (6.26) and (6.27), we get the relation

$$\sum_{n=0}^{\infty} \binom{x+w+nz}{n} u^n = \left[\sum_{k=0}^{\infty} \frac{x}{x+kz}\binom{x+kz}{k}u^k\right]\left[\sum_{j=0}^{\infty}\binom{w+jz}{j}u^j\right]$$

which, according to (6.10), implies *Gould's formula*:

$$(x+w+nz)_n = \sum_{k=0}^{n} \binom{n}{k} x(x+kz-1)_{k-1}(w+nz-kz)_{n-k}.$$

Putting $y = w + nz$, it yields

$$(x+y)_n = \sum_{k=0}^{n} \binom{n}{k} x(x+kz-1)_{k-1}(y-kz)_{n-k}.$$

Note that Vandermonde's formula is a particular case of Gould's formula, corresponding to $z = 0$. □

6.3   COMBINATIONS AND PERMUTATIONS

Let us consider a finite set $W_n = \{w_1, w_2, \ldots, w_n\}$ and the set $\mathcal{C}_k(W_n)$ of the k-combinations of n and attach the indicator x_j to the element w_j, $j = 1, 2, \ldots, n$. Then the homogeneous product of order k, $x_1^{r_1} x_2^{r_2} \cdots x_n^{r_n}$, with $r_j \geq 0$, $j = 1, 2, \ldots, n$, and $r_1 + r_2 + \cdots + r_n = k$, uniquely determines the k-combination of n, $\{a_1, a_2, \ldots, a_k\}$, $a_j \in W_n$, $j = 1, 2, \ldots, k$, in which $r_j \geq 0$ elements are w_j, $j = 1, 2, \ldots, n$, with $r_1 + r_2 + \cdots + r_n = k$. Further, the homogeneous product sum

$$h_k(x_1, x_2, \ldots, x_n) = \sum x_1^{r_1} x_2^{r_2} \cdots x_n^{r_n}, \quad k = 0, 1, \ldots, \tag{6.28}$$

where the summation is extended over all $r_j \geq 0$, $j = 1, 2, \ldots, n$, such that $r_1 + r_2 + \cdots + r_n = k$, exhibits all k-combinations of n. The advantage of this correspondence is that the homogeneous product sum (6.28) can be obtained algebraically. Specifically,

$$\prod_{j=1}^{n}(1 + x_j t + x_j^2 t^2 + \cdots) = \sum_{k=0}^{\infty} h_k(x_1, x_2, \ldots, x_n) t^k. \qquad (6.29)$$

Indeed, if the operations are executed by selecting the term t^{r_j} from the j-th factor of the product and forming the product $t^{r_1} t^{r_2} \cdots t^{r_n} = t^{r_1 + r_2 + \cdots + r_n}$, with $r_1 + r_2 + \cdots + r_n = k$, the k-th power of t, t^k, is turned up with coefficient equal to the sum of the homogeneous products of order k, $x_1^{r_1} x_2^{r_2} \cdots x_n^{r_n}$, extended over all $r_j \geq 0$, $j = 1, 2, \ldots, n$, such that $r_1 + r_2 + \cdots + r_n = k$.

Note that, in the j-th factor of the product in (6.29),

$$A(t, x_j) = 1 + x_j t + x_j^2 t^2 + \cdots, \quad j = 1, 2, \ldots, n, \qquad (6.30)$$

the term $x_j^{r_j} t^{r_j}$ indicates that the element w_j may appear in any combination r_j times, $r_j = 0, 1, \ldots$. Moreover, any restrictions on the appearance of any elements of W_n modify the factors (6.30) accordingly. Thus, if the element w_j is allowed to appear in any combination at least m_j and at most n_j times, then (6.30) becomes

$$A_{m_j, n_j}(t, x_j) = x_j^{m_j} t^{m_j} + x_j^{m_j + 1} t^{m_j + 1} + \cdots + x_j^{n_j} t^{n_j}, \qquad (6.31)$$

for $j = 1, 2, \ldots, n$ and expression (6.29) is modified accordingly. Putting $x_j = 1$, $j = 1, 2, \ldots, n$ in the resulting expression, the generating function of the number of k-combinations of n is deduced.

Note that, in the particular case of k-combinations of n (without repetitions), the homogeneous product sum of order k, (6.28), coincides with the elementary symmetric function of order k, (6.2), since $r_j = 1$ for k indices, $\{i_1, i_2, \ldots, i_k\}$, out of the n indices $\{1, 2, \ldots, n\}$, and $r_j = 0$ for the remaining $n - k$ indices, $\{1, 2, \ldots, n\} - \{i_1, i_2, \ldots, i_k\}$. Also, the factor (6.31) reduces to $(1 + x_j t)$ and expression (6.29), modified accordingly, coincides with (6.1).

Since the ordinary multiplication is commutative, whence $x_1 x_2$ and $x_2 x_1$ are not distinguishable, it is not possible to construct a function exhibiting all permutations of the set $\mathcal{P}_k(W_n)$, of the k-permutations of n. For this reason, to the homogeneous product of order k, $x_1^{r_1} x_2^{r_2} \cdots x_n^{r_n}$, with $r_j \geq 0$, $j = 1, 2, \ldots, n$, and $r_1 + r_2 + \cdots + r_n = k$, in addition to the k-permutation of n, $(a_1, a_2, \ldots, a_k)$, $a_j \in W_n$, $j = 1, 2, \ldots, k$, in which the first $r_1 \geq 0$ elements are w_1, the next $r_2 \geq 0$ elements are w_2 and finally the last $r_n \geq 0$ elements are w_n, there correspond all its permutations, which (including the initial), according to Theorem 2.4, are equal to

$$\frac{k!}{r_1! r_2! \cdots r_n!}, \quad k = r_1 + r_2 + \cdots + r_n.$$

In consequence, to the set $\mathcal{P}_k(W_n)$ of the k-permutations of n, we correspond the sum

$$g_k(x_1, x_2, \ldots, x_n) = \sum \frac{k!}{r_1! r_2! \cdots r_n!} x_1^{r_1} x_2^{r_2} \cdots x_n^{r_n}, \quad k = 0, 1, \ldots, (6.32)$$

where the summation is extended over all $r_j \geq 0$, $j = 1, 2, \ldots, n$, such that $r_1 + r_2 + \cdots + r_n = k$. This sum, according to the multinomial Theorem 3.7, equals

$$g_k(x_1, x_2, \ldots, x_n) = (x_1 + x_2 + \cdots + x_n)^k$$

and may be obtained algebraically as follows:

$$\prod_{j=1}^{n} \left(1 + x_j t + \frac{x_j^2 t^2}{2!} + \cdots \right) = \sum_{k=0}^{\infty} g_k(x_1, x_2, \ldots, x_n) \frac{t^k}{k!}, \qquad (6.33)$$

since

$$\sum_{k=0}^{\infty} \frac{x^k t^k}{k!} = e^{xt}.$$

Moreover, any restrictions on the appearance of any elements in permutations of W_n modify accordingly the corresponding factors,

$$E(t, x_j) = 1 + x_j t + \frac{x_j^2 t^2}{2!} + \cdots, \quad j = 1, 2, \ldots, n, \qquad (6.34)$$

of the product in (6.33). Thus, if the element w_j is allowed to appear in any permutation at least m_j and at most n_j times, then (6.34) becomes

$$E_{m_j, n_j}(t, x_j) = \frac{x_j^{m_j} t^{m_j}}{m_j!} + \frac{x_j^{m_j+1} t^{m_j+1}}{(m_j + 1)!} + \cdots + \frac{x_j^{n_j} t^{n_j}}{n_j!}, \qquad (6.35)$$

for $j = 1, 2, \ldots, n$ and expression (6.33) is modified accordingly. Putting $x_j = 1$, $j = 1, 2, \ldots, n$ in the resulting expression, the exponential generating function of the number of k-permutations of n is deduced.

Example 6.11 Combinations with repetition

The generating function of the number of k-combinations of n with (unrestricted) repetition may be obtained as follows. The enumerator of the appearance of the j-th element in any combination, since it is allowed to appear r_j times, with $r_j = 0, 1, \ldots$, is

$$A_0(t, x_j) = 1 + x_j t + x_j^2 t^2 + \cdots = (1 - x_j t)^{-1}, \quad j = 1, 2, \ldots, n.$$

Therefore, the enumerator of k-combinations of n is given by

$$\prod_{j=1}^{n} A_0(t, x_j) = \prod_{j=1}^{n} (1 - x_j t)^{-1} = \sum_{k=0}^{\infty} h_k(x_1, x_2, \ldots, x_n) t^k,$$

which, with $x_j = 1$, $j = 1, 2, \ldots, n$, yields the generating function of the number of k-combinations of n with repetition:

$$A_0(t) = (1-t)^{-n} = \sum_{k=0}^{\infty} \binom{-n}{k} (-t)^k = \sum_{k=0}^{\infty} \binom{n+k-1}{k} t^k.$$

The coefficient of t^k,

$$E(n, k) = \binom{n+k-1}{k},$$

gives the number of k-combinations of n with repetition (see Theorem 2.7). ▯

Example 6.12

In the case of k-combinations of n with repetition, in which every element appears at least once, the enumerator of the appearance of the j-th element, since it is allowed to appear r_j times, with $r_j = 1, 2, \ldots$, is

$$A_1(t, x_j) = x_j t + x_j^2 t^2 + \cdots = x_j t (1 - x_j t)^{-1}, \quad j = 1, 2, \ldots, n.$$

The generating function of the number of k-combinations of n with repetition, in which every element appears at least once, is deduced from

$$\prod_{j=1}^{n} A_1(t, x_j) = \prod_{j=1}^{n} x_j t (1 - x_j t)^{-1},$$

by setting $x_j = 1$, $j = 1, 2, \ldots, n$. Expanding the resulting generating function into powers of t, we get

$$A_1(t) = t^n (1-t)^{-n} = \sum_{r=0}^{\infty} (-1)^r \binom{-n}{r} t^{r+n} = \sum_{r=0}^{\infty} \binom{n+r-1}{r} t^{r+n}$$

$$= \sum_{k=n}^{\infty} \binom{k-1}{k-n} t^k = \sum_{k=n}^{\infty} \binom{k-1}{n-1} t^k.$$

Therefore, the number of k-combinations of n with repetitions, in which every element appears at least once, is zero for $k < n$ and

$$\binom{k-1}{n-1}$$

for $k \geq n$. ▯

Example 6.13 **Combinations with limited repetition**

The generating function of the number of k-combinations of n with repetition and the restriction that each element appears, at most, s times is deduced from the enumerator:

$$\prod_{j=1}^{n}(1 + x_j t + x_j^2 t^2 + \cdots + x_j^s t^s) = \prod_{j=1}^{n}(1 - x_j^{s+1} t^{s+1})(1 - x_j t)^{-1},$$

by putting $x_j = 1$, $j = 1, 2, \ldots, n$. Thus,

$$A_{n,s}(t) = (1 - t^{s+1})^n (1 - t)^{-n}.$$

The determination of the number $E_s(n, k)$, of k-combinations of n with repetition and the restriction that each element appears, at most, s times requires the expansion of the generating function into powers of t,

$$A_{n,s}(t) = \left[\sum_{r=0}^{n}(-1)^r \binom{n}{r} t^{r(s+1)}\right] \cdot \left[\sum_{j=0}^{\infty} \binom{n+j-1}{j} t^j\right].$$

Note that, in the case $k \leq s$, the first factor of this product in addition to the first term, which is equal to 1, contains powers of t of order greater than k. Consequently, the coefficient of t^k is equal to the coefficient of this power in the second factor of the product, that is,

$$E_s(n, k) = \binom{n+k-1}{k}, \quad k \leq s.$$

In the case of $k \geq s + 1$, the term t^k is formed by multiplying the term $t^{r(s+1)}$ of the first factor with the term $t^{k-r(s+1)}$ of the second factor. Consequently, the coefficient of t^k is equal to the sum of the product of the coefficients of $t^{r(s+1)}$ from the first factor and the coefficients of $t^{k-r(s+1)}$ from the second factor, for $r = 0, 1, \ldots, n$, whence

$$E_s(n, k) = \sum_{r=0}^{n}(-1)^r \binom{n}{r}\binom{n+k-r(s+1)-1}{n-1}, \quad k \geq s + 1.$$

The next two particular cases are worth noticing. For $s = 1$ we have

$$A_{n,1}(t) = (1 + t)^n = \sum_{k=0}^{n}\binom{n}{k} t^k,$$

and so

$$E_1(n, k) = \binom{n}{k},$$

while, for $s \to \infty$, we have

$$\lim_{s \to \infty} A_{n,s}(t) = \left(\sum_{r=0}^{\infty} t^r\right)^n = (1 - t)^{-n} = \sum_{k=0}^{\infty}\binom{n+k-1}{k} t^k.$$

and so

$$\lim_{s \to \infty} E_s(n, k) = \binom{n + k - 1}{k}. \quad \square$$

Example 6.14 Permutations with repetition

The generating function of the number of k-permutations of n with (unrestricted) repetition is deduced from the function

$$\prod_{j=1}^{n} \left(1 + x_j t + \frac{x_j^2 t^2}{2!} + \cdots \right),$$

by putting $x_j = 1$, $j = 1, 2, \ldots, n$. Hence, we get

$$E_0(t) = \left(1 + t + \frac{t^2}{2!} + \cdots \right)^n = e^{nt}.$$

Expanding it into powers of t,

$$E_0(t) = e^{nt} = \sum_{k=0}^{\infty} n^k \frac{t^k}{k!},$$

we conclude the number of k-permutations of n with repetition equals

$$U(n, k) = n^k. \quad \square$$

Example 6.15

The generating function of the number of k-permutations of n, with repetition, in which every element appears at least once, is deduced from the function

$$\prod_{j=1}^{n} \left(x_j t + \frac{x_j^2 t^2}{2!} + \cdots \right),$$

by setting $x_j = 1$, $j = 1, 2, \ldots, n$. Expanding the resulting generating function into powers of t, we find

$$E_1(t) = \left(t + \frac{t^2}{2!} + \cdots \right) = (e^t - 1)^n = \sum_{r=0}^{n} (-1)^{n-r} \binom{n}{r} e^{rt}$$

$$= \sum_{r=0}^{n} (-1)^{n-r} \binom{n}{r} \sum_{k=0}^{\infty} r^k \frac{t^k}{k!} = \sum_{k=0}^{\infty} \left\{ \sum_{r=0}^{n} (-1)^{n-r} \binom{n}{r} r^k \right\} \frac{t^k}{k!}.$$

Therefore, the number of k-permutations of n, with repetition, in which every element appears at least once, is given by

$$U_1(n, k) = \sum_{r=0}^{n} (-1)^{n-r} \binom{n}{r} r^k = n! S(k, n),$$

where $S(k, n)$ is the Stirling number of the second kind.   ⬜

Example 6.16 Permutations with limited repetition

The generating function of the number of k-permutations of the set $W_n = \{w_1, w_2, \ldots, w_n\}$, with repetition, in which the element w_j is allowed to appear at most k_j times, $j = 1, 2, \ldots, n$, where $k_1 + k_2 + \cdots + k_n = r \geq k$, is deduced from the function

$$\prod_{j=1}^{n} \left(1 + x_j t + \frac{x_j^2 t^2}{2!} + \cdots + \frac{x_j^{k_j} t^{k_j}}{k_j!}\right)$$

by setting $x_j = 1$, $j = 1, 2, \ldots, n$. Then we get the expression

$$F(t) = \prod_{j=1}^{n} \left(1 + t + \frac{t^2}{2!} + \cdots + \frac{t^{k_j}}{k_j!}\right),$$

which, after executing the multiplication and arranging the terms in ascending order of powers of t, takes the form

$$F(t) = \sum_{k=0}^{r} U(k_1, k_2, \ldots, k_n; k) \frac{t^k}{k!}.$$

The coefficient of $t^k/k!$ is the number of k-permutations of n with repetition, in which the j-th element is allowed to appear at most k_j times, $j = 1, 2, \ldots, n$.

In the particular case of $k = r$, the number $U(k_1, k_2, \ldots, k_n; r) \equiv M(k_1, k_2, \ldots, k_n)$ of r-permutations of n with repetitions, in which the j-th element is allowed to appear at most k_j times, $j = 1, 2, \ldots, n$, where $k_1 + k_2 + \cdots + k_n = r$, equals

$$M(k_1, k_2, \ldots, k_n) = \frac{r!}{k_1! k_2! \cdots k_n!}, \quad r = k_1 + k_2 + \cdots + k_n,$$

since the coefficient of $t^r = t^{k_1 + k_2 + \cdots + k_n}$ is equal to the product

$$\frac{1}{k_1!} \frac{1}{k_2!} \cdots \frac{1}{k_n!},$$

where $1/k_j!$ is the coefficient of the last term t^{k_j} of the j-th factor of the generating function $F(t)$, $j = 1, 2, \ldots, n$.

In the general case, it similarly follows that

$$U(k_1, k_2, \ldots, k_n; k) = \sum \frac{k!}{s_1! s_2! \cdots s_n!},$$

where the summation is extended over all $s_j = 0, 1, \ldots, k_j$, $j = 1, 2, \ldots, n$, such that $s_1 + s_2 + \cdots + s_n = k$.   ⬜

6.4   MOMENT GENERATING FUNCTIONS

Let us consider a sequence of real numbers a_k, $k = 0, 1, \ldots$. In combinatorics, the general term a_k of such a sequence, usually, expresses a combinatorial number. Also, in probability theory and, in particular, when the total probability is concentrated at discrete points, such a sequence of probabilities defines a probability (density) function. In mechanics and particularly when the total mass is concentrated at discrete points, such a sequence of masses defines a density mass function. Among the characteristics of the sequence a_k, $k = 0, 1, \ldots$ of special interest are its moments defined as follows.

The *(power) moment of order r* of the sequence a_k, $k = 0, 1, \ldots$, denoted by μ_r', is defined by

$$\mu_r' = \sum_{k=0}^{\infty} k^r a_k, \quad r = 0, 1, \ldots, \tag{6.36}$$

while the *factorial moment of order r* of the same sequence, denoted by $\mu_{(r)}$, is defined by

$$\mu_{(r)} = \sum_{k=0}^{\infty} (k)_r a_k, \quad r = 0, 1, \ldots. \tag{6.37}$$

It should be noted that the absolute convergence of the series (6.36) and (6.37) is an essential requirement for the existence of the moments μ_r' and $\mu_{(r)}$. Note, also, that the factorial moment of order r, $\mu_{(r)}$, is closely connected with the *binomial moment of order r*, $b_{(r)}$, which is defined by

$$b_{(r)} = \sum_{k=0}^{\infty} \binom{k}{r} a_k, \quad r = 0, 1, \ldots, \tag{6.38}$$

whence $b_{(r)} = \mu_{(r)}/r!$, $r = 0, 1, \ldots$.

REMARK 6.1   The moments μ_r', $\mu_{(r)}$ and $b_{(r)}$ are particular cases of the *weighted moment of order r, m_r*, defined by

$$m_r = \sum_{k=0}^{\infty} c_r(k) a_k, \quad r = 0, 1, \ldots,$$

where, for fixed r, $c_r(k)$, $k = 0, 1, \ldots$, is a sequence of weights. When a_k, $k = 0, 1, \ldots$ is a sequence of probabilities, $\mu \equiv \mu_1' = \mu_{(1)}$ is the *mean* and

$\sigma^2 = \mu_2' - \mu_1'^2 = \mu_{(2)} - \mu_{(1)} - \mu_{(1)}^2$ is the *variance*; when it is a sequence of masses, μ is the *center of gravity* and σ^2 is the *moment of inertia*. ∎

The calculation of the moments μ_r', $\mu_{(r)}$ and $b_{(r)}$, $r = 0, 1, \ldots$, when the sequence a_k, $k = 0, 1, \ldots$, is known and vice versa, is facilitated by using suitable generating functions. Such generating functions are the *moment generating function*, defined by

$$M(t) = \sum_{r=0}^{\infty} \mu_r' \frac{t^r}{r!}, \tag{6.39}$$

and the *binomial* or *factorial moment generating function*, defined by

$$B(t) = \sum_{r=0}^{\infty} b_{(r)} t^r = \sum_{r=0}^{\infty} \mu_{(r)} \frac{t^r}{r!}. \tag{6.40}$$

These generating functions are connected with the ordinary generating functions of the sequence a_k, $k = 0, 1, \ldots$,

$$A(t) = \sum_{k=0}^{\infty} a_k t^k. \tag{6.41}$$

Indeed, from (6.36) and (6.37), we have

$$M(t) = \sum_{r=0}^{\infty} \sum_{k=0}^{\infty} k^r a_k \frac{t^r}{r!} = \sum_{k=0}^{\infty} a_k \left\{ \sum_{r=0}^{\infty} \frac{(kt)^r}{r!} \right\} = \sum_{k=0}^{\infty} a_k e^{kt}$$

and, using (6.41) with e^t instead of t, we get

$$M(t) = A(e^t), \quad A(u) = M(\log u), \tag{6.42}$$

while, from (6.39) and (6.40), we find

$$B(t) = \sum_{r=0}^{\infty} \sum_{k=0}^{\infty} \binom{k}{r} a_k t^r = \sum_{k=0}^{\infty} a_k \left\{ \sum_{r=0}^{\infty} \binom{k}{r} t^r \right\} = \sum_{k=0}^{\infty} a_k (1+t)^k$$

and so

$$B(t) = A(1+t), \quad A(u) = B(u-1). \tag{6.43}$$

From (6.42) and (6.43) it follows that

$$M(t) = B(e^t - 1), \quad B(t) = M(\log(1+t)). \tag{6.44}$$

Expanding the second of (6.43) into powers of u, using (6.40) and (6.41), we get

$$\sum_{k=0}^{\infty} a_k u^k = \sum_{r=0}^{\infty} \mu_{(r)} \frac{(u-1)^r}{r!} = \sum_{r=0}^{\infty} \frac{\mu_{(r)}}{r!} \sum_{k=0}^{r} (-1)^{r-k} \binom{r}{k} u^k$$

$$= \sum_{k=0}^{\infty} \left\{ \frac{1}{k!} \sum_{r=k}^{\infty} \frac{(-1)^{r-k}}{(r-k)!} \mu_{(r)} \right\} u^k$$

and thus

$$a_k = \frac{1}{k!} \sum_{r=k}^{\infty} \frac{(-1)^{r-k}}{(r-k)!} \mu_{(r)}.$$

Introducing the binomial moments $b_{(r)} = \mu_{(r)}/r!$, $r = 0, 1, 2, \ldots$, this expression takes the form

$$a_k = \sum_{r=k}^{\infty} (-1)^{r-k} \binom{r}{k} b_{(r)}.$$

Note that the last expression is similar to (4.14), which expresses the number $N_{n,k}$, $k = 0, 1, \ldots, n$, of elements that are contained in k among n subsets, $A_1, A_2, \ldots, A_n$, of a finite set Ω, in terms of the numbers $S_{n,r} = \sum N(A_{i_1} A_{i_2} \cdots A_{i_r})$, $r = 0, 1, \ldots, n$, where the summation is extended over all r-combinations, $\{i_1, i_2, \ldots, i_r\}$, of the n indices $\{1, 2, \ldots, n\}$.

Example 6.17

Let us determine the factorial moments $\mu_{(r)}$, $r = 0, 1, \ldots$, of the sequence of geometric probabilities $p_k = p q^k$, $k = 0, 1, \ldots$, where $0 < p < 1$ and $q = 1 - p$.

The probability generating function, on using the geometric series, is obtained as

$$A(t) = \sum_{k=0}^{\infty} p_k t^k = p \sum_{k=0}^{\infty} (qt)^k = \frac{p}{1 - qt}, \quad |t| < 1/q.$$

Therefore, by the first of (6.43), the factorial moment generating function is given by

$$B(t) = \frac{1}{1 - (q/p)t}, \quad |t| < p/q$$

which, expanded into powers of t,

$$B(t) = \frac{1}{1 - (q/p)t} = \sum_{r=0}^{\infty} (q/p)^r t^r,$$

by virtue of (6.40), yields

$$\mu_{(r)} = r!(q/p)^r, \quad r = 0, 1, \ldots. \qquad \square$$

Example 6.18

Let us determine the sequence of probabilities p_k, $k = 0, 1, \ldots$, with binomial moments

$$b_{(r)} = \binom{n}{r} p^r, \quad r = 0, 1, \ldots,$$

where $0 < p < 1$ and n is a positive integer.

The binomial moment generating function, on using Newton's binomial formula, is obtained as

$$B(t) = \sum_{r=0}^{\infty} b_{(r)} t^r = \sum_{r=0}^{n} \binom{n}{r} (pt)^r = (1 + pt)^n.$$

Consequently, by virtue of the second of (6.43), the generating function of the sequence p_k, $k = 0, 1, \ldots$, is given by

$$A(u) = (1 - p + pu)^n = \sum_{k=0}^{n} \binom{n}{k} p^k (1 - p)^{n-k} u^k,$$

and so

$$p_k = \binom{n}{k} p^k (1 - p)^{n-k}, \quad k = 0, 1, \ldots, n. \quad \square$$

Example 6.19

Let us consider the number $N_{n,k}$, $k = 0, 1, \ldots, n$, of elements that are contained in k among n subsets $A_1, A_2, \ldots, A_n$, of a finite set Ω, and

$$S_{n,r} = \sum N(A_{i_1} A_{i_2} \cdots A_{i_r}), \quad r = 0, 1, \ldots, n,$$

where the summation is extended over all r-combinations $\{i_1, i_2, \ldots, i_r\}$ of the n indices $\{1, 2, \ldots, n\}$. Then, according to (4.14) and (4.20),

$$N_{n,k} = \sum_{r=k}^{n} (-1)^{r-k} \binom{r}{k} S_{n,r}, \quad S_{n,r} = \sum_{k=r}^{n} \binom{k}{r} N_{n,k}.$$

Consequently, $S_{n,r}$, $r = 0, 1, \ldots, n$, is the sequence of binomial moments of the sequence $N_{n,k}$, $k = 0, 1, \ldots, n$. Thus, if

$$N_n(t) = \sum_{k=0}^{n} N_{n,k} t^k, \quad S_n(t) = \sum_{r=0}^{n} S_{n,r} t^r, \tag{6.45}$$

then, according to (6.43),

$$S_n(t) = N_n(t + 1), \quad N_n(t) = S_n(t - 1). \tag{6.46}$$

In the particular case with Ω the set of permutations $(j_1, j_2, \ldots, j_n)$ of $\{1, 2, \ldots, n\}$ and A_s the subsets of these permutations for which the point j_s is a fixed point, $s = 1, 2, \ldots, n$, we have

$$N_{n,k} = D_{n,k}, \quad S_{n,r} = n!/r!,$$

where $D_{n,k}$ is the coincidence number (see Section 5.2). The generating function

$$D_n(t) = \sum_{k=0}^{n} D_{n,k} t^k,$$

according to (6.45) and (6.46), is written as

$$D_n(t) = \sum_{r=0}^{n} \frac{n!}{r!} (t-1)^r.$$

Differentiating it we get the relation

$$\frac{d}{dt} D_n(t) = n D_{n-1}(t).$$

Also, we find

$$D_n(t) = n D_{n-1}(t) + (t-1)^n.$$

These relations imply the recurrence relations:

$$k D_{n,k} = n D_{n-1,k-1}$$

and

$$D_{n,k} = n D_{n-1,k} + (-1)^{n-k} \binom{n}{k},$$

respectively. □

6.5 MULTIVARIATE GENERATING FUNCTIONS

The notion of a generating function can be extended to cover the case of a double-index and, more generally, a multi-index sequence. Specifically, the following definitions are introduced.

DEFINITION 6.4 *Let $a_{n,k}$, $k = 0, 1, \ldots$, $n = 0, 1, \ldots$ be a sequence of real numbers. The sum*

$$A(t, u) = \sum_{n=0}^{\infty} \sum_{k=0}^{\infty} a_{n,k} t^k u^n \tag{6.47}$$

is called (ordinary) generating function, while the sum

$$E(t, u) = \sum_{n=0}^{\infty} \sum_{k=0}^{\infty} a_{n,k} \frac{t^k}{k!} \frac{u^n}{n!} \qquad (6.48)$$

is called exponential generating function of the sequence $a_{n,k}$, $k = 0, 1, \ldots$, $n = 0, 1, \ldots$.

It should be noted that the absolute convergence of the series (6.47) and (6.48) is required for the existence of the corresponding generating functions.

REMARK 6.2   Both generating functions $A(t, u)$ and $E(t, u)$ are particular cases of

$$G(t, u) = \sum_{n=0}^{\infty} \sum_{k=0}^{\infty} a_{n,k} f_k(t) g_n(u), \qquad (6.49)$$

where the functions $f_k(t)$, $k = 0, 1, \ldots$, as well as the functions $g_n(u)$, $n = 0, 1, \ldots$, are linearly independent and so (6.49) is uniquely defined. It is worth noting that, with

$$F_n(t) = \sum_{k=0}^{\infty} a_{n,k} f_k(t), \quad G_k(u) = \sum_{n=0}^{\infty} a_{n,k} g_n(u),$$

(6.49) may be written in the form

$$G(t, u) = \sum_{k=0}^{\infty} G_k(u) f_k(t),$$

or in the form

$$G(t, u) = \sum_{n=0}^{\infty} F_n(t) g_n(u).$$

Note that the last two expressions are of the general form (6.13) of a generating function. With this remark, the study of generating functions of double-index sequences is reduced to the previous study of generating functions of single-index sequences.

In the particular case of *triangular* double-index sequences $a_{n,k}$, $k = 0, 1, \ldots$, $n = 0, 1, \ldots$, for which $a_{n,k} = 0$ for $k > n$, it is more convenient to use the generating function

$$\Phi(t, u) = \sum_{n=0}^{\infty} \sum_{k=0}^{n} a_{n,k} t^k \frac{u^n}{n!}. \qquad (6.50)$$

In an analogous way, a multivariate generating function $G(t_1, t_2, \ldots, t_r)$ of a multi-index sequence $a_{k_1, k_2, \ldots, k_r}$, $k_i = 0, 1, \ldots, i = 1, 2, \ldots, r$, may be introduced and studied. ∎

Example 6.20

Let us consider the double-index sequence:

$$a_{n,k} = \binom{n}{k}, \quad k = 0, 1, \ldots, n, \ n = 0, 1, \ldots.$$

For fixed n, using Newton's binomial formula, the ordinary generating function of this sequence is readily deduced as

$$A_n(t) = \sum_{k=0}^{n} \binom{n}{k} t^k = (1+t)^n.$$

Consequently, the bivariate generating function of the double-index sequence $a_{n,k}$, $k = 0, 1, \ldots, n, n = 0, 1, \ldots$, is obtained as

$$A(t, u) = \sum_{n=0}^{\infty} \sum_{k=0}^{n} a_{n,k} t^k u^n = \sum_{n=0}^{\infty} A_n(t) u^n$$

$$= \sum_{n=0}^{\infty} [(1+t)u]^n = [1 - (1+t)u]^{-1}.$$

Since the considered double-index sequence is triangular, its generating function of the form (6.50) is derived as

$$\Phi(t, u) = \sum_{n=0}^{\infty} \sum_{k=0}^{n} a_{n,k} t^k \frac{u^n}{n!} = \sum_{n=0}^{\infty} A_n(t) \frac{u^n}{n!}$$

$$= \sum_{n=0}^{\infty} \frac{[(1+t)u]^n}{n!} = e^{(1+t)u}.$$

The exponential generating function of the same sequence is quite involved and thus not of any interest. ▯

Example 6.21

Consider the coincidence number (see Theorem 5.3),

$$D_{n,k} = \frac{n!}{k!} \sum_{j=0}^{n-k} \frac{(-1)^j}{j!}, \quad k = 0, 1, \ldots, n, \ n = 0, 1, \ldots.$$

For fixed n, the generating function of this sequence is (see Example 6.19),

$$D_n(t) = \sum_{k=0}^{n} D_{n,k} t^k = n! \sum_{r=0}^{n} \frac{(t-1)^r}{r!}.$$

Then, the bivariate generating function

$$D(t, u) = \sum_{n=0}^{\infty} \sum_{k=0}^{n} D_{n,k} t^k \frac{u^n}{n!},$$

is obtained as

$$D(t, u) = \sum_{r=0}^{\infty} \frac{[u(t-1)]^r}{r!} \sum_{n=r}^{\infty} u^{n-r} = (1-u)^{-1} e^{u(t-1)}. \qquad \square$$

Example 6.22
Let us consider the multi-index sequence

$$C(n, k_1, k_2, \ldots, k_{r-1}) = \binom{n}{k_1, k_2, \ldots, k_{r-1}} = \frac{n!}{k_1! k_2! \cdots k_{r-1}! k_r!},$$

$k_i = 0, 1, \ldots, n, i = 1, 2, \ldots, r-1, k_r = n-(k_1+k_2+\cdots+k_{r-1}), n = 0, 1, \ldots$.
For fixed n, the multivariate generating function of this sequence, on using the multinomial formula (see Theorem 3.7), is obtained as

$$A_n(t_1, t_2, \ldots, t_{r-1}) = \sum \binom{n}{k_1, k_2, \ldots, k_{r-1}} t_1^{k_1} t_2^{k_2} \cdots t_{r-1}^{k_{r-1}}$$
$$= (1 + t_1 + t_2 + \cdots + t_{r-1})^n.$$

Consequently,

$$A(t_1, t_2, \ldots, t_{r-1}, u) = \sum_{n=0}^{\infty} \sum_{}^{\infty} \binom{n}{k_1, k_2, \ldots, k_{r-1}} t_1^{k_1} t_2^{k_2} \cdots t_{r-1}^{k_{r-1}} u^n$$
$$= \sum_{n=0}^{\infty} [(1 + t_1 + t_2 + \cdots + t_{r-1})u]^n$$
$$= [1 - (1 + t_1 + t_2 + \cdots + t_{r-1})u]^{-1}$$

and

$$\Phi(t_1, t_2, \ldots, t_{r-1}, u) = \sum_{n=0}^{\infty} \sum_{}^{\infty} \binom{n}{k_1, k_2, \ldots, k_{r-1}} t_1^{k_1} t_2^{k_2} \cdots t_{r-1}^{k_{r-1}} \frac{u^n}{n!}$$
$$= \sum_{n=0}^{\infty} \frac{[(1 + t_1 + t_2 + \cdots + t_{r-1})u]^n}{n!}$$
$$= e^{(1+t_1+t_2+\cdots+t_{r-1})u}.$$

Other bivariate generating functions are discussed in the exercises. $\qquad \square$

6.6   BIBLIOGRAPHIC NOTES

The roots of generating functions may be traced in the work of Abraham De Moivre (1718). Also, generating functions were used by L. Euler (1746) in his study on the partitions of integers. The development of the theory of generating functions arose in conjunction with the calculus of probability. P. S. Laplace (1812) devoted half of his book to a systematic and complete treatment of generating functions and especially the moment generating functions; he is considered the inventor of the method of generating functions. P. MacMahon (1915, 1916) extensively used enumerating generating functions in his treatise on combinatorial analysis. The basic elements of the symbolic calculus included in this chapter are based on the paper of E. T. Bell (1940) and the book of J. Riordan (1958). The exponential Bell numbers were examined in E. T. Bell (1934a,b). Abel's generalization of Newton's binomial formula first appeared in N. H. Abel (1826) and Gould's generalization of Vandermonde's formula was derived in H. W. Gould (1956). An illuminating historical coverage of the use of generating functions in probability theory is given by H. L. Seal (1949). A wealth of information can be found in the survey article on generating function by R. P. Stanley (1978).

6.7   EXERCISES

1. Construct a generating function for the number a_k of k-combinations of the set $W_5 = \{w_1, w_2, \dots, w_5\}$, with repetition, in which the element w_1 appears at least once, each of the elements w_2, w_3 and w_4 appears at least twice and the element w_5 may appear without any restriction. Expanding the generating function, deduce the number a_k, $k = 7, 8, \dots$.

2. Suppose that three distinguishable dice are rolled. Construct a generating function of the number a_k of possible results in which the sum of the three numbers equals k. Derive the numbers a_9 and a_{10} (see Exercise 2.11).

3. Consider ten urns, each containing five numbered balls. The balls of each of the first five urns bear the even integers $\{0, 2, 4, 6, 8\}$, while the balls of each of the other five urns bear the odd integers $\{1, 3, 5, 7, 9\}$. Assume that one ball is drawn from each of the urns. Construct a generating function for the number a_k of drawings in which the sum of the two numbers equals k. Derive the numbers a_7, a_9 and a_{11}.

4. Consider a collection of 52 cards of four different kinds with 13 cards like from each kind. Construct a generating function for the number a_k of different sub-collections of k cards. Expanding it, deduce the numbers a_{13} and a_{26}.

5. Suppose that five letters $\{l_1, l_2, \ldots, l_5\}$ from an alphabet are given. Construct a generating function for the number of k-letter words that can be formed (a) without repeated letters and (b) without any restriction.

6. Calculate the number of k-permutations of the set $\{0, 1, \ldots, 9\}$ in which (a) each of the digits 0 and 5 appears an odd number of times and (b) each of the digits 0 and 5 appears zero or an even number of times, using suitable generating functions.

7. Calculate the number of k-permutations of the set $\{0, 1, \ldots, 9\}$ (a) with zero or an even number of 0s and odd number of 5s and (b) in which the total number of 0s and 5s is zero or even.

8. In the quaternary number system each number is represented by an ordered sequence of the digits from the set $\{0, 1, 2, 3\}$. Using suitable generating functions, calculate the number of k-digit quaternary sequences in which (a) each of the digits 1 and 3 appears an even number of times and (b) the total number of appearance of the digits 1 and 3 is an even number.

9. Let $T(n, k)$ be the number of k-combinations of n with repetition and the restriction that each element may appear at most two times. For fixed n, construct a generating function for the sequence $T(n, k)$, $k = 0, 1, \ldots$, and derive the expressions

$$T(n, k) = \sum_{j=0}^{[k/2]} \binom{n}{j}\binom{n-j}{k-2j}$$

and

$$T(n, k) = \sum_{j=0}^{[k/3]} (-1)^j \binom{n}{j}\binom{n+k-3j-1}{n-1}.$$

10. (*Continuation*). Show that

$$T(n, k) = T(n-1, k) + T(n-1, k-1) + T(n-1, k-2),$$

for $k = 2, 3, \ldots, 2n$ and $n = 1, 2, \ldots$, with initial conditions

$$T(n, 0) = 1, \quad T(n, 1) = n, \quad T(n, k) = 0, \quad k > 2n.$$

11. (*Continuation*). Show that

$$kT(n, k) = nT(n-1, k-1) + 2nT(n-1, k-2),$$

for $k = 2, 3, \ldots, 2n$, $n = 1, 2, \ldots$, and

$$k(k-1)T(n, k) = 2n(2n-1)T(n-1, k-2) + 3n(n-1)T(n-2, k-2),$$

for $k = 2, 3, \ldots, 2n$ and $n = 2, 3, \ldots$, with initial conditions

$$T(n, 0) = 1, \quad T(n, 1) = n, \quad T(n, k) = 0, \quad k > 2n.$$

12. (*Continuation*). Let $T_n \equiv T(n, n)$ be the number of n-combinations of n with repetition and the restriction that each element may appear, at most, twice. Show that

$$nT_n - (2n-1)T_{n-1} - 3(n-1)T_{n-2} = 0, \quad n = 2, 3, \ldots, \quad T_0 = T_1 = 1$$

and

$$T(t) = \sum_{n=0}^{\infty} T_n t^n = (1 - 2t - 3t^2)^{-1/2}.$$

13*. (*Continuation*). From the generating function

$$T_n(t) = \sum_{k=0}^{2n} T(n, k)t^k$$

and using Lagrange formula,

$$\sum_{n=0}^{\infty} \left[\frac{d^n}{dt^n} f(t) g^n(t) \right]_{t=0} \cdot \frac{u^n}{n!} = f(t) \left(1 - u \frac{dg(t)}{dt} \right)^{-1}, \quad u = \frac{t}{g(t)},$$

deduce that

$$T(u) = \sum_{n=0}^{\infty} T_n u^n = (1 - 2u - 3u^2)^{-1/2}.$$

14. (*Continuation*). Let $A(n, r, k)$ be the number of k-combinations of $n + r$ with repetition and the restriction that each of n specified elements may appear at most twice, while each of the other r elements may appear at most once. For fixed n and r, construct a generating function for the sequence $A(n, r, k)$, $k = 0, 1, \ldots$, and derive the expressions

$$A(n, r, k) = \sum_{j=0}^{[k/2]} \binom{n}{j} \binom{n+r-j}{k-2j},$$

and

$$A(n, r, k) = \sum_{i=0}^{k} \binom{r}{i} T(n, k-i).$$

15. Let $G(n, k)$ be the number of k-combinations of n with repetition and the restriction that each element appears zero or an even number of times. For fixed n, construct a generating function for the sequence $G(n, k)$, $k = 0, 1, \ldots$ and, expanding it, show that

$$G(n, 2j) = \binom{n+j-1}{j}, \quad j = 0, 1, \ldots, \quad G(n, 2j+1) = 0, \quad j = 0, 1, \ldots.$$

16. Let $H(n, k)$ be the number of k-combinations of n with repetition and the restriction that each element appears an odd number of times. For fixed n, construct a generating function for the sequence $H(n, k)$, $k = 0, 1, \ldots.$ If $n = 2r$, conclude that

$$H(2r, 2j) = \binom{r+j-1}{j-r}, \quad j = r, r+1, \ldots,$$

$$H(2r, 2j) = 0, \quad j = 0, 1, \ldots, r-1, \quad H(2r, 2j+1) = 0, \quad j = 0, 1, \ldots,$$

while if $n = 2r + 1$, conclude that

$$H(2r+1, 2j+1) = \binom{r+j}{j-r}, \quad j = r, r+1, \ldots,$$

$$H(2r+1, 2j+1) = 0, \quad j = 0, 1, \ldots, r-1, \quad H(2r+1, 2j) = 0, \quad j = 0, 1, \ldots.$$

17. Let $B(n, k; s)$ be the number of k-combinations of the set $W_{n+1} = \{w_0, w_1, \ldots, w_n\}$ with repetition and the restriction that the element w_0 may appear at most s times and each of the other elements at most once. Using a suitable generating function, show that

$$B(n, k; s) = \sum_{j=0}^{m} \binom{n}{k-j}, \quad m = \min\{k, s\}$$

and

$$B(n, k; s) - B(n, k-1; s) = \binom{n}{k} - \binom{n}{k-s-1},$$

with $B(n, 0; s) = B(n, n+s; s) = 1$.

18. Let $E_s(n, k)$ be the number of k-combinations of n with repetition and the restriction that each element may appear at most s times. Using a suitable generating function, show that

$$E_s(n, k) = E_s(n-1, k) + E_s(n-1, k-1) + \cdots + E_s(n-1, k-s), \quad k \geq s,$$

$$E_s(n, k) = \binom{n+k-1}{k}, \quad k \leq s.$$

19. (*Continuation*). Show that

$$\sum_{j=0}^{k} E_s(r,j)E_s(n-r,k-j) = E_s(n,k),$$

and

$$\sum_{j=m}^{k} \binom{n}{j} E_{s-1}(n-j,k-n+j) = E_s(n,k), \quad m = \max\{0, n-k\}.$$

20. Let P_n be the (total) number of permutations of n:

$$P_n = \sum_{k=0}^{n} (n)_k, \quad n = 0, 1, \dots .$$

Show that

$$P(t) = \sum_{n=0}^{\infty} P_n \frac{t^n}{n!} = (1-t)^{-1} e^t$$

and

$$P_n = nP_{n-1} + 1, \quad n = 1, 2, \dots, \quad P_0 = 1.$$

21. Let $A(n,k)$ be the number of k-permutations of n with repetition and the restriction that each element appears zero or an even number of times. Using a suitable generating function, show that

$$A(n, 2s) = 2^{-n} \sum_{j=0}^{n} \binom{n}{j} (n-2j)^{2s}, \quad A(n, 2s+1) = 0, \quad s = 0, 1, \dots$$

and, for $A_1(r,s) = A(2r+1, 2s)$, $A_2(r,s) = A(2r, 2s)$, that

$$A_1(r,s) = (2r+1)\{(2r+1)A_1(r, s-1) - 2rA_1(r-1, s-1)\},$$

$$A_2(r,s) = 2r\{2rA_2(r, s-1) - (2r-1)A_2(r-1, s-1)\}.$$

22. Let $B(n,k)$ be the number of k-permutations of n with repetition and the restriction that each element appears an odd number of times. Using a suitable generating function, show that

$$B(2r, 2s) = 2^{-2(r-s)+1} \sum_{j=0}^{r} (-1)^j \binom{2r}{j} (r-j)^{2s},$$

$$B(2r, 2s+1) = 0,$$

and

$$B(2r + 1, 2s + 1) = 2^{-2(r-s)+1} \sum_{j=0}^{r} (-1)^j \binom{2r + 1}{j} (r - 2j + 1/2)^{2s+1},$$

$$B(2r + 1, 2s) = 0.$$

23. (*Continuation*). For $B_1(r, s) = B(2r + 1, 2s + 1)$ and $B_2(r, s) = B(2r, 2s)$, show that

$$B_1(r, s) = (2r + 1)\{(2r + 1)B_1(r, s - 1) + 2rB_1(r - 1, s - 1)\},$$

and

$$B_2(r, s) = 2r\{2rB_2(r, s - 1) + (2r - 1)B_2(r - 1, s - 1)\}.$$

24. Let $R(n, k)$ be the number of k-permutations of n with repetition and the restriction that each element appears at most twice. Using a suitable generating function, deduce the recurrence relations

$$R(n + 1, k) = R(n, k) + kR(n, k - 1) + \binom{k}{2}R(n, k - 2),$$

$$R(n, k + 1) = nR(n, k) - n\binom{k}{2}R(n - 1, k - 2)$$

and

$$R(n, k + 2) = n(2n - 1)R(n - 1, k) - n(n - 1)R(n - 2, k)$$

with $R(n, 0) = 1$, $R(n, 1) = 1$, $R(n, 2) = n^2$.

25. (*Continuation*). Let $R(n)$ be the (total) number of permutations of n with repetition and the restriction that each element appears at most twice:

$$R(n) = \sum_{k=0}^{2n} R(n, k), \quad n = 0, 1, \dots .$$

Show that

$$R(n) - n(2n - 1)R(n - 1) + n(n - 1)R(n - 2) = n + 1,$$

with $R(0) = 1$, $R(1) = 3$.

26. *Parcelling out procedures.* Let K_n be the number of ways of parcelling out a stick of n units length into n unitary parts when, at each step, all the parts of length greater than one unit are parceled out into two parts. Show that

$$K_n = \sum_{j=1}^{n-1} K_j K_{n-j}, \quad n = 2, 3, \dots, \quad K_1 = 1$$

and conclude that

$$K_n = C_n = \frac{1}{n}\binom{2n-2}{n-1}, \quad n = 1, 2, \ldots,$$

where C_n, $n = 1, 2, \ldots$, are the *Catalan numbers*.

27*. *Triangulation of convex polygons.* Let T_n be the number of triangulations of a convex n-gon (ways to cut up a convex polygon of n vertices into $n - 2$ triangles by means of $n - 3$ non-intersecting diagonals). Show that

$$T_n = \sum_{k=2}^{n-1} T_k T_{n-k+1}, \quad n = 3, 4, \ldots, \quad T_2 = 1.$$

Deduce the generating function

$$T(t) = \sum_{n=2}^{\infty} T_n t^n = \frac{t}{2}\left(1 - \sqrt{1 - 4t}\right)$$

and conclude that

$$T_n = C_{n-1} = \frac{1}{n-1}\binom{2n-4}{n-2}, \quad n = 2, 3, \ldots,$$

where C_n, $n = 1, 2, \ldots$, are the *Catalan numbers*.

28. *Convolution of Catalan numbers.* Let C_n, $n = 1, 2, \ldots$, be the sequence of Catalan numbers. For fixed $k = 2, 3, \ldots$, the sequence

$$C_n^{(k)} = \sum_{j=1}^{n-1} C_j C_{n-j}^{(k-1)}, \quad n = 1, 2, \ldots,$$

with $C_n^{(1)} = C_n$, $n = 1, 2, \ldots$, is called *k-fold convolution* of the sequence C_n, $n = 1, 2, \ldots$ (a) Derive the generating function

$$C_k(t) = \sum_{n=1}^{\infty} C_n^{(k)} t^n = 2^{-k}(1 - \sqrt{1 - 4t})^k$$

and (b) deduce the recurrence

$$C_n^{(k)} = C_n^{(k+1)} + C_{n-1}^{(k-1)}, \quad k = 2, 3, \ldots, \quad n = 1, 2, \ldots,$$

with $C_n^{(1)} = C_n^{(2)} = C_n$. (c) Show that

$$C_n^{(k)} = \psi_{n-1, n-k} = \frac{k}{n}\binom{2n-k-1}{n-1}, \quad n = 1, 2, \ldots, \quad k = 2, 3, \ldots,$$

where $\psi_{n,k}$, $k = 0, 1, \ldots, n$, $n = 0, 1, \ldots$, are the *ballot numbers*.

29*. Let S_n be the number of permutations of $\{1, 2, \ldots, n\}$ without a succession. If $S_{n,k}$ is the number of permutations of $\{1, 2, \ldots, n\}$ with k successions, then (see Theorem 5.6)

$$S_{n,k} = \binom{n-1}{k} S_{n-k}, \quad k = 0, 1, \ldots, n-1, \quad n = 1, 2, \ldots.$$

Using this relation and the symbolic calculus, show that

$$S(t) = \sum_{n=0}^{\infty} S_{n+1} \frac{t^n}{n!} = (1-t)^{-2} e^{-t}.$$

30*. (*Continuation*). Show that

$$S(t, u) = \sum_{n=0}^{\infty} \sum_{k=0}^{n} S_{n+1,k} t^k \frac{u^n}{n!} = (1-u)^{-2} e^{u(t-1)}.$$

31. Determine the sequence of probabilities p_k, $k = 0, 1, \ldots$, with factorial moments $\mu_{(r)} = \lambda^r$, $r = 0, 1, \ldots$, where $\lambda > 0$.

32. Let

$$S_{n,k}(x, z) = \sum_{r=0}^{k} (x + rz)^n, \quad n = 0, 1, \ldots, \quad k = 0, 1, \ldots$$

Show that

$$S_k(u, x, z) = \sum_{n=0}^{\infty} S_{n,k}(x, z) \frac{u^n}{n!} = \frac{e^{xu}(e^{(k+1)zu} - 1)}{e^{zu} - 1}$$

and

$$S(t, u, x, z) = \sum_{n=0}^{\infty} \sum_{k=0}^{\infty} S_{n,k}(x, z) \frac{t^k}{k!} \frac{u^n}{n!} = \frac{e^{xu}(e^{zu+te^{zu}} - e^t)}{e^{zu} - 1}.$$

33. Let

$$C_{n,k}(x, z) = \sum_{r=0}^{k} (x + rz)_n, \quad n = 0, 1, \ldots, \quad k = 0, 1, \ldots.$$

Show that

$$C_k(u, x, z) = \sum_{n=0}^{\infty} C_{n,k}(x, z) \frac{u^n}{n!} = \frac{(1+u)^x [(1+u)^{(k+1)z} - 1]}{(1+u)^z - 1}.$$

and

$$C(t, u, x, z) = \sum_{n=0}^{\infty} \sum_{k=0}^{\infty} C_{n,k}(x, z) \frac{t^k}{k!} \frac{u^n}{n!} = \frac{(1+u)^x [(1+u)^z e^{t(1+u)^z} - e^t]}{(1+u)^z - 1}.$$

34. Let $a_{n,k} = \min\{n, k\}$, $k = 1, 2, \ldots$, $n = 1, 2, \ldots$. Show that

$$A(t, u) = \sum_{n=1}^{\infty} \sum_{k=1}^{\infty} a_{n,k} t^k u^n = \frac{tu}{(1-t)(1-u)(1-tu)}.$$

35*. *Generating function of the cycles of binomial coefficients.* The product

$$a_{k,r} = \binom{r+k}{k} \binom{k+r}{r}, \quad k = 0, 1, \ldots, \quad r = 0, 1, \ldots,$$

is called *cycle of binomial coefficients of length* 2. Show that

$$A(t, u) = \sum_{r=0}^{\infty} \sum_{k=0}^{\infty} \binom{r+k}{k} \binom{k+r}{r} t^k u^r = \sum_{r=0}^{\infty} \sum_{k=0}^{\infty} \binom{r+k}{k}^2 t^k u^r$$

$$= [(1-t+u)^2 - 4u]^{-1/2}.$$

Chapter 7

RECURRENCE RELATIONS

7.1 INTRODUCTION

Recurrence relations were introduced in Chapter 1, on the basic counting principles, and encountered in several places in subsequent chapters. As has been already noted, in certain enumeration problems, the number of configurations satisfying specified conditions can only be expressed recursively. Also, even when the direct expression of this number in a closed form is possible, a recurrence relation is useful at least for tabulation purposes.

This chapter presents the basic methods of solving linear recurrence relations. Specifically, after the introduction of the basic notions of linear recurrence relations, the iteration method is employed to derive the solutions of linear recurrence relations of the first order. This recursive, step-by-step, derivation of the solutions contributes to the understanding of the term *recurrence relation*. Then, the method of characteristic roots for the solution of linear recurrence relations with constant coefficients is presented. The last section is devoted to the use of generating functions in solving linear recurrence relations with constant or variable coefficients.

7.2 BASIC NOTIONS

Consider a sequence of numbers a_n, $n = 0, 1, \ldots$, and let

$$b_0(n)a_{n+r} + b_1(n)a_{n+r-1} + \cdots + b_r(n)a_n = u(n), \quad n = 0, 1, \ldots, \quad (7.1)$$

where $u(n)$ and the coefficients $b_j(n)$, $j = 0, 1, \ldots, r$, with $b_0(n) \neq 0$ and $b_r(n) \neq 0$, are given functions of n. Recurrence relation (7.1) is called *linear recurrence relation of order* r. If $u(n) = 0$, $n = 0, 1, \ldots$, (7.1) is called *homogeneous*; otherwise is called *complete*. If the coefficients are constant (independent of n), $b_j(n) = b_j$, $j = 0, 1, \ldots, r$, $n = 0, 1, \ldots$, (7.1) is called *linear recurrence relation of order r with constant coefficients*. In the case of a double-index sequence $a_{n,k}$, $n = 0, 1, \ldots$, $k = 0, 1, \ldots$, the recurrence relation

$$b_{0,0}(n,k)a_{n+r,k+s} + b_{0,1}(n,k)a_{n+r,k+s-1} + \cdots + b_{r,s}(n,k)a_{n,k} = u(n,k),$$
$$(7.2)$$

$n = 0, 1, \ldots$, $k = 0, 1, \ldots$, where $u(n,k)$ and the coefficients $b_{i,j}(n,k)$, $i = 0, 1, \ldots, r$, $j = 0, 1, \ldots, s$, with $b_{0,0}(n,k) \neq 0$ and $b_{r,s}(n,k) \neq 0$, are given functions of n and k, is called *linear recurrence relation of order (r, s)*.

Solution of recurrence relation (7.1) or (7.2) in a set S is called a sequence that makes this equation an identity in S. A general solution of recurrence relation (7.1) includes r arbitrary constants. The knowledge of the r *initial conditions* (values) $a_0, a_1, \ldots, a_{r-1}$ specifies the constants and makes the solution unique. In the case of recurrence relation (7.2), the knowledge of the $r + s$ *initial conditions* (sequences) $a_{0,k}, a_{1,k}, \ldots, a_{r-1,k}$ and $a_{n,0}, a_{n,1}, \ldots, a_{n,s-1}$ guarantees the uniqueness of its solution.

Let us now consider the homogeneous linear recurrence relation of order r corresponding to (7.1):

$$b_0(n)a_{n+r} + b_1(n)a_{n+r-1} + \cdots + b_r(n)a_n = 0, \quad n = 0, 1, \ldots, \quad (7.3)$$

where the coefficients $b_j(n)$, $j = 0, 1, \ldots, r$, with $b_0(n) \neq 0$ and $b_r(n) \neq 0$, are given functions of n. Note that, if $a_1(n)$ and $a_2(n)$ are any solutions of (7.3), then $c_1 a_1(n) + c_2 a_2(n)$, where c_1 and c_2 are arbitrary constants, is also a solution of (7.3). In general, it can be shown that the set of solutions of (7.3) constitutes a linear r dimensional vector space. Note that r solutions $a_1(n), a_2(n), \ldots, a_r(n)$ of (7.3) are linearly independent if and only if their Wronski determinant

$$W_r(n) = \begin{vmatrix} a_1(n) & a_2(n) & \cdots & a_r(n) \\ a_1(n+1) & a_2(n+1) & \cdots & a_r(n+1) \\ \cdots\cdots\cdots\cdots\cdots\cdots\cdots\cdots\cdots\cdots\cdots\cdots\cdots\cdots \\ a_1(n+r-1) & a_2(n+r-1) & \cdots & a_r(n+r-1) \end{vmatrix}$$

is different from zero for some index $n = m$. Consequently, if the r solutions $a_1(n), a_2(n), \ldots, a_r(n)$ of the homogeneous linear recurrence relation of order r (7.3) are linearly independent, then they constitute a base for the r dimensional linear vector space of its solutions. Further, every solution of (7.3) is of the form

$$a_n = c_1 a_1(n) + c_2 a_2(n) + \cdots + c_r a_r(n), \quad\quad (7.4)$$

where $c_1, c_2, \ldots, c_r$ are arbitrary constants. The solution (7.4) is called *general solution* of (7.3). In the case where r values $a_m, a_{m+1}, \ldots, a_{m+r-1}$

are given, the system

$$c_1 a_1(m) + c_2 a_2(m) + \cdots + c_r a_r(m) = a_m,$$
$$c_1 a_1(m+1) + c_2 a_2(m+1) + \cdots + c_r a_r(m+1) = a_{m+1},$$
$$\cdots\cdots\cdots\cdots\cdots\cdots\cdots\cdots\cdots\cdots\cdots\cdots\cdots\cdots\cdots\cdots\cdots\cdots$$
$$c_1 a_1(m+r-1) + c_2 a_2(m+r-1) + \cdots + c_r a_r(m+r-1) = a_{m+r-1},$$

since $W_r(m) \neq 0$, has a unique solution with respect to $c_1, c_2, \dots, c_r$. Introducing this solution into (7.4), the unique solution of (7.3) is deduced. Further, if $w(n)$ is a *particular solution* of the complete linear recurrence relation of order r (7.1), then, according to the preceding analysis, it follows directly that

$$a_n = c_1 a_1(n) + c_2 a_2(n) + \cdots + c_r a_r(n) + w(n)$$

is the *general solution* of (7.1).

7.3 RECURRENCE RELATIONS OF THE FIRST ORDER

Let us first consider the *complete linear recurrence relation of the first order with variable coefficient,*

$$a_{n+1} - b(n)a_n = u(n), \quad n = m, m+1, \dots, \tag{7.5}$$

where $b(n) \neq 0$, $n = m, m+1, \dots$ and m is a given nonnegative integer. The solution of this recurrence relation is deduced in the following theorem by employing the *iteration method*. This recursive, step-by-step, derivation of the solution serves to the better understanding of the term *recurrence relation*.

THEOREM 7.1
The solution of the linear recurrence relation (7.5), with a_m a given initial condition, is

$$a_n = a_m b_{n-1,m} + \sum_{r=m}^{n-2} u(r) b_{n-1,r+1} + u(n-1), \tag{7.6}$$

for $n = m+1, m+2, \dots$, where

$$b_{n,r} = \prod_{k=r}^{n} b(k). \tag{7.7}$$

PROOF   Introducing the transformation

$$h_n = a_n/b_{n-1,m}, \quad n = m+1, m+2, \ldots, \quad h_m = a_m$$

and setting $w(n) = u(n)/b_{n-1,m}, n = m, m+1, \ldots,$ recurrence relation (7.5) is transformed to the recurrence relation

$$h_{n+1} = h_n + w(n), \quad n = m, m+1, \ldots,$$

where $h_m = a_m$ is a given initial condition. Iterating (applying repeatedly) this recurrence, we get

$$h_{m+1} = h_m + w(m),$$

$$h_{m+2} = h_{m+1} + w(m+1) = h_m + w(m) + w(m+1)$$

and

$$h_{n+1} = h_n + w(n) = \{h_m + w(m) + w(m+1) + \cdots + w(n-1)\} + w(n).$$

Consequently,

$$h_{n+1} = h_m + \sum_{r=m}^{n} w(r), \quad n = m, m+1, \ldots.$$

Returning to the sequence $a_n, n = m, m+1, \ldots,$ the last expression, upon using the inverse transformation

$$a_n = h_n b_{n-1,m}, \quad n = m+1, m+2, \ldots, \quad a_m = h_m,$$

and since $u(n) = w(n)b_{n-1,m}, n = m, m+1, \ldots,$ implies

$$a_{n+1} = a_m b_{n,m} + \sum_{r=m}^{n-1} u(r)b_{n,r+1} + u(n), \quad n = m, m+1, \ldots,$$

which is the required expression with $n+1$ instead of n.   ∎

The solution of the complete linear recurrence relation of the first order with constant coefficient,

$$a_{n+1} - ba_n = u(n), \quad n = m, m+1, \ldots, \tag{7.8}$$

where $b \neq 0, n = m, m+1, \ldots$ and m is a given nonnegative integer, is readily deduced from Theorem 7.1 by setting $b(n) = b, n = m, m+1, \ldots.$ Also, its particular case with $u(n) = u$ constant for all $n = m, m+1, \ldots,$ upon using the geometric progression summation formula, is concluded. These solutions of (7.8) are given in the following corollary.

COROLLARY 7.1
The solution of the linear recurrence relation (7.8), with a_m a given initial condition, is

$$a_n = a_m b^{n-m} + \sum_{r=m}^{n-1} u(r) b^{n-r-1}, \quad n = m+1, m+2, \ldots. \qquad (7.9)$$

In particular, with $u(n) = u$ constant for all $n = m, m+1, \ldots,$

$$a_n = \begin{cases} a_m b^{n-m} + u \cdot (1 - b^{n-m})/(1-b), & b \neq 1, \\ \\ a_m + u \cdot (n-m), & b = 1, \end{cases} \qquad (7.10)$$

for $n = m+1, m+2, \ldots.$

REMARK 7.1   If the term a_m is not given, (7.6) with $a_m = c$, an arbitrary constant, constitutes a family of solutions, which is the *general solution* of recurrence relation (7.5). Similarly, (7.9) and (7.10) with $a_m = c$ are the respective general solutions of (7.8).   ∎

Example 7.1   Tennis tournament
Let $2n$ players participate in a singles tennis tournament. Determine, by the aid of a recurrence relation, the number a_n of different pairs that can be formed for the n matches of the first round.

Consider the player with the number $2n$ in the list. This player can be paired with any of the other $2n - 1$ players. Since a_{n-1} different pairs can be formed by the remaining $2(n-1)$ players for the other $n-1$ matches of the first round, it follows that
$$a_n = (2n-1)a_{n-1}, \quad n = 2, 3, \ldots, \quad a_1 = 1.$$
This is a homogeneous linear recurrence relation of the first order with variable coefficient. Thus from (7.6), with $m = 1$, $u(n) = 0$, $b(n) = 2n+1$, $n = 1, 2, \ldots,$ and $a_1 = 1$, we deduce the expression

$$a_n = 1 \cdot 3 \cdots (2n-1), \quad n = 1, 2, \ldots.$$

Multiplying it by $2 \cdot 4 \cdots 2n = 2^n n!$ and dividing the resulting expression by the same number, we find the following equivalent expression for a_n:

$$a_n = \frac{1 \cdot 2 \cdot 3 \cdots (2n-1)(2n)}{2^n n!} = \frac{(2n)!}{2^n n!}, \quad n = 1, 2, \ldots. \qquad ∎$$

Example 7.2   Regions of a plane
Determine, using a recurrence relation, the number a_n of regions into which a plane is separated by n circles, with each pair of circles intersecting in exactly two points and with no triple of circles having a common intersecting point.

Consider n circles on a plane, with each pair of circles intersecting in exactly two points and no triple of circles having a common intersecting point. Then these circles separate the plane into a_n regions. Now add another circle, which intersects each of the n circles at exactly two points and does not intersect any pair of circles at any of their intersecting points. Thus, this circle intersects the n circles at a total of $2n$ points, passing through $2n$ regions. Each of these regions is separated into two regions. Consequently, the addition of the $(n + 1)$st circle increases the number of regions by $2n$ and so

$$a_{n+1} = a_n + 2n, \quad n = 1, 2, \ldots, \quad a_1 = 2.$$

This is a complete linear recurrence relation of the first order with constant coefficient. Hence, from (7.9), with $m = 1, b = 1, a_1 = 2$ and $u(r) = 2r$, we deduce the expression

$$a_n = 2 + 2 \sum_{r=1}^{n-1} r,$$

which, upon using the arithmetic progression summation formula, reduces to

$$a_n = n(n - 1) + 2, \quad n = 1, 2, \ldots. \qquad \square$$

Example 7.3 The transfer of the Hanoi tower

Consider three pegs A, B and C and n cyclic discs of different diameters $\{d_1, d_2, \ldots, d_n\}$, with $d_1 < d_2 < \cdots < d_n$. Initially, the discs are placed on peg A in decreasing order of size from the bottom to the top. Let a_n be the minimum number of movements of discs required for the transfer of the tower from peg A to peg B, using peg C as an auxiliary, under the restriction that, in each movement, only one disc is transferred and no disc is placed over a disc of smaller size.

In order to find a recurrence relation for a_n, note that the $n - 1$ discs $\{d_1, d_2, \ldots, d_{n-1}\}$ of the tower can be transferred to peg C after a_{n-1} movements. Then, in one movement, the last disc d_n is transferred from peg C to peg B. Finally, the tower of the $n - 1$ discs is transferred from peg C to peg B, over the disc d_n, after a_{n-1} movements. Consequently,

$$a_n = 2a_{n-1} + 1, \quad n = 2, 3, \ldots, \quad a_1 = 1.$$

This is a complete recurrence relation of the first order with constant coefficient. Hence, from (7.10), with $m = 1, a_1 = 1$ and $u = 1$, we get

$$a_n = 2^{n-1} - (1 - 2^{n-1}) = 2^n - 1.$$

According to legend, at the creation God established the tower of Hanoi at the temple of Benares with $n = 64$ gold discs. Since then, the tower is being transferred by priests, with the prediction that when the transfer of the tower is completed the world will vanish. Note that $a_2 = 3$, $a_3 = 7$, $a_4 = 15$, $a_{16} = 65{,}535$, $a_{32} = 4{,}294{,}967{,}295$ and $a_{64} = 18{,}446{,}744{,}073{,}709{,}551{,}615$. $\qquad \square$

7.4 THE METHOD OF CHARACTERISTIC ROOTS

Let us first consider the *homogeneous linear recurrence relation of the second order with constant coefficients*:

$$a_{n+2} + b_1 a_{n+1} + b_2 a_n = 0, \quad n = m, m+1, \dots, \qquad (7.11)$$

where $b_2 \neq 0$ and m is a given nonnegative integer. Clearly, a sequence of the form $a(n) = \rho^n$, $n = m, m+1, \dots$, is a solution of (7.11) if and only if ρ is a root of the equation

$$t^{n+2} + b_1 t^{n+1} + b_2 t^n = 0.$$

Dividing both sides of this equation by t^n, for $t \neq 0$, we obtain the following equation of the second order,

$$t^2 + b_1 t + b_2 = 0, \qquad (7.12)$$

which is called the *characteristic equation* of the recurrence relation (7.11). Further, the *discriminant* $b_1^2 - 4b_2$ of the characteristic equation determines the kind of its roots. Specifically, if $b_1^2 - 4b_2 > 0$, (7.12) has two distinct real roots,

$$\rho_1 = \frac{-b_1 - \sqrt{b_1^2 - 4b_2}}{2}, \quad \rho_2 = \frac{-b_1 + \sqrt{b_1^2 - 4b_2}}{2},$$

while, if $b_1^2 - 4b_2 = 0$, it has one double root $\rho = -b_1/2$. Also, if $b_1^2 - 4b_2 < 0$, (7.12) has two conjugate complex roots

$$\rho_1 = \frac{-b_1 - i\sqrt{4b_2 - b_1^2}}{2}, \quad \rho_2 = \frac{-b_1 + i\sqrt{4b_2 - b_1^2}}{2},$$

where $i = \sqrt{-1}$ is the imaginary unit. These complex numbers, on using the length $\rho = b_2$ and argument θ, with $\tan\theta = -\sqrt{4b_1 - b_1^2}/b_1$, may be written in trigonometric form as

$$\rho_1 = \rho(\cos\theta + i\sin\theta), \quad \rho_2 = \rho(\cos\theta - i\sin\theta).$$

The general solution of recurrence relation (7.11) depends on the kind of roots of the characteristic equation (7.12). More precisely, we prove the following theorem.

THEOREM 7.2

The general solution of the homogeneous linear recurrence relation of the second order with constant coefficients, (7.11), is given by

$$a_n = c_1 a_1(n) + c_2 a_2(n), \quad n = m, m+1, \dots, \qquad (7.13)$$

where c_1 and c_2 are arbitrary constants. Further, if the characteristic equation (7.12) has:

(a) *two distinct real roots ρ_1 and ρ_2, then*

$$a_1(n) = \rho_1^n, \quad a_2(n) = \rho_2^n, \tag{7.14}$$

(b) *one double root ρ, then*

$$a_1(n) = \rho^n, \quad a_2(n) = n\rho^n, \tag{7.15}$$

(c) *two conjugate complex roots ρ_1 and $\rho_2 = \bar{\rho}_1$, with length ρ and argument θ and $-\theta$, respectively, then*

$$a_1(n) = \rho^n \cos(n\theta), \quad a_2(n) = \rho^n \sin(n\theta). \tag{7.16}$$

PROOF   (a) Since ρ_1 and ρ_2 are roots of the characteristic equation (7.12), $a_1(n) = \rho_1^n$ and $a_2(n) = \rho_2^n$ are solutions of (7.11). The Wronski determinant of these solutions is

$$W_2(n) = \begin{vmatrix} \rho_1^n & \rho_2^n \\ \rho_1^{n+1} & \rho_2^{n+1} \end{vmatrix} = (\rho_1\rho_2)^n(\rho_2 - \rho_1),$$

with $\rho_1\rho_2 \neq 0$, since $b_2 \neq 0$, and $\rho_1 \neq \rho_2$. Hence, $W_2(n) \neq 0$ and so the solutions in (7.14) constitute a base of the two-dimensional linear vector space of the solutions of (7.11).

(b) Since ρ is a solution of the characteristic equation (7.12), $a(n) = \rho^n$ is a solution of (7.11). A base of the two-dimensional vector space of the solutions of (7.11) is composed of two linearly independent solutions. In order to determine them, we set

$$a_n = h_n\rho^n, \quad n = m, m + 1, \ldots, \tag{7.17}$$

where $h_n, n = m, m + 1, \ldots$, is a sequence to be determined. Introducing it into (7.11), we get

$$\rho^2 h_{n+2} + b_1\rho h_{n+1} + b_2 h_n = 0,$$

and, since $b_1 = -2\rho$, $b_2 = \rho^2$, with $\rho \neq 0$, we deduce for h_n, $n = m$, $m + 1, \ldots$, the recurrence relation

$$h_{n+2} - 2h_{n+1} + h_n = 0, \quad n = m, m + 1, \ldots.$$

Putting

$$h_{n+1} - h_n = g_n, \quad n = m, m + 1, \ldots,$$

we get the homogeneous linear recurrence relation of the first order

$$g_{n+1} - g_n = 0, \quad n = m, m + 1, \ldots,$$

the general solution of which, according to Remark 7.1, is given by

$$g_n = c_1, \quad n = m, m+1, \ldots .$$

Consequently,

$$h_{n+1} - h_n = c_1, \quad n = m, m+1, \ldots .$$

The general solution of this recurrence relation, again according to Remark 7.1, is given by

$$h_n = c_1 + c_2 n, \quad n = m, m+1, \ldots .$$

Introducing this general solution into (7.17), we deduce the required expression of the general solution of (7.11). Note that the Wronski determinant of the solutions $a_1(n) = \rho^n$ and $a_2(n) = n\rho^n$ is given by

$$W_2(n) = \begin{vmatrix} \rho^n & n\rho^n \\ \rho^{n+1} & (n+1)\rho^{n+1} \end{vmatrix} = \rho^{2n+1}$$

and since $\rho \neq 0$, $W_2(n) \neq 0$.

(c) Since ρ and $\rho_2 = \bar{\rho}_1$ are conjugate complex roots of the characteristic equation (7.12), $a_n = \rho_1^n$ and $\bar{a}_n = \bar{\rho}_1^n$ are conjugate complex solutions of (7.11), which may be written in trigonometric form as

$$a_n = \rho^n \cos(n\theta) + i\rho^n \sin(n\theta), \quad \bar{a}_n = \rho^n \cos(n\theta) - i\rho^n \sin(n\theta).$$

Interested only in the real solutions of recurrence relation (7.11), it is necessary to isolate them. In this respect, note that, if $a_n = a_1(n) + ia_2(n)$ is a complex solution of the characteristic equation (7.12), then each of $a_1(n)$ and $a_2(n)$ is also a solution of the same equation since

$$[a_1(n+2) + b_1 a_1(n+1) + b_2 a_1(n)]$$
$$+ i[a_2(n+2) + b_1 a_2(n+1) + b_2 a_2(n)] = 0, \quad n = m, m+1, \ldots ,$$

implies

$$a_1(n+2) + b_1 a_1(n+1) + b_2 a_1(n) = 0, \quad n = m, m+1, \ldots ,$$

and

$$a_2(n+2) + b_1 a_2(n+1) + b_2 a_2(n) = 0, \quad n = m, m+1, \ldots .$$

Consequently,

$$a_1(n) = \rho^n \cos(n\theta), \quad a_2(n) = \rho^n \sin(n\theta)$$

are two real solutions of (7.11). The Wronski determinant of these solutions is given by

$$W_2(n) = \begin{vmatrix} \rho^n \cos(n\theta) & \rho^n \sin(n\theta) \\ \rho^{n+1} \cos(n\theta + \theta) & \rho^{n+1} \sin(n\theta + \theta) \end{vmatrix} .$$

Thus,

$$W_2(n) = \rho^{2n+1}[\cos(n\theta)\sin(n\theta + \theta) - \sin(n\theta)\cos(n\theta + \theta)]$$

and since

$$\sin(n\theta + \theta) = \sin(n\theta)\cos\theta + \cos(n\theta)\sin\theta,$$

$$\cos(n\theta + \theta) = \cos(n\theta)\cos\theta - \sin(n\theta)\sin\theta,$$

the Wronski determinant reduces to

$$W_2(n) = \rho^{2n+1}\sin\theta[\cos^2(n\theta) + \sin^2(n\theta)] = \rho^{2n+1}\sin\theta,$$

which, by virtue of $\rho \neq 0$ and $\sin\theta \neq 0$, implies $W_2(n) \neq 0$. Hence, the solutions in (7.16) constitute a base of the two-dimensional linear vector space of the solutions of (7.11) and the proof of the theorem is completed. ∎

Example 7.4 Gambler's ruin

A gambler plays a series of games of chance against a casino (or against an adversary). In any game the probability of the gambler to win \$1 is p and to loose \$1 is $q = 1 - p$. Assume that initially the gambler possesses n dollars and the casino (or his adversary) possesses $k - n$ dollars. Find the probability p_n of the gambler's ruin.

In the first game, the gambler may either win or loose \$1, with probabilities p and q, respectively. If he wins the first game, then his fortune is increased to $n + 1$ and so his ruin probability becomes p_{n+1}. If he looses the first game, his fortune is decreased to $n - 1$ and so his ruin probability becomes p_{n-1}. Hence

$$p_n = pp_{n+1} + qp_{n-1}, \quad n = 1, 2, \ldots, k - 1$$

or

$$pp_{n+2} - p_{n+1} + qp_n = 0, \quad n = 0, 1, \ldots, k - 2,$$

with

$$p_0 = 1, \quad p_k = 0.$$

This is a linear recurrence relation with constant coefficients. The roots of its characteristic equation,

$$pt^2 - t + q = 0,$$

are $\rho_1 = 1$ and $\rho_2 = q/p$. If $p \neq 1/2$, whence $q \neq 1/2$, these two roots are distinct, while if $p = 1/2$, whence $q = 1/2$, $\rho_2 = \rho_1 = 1$. Thus, according to Theorem 7.2, the general solution of the recurrence relation: (a) for $p \neq 1/2$ is

$$p_n = c_1 + c_2(q/p)^n, \quad n = 0, 1, \ldots, k,$$

while (b) for $p = 1/2$ is

$$p_n = c_1 + c_2 n, \quad n = 0, 1, \ldots, k.$$

Using the initial conditions we get: (a) for $p \neq 1/2$,

$$c_1 + c_2 = 1, \quad c_1 + c_2(q/p)^k = 0,$$

whence

$$c_1 = \frac{(q/p)^k}{1 - (q/p)^k}, \quad c_2 = \frac{1}{1 - (q/p)^k},$$

while (b) for $p = 1/2$,

$$c_1 = 1, \quad c_1 + c_2 k = 0,$$

whence

$$c_1 = 1, \quad c_2 = -\frac{1}{k}.$$

Consequently, the solution of the recurrence relation that satisfies the initial conditions: (a) for $p \neq 1/2$ is

$$p_n = \frac{(q/p)^n - (q/p)^k}{1 - (q/p)^k}, \quad n = 0, 1, \ldots, k$$

and (b) for $p = 1/2$ is

$$p_n = \frac{k - n}{k}, \quad n = 0, 1, \ldots, k. \qquad \square$$

Example 7.5

Determine the sequence a_n, $n = 0, 1, \ldots$, for which the general term is the arithmetic mean of its two preceding terms and the first two terms are 0 and 1.

The sequence a_n, $n = 0, 1, \ldots$, according to its definition, satisfies the linear recurrence relation

$$a_n = \frac{1}{2}(a_{n-1} + a_{n-2}), \quad n = 2, 3, \ldots,$$

or

$$2a_{n+2} - a_{n+1} - a_n = 0, \quad n = 0, 1, \ldots,$$

with initial conditions $a_0 = 0$ and $a_1 = 1$. The roots of the characteristic equation,

$$2t^2 - t - 1 = 0,$$

are $\rho_1 = 1$ and $\rho_2 = -1/2$. Thus, according to Theorem 7.2, the general solution of the recurrence relation is

$$a_n = c_1 + c_2 \frac{(-1)^n}{2^n}, \quad n = 0, 1, \ldots.$$

The initial conditions require that

$$c_1 + c_2 = 0, \quad c_1 - \frac{1}{2}c_2 = 1.$$

Hence

$$c_1 = \frac{2}{3}, \quad c_2 = -\frac{2}{3}$$

and

$$a_n = \frac{2}{3}\left[1 - \frac{(-1)^n}{2^n}\right], \quad n = 0, 1, \dots . \quad \square$$

Let us now consider the *complete linear recurrence relation of the second order with constant coefficients*:

$$a_{n+2} + b_1 a_{n+1} + b_2 a_n = u(n), \quad n = m, m+1, \dots , \tag{7.18}$$

where $b_2 \neq 0$ and m is a given nonnegative integer. According to what was stated in Section 7.2, if $c_1 a_1(n) + c_2 a_2(n)$, $n = m, m+1, \dots$, is the general solution of the corresponding homogeneous linear recurrence relation (7.11) and $w(n)$, $n = m, m+1, \dots$, is a particular solution of (7.18), then the general solution of (7.18) is given by

$$a_n = c_1 a_1(n) + c_2 a_2(n) + w(n), \quad n = m, m+1, \dots .$$

Consequently, the derivation of a particular solution of (7.18) is what remains to be done. Consider the case

$$u(n) = b^n \sum_{j=0}^{s} u_j \binom{n}{j}, \quad n = m, m+1, \dots , \tag{7.19}$$

where b and u_j, $j = 0, 1, \dots , s$, are constants. The method of arbitrary constants for the derivation of a particular solution of (7.18) is stated in the following theorem.

THEOREM 7.3
A particular solution of the complete linear recurrence relation of the second order with constant coefficients (7.18) in the case the function $u(n)$ is given by (7.19), with b a root of the characteristic polynomial $\phi(t) = t^2 + b_1 t + b_2$ of multiplicity $k \geq 0$, is given by

$$w(n) = b^{n-k} \sum_{j=k}^{k+s} w_j \binom{n}{j}, \quad n = m, m+1, \dots , \quad k = 0, 1, 2, \tag{7.20}$$

where the coefficients w_j, $j = k, k+1, \dots , k+s$, are determined by the system of the $s+1$ equations:
    (a) for $k = 0$,

$$\phi(b)w_s = u_s,$$

$$\phi(b)w_{s-1} + b\phi'(b)w_s = u_{s-1}, \tag{7.21}$$

$$\phi(b)w_j + b\phi'(b)w_{j+1} + b^2 w_{j+2} = u_j, \quad j = 0, 1, \ldots, s - 2,$$

(b) *for* $k = 1$,

$$\phi'(b)w_{s+1} = u_s,$$

$$\phi'(b)w_{j+1} + bw_{j+2} = u_j, \quad j = 0, 1, \ldots, s - 1, \tag{7.22}$$

(c) *for* $k = 2$,

$$w_{j+2} = u_j, \quad j = 0, 1, \ldots, s, \tag{7.23}$$

with $\phi'(b)$ *the derivative of the characteristic polynomial* $\phi(t)$ *at the point* $t = b$.

PROOF   Introducing (7.19) into (7.18) and requiring (7.20) to satisfy the resulting recurrence relation, we get

$$b^2 \sum_{j=k}^{k+s} w_j \binom{n+2}{j} + b_1 b \sum_{j=k}^{k+s} w_j \binom{n+1}{j} + b_2 \sum_{j=k}^{k+s} w_j \binom{n}{j} = b^k \sum_{j=0}^{s} u_j \binom{n}{j}.$$

Using the recurrence relations

$$\binom{n+1}{j} = \binom{n}{j} + \binom{n}{j-1}$$

and

$$\binom{n+2}{j} = \binom{n}{j} + 2\binom{n}{j-1} + \binom{n}{j-2},$$

we deduce the relation

$$b^2 \sum_{j=k}^{k+s} w_j \binom{n}{j} + 2b^2 \sum_{j=k}^{k+s} w_j \binom{n}{j-1} + b^2 \sum_{j=k}^{k+s} w_j \binom{n}{j-2}$$

$$+ b_1 b \sum_{j=k}^{k+s} w_j \binom{n}{j} + b_1 b \sum_{j=k}^{k+s} w_j \binom{n}{j-1} + b_2 \sum_{j=k}^{k+s} w_j \binom{n}{j} = b^k \sum_{j=0}^{s} u_j \binom{n}{j}.$$

Introducing the characteristic polynomial $\phi(b) = b^2 + b_1 b + b_2$ and its derivative $\phi'(b) = 2b + b_1$, we get

$$\sum_{j=k}^{k+s} \phi(b)w_j \binom{n}{j} + \sum_{j=k}^{k+s} b\phi'(b)w_j \binom{n}{j-1} + \sum_{j=k}^{k+s} b^2 w_j \binom{n}{j-2} = b^k \sum_{j=0}^{s} u_j \binom{n}{j}.$$

Equating the coefficients of the binomials $\binom{n}{j}$ of both sides of this expression and since (a) for $k = 0$, $\phi(b) \neq 0$, (b) for $k = 1$, $\phi(b) = 0$, $\phi'(b) \neq 0$ and

(c) for $k = 2$, $\phi(b) = 0$, $\phi'(b) = 0$, we conclude (7.21), (7.22) and (7.23), respectively. ∎

Example 7.6 Expected time to gambler's ruin

Consider the gambler's ruin problem of Example 7.4 and let d_n be the expected value of the number of games until the gambler is ruined. Derive a recurrence relation for d_n and from that deduce an explicit expression for the expected time to gambler's ruin.

In the first game, the gambler may either win or loose $\$\,1$, with probabilities p and q, respectively. If he wins the first game, then his fortune is increased to $n+1$ and, after that game, the expected time to his ruin becomes d_{n+1}. If he looses the first game, his fortune is decreased to $n-1$ and, after that game, the expected time to his ruin becomes d_{n-1}. In both cases, adding the first game we deduce the following recurrence relation

$$d_n = p(d_{n+1} + 1) + q(d_{n-1} + 1), \quad n = 1, 2, \ldots, k-1,$$

or

$$d_{n+2} - (1/p)d_{n-1} + (q/p)d_n = -1/p, \quad n = 0, 1, \ldots, k-2,$$

with initial conditions

$$d_0 = 0, \quad d_k = 0.$$

This is a complete linear recurrence relation with constant coefficients. The general solution of the corresponding homogeneous recurrence relation,

$$d_{n+2} - (1/p)d_{n+1} + (q/p)d_n = 0, \quad n = 0, 1, \ldots, k-2,$$

according to Example 7.4: (a) for $p \neq 1/2$ is

$$d_n = c_1 + c_2(q/p)^n, \quad n = 0, 1, \ldots, k,$$

while (b) for $p = 1/2$ is

$$d_n = c_1 + c_2 n, \quad n = 0, 1, \ldots, k.$$

The function $u(n)$ of the complete linear recurrence relation is of the form (7.19), with $b = 1$, $s = 0$ and $u_0 = -1/p$. Note that (a) if $p \neq 1/2$, $b = 1$ is a simple root ($k = 1$), while (b) if $p = 1/2$, $b = 1$ is a double root ($k = 2$) of the characteristic polynomial $\phi(t) = t^2 - (1/p)t + (q/p)$. Thus, according to Theorem 7.3, a particular solution of the complete linear recurrence relation with constant coefficients: (a) for $p \neq 1/2$ is given by $w(n) = w_1 n$, where, by (7.22) and since $\phi'(1) = (2p-1)/p$, we get $w_1 = 1/(1-2p)$ and so

$$w(n) = \frac{n}{1-2p}.$$

Also, a particular solution (b) for $p = 1/2$ is given by $w(n) = w_2 n(n-1)/2$, where, by (7.23), $w_2 = -2$ and so

$$w(n) = -n(n-1).$$

Consequently, the general solution of the complete linear recurrence relation with constant coefficients: (a) for $p \neq 1/2$ is given by

$$d_n = c_1 + c_2 \left(\frac{q}{p}\right)^n + \frac{n}{1 - 2p}, \quad n = 0, 1, \ldots, k,$$

while (b) for $p = 1/2$ is given by

$$d_n = c_1 + c_2 n - n(n-1), \quad n = 0, 1, \ldots, k.$$

Using the initial conditions, we get (a) for $p \neq 1/2$,

$$c_1 + c_2 = 0, \quad c_1 + c_2 \left(\frac{q}{p}\right)^k + \frac{k}{1 - 2p} = 0,$$

whence

$$c_1 = -\frac{k}{1 - 2p} \cdot \frac{1}{1 - (q/p)^k}, \quad c_2 = \frac{k}{1 - 2p} \cdot \frac{1}{1 - (q/p)^k},$$

while (b) for $p = 1/2$,

$$c_1 = 0, \quad c_2 k - k(k-1) = 0,$$

whence

$$c_1 = 0, \quad c_2 = (k-1).$$

Therefore, the unique solution of the complete linear recurrence relation: (a) for $p \neq 1/2$ is given by

$$d_n = \frac{n}{1 - 2p} - \frac{k}{1 - 2p} \cdot \frac{1 - (q/p)^n}{1 - (q/p)^k}, \quad n = 0, 1, \ldots, k$$

and (b) for $p = 1/2$ is

$$d_n = n(k-n), \quad n = 0, 1, \ldots, k. \qquad \square$$

Let us finally consider the general case of the linear recurrence relation of the r-th order with constant coefficients:

$$a_{n+r} + b_1 a_{n+r-1} + \cdots + b_r a_n = u(n), \quad n = m, m+1, \ldots, \qquad (7.24)$$

where $b_r \neq 0$ and m is a given nonnegative integer. Let

$$u(n) = b^n \sum_{j=0}^{s} u_j \binom{n}{j}, \quad n = m, m+1, \ldots . \tag{7.25}$$

The corresponding homogeneous linear recurrence relation of the r-th order with constant coefficients is given by

$$a_{n+r} + b_1 a_{n+r-1} + \cdots + b_r a_n = 0, \quad n = m, m+1, \ldots, \tag{7.26}$$

where $b_r \neq 0$ and m is a given nonnegative integer. The characteristic polynomial of this recurrence relation is

$$\phi(t) = t^r + b_1 t^{r-1} + \cdots + b_r. \tag{7.27}$$

The general solution of (7.26) and a particular solution of (7.24) in the case where $u(n)$ is given by (7.25) are stated in the following two theorems. The proofs, similar to the proofs of Theorems 7.2 and 7.3, respectively, are omitted.

THEOREM 7.4

The general solution of the homogeneous linear recurrence relation of order r with constant coefficients, (7.26), is given by

$$a_n = \sum_{j=1}^{\nu_1} (c_{j,1} + c_{j,2} n + \cdots + c_{j,k_j} n^{k_j-1}) \rho_j^n$$

$$+ \sum_{j=\nu+1}^{\nu_2} (c_{j,1} + c_{j,2} n + \cdots + c_{j,k_j} n^{k_j-1}) \rho_j^n \cos(n\theta_j)$$

$$+ \sum_{j=\nu+1}^{\nu_2} (d_{j,1} + d_{j,2} n + \cdots + d_{j,k_j} n^{k_j-1}) \rho_j^n \sin(n\theta_j),$$

where ρ_j is a real root of the characteristic polynomial (7.27) of multiplicity $k_j \geq 0$, $j = 1, 2, \ldots, \nu_1$, with $k_1 + k_2 + \cdots + k_{\nu_1} = \nu$ and $\rho_j(\cos\theta_j + i\sin\theta_j)$ is a complex root of multiplicity $k_j \geq 0$, $j = \nu+1, \nu+2, \ldots, \nu_2$, with $2(k_{\nu+1} + k_{\nu+2} + \cdots + k_{\nu_2}) = r - \nu$.

THEOREM 7.5

A particular solution of the complete linear recurrence relation of order r with constant coefficients, (7.24), in the case the function $u(n)$ is given by (7.25), with b a root of the characteristic polynomial (7.27) of multiplicity $k \geq 0$, is given by

$$w(n) = b^{n-k} \sum_{j=k}^{k+s} w_j \binom{n}{j}, \quad n = m, m+1, \ldots, \quad k = 0, 1, \ldots, r,$$

where the coefficients w_j, $j = k, k + 1, \ldots, k + s$ are determined by the system of the $s + 1$ equations

$$\sum_{\nu=j}^{s} \frac{b^{\nu-j}}{(\nu + k - j)!} \phi^{(\nu+k-j)}(b) w_\nu = u_j, \quad j = 0, 1, \ldots, s,$$

with $\phi^{(\nu+k-j)}(b)$ denoting the derivative of order $\nu + k - j$ of the characteristic polynomial $\phi(t)$ at the point $t = b$.

Example 7.7

Determine the general solution of the recurrence relation

$$a_{n+4} + 2a_{n+2} + a_n = 2(a_{n+3} + a_{n+1}) + (5n - 2)2^n, \quad n = 0, 1, \ldots.$$

This relation is written as

$$a_{n+4} - 2a_{n+3} + 2a_{n+2} - 2a_{n+1} + a_n = (5n - 2)2^n, \quad n = 0, 1, \ldots,$$

which is a complete linear recurrence relation of the fourth order with constant coefficients. The corresponding homogeneous recurrence relation is

$$a_{n+4} - 2a_{n+3} + 2a_{n+2} - 2a_{n+1} + a_n = 0, \quad n = 0, 1, \ldots.$$

Its characteristic polynomial is given by

$$\phi(t) = t^4 - 2t^3 + 2t^2 - 2t + 1.$$

Note that $\phi(1) = 0$ and $\phi'(t) = 2(2t^3 - 3t^2 + 2t - 1)$, whence $\phi'(1) = 0$. Thus $\rho = 1$ is a double real root of the characteristic polynomial. Dividing it by $(t-1)^2$ we deduce the expression

$$\phi(t) = (t - 1)^2(t^2 + 1),$$

which implies that the characteristic polynomial also has the conjugate complex roots $\rho_1 = i$ and $\rho_2 = -i$, with $i = \sqrt{-1}$. The complex root $\rho_1 = i$ has length 1 and argument $\theta = \pi/2$. The general solution of the homogeneous recurrence relation, according to Theorem 7.4, is given by

$$a_n = c_1 + c_2 n + c_3 \cos(n\pi/2) + c_4 \sin(n\pi/2), \quad n = 0, 1, \ldots.$$

Further, the function $u(n) = (5n - 2)2^n$ of the complete recurrence relation is of the form (7.25), with $b = 2$, $s = 1$, $u_0 = -2$ and $u_1 = 5$. Thus, from Theorem 7.5 and since $b = 2$ is not root of the characteristic polynomial, whence $k = 0$, we conclude that

$$w(n) = (w_0 + w_1 n)2^n, \quad n = 0, 1, \ldots$$

is a particular solution of the complete recurrence relation. The coefficients w_0 and w_1 are determined from the equations:

$$\phi(2)w_1 = u_1, \quad \phi(2)w_0 + 2\phi'(2)w_1 = u_0.$$

Since $u_0 = -2$, $u_1 = 5$, $\phi(2) = 5$ and $\phi'(2) = 14$, these equations become

$$5w_1 = 5, \quad 5w_0 + 28w_1 = -2$$

and so $w_1 = 1$ and $w_0 = -6$. Therefore, $w(n) = (n - 6)2^n$ and the general solution of the complete recurrence relation is given by

$$a_n = c_1 + c_2 n + c_3 \cos(n\pi/2) + c_4 \sin(n\pi/2) + (n - 6)2^n,$$

for $n = 0, 1, \ldots$. $\square$

7.5 THE METHOD OF GENERATING FUNCTIONS

In Chapter 6, we have examined how generating functions may be used in coping with combinatorial problems. The direct determination of generating functions by invoking the assumptions and restrictions imposed by these problems was a crucial characteristic of many of them. However, a direct determination of a suitable generating function is not always possible. In many problems, a linear recurrence relation for the sequence of the numbers under consideration can be derived by appealing to the assumptions and restrictions imposed by them. We have already dealt with several such problems. In this way, the interest is converted to the solution of a linear recurrence relation.

In this section, generating functions are used in solving a linear recurrence relation of order r with constant or variable coefficients and a bivariate linear recurrence relation of order (r, s) with constant coefficients.

Let us, first, consider the linear recurrence relation with constant coefficients,

$$b_0 a_{n+r} + b_1 a_{n+r-1} + \cdots + b_r a_n = u(n), \quad n = 0, 1, \ldots, \qquad (7.28)$$

where $b_0 \neq 0$, $b_r \neq 0$ and $u(n)$, $n = 0, 1, \ldots$, is a given sequence (function) of n. The determination of a sequence a_k, $k = 0, 1, \ldots$, satisfying this recurrence relation requires the knowledge of r *initial values* $a_0, a_1, \ldots, a_{r-1}$. Since the coefficients $b_0, b_1, \ldots, b_{r-1}$ are known, the numbers

$$c_k = \sum_{j=0}^{k} a_j b_{k-j}, \quad k = 0, 1, \ldots, r - 1 \qquad (7.29)$$

can be calculated. The derivation of the solution a_k, $k = 0, 1, \ldots$, of (7.28) by using a generating function is carried out in two steps. At the first step, on using (7.28) and its initial values, the generating function of the sequence a_k, $k = 0, 1, \ldots$ is derived in the next theorem.

THEOREM 7.6
Let a_k, $k = 0, 1, \ldots$, be a sequence satisfying the linear recurrence relation (7.28). Then the generating function

$$A(t) = \sum_{k=0}^{\infty} a_k t^k, \tag{7.30}$$

is given by

$$A(t) = \frac{C(t) + t^r U(t)}{B_r(t)}, \tag{7.31}$$

where

$$B_r(t) = \sum_{k=0}^{r} b_k t^k, \quad C(t) = \sum_{k=0}^{r-1} c_k t^k \tag{7.32}$$

and

$$U(t) = \sum_{n=0}^{\infty} u(n) t^n. \tag{7.33}$$

PROOF Multiplying (7.28) by t^{n+r} and summing for $n = 0, 1, \ldots$, we get

$$b_0 \sum_{n=0}^{\infty} a_{n+r} t^{n+r} + b_1 t \sum_{n=0}^{\infty} a_{n+r-1} t^{n+r-1} + \cdots + b_r t^r \sum_{n=0}^{\infty} a_n t^n = t^r \sum_{n=0}^{\infty} u(n) t^n.$$

Since

$$\sum_{n=0}^{\infty} a_{n+s} t^{n+s} = A(t) - \sum_{k=0}^{s-1} a_k t^k, \quad s = 1, 2, \ldots, r,$$

upon introducing the generating functions (7.30) and (7.33), we deduce the relation

$$\sum_{j=0}^{r-1} b_j t^j \left(A(t) - \sum_{k=0}^{r-j-1} a_k t^k \right) + b_r t^r A(t) = t^r U(t)$$

and so

$$A(t) \left(\sum_{k=0}^{r} b_k t^k \right) - \sum_{k=0}^{r-1} \left(\sum_{j=0}^{k} a_j b_{k-j} \right) t^k = t^r U(t).$$

Clearly, on using (7.29) and (7.32), we deduce expression (7.31). ∎

After the derivation of expression (7.31) for the generating function $A(t)$, the next step is the determination of the sequence a_k, $k = 0, 1, \ldots$, which requires the expansion of the right-hand side of (7.31) into powers of t. Note that the polynomial $B_r(t)$ is connected with the characteristic polynomial

$$\phi_r(t) = \sum_{k=0}^{r} b_k t^{r-k}$$

by the relation $B_r(t) = t^r \phi_r(1/t)$. Further, when the r initial values $a_0, a_1, \ldots, a_{r-1}$ are not given, the numbers c_k, $k = 0, 1, \ldots, r-1$, defined by (7.29), enter in expression (7.31) of the generating function and, consequently, in the solution of (7.28) as r arbitrary constants. In order to expand the generating function (7.31), we express it as a sum of algebraic fractions of the form, $c/(1 - \rho t)^k$, by the aid of the method of partial fractions. Specifically, if $\rho_1, \rho_2, \ldots, \rho_s$ are the roots of the characteristic polynomial $\phi_r(t)$, with multiplicities $m_1, m_2, \ldots, m_s$, respectively, so that $m_1 + m_2 + \cdots + m_s = r$, then

$$\phi_r(t) = b_0(t - \rho_1)^{m_1}(t - \rho_2)^{m_2} \cdots (t - \rho_s)^{m_s}$$

and the denominator in (7.31) is factored as follows:

$$B_r(t) = b_0(1 - \rho_1 t)^{m_1}(1 - \rho_2 t)^{m_2} \cdots (1 - \rho_s t)^{m_s}.$$

Therefore, the generating function (7.31) may be written in the form

$$A(t) = \sum_{j=1}^{s} \sum_{k=1}^{m_j} \frac{c_{k,j}}{(1 - \rho_j t)^k} + b_0^{-1} t^r U(t) \prod_{j=1}^{s} \frac{1}{(1 - \rho_j t)^{m_j}},$$

where $c_{k,j}$, $k = 1, 2, \ldots, m_j$, $j = 1, 2, \ldots, s$, are constants to be determined from the system of r equations that follows from the equation of the coefficients of t^k, for $k = 0, 1, \ldots, r-1$, in both members of the identity

$$\frac{C(t)}{B_r(t)} = \sum_{j=1}^{s} \sum_{k=1}^{m_j} \frac{c_{k,j}}{(1 - \rho_j t)^k}.$$

The generating function $A(t)$, on using Newton's general binomial formula, may be expanded into powers of t as

$$A(t) = \sum_{j=1}^{s} \sum_{k=1}^{m_j} c_{k,j} \sum_{n=0}^{\infty} \binom{k+n-1}{k-1} \rho_j^n t^n$$

$$+ b_0^{-1} \left[\sum_{n=0}^{\infty} u(n) t^{n+r} \right] \prod_{j=1}^{s} \left[\sum_{n_j=0}^{\infty} \binom{m_j + n_j - 1}{m_j - 1} \rho_j^{n_j} t^{n_j} \right].$$

Thus

$$a_n = \sum_{j=1}^{s} \sum_{k=1}^{m_j} c_{k,j} \binom{k+n-1}{k-1} \rho_j^n$$

$$+ b_0^{-1} \sum_{k=0}^{n-r} u(n-k-r) \sum \prod_{j=1}^{s} \binom{m_j + n_j - 1}{m_j - 1} \rho_j^{n_j}, \qquad (7.34)$$

where, in the inner sum of the second summand, the summation is extended over all $n_j = 0, 1, \ldots, k$, $j = 1, 2, \ldots, s$, with $n_1 + n_2 + \cdots + n_s = k$.

The use of generating functions in solving linear recurrence relations is illustrated in the following examples.

Example 7.8 Fibonacci numbers revisited

Consider n points, each being either 0 or 1, and let Q_n be the number of arrangements of them in a row so that no two zeros are consecutive. The numbers $f_0 = 1$, $f_1 = 1$, $f_n = Q_{n-1}$, $n = 2, 3, \ldots$, are the *Fibonacci numbers* (see Example 2.28). Derive a recurrence relation for the numbers f_n, $n = 0, 1, \ldots$, and solve it by using a generating function.

In order to derive a recurrence relation for the numbers f_n, $n = 0, 1, \ldots$, note that, in an arrangement of n points in a row, each being either 0 or 1 so that no two zeros are consecutive, the first position is occupied either by 1 or by 0. If it is occupied by 1, then there are f_n arrangements of the remaining $n - 1$ points so that no two zeros are consecutive. If the first position is occupied by 0, the second position is necessarily occupied by 1 and then there are f_{n-1} arrangements of the remaining $n - 2$ points so that no two zeros are consecutive. Therefore, according to the addition principle, we deduce the following recurrence relation

$$f_{n+1} = f_n + f_{n-1}, \quad n = 3, 4, \ldots .$$

In particular, for $n = 1, 2$ we have the initial conditions $f_2 = 2$, $f_3 = 3$. Note that these initial conditions can be replaced by the values of f_0 and f_1, which it should be noted that they do not have any combinatorial meaning. The values f_0 and f_1 are deduced from the extension of the recurrence relation at the point $n = 1$: $f_3 - f_2 - f_1 = 0$, whence $f_1 = f_3 - f_2 = 3 - 2 = 1$ and at the point $n = 2$: $f_2 - f_1 - f_0 = 0$, whence $f_0 = f_2 - f_1 = 2 - 1 = 1$. Thus, we may have the second order recurrence relation with constant coefficients,

$$f_{n+2} - f_{n+1} - f_n = 0, \quad n = 0, 1, \ldots \qquad (7.35)$$

and initial conditions $f_0 = 1$ and $f_1 = 1$.

Using recurrence relation (7.35) and its initial conditions, Table 7.1 of Fibonacci numbers f_n, $n = 0, 1, \ldots, 14$, is constructed.

Table 7.1 *Fibonacci Numbers*

n	0	1	2	3	4	5	6	7	8	9	10	11	12	13	14
f_n	1	1	2	3	5	8	13	21	34	55	89	144	233	377	610

The generating function

$$F(t) = \sum_{n=0}^{\infty} f_n t^n$$

is determined by multiplying the recurrence relation (7.35) by t^{n+2} and summing for $n = 0, 1, \ldots$. Then

$$\sum_{n=0}^{\infty} f_{n+2} t^{n+2} - t \sum_{n=0}^{\infty} f_{n+1} t^{n+1} - t^2 \sum_{n=0}^{\infty} f_n t^n = 0$$

and

$$[F(t) - 1 - t] - t[F(t) - 1] - t^2 F(t) = 0,$$

whence

$$F(t) = \frac{1}{1 - t - t^2}. \tag{7.36}$$

In order to expand it into powers of t, we calculate the roots of the characteristic polynomial $\phi_2(t) = t^2 - t - 1$. These roots are

$$\rho_1 = \frac{1 + \sqrt{5}}{2}, \quad \rho_2 = \frac{1 - \sqrt{5}}{2}$$

and so

$$B_2(t) = t^2 \phi_2(1/t) = 1 - t - t^2 = \left(1 - \frac{1 + \sqrt{5}}{2} t\right)\left(1 - \frac{1 - \sqrt{5}}{2} t\right).$$

Then, the generating function (7.36) is written in the form

$$F(t) = \frac{1}{(1 - \rho_1 t)(1 - \rho_2 t)} = \frac{c_1}{1 - \rho_1 t} + \frac{c_2}{1 - \rho_2 t},$$

where c_1 and c_2 are constants to be determined. Since

$$(c_1 + c_2) - (c_1 \rho_2 + c_2 \rho_1)t = 1,$$

these constants satisfy the following system of equations:

$$c_1 + c_2 = 1, \quad c_1 \rho_2 + c_2 \rho_1 = 0.$$

Therefore,

$$c_1 = \frac{\rho_1}{\rho_1 - \rho_2} = \frac{\rho_1}{\sqrt{5}}, \quad c_2 = -\frac{\rho_2}{\rho_1 - \rho_2} = -\frac{\rho_2}{\sqrt{5}}$$

and

$$F(t) = \frac{1}{\sqrt{5}} \left(\frac{\rho_1}{1 - \rho_1 t} - \frac{\rho_2}{1 - \rho_2 t} \right) = \frac{1}{\sqrt{5}} \sum_{n=0}^{\infty} (\rho_1^{n+1} - \rho_2^{n+1}) t^n,$$

whence

$$f_n = \frac{1}{\sqrt{5}} \left\{ \left(\frac{1 + \sqrt{5}}{2} \right)^{n+1} - \left(\frac{1 - \sqrt{5}}{2} \right)^{n+1} \right\}, \quad n = 0, 1, \ldots . \quad \square$$

Example 7.9 Partial sums

Consider a sequence a_k, $k = 0, 1, \ldots$, and let

$$A(t) = \sum_{k=0}^{\infty} a_k t^k.$$

The sequence of partial sums

$$s_n = \sum_{k=0}^{n} a_k, \quad n = 0, 1, \ldots ,$$

is of interest, especially in probability theory. Clearly, this sequence satisfies the relation

$$s_n - s_{n-1} = a_n, \quad n = 1, 2, \ldots ,$$

which is a first order linear recurrence relation with constant coefficients. Multiplying it by t^n and summing for $n = 1, 2, \ldots$, we get

$$\sum_{n=1}^{\infty} s_n t^n - t \sum_{n=1}^{\infty} s_{n-1} t^{n-1} = \sum_{n=1}^{\infty} a_n t^n.$$

Thus, the generating function

$$S(t) = \sum_{n=0}^{\infty} s_n t^n$$

satisfies the relation $S(t) - s_0 - tS(t) = A(t) - a_0$ and since $s_0 = a_0$,

$$S(t) = (1 - t)^{-1} A(t). \tag{7.37}$$

Let us apply this result to the sequence

$$a_k = \binom{x + k - 1}{k}, \quad k = 0, 1, \ldots ,$$

where x is a real number. Then, according to Newton's general binomial formula,

$$A(t) = \sum_{k=0}^{\infty} \binom{x+k-1}{k} t^k = \sum_{k=0}^{\infty} \binom{-x}{k} (-t)^k = (1-t)^{-x},$$

the generating function of the partial sum

$$s_n = \sum_{k=0}^{n} \binom{x+k-1}{k}, \quad n = 0, 1, \dots ,$$

on using (7.37), is obtained as

$$S(t) = \sum_{n=0}^{\infty} s_n t^n = (1-t)^{-x-1}.$$

This generating function is expanded into powers of t as

$$S(t) = \sum_{n=0}^{\infty} \binom{-(x+1)}{n} (-t)^n = \sum_{n=0}^{\infty} \binom{x+n}{n} t^n$$

and so

$$s_n = \sum_{k=0}^{n} \binom{x+k-1}{k} = \binom{x+n}{n}. \quad \square$$

Consider now the linear recurrence relation of order r with variable coefficients:

$$b_0(n)a_{n+r} + b_1(n)a_{n+r-1} + \cdots + b_r(n)a_n = u(n), \quad n = 0, 1, \dots , \quad (7.38)$$

where $u(n)$ and the coefficients $b_j(n)$, $j = 0, 1, \dots , r$, with $b_0(n) \neq 0$ and $b_r(n) \neq 0$, are given functions of n. The derivation of the solution a_k, $k = 0, 1, \dots$, of this recurrence, by using a generating function, when the coefficients $b_j(n)$ are polynomials in n of order s_j, $j = 0, 1, \dots , r$, may be carried out as follows. Multiplying (7.38) by t^{n+r} and summing for $n = 0, 1, \dots$, we get for the generating function

$$A(t) = \sum_{k=0}^{\infty} a_k t^k,$$

a linear differential equation of order at most $s = \max\{s_0, s_1, \dots , s_r\}$. The solution of this equation gives $A(t)$ and its expansion into powers of t yields a_k, $j = 0, 1, \dots$. This method is illustrated in the following examples.

Example 7.10

Let a_k, $k = 0, 1, \ldots, n$, be the number of k-combinations of the set $W_n = \{w_1, w_2, \ldots, w_n\}$. A recurrence relation for the sequence a_k, $k = 0, 1, \ldots, n$ may be obtained as follows. Attaching in every k-combination $\{w_{i_1}, w_{i_2}, \ldots, w_{i_k}\}$ of W_n any one of the $n - k$ elements of $W_n - \{w_{i_1}, w_{i_2}, \ldots, w_{i_k}\}$, we get all the $(k + 1)$-combinations $\{w_{i_1}, w_{i_2}, \ldots, w_{i_{k+1}}\}$ of W_n, $k + 1$ times each. Specifically, the $(k + 1)$-combination $\{w_{i_1}, w_{i_2}, \ldots, w_{i_{k+1}}\}$ of W_n, is obtained from each of the k-combinations

$$\{w_{i_2}, w_{i_3}, \ldots, w_{i_{k+1}}\}, \{w_{i_1}, w_{i_3}, \ldots, w_{i_{k+1}}\}, \ldots, \{w_{i_1}, w_{i_2}, \ldots, w_{i_k}\}$$

by attaching the elements $w_{i_1}, w_{i_2}, \ldots, w_{i_{k+1}}$, respectively. Thus, for a fixed integer n, we get

$$(k + 1)a_{k+1} - (n - k)a_k = 0, \quad k = 0, 1, \ldots, n - 1, \quad a_0 = 1,$$

which is a first order linear recurrence relation with variable coefficients. Its solution may be obtained by using a generating function

$$A(t) = \sum_{k=0}^{n} a_k t^k,$$

as follows. Multiplying the recurrence relation by t^k and summing for $k = 0, 1, \ldots, n - 1$, we get

$$\sum_{k=0}^{n-1} (k + 1)a_{k+1} t^k - n \sum_{k=0}^{n} a_k t^k + t \sum_{k=1}^{n} k a_k t^{k-1} = 0.$$

Further,

$$\frac{d}{dt} A(t) = \sum_{k=1}^{n} k a_k t^{k-1} = \sum_{r=0}^{n-1} (r + 1)a_{r+1} t^r$$

and so

$$(1 + t)\frac{d}{dt} A(t) = nA(t).$$

Therefore

$$\int_0^1 \frac{dA(s)}{A(s)} = n \int_0^1 \frac{ds}{1 + s}$$

and

$$\log A(t) - \log A(0) = n \log(1 + t) - n \log 1.$$

Since $A(0) = a_0 = 1$ and $\log 1 = 0$, it follows that

$$A(t) = (1 + t)^n = \sum_{k=0}^{n} \binom{n}{k} t^k$$

and so

$$a_k = \binom{n}{k}, \quad k = 0, 1, \ldots, n. \quad \Box$$

Example 7.11

Consider the sequence $a_0 = 1$, $a_n = \binom{2n}{n}$, $n = 1, 2, \ldots$, and let

$$A(t) = \sum_{n=0}^{\infty} a_n t^n = 1 + \sum_{n=1}^{\infty} \binom{2n}{n} t^n.$$

In order to determine this generating function, note that

$$a_{n+1} = \binom{2n+2}{n+1} = \frac{2(2n+1)}{n+1}\binom{2n}{n} = \frac{2(2n+1)}{n+1}a_n.$$

Thus, the sequence a_n, $n = 0, 1, \ldots$, satisfies the first order linear recurrence relation with variable coefficients:

$$(n+1)a_{n+1} - 2(2n+1)a_n = 0, \quad n = 0, 1, \ldots, \quad a_0 = 1.$$

Multiplying it by t^n and summing for $n = 0, 1, \ldots$, since

$$\frac{d}{dt}A(t) = \sum_{n=1}^{\infty} n a_n t^{n-1} = \sum_{k=0}^{\infty}(k+1)a_{k+1}t^k,$$

we get the first order differential equation

$$(1 - 4t)\frac{d}{dt}A(t) = 2A(t).$$

Hence

$$\int_0^1 \frac{dA(s)}{A(s)} = \int_0^1 \frac{2ds}{1 - 4s}$$

and so

$$\log A(t) - \log A(0) = -\frac{1}{2}\{\log(1 - 4t) - \log 1\}.$$

Since $A(0) = a_0 = 1$ and $\log 1 = 0$, it follows that

$$A(t) = (1 - 4t)^{-1/2}. \quad \Box$$

Let us finally consider the bivariate linear recurrence relation of order (r, s) with constant coefficients:

$$b_{0,0}a_{n+r,k+s} + b_{0,1}a_{n+r,k+s-1} + b_{1,0}a_{n+r-1,k+s} + \cdots + b_{r,s}a_{n,k} = w(n,k),$$
$$(7.39)$$

where $w(n, k)$ is a given function and the coefficients $b_{i,j}$, $i = 0, 1, \ldots, n$, $j = 0, 1, \ldots, k$, are given constants. The determination of the double-index sequence $a_{n,k}$, $n = 0, 1, \ldots$, $k = 0, 1, \ldots$, that satisfies the recurrence relation (7.39) requires the knowledge of $r + s$ independent sequences,

$$a_{i,k}, \quad k = 0, 1, \ldots, \quad i = 0, 1, \ldots, r - 1, \tag{7.40}$$

$$a_{n,j}, \quad n = 0, 1, \ldots, \quad j = 0, 1, \ldots, s - 1, \tag{7.41}$$

the *initial conditions*.

The derivation of the solution of (7.39) by using a generating function is carried out in two steps. In the first step, we determine the bivariate generating function

$$A(t, u) = \sum_{n=0}^{\infty} \sum_{k=0}^{\infty} a_{n,k} t^k u^n,$$

which, on using the sequence of generating functions

$$A_n(t) = \sum_{k=0}^{\infty} a_{n,k} t^k, \quad n = 0, 1, \ldots,$$

is written in the form

$$A(t, u) = \sum_{n=0}^{\infty} A_n(t) u^n.$$

For the derivation of the sequence of generating functions $A_n(t)$, $n = 0, 1, \ldots$, we multiply (7.39) by t^{k+s} and sum for $k = 0, 1, \ldots$. Then

$$\sum_{i=0}^{r} \sum_{j=0}^{s} b_{i,j} t^j \sum_{k=0}^{\infty} a_{n+r-i,k+s-j} t^{k+s-j} = t^s \sum_{k=0}^{\infty} w(n, k) t^k, \quad n = 0, 1, \ldots$$

and, since

$$\sum_{k=0}^{\infty} a_{n+r-i,k+s-j} t^{k+s-j} = A_{n+r-i}(t) - \sum_{m=j}^{s-1} a_{n+r-i,m-j} t^{m-j},$$

for $j = 0, 1, \ldots, s - 1$, we deduce the relation

$$\sum_{i=0}^{r} \sum_{j=0}^{s} b_{i,j} t^j A_{n+r-i}(t) - \sum_{i=0}^{r} \sum_{j=0}^{s-1} \sum_{m=j}^{s-1} a_{n+r-i,m-j} t^m = t^s \sum_{k=0}^{\infty} w(n, k) t^k,$$

for $n = 0, 1, \ldots$. Introducing

$$b_i(t) = \sum_{j=0}^{s} b_{i,j} t^j, \quad i = 0, 1, \ldots, r,$$

$$C_n(t) = \sum_{m=0}^{s-1} \left\{ \sum_{i=0}^{r} \sum_{j=0}^{m} a_{n+r-i,m-j} b_{i,j} \right\} t^m$$

and

$$W_n(t) = \sum_{k=0}^{\infty} w(n,k) t^k,$$

this relation may be written in the form

$$b_0(t) A_{n+r}(t) + b_1(t) A_{n+r-1}(t) + \cdots + b_r(t) A_n(t) = C_n(t) + t^s W_n(t),$$
(7.42)

for $n = 0, 1, \ldots$. The last relation is a linear recurrence relation of order r for the sequence of generating functions $A_n(t)$, $n = 0, 1, \ldots$ with constant, with respect to n, coefficients $b_i(t)$, $i = 0, 1, \ldots, r$. Note that the generating functions

$$A_i(t) = \sum_{k=0}^{\infty} a_{i,k} t^k, \quad i = 0, 1, \ldots, r-1,$$
(7.43)

which, according to the initial conditions (7.40), are known, constitute the *initial conditions* of (7.42). Considering t as a parameter, the recurrence relation (7.42) is solved in exactly the same way as the recurrence relation (7.28). So, multiplying (7.42) by u^{n+r} and summing for $n = 0, 1, \ldots$, we deduce the relation

$$A(t, u) = \frac{C(t, u) + u^r t^s W(t, u)}{B_{r,s}(t, u)},$$
(7.44)

where

$$B_{r,s}(t, u) = \sum_{i=0}^{r} b_i(t) u^i = \sum_{i=0}^{r} \sum_{j=0}^{s} b_{i,j} t^j u^i,$$

$$W(t, u) = \sum_{n=0}^{\infty} W_n(t) u^n = \sum_{n=0}^{\infty} \sum_{k=0}^{\infty} w(n,k) t^k u^n$$

and

$$C(t, u) = \sum_{n=0}^{r-1} \sum_{i=0}^{n} A_{n-i}(t) b_i(t) u^n + \sum_{m=0}^{s-1} \sum_{j=0}^{m} A_{m-j}^*(u) b_j^*(u) t^m$$

$$- \sum_{n=0}^{r-1} \sum_{m=0}^{s-1} \sum_{i=0}^{n} \sum_{j=0}^{n} a_{n-i,m-j} b_{i,j} t^m u^n,$$

with

$$A_j^*(u) = \sum_{n=0}^{\infty} a_{n,j} u^n, \quad j = 0, 1, \ldots, s-1,$$

known generating functions, according to the initial conditions (7.41), and

$$b_j^*(u) = \sum_{i=0}^{r} b_{i,j} u, \quad j = 0, 1, \ldots, s.$$

After the derivation of the generating function $A(t, u)$, the next step is the determination of the sequence $a_{n,k}$, $k = 0, 1, \ldots$, $n = 0, 1, \ldots$, which requires the expansion of the right-hand member of (7.44) into powers of t and u. For this purpose, considering t as a parameter, we expand this function into powers of u, and then we expand the resulting expression into powers of t. This procedure is further clarified in the following examples.

Example 7.12 Sums of sums

Let us consider a sequence a_k, $k = 0, 1, \ldots$, with generating function

$$A(t) = \sum_{k=0}^{\infty} a_k t^k.$$

Further, consider the sum

$$s_{1,j} = \sum_{i=0}^{j} a_i, \quad j = 0, 1, \ldots,$$

the double sum

$$s_{2,r} = \sum_{j=0}^{r} \sum_{i=0}^{j} a_i = \sum_{j=0}^{r} s_{1,j}, \quad r = 0, 1, \ldots$$

and, generally, the n-tuple sum

$$s_{n,k} = \sum_{r=0}^{k} s_{n-1,r}, \quad k = 0, 1, \ldots, \quad n = 0, 1, \ldots.$$

Clearly,

$$s_{n,k} - s_{n,k-1} = s_{n-1,k}, \quad k = 1, 2, \ldots, \quad n = 1, 2, \ldots.$$

Multiplying this relation by t^k and summing for $k = 1, 2, \ldots$, we get

$$\sum_{k=1}^{\infty} s_{n,k} t^k - t \sum_{k=1}^{\infty} s_{n,k-1} t^{k-1} = \sum_{k=1}^{\infty} s_{n-1,k} t^k$$

and, since $s_{n,0} = s_{n-1,0}$, we deduce for the generating function

$$S_n(t) = \sum_{k=0}^{\infty} s_{n,k} t^k, \quad n = 0, 1, \ldots,$$

the first order linear recurrence relation with constant, with respect to n, coefficients

$$(1 - t)S_n(t) = S_{n-1}(t), \quad n = 1, 2, \ldots,$$

and initial condition

$$S_0(t) = \sum_{k=0}^{\infty} s_{0,k} t^k = \sum_{k=0}^{\infty} a_k t^k = A(t).$$

Multiplying it by u^n and summing for $n = 1, 2, \ldots$, we get

$$(1 - t) \sum_{n=1}^{\infty} S_n(t) u^n = u \sum_{n=1}^{\infty} S_{n-1}(t) u^{n-1}.$$

Further, using the initial condition $S_0(t) = A(t)$, we deduce for the double (bivariate) generating function

$$S(t, u) = \sum_{n=0}^{\infty} \sum_{k=0}^{\infty} s_{n,k} t^k u^n = \sum_{n=0}^{\infty} S_n(t) u^n,$$

the expression

$$S(t, u) = (1 - t)(1 - t - u)^{-1} A(t).$$

Expanding the right-hand side of this expression into powers of u, we find

$$S_n(t) = (1 - t)^{-n} A(t)$$

and since

$$(1 - t)^{-n} = \sum_{k=0}^{\infty} \binom{n + k - 1}{k} t^k,$$

we conclude that

$$s_{n,k} = \sum_{j=0}^{k} \binom{n + j - 1}{j} a_{k-j}. \qquad \square$$

Example 7.13 Distribution of shares

Let us consider two players $\mathcal{K}$ and $\mathcal{R}$ playing in a series of games in which winner is declared the one who first wins n games. Assume that the probability of $\mathcal{K}$ to win in a game is p and so that of $\mathcal{R}$ is $q = 1 - p$. Further, suppose that, for some reason, the series of games is interrupted when $\mathcal{K}$ has won $n - k$ games and $\mathcal{R}$ $n - r$ games, with $k, r < n$. In this case, what should be the shares of the two players from a total stake of s dollars?

The total stake should be distributed to the two players in shares proportional to the probability that each one has to win the series of games if it is continued.

So, let $p_{k,r}$ be the probability of $\mathcal{K}$ to win the series of games, when k wins of $\mathcal{K}$ are required before r wins of $\mathcal{R}$. Note that $\mathcal{K}$ may win or lose the next game with probability p or $q = 1 - p$, respectively. If $\mathcal{K}$ wins the next game, then the probability of winning the series is $p_{k-1,r}$, while if $\mathcal{K}$ loses (whence $\mathcal{R}$ wins) the next game, then the probability of winning the series is $p_{k,r-1}$. Consequently, the probability $p_{k,r}$, $r = 1, 2, \ldots, k = 1, 2, \ldots$, satisfies the recurrence relation

$$p_{k,r} = pp_{k-1,r} + qp_{k,r-1}, \quad r = 1, 2, \ldots, \quad k = 1, 2, \ldots,$$

with initial conditions

$$p_{0,r} = 1, \quad r = 1, 2, \ldots, \quad p_{k,0} = 0, \quad k = 1, 2, \ldots.$$

Multiplying this recurrence relation by t^r and summing for $r = 1, 2, \ldots$, we get

$$\sum_{r=1}^{\infty} p_{k,r}t^r = p \sum_{r=1}^{\infty} p_{k-1,r}t^r + qt \sum_{r=1}^{\infty} p_{k,r-1}t^{r-1}, \quad k = 1, 2, \ldots$$

and, since $p_{k,0} = 0$, we deduce for the sequence of generating functions

$$P_k(t) = \sum_{r=1}^{\infty} p_{k,r}t^r, \quad k = 1, 2, \ldots,$$

the first order recurrence relation with constant, with respect to k, coefficients:

$$(1 - qt)P_k(t) = pP_{k-1}(t), \quad k = 1, 2, \ldots$$

and initial condition

$$P_0(t) = \sum_{r=1}^{\infty} p_{0,r}t^r = \sum_{r=1}^{\infty} t^r = t(1 - t)^{-1}.$$

Multiplying this recurrence relation by u^k and summing for $k = 1, 2, \ldots$, we get

$$(1 - qt) \sum_{k=1}^{\infty} P_k(t)u^k = pu \sum_{k=1}^{\infty} P_{k-1}(t)u^{k-1}$$

and using the initial condition $P_0(t) = t(1 - t)^{-1}$, we deduce for the double generating function

$$P(t, u) = \sum_{k=1}^{\infty} P_k(t)u^k = \sum_{k=1}^{\infty} \sum_{r=1}^{\infty} p_{k,r}t^r u^k,$$

the expression

$$P(t, u) = t(1 - t)^{-1}(1 - qt - pu)^{-1}pu.$$

Expanding the right-hand side of this expression into powers of u, we get

$$P(t,u) = t(1-t)^{-1}(1-p(1-qt)^{-1}u)^{-1}p(1-qt)^{-1}u$$

$$= t(1-t)^{-1}\sum_{k=1}^{\infty} p^k(1-qt)^{-k}u^k$$

and so

$$P_k(t) = t(1-t)^{-1}p^k(1-qt)^{-k}.$$

Finally, expanding this generating function into powers of t, by using Newton's negative binomial formula, we deduce for the probability $p_{k,r}$ the expression

$$p_{k,r} = p^k \sum_{j=0}^{r-1} \binom{k+j-1}{k-1} q^j.$$

Consequently, from the total stake of s dollars, player $\mathcal{K}$ might get a share of $sp_{k,r}$ and player $\mathcal{R}$ a share of $s(1-p_{k,r})$ dollars. Note that, for $p = q = 1/2$, the probability $p_{k,r}$ reduces to the probability derived in Section 2.7.1. $\square$

7.6 BIBLIOGRAPHIC NOTES

The first recurrence relation was given in 1202 by Leonardo Fibonacci in his book on abacus (*Liber Abaci*). This recurrence relation of the Fibonacci numbers is deduced and solved in Example 7.8. The derivation of the solution by using a generating function is due to Abraham De Moivre (1718). E. Lucas (1891) named this sequence of numbers after Fibonacci and also examined the related sequence of Lucas numbers. An extensive coverage of the Fibonacci and Lucas numbers and some of their extensions and generalizations is provided by V. E. Hoggatt (1969). The interest of E. Lucas (1891) in the problem of the Hanoi tower contributed the most in its popularization. The gambler's ruin is an old problem; in 1657, Christian Huyghens' treatise, which was included in the book *Ars Conjectandi* of J. Bernoulli, discussed the particular case of $n = k - n = 12$ tokens and $p/q = 5/7$. Bernoulli considered and solved the general case.

7.7 EXERCISES

1. An amount of $a_0 = \$10,000$ is deposited in a bank at the beginning of an interest period at interest rate r. If the interest is compounded each period, show that the amount a_n on deposit after n periods satisfies the first order linear recurrence relation

$$a_{n+1} = (1+r)a_n, \quad n = 0, 1, \ldots .$$

Iterate it to derive a_n.

2. Let a_n be the number of n-digit nonnegative integers in which no two same digits are consecutive. Show that

$$a_{n+1} = 9a_n, \quad n = 1, 2, \ldots ,$$

with $a_1 = 10$. Iterate this recurrence relation to derive a_n.

3. A player decides to play a series of games against a casino (or against an adversary) until he wins. In any game, if he wins he gets $b + 1$ times the amount of his stake, while if he looses he stakes a new amount. Let a_n be the player's stake on the n-th game, $n = 1, 2, \ldots$. Provided that, if the player wins the n-th game, he gets back not only the stakes he lost in the $n-1$ previous games but also an amount a, which is fixed in advance, show that

$$ba_n = (b+1)a_{n-1}, \quad n = 2, 3, \ldots ,$$

with $a_1 = a/b$. Iterate this recurrence to find a_n.

4. Consider the set of sequences of flips of a coin that are terminated when heads appear for the second time. Let a_n be the number of such sequences that are terminated before or at the n-th flip. Show that

$$a_{n+1} = a_n + n, \quad n = 2, 3, \ldots ,$$

with $a_2 = 1$. Iterate this recurrence relation to derive a_n.

5. Let R_n be the number of regions in which a plane can be divided by n lines, each two of them having a point in common but no three of them having a point in common. Show that

$$R_n = R_{n-1} + n, \quad n = 2, 3, \ldots$$

with $R_1 = 2$. Further, show that the unique solution of this recurrence relation is given by

$$R_n = 1 + \binom{n+1}{2} = \frac{n^2 + n + 2}{2}.$$

6. Let a_n be the number of n-permutations of the set $\{0, 1, 2\}$ with repetition that include an even number of zeros. Show that

$$a_{n+1} = a_n + 3^n, \quad n = 1, 2, \ldots ,$$

with $a_1 = 2$. Iterate this recurrence relation to derive a_n.

7. Assume that the sequence a_n, $n = 0, 1, \ldots$, satisfies the recurrence relation

$$2(n+1)a_{n+1} = (2n+1)a_n, \quad n = 0, 1, \ldots,$$

with initial condition $a_0 = 1$. Show that its unique solution is given by

$$a_n = \frac{1}{2^{2n}}\binom{2n}{n}, \quad n = 0, 1, \ldots.$$

8. Let a_n and b_n be the numbers of n-permutations of the set $\{0, 1, \ldots, 9\}$ with repetition that include an odd and even number of 5s, respectively. Show that

$$a_{n+2} - 18a_{n+1} + 80a_n = 0,$$

with $a_1 = 1$, $a_2 = 18$ and

$$b_{n+2} - 18b_{n+1} + 80b_n = 0,$$

with $b_1 = 9$, $b_2 = 82$. Note that both sequences a_n, $n = 1, 2, \ldots$, and b_n, $n = 1, 2, \ldots$, satisfy the same recurrence relation but with different initial conditions. Further, derive the unique solutions

$$a_n = \frac{1}{2}(10^n - 8^n), \quad n = 1, 2, \ldots$$

and

$$b_n = \frac{1}{2}(10^n + 8^n), \quad n = 1, 2, \ldots.$$

9. Let a_n be the number of n-digit nonnegative integers in which no three same digits are consecutive. Show that

$$a_{n+2} = 9a_{n+1} + 9a_n, \quad n = 1, 2, \ldots,$$

with $a_1 = 10$ and $a_2 = 100$. Further, show that the unique solution of this recurrence relation is given by

$$a_n = \frac{5}{9\sqrt{13}}\left(\frac{3}{2}\right)^n \left\{\left(3 + \sqrt{13}\right)^{n+1} - \left(3 - \sqrt{13}\right)^{n+1}\right\}.$$

10. Let a_n be the number of n-permutations of the set $\{0, 1, 2\}$ with repetition and the restriction that no two zeros are consecutive. Show that

$$a_{n+2} = 2a_{n+1} + 2a_n, \quad n = 0, 1, \ldots,$$

with $a_0 = 1$ and $a_1 = 3$. Further, show that the unique solution of this recurrence relation is given by

$$a_n = \frac{1}{4\sqrt{3}}\left\{ \left(1 + \sqrt{3}\right)^{n+2} - \left(1 - \sqrt{3}\right)^{n+2} \right\}.$$

11. Let a_n be the number of n-permutations of the set $\{0, 1, 2\}$ with repetition and the restriction that no two zeros and no two ones are consecutive. Show that

$$a_{n+2} = 2a_{n+1} + a_n, \quad n = 0, 1, \ldots,$$

with $a_0 = 1$ and $a_1 = 3$. Further, show that the unique solution of this recurrence relation is given by

$$a_n = \frac{1}{2}\left\{ \left(1 + \sqrt{2}\right)^{n+1} - \left(1 - \sqrt{2}\right)^{n+1} \right\}.$$

12. Let

$$a_n = \sum_{k=0}^{n}(-1)^k 2^{2n-2k}\binom{2n - k + 1}{k}, \quad b_n = \sum_{k=0}^{n}(-1)^k 2^{2n-2k}\binom{2n - k}{k}.$$

Show that

$$a_n = b_n - a_{n-1}, \quad b_n = 4a_{n-1} - b_{n-1}, \quad n = 1, 2, \ldots$$

and thus conclude the recurrence relations:

$$a_n = 2a_{n-1} - a_{n-2}, \quad n = 2, 3, \ldots,$$

with $a_0 = 1$, $a_1 = 2$, and

$$b_n = 2b_{n-1} - b_{n-2}, \quad n = 2, 3, \ldots,$$

with $b_0 = 1$, $b_1 = 3$. Solve these recurrence relations to get

$$a_n = n + 1, \quad b_n = 2n + 1, \quad n = 0, 1, \ldots.$$

13. Find the unique solution of the recurrence relation

$$a_{n+1} = \frac{1}{4}(a_{n+2} - a_n), \quad n = 0, 1, \ldots,$$

with initial conditions $a_0 = 1$ and $a_1 = 4$.

14. Find the general solution of the recurrence relation

$$a_{n+2} + a_{n+1} - 2a_n = 2^n, \quad n = 0, 1, \ldots.$$

15. Find the general solution of the recurrence relation

$$a_{n+3} - 5a_{n+2} + 7a_{n+1} - 3a_n = 3^n, \quad n = 0, 1, \dots.$$

16*. Let $Q_r(n, k)$ be the number of k-permutations of n with repetition in which no r like elements are consecutive $r = 2, 3, \dots$. Show that

$$Q_r(n, k) = (n - 1) \sum_{j=1}^{r-1} Q_r(n, k - j), \quad k \geq r, \quad Q_r(n, k) = n^k, \quad k < r$$

and

$$Q_{n,r}(t) = \sum_{k=1}^{\infty} Q_r(n, k)t^n = \frac{nt(1 + t + \cdots + t^{r-2})}{1 - (n - 1)t(1 + t + \cdots + t^{r-2})}.$$

17*. Let $f_{n+1}(s)$ be the number of n-permutations of the set $\{0, 1, \dots, s\}$ with repetition and the restriction that no two zeros are consecutive. (a) Derive the recurrence relation

$$f_{n+1}(s) = s\{f_n(s) + f_{n-1}(s)\}, \quad n = 1, 2, \dots,$$

with initial conditions $f_0(s) = 1/s$, $f_1(s) = 1$. (b) Show that

$$f_{2k+1}(s) = s^k \sum_{j=0}^{k} \binom{k}{j} f_{k-j+1}(s), \quad k = 0, 1, \dots,$$

and

$$f_{2k+2}(s) = s^k \sum_{j=0}^{k} \binom{k}{j} f_{k-j+2}(s), \quad k = 0, 1, \dots.$$

(c) Derive the generating function

$$F(t; s) = \sum_{n=0}^{\infty} f_n(s)t^n = \frac{1}{s(1 - st - st^2)}.$$

18*. (*Continuation*). Show that

$$f_n(s) = \sum_{k=0}^{[n/2]} \binom{n-k}{k} s^{n-k+1}, \quad n = 0, 1, \dots.$$

19*. (*Continuation*). Show that

$$f_n(s) = \frac{1}{s\sqrt{s(s+4)}} \left\{ \left(\frac{s + \sqrt{s(s+4)}}{2} \right)^{n+1} - \left(\frac{s - \sqrt{s(s+4)}}{2} \right)^{n+1} \right\}$$

and conclude that

$$f_n(s) = 2^{-n} \sum_{k=0}^{[n/2]} \binom{n+1}{2k+1} s^{n-k-1} (s+4)^k, \quad n = 0, 1, \ldots .$$

Note that $f_n(1) = f_n$, $n = 0, 1, \ldots$, are the *Fibonacci numbers*.

20*. Let $f_{n+1}(r, s)$ be the number of n-permutations of the set $W = W_0 + W_1$, where $N(W_0) = r + 1$, $N(W_1) = s$, with repetition and the restriction that no two like elements of W_0 are consecutive. (a) Derive the recurrence relation

$$f_{n+1}(r, s) = (r+s) f_n(r, s) + s f_{n-1}(s), \quad n = 1, 2, \ldots ,$$

with initial conditions $f_0(r, s) = 1/s$, $f_1(r, s) = 1$ and (b) deduce the generating function

$$F(t; r, s) = \sum_{n=0}^{\infty} f_n(r, s) t^n = \frac{1 - rt}{s[1 - (r+s)t - st^2]}.$$

Note that $f_n(0, 1) = f_n$, $n = 0, 1, \ldots$, are the *Fibonacci numbers*.

21*. *Fibonacci numbers of order* s. Let $F_{n+1,s}$ be the number of n-permutations of the set $\{0, 1\}$ with repetition and the restriction that no s zeros are consecutive. Show that (a)

$$F_{n,s} = \sum_{j=1}^{m} F_{n-j,s}, \quad m = \min\{n, s\}, \quad n = 1, 2, \ldots , \quad F_{0,s} = F_{1,s} = 1$$

and (b)
$$F_{n,s} = 2^{n-1}, \quad n = 1, 2, \ldots , s,$$

$$F_{n,s} = 2 F_{n-1,s} - F_{n-s-1,s}, \quad n = s+1, s+2, \ldots .$$

(c) Derive the generating function

$$F_s(t) = \sum_{n=0}^{\infty} F_{n,s} t^n = \frac{1}{1 - t - t^2 - \cdots - t^s}.$$

22. (*Continuation*). Show that

$$F_{n,s} = \sum_{k=0}^{[n-n/k]} E_{s-1}(n - k, k), \quad n = 0, 1, \ldots ,$$

where $E_{s-1}(r, k)$ is the number of k-combinations of r with repetition and the restriction that each element appears at most $s - 1$ times.

23. *Lucas numbers revisited.* Let $g_0 = 2$ and g_n, $n = 1, 2, \ldots$, be the number of combinations of the n numbers $\{1, 2, \ldots, n\}$, displayed on a circle (whence $(n, 1)$ is a pair of consecutive elements), which include no pair of consecutive elements. These numbers satisfy the recurrence relation

$$g_{n+2} - g_{n+1} - g_n = 0, \quad n = 0, 1, \ldots$$

with initial conditions $g_0 = 2$, $g_1 = 1$. Show that

$$G(t) = \sum_{n=0}^{\infty} g_n t^n = \frac{2 - t}{1 - t - t^2}$$

and conclude that

$$g_n = \left(\frac{1 + \sqrt{5}}{2}\right)^n + \left(\frac{1 - \sqrt{5}}{2}\right)^n, \quad n = 0, 1, \ldots.$$

24*. *Lucas numbers of order s.* Let $G_{n,s}$ be the number of cyclic n-permutations of $\{0, 1\}$ with repetition and with one element marked by a star and the restriction that no s zeros are consecutive. Note that, for $s = 2$, $G_{n,2} = g_n$, $n = 0, 1, 2, \ldots$, are the *Lucas numbers*. Show that (a)

$$G_{n,s} = \sum_{j=1}^{s} G_{n-j,s}, \quad n = s+1, s+2, \ldots, \quad G_{n,s} = 2^n - 1, \quad n = 1, 2, \ldots, s$$

and (b)

$$G_{n,s} = 2G_{n-2,s} - G_{n-s-1,s}, \quad n = s+2, s+3, \ldots.$$

(c) Derive the generating function

$$G_s(t) = \sum_{n=1}^{\infty} G_{n,s} t^n = \frac{t + 2t^2 + \cdots + st^s}{1 - t - t^2 - \cdots - t^s}.$$

25. *Generalized Fibonacci numbers.* Let $f_{n+s,s}$ be the number of combinations of the n numbers $\{1, 2, \ldots, n\}$ possessing the property: between any two elements belonging to such a combination there exist at least s elements of n that do not belong to it. Show that the numbers $f_{n,s}$, $n = s, s+1, \ldots$ satisfy the recurrence relation

$$f_{n+s,s} = f_{n+s-1,s} + f_{n-1,s}, \quad n = s+1, s+2, \ldots$$

with initial conditions

$$f_{n+s,s} = n + 1, \quad n = 1, 2, \ldots, s.$$

Putting $f_{n,s} = 1$, $n = 0, 1, 2, \ldots, s$, show that

$$\Phi_s(t) = \sum_{n=0}^{\infty} f_{n,s} t^n = \frac{1}{1 - t - t^{s+1}}.$$

The numbers $f_{n,s}$, $n = 0, 1, \ldots$, which for $s = 1$ reduce to the Fibonacci numbers, are called *generalized Fibonacci numbers*.

26. *Generalized Lucas numbers.* Let $g_{n,s}$ be the number of combinations of the n numbers $\{1, 2, \ldots, n\}$, displayed on a circle, possessing the following property: between any two elements belonging to such a combination there exist at least s elements of n that do not belong to it. Show that the numbers $g_{n,s}$, $n = 1, 2, \ldots$, $g_{0,s} = s + 1$, satisfy the recurrence relation

$$g_{n,s} = g_{n-1,s} + g_{n-s-1,s}, \quad n = s+1, s+2, \ldots$$

with initial conditions

$$g_{0,s} = s + 1, \quad g_{n,s} = 1, \quad n = 1, 2, \ldots, s$$

and

$$G_s(t) = \sum_{n=0}^{\infty} g_{n,s} t^n = \frac{s + 1 - st}{1 - t - t^{s+1}}.$$

The numbers $g_{n,s}$, $n = 0, 1, \ldots$, which for $s = 1$ reduce to the Lucas numbers, are called *generalized Lucas numbers* and are connected with the generalized Fibonacci numbers by the relation

$$g_{n,s} = f_{n,s} + s f_{n-s-1,s}, \quad n = s+1, s+2, \ldots.$$

27. *Convolution of Fibonacci numbers.* Let f_n, $n = 0, 1, \ldots$, be the sequence of Fibonacci numbers with $f_0 = f_1 = 1$. The sequence

$$f_n^{(2)} = \sum_{j=0}^{n} f_j f_{n-j}, \quad n = 0, 1, \ldots$$

is called *convolution* of the sequence f_n, $n = 0, 1, \ldots$. In general, the sequence

$$f_n^{(k)} = \sum_{j=0}^{n} f_j f_{n-j}^{(k-1)}, \quad n = 0, 1, \ldots, \quad k = 2, 3, \ldots,$$

with $f_n^{(1)} = f_n$, $n = 0, 1, \ldots$, is called *k-fold convolution* of the sequence f_n, $n = 0, 1, \ldots$. (a) Derive the generating function.

$$F_k(t) = \sum_{n=0}^{\infty} f_n^{(k)} t^n = (1 - t - t^2)^{-k}$$

and deduce the expression

$$f_n^{(k)} = \sum_{r=0}^{[n/2]} \binom{k+n-r-1}{k-1} \binom{n-r}{r}.$$

(b) Show that

$$F(t,u) = \sum_{k=0}^{\infty} \sum_{n=0}^{\infty} f_n^{(k)} t^n u^k = \frac{1-t-t^2}{1-t-t^2-u}.$$

28. Let $R_{n,k}$ be the number of regions in which a plane can be divided by n lines, k of them being parallel to each other but no three of them having a point in common. Derive the recurrence relation

$$R_{n,k} = R_{n-1,k-1} + n - k + 1, \quad k = 1, 2, \dots, n, \quad n = 2, 3, \dots,$$

with $R_{1,0} = R_{1,1} = 2$ and show that its solution is given by

$$R_{n,k} = 1 + (n-k+1)k + \binom{n-k+1}{2}.$$

29. Let S_n be the number of permutations of $\{1, 2, \dots, n\}$ without a succession. Using the recurrence relation

$$S_n = (n-1)S_{n-1} + (n-2)S_{n-2}, \quad n = 3, 4, \dots, \quad S_1 = S_2 = 1,$$

show that

$$S(t) = \sum_{n=0}^{\infty} S_{n+1} \frac{t^n}{n!} = (1-t)^{-2} e^{-t}.$$

30*. Consider the sum of inverse binomial coefficients:

$$S_n = \sum_{k=0}^{n} \binom{n}{k}^{-1}, \quad n = 0, 1, \dots.$$

(a) Derive the recurrence relation

$$2nS_n - (n+1)S_{n-1} = 2n, \quad n = 1, 2, \dots,$$

with initial condition $S_0 = 1$. (b) Show that

$$S(t) = \sum_{n=0}^{\infty} S_n t^n = \left(1 - \frac{t}{2}\right)^{-1} (1-t)^{-1} - \frac{1}{2}\left(1 - \frac{t}{2}\right)^{-2} \log(1-t),$$

and (c) conclude that

$$S_n = \frac{n+1}{2^n} \sum_{k=0}^{n} \frac{2^k}{k+1}.$$

31. Consider the sequences

$$a_n = \sum_{i=0}^{n} \binom{n+i}{2i}, \quad n = 0, 1, \dots, \quad b_n = \sum_{i=0}^{n-1} \binom{n+i}{2i+1}, \quad n = 1, 2, \dots$$

and let

$$A(t) = \sum_{n=0}^{\infty} a_n t^n, \quad B(t) = \sum_{n=1}^{\infty} b_n t^n.$$

Show that

$$A(t) - 1 = tA(t) + B(t), \quad B(t) = tA(t) + tB(t)$$

and thus, conclude that

$$A(t) = \frac{1-t}{(1-t)^2 - t}, \quad B(t) = \frac{t}{(1-t)^2 - t}.$$

32*. Let B_n be the number of partitions of a finite set of n elements. Show that

$$B(t) = \sum_{n=0}^{\infty} B_n \frac{t^n}{n!} = \exp(e^t - 1)$$

and

$$B_n = e^{-1} \sum_{r=0}^{\infty} \frac{r^n}{r!}, \quad n = 0, 1, \dots .$$

33*. Let A_n be the number of partitions of a finite set of n elements into subsets with an even number of elements. Show that

$$A_n = \sum_{k=1}^{[n/2]} \binom{n-1}{2k-1} A_{n-2k}, \quad n = 2, 3, \dots, \quad A_0 = 1, \quad A_1 = 0$$

and

$$A(t) = \sum_{n=0}^{\infty} A_n \frac{t^n}{n!} = \exp(\cosh t - 1),$$

where $\cosh t = (e^t + e^{-t})/2$ is the hyperbolic cosine of t.

34. (*Continuation*). Show that

$$A_n = e^{-1} \sum_{r=0}^{\infty} \frac{G(r, n)}{r!}, \quad n = 0, 1, \dots,$$

where $G(r, n)$ is the number of n-permutations of r with repetition and the restriction that each element appears zero or an even number of times.

35*. (*Continuation*). Let E_n be the number of partitions of a finite set of n elements into subsets with an odd number of elements. Show that

$$E_n = \sum_{k=0}^{[(n-1)/2]} \binom{n-1}{2k} E_{n-2k-1}, \quad n = 1, 2, \dots, \quad E_0 = 1$$

and

$$E(t) = \sum_{n=0}^{\infty} E_n \frac{t^n}{n!} = \exp(\sinh t),$$

where $\sinh t = (e^t - e^{-t})/2$ is the hyperbolic sine of t.

36. (*Continuation*). Show that

$$E_n = \sum_{r=0}^{\infty} \frac{H(r, n)}{r!}, \quad n = 0, 1, \dots,$$

where $H(r, n)$ is the number of n-permutations of r with repetition and the restriction that each element appears an odd number of times.

37*. *Ballot numbers revisited.* In a ballot between two candidates $\mathcal{N}$ and $\mathcal{K}$, $\mathcal{N}$ receives n votes and $\mathcal{K}$ receives k votes, with $n \geq k$. Let x_r and y_r be the numbers of votes for $\mathcal{N}$ and $\mathcal{K}$, respectively, after the counting of r votes, $r = 1, 2, \dots, n + k$ and $\psi_{n,k}$ be the number of ways of counting the votes in which $x_r \geq y_r$, $r = 1, 2, \dots, n + k$. Show that

$$\psi_{n,k} = \psi_{n,k-1} + \psi_{n-1,k}, \quad k = 1, 2, \dots, n - 1, \quad n = 1, 2, \dots,$$

with $\psi_{n,n} = \psi_{n,n-1}$, $\psi_{n,0} = 1$, and

$$\psi(t, u) = \sum_{n=0}^{\infty} \sum_{k=0}^{n} \psi_{n,k} t^k u^n = \frac{(1 - 4tu)^{1/2} + 2u - 1}{2(1 - t - u)u}.$$

38*. *Delannoy numbers.* Let us consider an electoral district in which each voter may vote for, at most, two candidates of the same party and assume that two candidates of the same party $\mathcal{N}$ and $\mathcal{K}$ receive n and k

votes, respectively. The total number $D(n, k)$ of ways of counting the votes is called *Delannoy number*. (a) Show that

$$D(n, k) = D(n, k - 1) + D(n - 1, k - 1) + D(n - 1, k),$$

for $k = 1, 2, \ldots$, $n = 1, 2, \ldots$, with $D(n, 0) = D(0, k) = 1$. (b) Derive the generating function

$$G(t, u) = \sum_{n=0}^{\infty} \sum_{k=0}^{\infty} D(n, k) t^k u^n = (1 - t - u - tu)^{-1}$$

and conclude that

$$D(n, k) = \sum_{r=0}^{m} \binom{k}{r} \binom{n + k - r}{k} = \sum_{r=0}^{m} 2^r \binom{k}{r} \binom{n}{k}, \quad m = \min\{n, k\}.$$

39*. (*Continuation*). Let $D(n) \equiv D(n, n)$, $n = 0, 1, \ldots$. From the generating function

$$G_n(t) = \sum_{k=0}^{\infty} D(n, k) t^k$$

and using Lagrange formula,

$$\sum_{n=0}^{\infty} \left[\frac{d^n}{dt^n} f(t) g^n(t) \right]_{t=0} \cdot \frac{u^n}{n!} = f(t) \left(1 - u \frac{dg(t)}{dt} \right)^{-1}, \quad u = \frac{t}{g(t)},$$

deduce that

$$G(u) = \sum_{n=0}^{\infty} D(n) u^n = (1 - 6u + u^2)^{-1/2}$$

and conclude that

$$nD(n) - (6n - 4)D(n - 1) + (n - 5)D(n - 2) = 0, \quad n = 2, 3, \ldots,$$

with $D(0) = 1$ and $D(1) = 3$.

40*. *Leibnitz numbers.* The numbers

$$L(n, k) = \left[(n + 1) \binom{n}{k} \right]^{-1} = \frac{k!}{(n + 1)_{k+1}}, \quad k = 0, 1, \ldots, n, \quad n = 0, 1, \ldots$$

are called *Leibnitz numbers* (see Exercise 3.6). Show that

$$L(n, k) + L(n, k - 1) = L(n - 1, k - 1), \quad k = 1, 2, \ldots, n, \quad n = 1, 2, \ldots$$

and

$$G(t, u) = \sum_{n=0}^{\infty} \sum_{k=0}^{n} L(n, k) t^k u^n = \frac{-\log\{(1 - u)(1 - tu)\}}{u\{1 + t(1 - u)\}}.$$

Chapter 8

STIRLING NUMBERS

8.1 INTRODUCTION

The Stirling numbers of the first and second kind, which are the coefficients of the expansions of the factorials into powers and of the powers into factorials, respectively, were introduced by James Stirling in 1730. Since the factorials occupy the same central position in the calculus of finite differences as the powers in the infinitesimal calculus, the Stirling numbers constitute a part of the bridge connecting these two calculi. In the classical occupancy problem, the number of ways of distributing n distinguishable balls into k distinguishable urns so that no urn remains empty was expressed by Abraham De Moivre, in 1718, in the form of a simple sum of elementary terms with alternating sign; it is essentially the Stirling number of the second kind multiplied by $k!$. The number of different results of a tossing of n distinguishable dice in which each of the $k = 6$ faces appears (at least once), derived by P. R. Montmort in 1708, probably inspired De Moivre's more general result.

The Stirling numbers under different names attracted the attention of several other well-known mathematicians of the 18th and 19th centuries. The classical book of Ch. Jordan (1939a) on the calculus of finite differences revived the interest in these numbers. A variety of applications of the Stirling numbers in combinatorics and in probability theory was provided. The coefficients of the expansion of the generalized factorials into factorials are connected with the Stirling numbers and have applications in combinatorics, in occupancy problems, and probability theory.

In this chapter, the Stirling numbers and the generalized factorial coefficients are thoroughly examined. Basic properties, generating functions, explicit expressions and recurrence relations are presented. In the examples many of their applications are discussed. The Stirling numbers have been generalized to several directions. The non-central Stirling numbers are briefly examined in the last section of this chapter.

8.2   STIRLING NUMBERS OF THE FIRST AND SECOND KIND

Consider the factorial of t of order n:

$$(t)_n = t(t-1)\cdots(t-n+1),\ n = 1,2\ldots,\ (t)_0 = 1. \qquad (8.1)$$

Clearly, this is a polynomial of t of degree n. Executing the multiplications and arranging the terms in ascending order of powers of t, we get

$$(t)_n = \sum_{k=0}^{n} s(n,k)t^k,\ n = 0,1,\ldots. \qquad (8.2)$$

Inversely, the n-th power of t may be expressed in the form of a polynomial of factorials of t of degree n. Specifically, using (8.1), we get successively the expressions

$$t^0 = (t)_0 = 1,\ t^1 = (t)_1,\ t^2 = t[1 + (t-1)] = (t)_1 + (t)_2,$$

$$t^3 = (t)_1 t + (t)_2 t = (t)_1[1 + (t-1)] + (t)_2[2 + (t-2)] = (t)_1 + 3(t)_2 + (t)_3$$

and, generally,

$$t^n = \sum_{k=0}^{n} S(n,k)(t)_k,\ n = 0,1,\ldots. \qquad (8.3)$$

Then, the following definition is introduced.

DEFINITION 8.1   *The coefficients $s(n,k)$ and $S(n,k)$ of the expansions (8.2) and (8.3) of the factorials into powers and of the powers into factorials are called Stirling numbers of the first and second kind, respectively.*

Clearly, this definition implies

$$s(n,k) = S(n,k) = 0,\ k > n,\ s(0,0) = S(0,0) = 1.$$

Further, replacing in (8.2) t by $-t$ and, since $(t + n - 1)_n = (-1)^n(-t)_n$, we deduce the expression

$$(t + n - 1)_n = \sum_{k=0}^{n} |s(n,k)|t^k,\ n = 0,1,\ldots, \qquad (8.4)$$

where

$$|s(n,k)| = (-1)^{n-k} s(n,k). \qquad (8.5)$$

Note that $|s(n, k)|$, according to (8.4), as sum of products of $n - k$ positive integers from the set $\{1, 2, \ldots, n - 1\}$ is a positive integer (see Theorem 8.1 that follows). The coefficient $|s(n, k)|$, in expansion (8.4), of the rising factorials into powers, is called *signless or absolute Stirling number of the first kind*.

Expansions (8.2) and (8.3) readily imply that the Stirling numbers of the first kind are derivatives of factorials and the Stirling numbers of the second kind are finite differences of powers. Specifically, a function $f(t)$ for which the derivatives at zero, $[D^k f(t)]_{t=0}$, $k = 0, 1, \ldots$, exist may be expanded, according to Maclaurin formula, into powers of t as

$$f(t) = \sum_{k=0}^{\infty} \left[\frac{1}{k!} D^k f(t) \right]_{t=0} \cdot t^k.$$

In the case of the n-th order factorial of t, we have $D^k (t)_n = 0$, $k > n$, and so

$$(t)_n = \sum_{k=0}^{n} \left[\frac{1}{k!} D^k (t)_n \right]_{t=0} \cdot t^k, \quad n = 0, 1, \ldots. \tag{8.6}$$

From (8.2) and (8.6) it follows that

$$s(n, k) = \left[\frac{1}{k!} D^k (t)_n \right]_{t=0}, \quad k = 0, 1, \ldots, n, \quad n = 0, 1, \ldots. \tag{8.7}$$

Similarly

$$|s(n, k)| = \left[\frac{1}{k!} D^k (t + n - 1)_n \right]_{t=0}, \quad k = 0, 1, \ldots, n, \quad n = 0, 1, \ldots. \tag{8.8}$$

Further, a function $f(t)$ for which the differences at zero, $[\Delta^k f(t)]_{t=0}$, $k = 0, 1, \ldots$, exist may be expanded, according to Newton's formula, into factorials of t as

$$f(t) = \sum_{k=0}^{\infty} \left[\frac{1}{k!} \Delta^k f(t) \right]_{t=0} \cdot (t)_k.$$

In the case of the n-th power t, we have $\Delta^k t^n = 0$, $k > n$, and so

$$t^n = \sum_{k=0}^{n} \left[\frac{1}{k!} \Delta^k t^n \right]_{t=0} \cdot (t)_k, \quad n = 0, 1, \ldots. \tag{8.9}$$

From (8.3) and (8.9) it follows that

$$S(n, k) = \left[\frac{1}{k!} \Delta^k t^n \right]_{t=0}, \quad k = 0, 1, \ldots, n, \quad n = 0, 1, \ldots. \tag{8.10}$$

THEOREM 8.1
The signless Stirling number of the first kind $|s(n,k)|$, $k = 1, 2, \ldots, n$, $n = 2, 3, \ldots$, *is given by the sum*

$$|s(n,k)| = \sum i_1 i_2 \cdots i_{n-k}, \qquad (8.11)$$

where the summation is extended over all $(n-k)$-*combinations* $\{i_1, i_2, \ldots, i_{n-k}\}$ *of the* $n - 1$ *positive integers* $\{1, 2, \ldots, n-1\}$.

PROOF   According to definition (8.4) of $|s(n,k)|$, we have for $n = 2, 3, \ldots$,

$$(t+1)(t+2) \cdots (t+n-1) = \sum_{k=1}^{n} |s(n,k)| t^{k-1}.$$

Note that the i-th factor of the product of the left-hand side,

$$p_i(t) = t + i, \; i = 1, 2, \ldots, n-1,$$

is a monomial with constant term i. Executing the multiplications, the $(k-1)$-th order power of t is formed by multiplying the constant terms of any $n - k$ factors $\{i_1, i_2, \ldots, i_{n-k}\}$, out of the $n - 1$ factors $\{1, 2, \ldots, n-1\}$, together with the first order terms t of the remaining $k - 1$ factors. Since the coefficient of the first order term t, in any factor, equals one, by the multiplication principle, (8.11) is deduced.   ∎

REMARK 8.1   Expression (8.11) of the signless Stirling number of the first kind $|s(n,k)|$ may be transformed as

$$|s(n,k)| = (n-1)! \sum \frac{1}{j_1 j_2 \cdots j_{k-1}}, \qquad (8.12)$$

where the summation is extended over all $(k-1)$-combinations $\{j_1, j_2, \ldots, j_{k-1}\}$ of the $n-1$ positive integers $\{1, 2, \ldots, n-1\}$. Indeed, to each $(n-k)$-combination $\{i_1, i_2, \ldots, i_{n-k}\}$ of $\{1, 2, \ldots, n-1\}$, there uniquely corresponds the $(k-1)$-combination $\{j_1, j_2, \ldots, j_{k-1}\} = \{1, 2, \ldots, n-1\} - \{i_1, i_2, \ldots, i_{n-k}\}$ and vice versa. Therefore

$$i_1 i_2 \cdots i_{n-k} = (n-1)! \frac{i_1 i_2 \cdots i_{n-k}}{1 \cdot 2 \cdots (n-1)} = (n-1)! \frac{1}{j_1 j_2 \cdots j_{k-1}}.$$

Introducing this expression into (8.11), (8.12) is deduced.   ∎

The Stirling numbers of the first and second kind constitute a pair of orthogonal bivariate sequences. This is shown in the next theorem.

THEOREM 8.2
The Stirling numbers of the first and second kind satisfy the following orthogonality relations:

$$\sum_{r=k}^{n} s(n,r)S(r,k) = \delta_{n,k}, \quad \sum_{r=k}^{n} S(n,r)s(r,k) = \delta_{n,k}, \qquad (8.13)$$

where $\delta_{n,k} = 1$, if $k = n$ and $\delta_{n,k} = 0$, if $k \neq n$ is the Kronecker delta.

PROOF   Expanding the n-th order factorial of t into powers of t, by using (8.2), and in the resulting expression expanding the powers of t into factorials of t, by using (8.3), we get the relation

$$(t)_n = \sum_{r=0}^{n} s(n,r)t^r = \sum_{r=0}^{n} s(n,r) \sum_{k=0}^{r} S(r,k)(t)_k$$

$$= \sum_{k=0}^{n} \left\{ \sum_{r=k}^{n} s(n,r)S(r,k) \right\} (t)_k,$$

which implies the first of (8.13). Similarly, expanding the n-th power of t into factorials of t and, in the resulting expression, expanding the factorials of t into powers of t, we deduce the second of (8.13).   ∎

Note that, for fixed n, the (usual) generating function of the sequence of Stirling numbers of the first kind $s(n,k)$, $k = 0, 1, \ldots, n$, according to (8.2), is given by

$$s_n(t) = \sum_{k=0}^{n} s(n,k)t^k = (t)_n, \ n = 0, 1, \ldots,$$

while, the factorial generating function of the sequence of Stirling numbers of the second kind $S(n,k)$, $k = 0, 1, \ldots, n$, according to (8.3), is given by

$$S_n(t) = \sum_{k=0}^{n} S(n,k)(t)_k = t^n, \ n = 0, 1, \ldots.$$

Further, from (8.2), using Newton's general binomial theorem, we deduce the double generating function

$$g(t,u) = \sum_{n=0}^{\infty} \sum_{k=0}^{n} s(n,k)t^k \frac{u^n}{n!} = (1+u)^t, \qquad (8.14)$$

while from (8.3) we get

$$f(t,u) = \sum_{n=0}^{\infty} \sum_{k=0}^{n} S(n,k)(t)_k \frac{u^n}{n!} = e^{tu}. \qquad (8.15)$$

The exponential generating functions of the sequences $s(n, k)$, $n = k$, $k + 1, \ldots$, and $S(n, k)$, $n = k, k + 1, \ldots$, for fixed k, are derived in the next theorem.

THEOREM 8.3

(a) *The exponential generating function of the Stirling numbers of the first kind* $s(n, k)$, $n = k, k + 1, \ldots$, *for fixed k, is given by*

$$g_k(u) = \sum_{n=k}^{\infty} s(n, k) \frac{u^n}{n!} = \frac{[\log(1 + u)]^k}{k!}, \quad k = 0, 1, \ldots. \qquad (8.16)$$

(b) *The exponential generating function of the Stirling numbers of the second kind* $S(n, k)$, $n = k, k + 1, \ldots$, *for fixed k, is given by*

$$f_k(u) = \sum_{n=k}^{\infty} S(n, k) \frac{u^n}{n!} = \frac{(e^u - 1)^k}{k!}, \quad k = 0, 1, \ldots. \qquad (8.17)$$

PROOF   (a) Interchanging the order of summation in (8.14), we get

$$g(t, u) = \sum_{k=0}^{\infty} \sum_{n=k}^{\infty} s(n, k) \frac{u^n}{n!} t^k = \sum_{k=0}^{\infty} g_k(u) t^k$$

and since

$$g(t, u) = (1 + u)^t = \exp\{t \log(1 + u)\} = \sum_{k=0}^{\infty} \frac{[\log(1 + u)]^k}{k!} t^k,$$

we deduce (8.16).

(b) Similarly, interchanging the order of summation in (8.15), we get

$$f(t, u) = \sum_{k=0}^{\infty} \sum_{n=k}^{\infty} S(n, k) \frac{u^n}{n!} (t)_k = \sum_{k=0}^{\infty} f_k(u)(t)_k$$

and since

$$f(t, u) = [1 + (e^u - 1)]^t = \sum_{k=0}^{\infty} \binom{t}{k} (e^u - 1)^k = \sum_{k=0}^{\infty} \frac{(e^u - 1)^k}{k!} (t)_k,$$

(8.17) is established.   ∎

The generating functions of the signless Stirling numbers of the first kind may be obtained from the corresponding generating functions of the numbers $s(n, k)$ by using (8.5). Thus, from the first part of Theorem 8.3, we deduce the following corollary.

COROLLARY 8.1
The exponential generating function of the signless Stirling numbers of the first kind $|s(n, k)|$, $n = k, k + 1, \ldots$, for fixed k, is given by

$$h_k(u) = \sum_{n=k}^{\infty} |s(n, k)| \frac{u^n}{n!} = \frac{[-\log(1 - u)]^k}{k!}, \quad k = 0, 1, \ldots. \quad (8.18)$$

In the following example, a probabilistic application of the signless Stirling numbers of the first kind is given.

Example 8.1 Bernoulli trials with varying success probability
Suppose that balls are successively drawn one after the other from an urn initially containing m white balls, according to the following scheme. After each trial the drawn ball is placed back in the urn along with s black balls. Determine (a) the probability $p(k; n)$ of drawing k white balls in n trials and (b) the probability $q(n; k)$ that n trials are required until the k-th white ball is drawn.

(a) Let A_j be the event of drawing a white ball at the j-th trial, $j = 1, 2, \ldots, n$. Then, setting $\theta = m/s$, we get

$$p_j = P(A_j) = \frac{\theta}{\theta + j - 1},$$

$$q_j = P(A'_j) = 1 - P(A_j) = \frac{j - 1}{\theta + j - 1}, \quad j = 1, 2, \ldots, n$$

and

$$P(A_{j_1} A_{j_2} \cdots A_{j_k} A'_{j_{k+1}} \cdots A'_{j_n})$$
$$= P(A_{j_1}) P(A_{j_2}) \cdots P(A_{j_k}) P(A'_{j_{k+1}}) \cdots P(A'_{j_n})$$
$$= \frac{\theta^k}{(\theta + n - 1)_n} (j_{k+1} - 1)(j_{k+2} - 1) \cdots (j_n - 1).$$

Therefore, the probability $p(k; n)$ of drawing k white balls in n trials is given by the sum

$$p(k; n) = \frac{\theta^k}{(\theta + n - 1)_n} \sum (j_{k+1} - 1)(j_{k+2} - 1) \cdots (j_n - 1),$$

where, since in the first trial the probability of drawing a black ball is $q_1 = 0$, the summation is extended over all $(n - k)$-combinations $\{j_{k+1}, j_{k+2}, \ldots, j_n\}$ of the

$n-1$ positive integers $\{2, 3, \ldots, n\}$. Putting $i_m = j_{k+m} - 1, m = 1, 2, \ldots, n-k$, and using (8.11), we get

$$p(k; n) = \frac{|s(n, k)|\theta^k}{(\theta + n - 1)_n}, \quad k = 1, 2, \ldots, n.$$

Note that, according to (8.4), these probabilities sum to unity.

(b) The probability $q(n; k)$ that n trials are required until the k-th white ball is drawn equals the probability $p(k - 1; n - 1)$ of drawing $k - 1$ white balls in $n - 1$ trials multiplied by the probability $p_n = P(A_n) = \theta/(\theta + n - 1)$ of drawing a white ball at the n-th trial. Therefore

$$q(n; k) = \frac{|s(n - 1, k - 1)|\theta^k}{(\theta + n - 1)_n}, \quad n = k, k + 1, \ldots. \qquad \square$$

An interesting application of the orthogonality property of the Stirling numbers is presented in the following example.

Example 8.2 Inverse relations

Consider a sequence of real numbers x_k, $k = 0, 1, \ldots$, and let

$$y_r = \sum_{k=0}^{r} a_{r,k} x_k, \quad r = 0, 1, \ldots,$$

where $a_{r,k}$, $k = 0, 1, \ldots, r$, $r = 0, 1, \ldots$, are given coefficients. If there exist coefficients $b_{n,r}$, $r = 0, 1, \ldots, n$, $n = 0, 1, \ldots$, orthogonal to the coefficients $a_{r,k}$, $k = 0, 1, \ldots, r$, $r = 0, 1, \ldots$,

$$\sum_{r=k}^{n} b_{n,r} a_{r,k} = \delta_{n,k},$$

then

$$\sum_{r=k}^{n} b_{n,r} y_r = \sum_{r=k}^{n} b_{n,r} \sum_{k=0}^{r} a_{r,k} x_k = \sum_{k=0}^{n} \left\{ \sum_{r=k}^{n} b_{n,r} a_{r,k} \right\} x_k$$

and so

$$x_n = \sum_{r=0}^{n} b_{n,r} y_r, \quad n = 0, 1, \ldots.$$

The expressions of y_r, $r = 0, 1, \ldots$, in terms of x_k, $k = 0, 1, \ldots, r$, and inversely of x_n, $n = 0, 1, \ldots$, in terms of y_r, $r = 0, 1, \ldots, n$ constitute a pair of inverse relations. Similarly if

$$u_r = \sum_{n=r}^{\infty} a_{n,r} w_n, \quad r = 0, 1, \ldots,$$

and

$$\sum_{r=k}^{n} a_{n,r} b_{r,k} = \delta_{n,k},$$

then

$$w_k = \sum_{r=k}^{\infty} b_{r,k} u_r, \quad k = 0, 1, \dots .$$

In conclusion, every pair of orthogonal sequences entails a pair of inverse relations. According to Theorem 8.2, the Stirling numbers of the first and second kind constitute a pair of orthogonal sequences. Consequently, if

$$y_r = \sum_{k=0}^{r} s(r, k) x_k, \quad r = 0, 1, \dots ,$$

then

$$x_n = \sum_{r=0}^{n} S(n, r) y_r, \quad n = 0, 1, \dots$$

and inversely. Also, if

$$u_r = \sum_{n=r}^{\infty} s(n, r) w_n, \quad r = 0, 1, \dots ,$$

then

$$w_k = \sum_{r=k}^{\infty} S(r, k) u_r, \quad k = 0, 1, \dots$$

and inversely. ☐

Example 8.3 Connection of ordinary and factorial moments

Consider a sequence a_j, $j = 0, 1, \dots$ (of numbers with a combinatorial interpretation or probabilities or masses). The r-th order (power) moment μ'_r and the r-th order factorial moment $\mu_{(r)}$ of this sequence are defined by

$$\mu'_r = \sum_{j=0}^{\infty} j^r a_j, \quad \mu_{(r)} = \sum_{j=0}^{\infty} (j)_r a_j, \quad r = 0, 1, \dots .$$

Using (8.2), the r-th order factorial moment $\mu_{(r)}$ may be expressed as

$$\mu_{(r)} = \sum_{j=0}^{\infty} \sum_{k=0}^{r} s(r, k) j^k a_j = \sum_{k=0}^{r} s(r, k) \sum_{j=0}^{\infty} j^k a_j$$

and so

$$\mu_{(r)} = \sum_{k=0}^{r} s(r, k) \mu'_k, \quad r = 0, 1, \dots .$$

In the same way, using (8.3), we conclude that

$$\mu'_n = \sum_{r=0}^{n} S(n,r)\mu_{(r)}, \quad n = 0, 1, \dots .$$

Note that these two relations constitute a pair of inverse relations.   □

Example 8.4 The operator $\Theta = tD$

In the infinitesimal calculus, besides the usual derivative operator D, the operator $\Theta = tD$ is frequently used. Express the operator Θ in terms of the operator D.

Clearly,

$$\Theta f(t) = tDf(t),$$

$$\Theta^2 f(t) = \Theta(\Theta f(t)) = tD(tDf(t)) = tDf(t) + t^2 D^2 f(t),$$

$$\Theta^3 f(t) = \Theta(\Theta^2 f(t)) = tD(tDf(t) + t^2 D^2 f(t))$$
$$= tDf(t) + 3t^2 D^2 f(t) + t^3 D^3 f(t)$$

and, generally,

$$\Theta^n f(t) = \sum_{r=1}^{n} C_{n,r} t^r D^r f(t),$$

where the coefficients $C_{n,r}, r = 1, 2, \dots, n$ are independent of the function $f(t)$. Thus, for their determination the most convenient function can be chosen. Let $f(t) = t^u$, whence

$$\Theta^n t^u = u^n t^u, \quad D^r t^u = (u)_r t^{u-r}.$$

Consequently,

$$u^n = \sum_{r=1}^{n} C_{n,r}(u)_r$$

and, according to (8.3), $C_{n,r} = S(n,r), r = 1, 2, \dots, n$. Thus

$$\Theta^n f(t) = \sum_{r=1}^{n} S(n,r) t^r D^r f(t).$$

The inverse of this relation (see Example 8.2) is

$$D^r f(t) = t^{-r} \sum_{k=1}^{r} s(r,k) \Theta^k f(t).   \quad □$$

Example 8.5 The operator $\Psi = t\Delta$

In the calculus of finite differences, besides the usual difference operator Δ, the operator $\Psi = t\Delta$ is frequently used. Express the operator Ψ in terms of the operator Δ.

Clearly,

$$\Psi f(t) = t\Delta f(t),$$

$$\Psi^2 f(t) = \Psi(\Psi f(t)) = t\Delta(t\Delta f(t)) = t\Delta f(t) + (t+1)_2 \Delta^2 f(t),$$

$$\Psi^3 f(t) = \Psi(\Psi^2 f(t)) = t\Delta(t\Delta f(t) + (t+1)_2 \Delta^2 f(t))$$
$$= t\Delta f(t) + (t+1)_2 \Delta^2 f(t) + (t+2)_3 \Delta^3 f(t)$$

and, generally,

$$\Psi^n f(t) = \sum_{r=1}^{n} B_{n,r}(t+r-1)_r \Delta^r f(t),$$

where the coefficients $B_{n,r}$, $r = 1, 2, \ldots, n$ are independent of the function $f(t)$. Thus, for their determination the most convenient function can be chosen. Let $f(t) = (t+u-1)_u$, whence

$$\Psi^n(t+u-1)_u = u^n(t+u-1)_u, \quad \Delta^r(t+u-1)_u = (u)_r(t+u-1)_{u-r}.$$

Consequently,

$$u^n = \sum_{r=1}^{n} B_{n,r}(u)_r$$

and, according to (8.3), $B_{n,r} = S(n,r)$, $r = 1, 2, \ldots, n$. Thus

$$\Psi^n f(t) = \sum_{r=1}^{n} S(n,r)(t+r-1)_r \Delta^r f(t).$$

The inverse of this relation (see Example 8.2) is

$$\Delta^r f(t) = \frac{1}{(t+r-1)_r} \sum_{k=1}^{r} s(r,k)\Psi^k f(t). \quad \square$$

Example 8.6 Convolution of a logarithmic distribution
The probability function of a logarithmic distribution is

$$p_n = [-\log(1-\theta)]^{-1}\frac{\theta^n}{n}, \quad n = 1, 2, \ldots, \quad 0 < \theta < 1.$$

The generating function of the sequence of probabilities p_n, $n = 1, 2, \ldots$, is readily obtained as

$$h(u) = \sum_{n=1}^{\infty} p_n t^n = [-\log(1-\theta)]^{-1} \sum_{n=1}^{\infty} \frac{(\theta u)^n}{n}$$
$$= [-\log(1-\theta)]^{-1}[-\log(1-\theta u)].$$

Therefore, if $p_n(k)$, $n = k, k+1, \ldots$, is the k-fold convolution of p_n, $n = 1, 2, \ldots$, then

$$h_k(u) = \sum_{n=k}^{\infty} p_n(k) t^n = [h(u)]^k = [-\log(1-\theta)]^{-k} [-\log(1-\theta u)]^k.$$

Using (8.14), we deduce the expression of the probability function

$$p_n(k) = [-\log(1-\theta)]^{-1} k! |s(n,k)| \frac{\theta^n}{n!}, \quad n = k, k+1, \ldots.$$

The exponential generating function

$$b_k(u) = \sum_{r=0}^{\infty} \mu_{(r)}(k) \frac{u^r}{r!}, \quad r = 0, 1, \ldots,$$

of the factorial moments

$$\mu_{(r)}(k) = \sum_{n=k}^{\infty} (n)_r p_n(k), \quad r = 0, 1, \ldots,$$

of the k-fold convolution $p_n(k)$, $n = k, k+1, \ldots$, of the logarithmic distribution, is connected with the generating function $h_k(u)$ by the relation $b_k(u) = h_k(u+1)$. Therefore, using (8.14), we successively get

$$b_k(u) = \left(1 + [-\log(1-\theta)]^{-1} [-\log(1-\theta(1-\theta)^{-1}u)]\right)^k$$

$$= \sum_{j=0}^{\infty} \binom{k}{j} [-\log(1-\theta)]^{-j} [-\log(1-\theta(1-\theta)^{-1}u)]^j$$

$$= \sum_{j=0}^{\infty} (k)_j [-\log(1-\theta)]^{-j} \sum_{r=j}^{\infty} s(r,j) \frac{\theta^r}{(1-\theta)^r} \frac{u^r}{r!}$$

$$= \sum_{r=0}^{\infty} \left\{ \frac{\theta^r}{(1-\theta)^r} \sum_{j=0}^{\min\{r,k\}} (k)_j |s(r,j)| [-\log(1-\theta)]^{-j} \right\} \frac{u^r}{r!}$$

and so

$$\mu_{(r)}(k) = \frac{\theta^r}{(1-\theta)^r} \sum_{j=0}^{\min\{r,k\}} (k)_j |s(r,j)| [-\log(1-\theta)]^{-j},$$

for $r = 0, 1, \ldots$. $\square$

8.3 EXPLICIT EXPRESSIONS AND RECURRENCE RELATIONS

An explicit expression of the Stirling numbers of the second kind is deduced in the following theorem.

THEOREM 8.4
The Stirling number of the second kind $S(n, k)$, $k = 0, 1, \ldots, n$, $n = 0, 1, \ldots,$ is given by the sum

$$S(n, k) = \frac{1}{k!} \sum_{r=0}^{k} (-1)^{k-r} \binom{k}{r} r^n. \tag{8.19}$$

PROOF   Expanding the generating function (8.17) into powers of u, we get the expression

$$\sum_{n=k}^{\infty} S(n, k) \frac{u^n}{n!} = \frac{1}{k!} \sum_{r=0}^{k} (-1)^{k-r} \binom{k}{r} e^{ru}$$

$$= \frac{1}{k!} \sum_{r=0}^{k} (-1)^{k-r} \binom{k}{r} \sum_{n=0}^{\infty} r^n \frac{u^n}{n!}$$

$$= \sum_{n=0}^{\infty} \left\{ \frac{1}{k!} \sum_{r=0}^{k} (-1)^{k-r} \binom{k}{r} r^n \right\} \frac{u^n}{n!},$$

which implies (8.19).   ∎

REMARK 8.2   The explicit expression (8.19) of the Stirling numbers of the second kind may also be deduced from (8.10), by using the expression of the k-th power of the difference operator in terms of the shift operator,

$$\Delta^k = \sum_{r=0}^{k} (-1)^{k-r} \binom{k}{r} E^r.$$

Thus

$$S(n, k) = \left[\frac{1}{k!} \Delta^k t^n \right]_{t=0} = \frac{1}{k!} \sum_{r=0}^{k} (-1)^{k-r} \binom{k}{r} [E^r t^n]_{t=0}$$

and, since $[E^r t^n]_{t=0} = r^n$, (8.19) is deduced.   ∎

The Stirling numbers of the second kind, according to Theorem 8.4, are expressed in the form of a single summation of elementary terms, which are products and quotients of factorials and powers. There does not exist an analogous expression for the Stirling numbers of the first kind. An expression of the Stirling numbers of the first kind in the form of a single summation of terms that are products of binomial coefficients and Stirling numbers of the second kind is derived in the next theorem due to Schlömilch (1852). For the Stirling numbers of the first kind, it leads to an expression in the form of a double summation of elementary terms.

THEOREM 8.5 Schlömilch's formula
The Stirling numbers of the first kind are expressed in terms of the Stirling numbers of the second kind by

$$s(n,k) = \sum_{r=0}^{n-k} (-1)^r \binom{n+r-1}{k-1} \binom{2n-k}{n-k-r} S(n-k+r,r). \quad (8.20)$$

PROOF   The power series $t = \phi^{-1}(u) = \log(1+u)$ is the inverse of the power series $u = \phi(t) = e^t - 1$, with $\phi(0) = 0$. Thus, by (8.16) and applying Lagrange inversion formula (see Theorem 11.11),

$$\frac{1}{k!}\left[\frac{d^n}{du^n}(\phi^{-1}(u))^k\right]_{u=0} = \binom{n-1}{k-1}\left[\frac{d^{n-k}}{dt^{n-k}}\left(\frac{\phi(t)}{t}\right)^{-n}\right]_{t=0},$$

we get

$$s(n,k) = \frac{1}{k!}\left[\frac{d^n}{du^n}(\log(1+u))^k\right]_{u=0} = \binom{n-1}{k-1}\left[\frac{d^{n-k}}{dt^{n-k}}\left(\frac{e^t-1}{t}\right)^{-n}\right]_{t=0}.$$

Since (see Remark 11.5)

$$\left[\frac{d^m}{dt^m}(h(t))^s\right]_{t=0} = \sum_{r=0}^{m}\binom{s}{r}\binom{m-s}{m-r}\left[\frac{d^m}{dt^m}(h(t))^r\right]_{t=0},$$

for s a real number and $h(t)$ a function with $h(0) = 1$, for which the derivatives at zero exist, we deduce the expression

$$s(n,k) = \binom{n-1}{k-1}\sum_{r=0}^{n-k}\binom{-n}{r}\binom{2n-k}{n-k-r}\left[\frac{d^{n-k}}{dt^{n-k}}\left(\frac{e^t-1}{t}\right)^r\right]_{t=0}.$$

Further, by (8.17), we have

$$\frac{(e^t-1)^r}{t^r} = \sum_{n=r}^{\infty} r!S(n,r)\frac{t^{n-r}}{n!} = \sum_{j=0}^{\infty} r!S(j+r,r)\frac{t^j}{(j+r)!},$$

whence

$$\left[\frac{d^{n-k}}{dt^{n-k}}\left(\frac{e^t-1}{t}\right)^r\right]_{t=0} = \frac{S(n-k+r,r)}{\binom{n-k+r}{r}}$$

and so, using the relation

$$\binom{n-1}{k-1}\binom{-n}{r} = (-1)^r\binom{n+r-1}{k-1}\binom{n-k+r}{r},$$

we conclude (8.20). ∎

REMARK 8.3   Schlömilch's formula (8.20), by replacing k by $n-k$ and since

$$(-1)^k s(n, n-k) = |s(n, n-k)|,$$

$$\binom{n+r-1}{n-k-1}\binom{k+n}{k-r} = (-1)^{k+r}\binom{k-n}{k+r}\binom{k+n}{k-r},$$

may be rewritten in the following symmetric form

$$|s(n, n-k)| = \sum_{r=0}^{k}\binom{k-n}{k+r}\binom{k+n}{k-r}S(k+r,r).$$

The inverse of this relation is given in Exercise 2. ∎

Introducing expression (8.19) into Schlömilch's formula (8.20), we conclude the following corollary.

COROLLARY 8.2
The Stirling number of the first kind $s(n,k)$, $k = 0, 1, \ldots, n$, $n = 0, 1, \ldots$, is given by the double sum

$$s(n,k) = \sum_{r=0}^{n-k}\sum_{j=0}^{r}(-1)^j\binom{r}{j}\binom{n+r-1}{k-1}\binom{2n-k}{n-k-r}\frac{j^{n-k+r}}{r!}. \tag{8.21}$$

Expressions of the Stirling numbers $s(n,k)$ and $S(n,k)$ as multiple sums over all compositions, as well as over all partitions of n into k parts, are derived in the following theorem.

THEOREM 8.6
The Stirling numbers of the first and second kind $s(n,k)$ and $S(n,k)$ are given by

$$|s(n,k)| = \frac{n!}{k!}\sum\frac{1}{r_1 \cdot r_2 \cdots r_k} \tag{8.22}$$

and

$$S(n, k) = \frac{n!}{k!} \sum \frac{1}{r_1! r_2! \cdots r_k!}, \tag{8.23}$$

respectively, where the summation in both sums is extended over all nonnegative integer solutions of the equation $r_1 + r_2 + \cdots + r_k = n$. Alternatively,

$$|s(n, k)| = \sum \frac{n!}{k_1! k_2! \cdots k_n!} \left(\frac{1}{1}\right)^{k_1} \left(\frac{1}{2}\right)^{k_2} \cdots \left(\frac{1}{n}\right)^{k_n} \tag{8.24}$$

and

$$S(n, k) = \sum \frac{n!}{k_1! k_2! \cdots k_n!} \left(\frac{1}{1!}\right)^{k_1} \left(\frac{1}{2!}\right)^{k_2} \cdots \left(\frac{1}{n!}\right)^{k_n}, \tag{8.25}$$

respectively, where the summation in both sums is extended over all nonnegative integer solutions of the equations $k_1 + 2k_2 + \cdots + nk_n = n$ and $k_1 + k_2 + \cdots + k_n = k$.

PROOF The exponential generating function of the singless Stirling numbers of the first kind (8.18), upon using the expansion $-\log(1 - u) = \sum_{r=1}^{\infty} u^r/r$ and then the Cauchy rule of products of series, may be written as

$$\sum_{n=k}^{\infty} |s(n, k)| \frac{u^n}{n!} = \frac{1}{k!} \left(\sum_{r=1}^{\infty} \frac{u^r}{r}\right)^k = \frac{1}{k!} \prod_{i=1}^{k} \left(\sum_{r_i=1}^{\infty} \frac{u^{r_i}}{r_i}\right)$$

$$= \sum_{n=k}^{\infty} \left\{\frac{n!}{k!} \sum \frac{1}{r_1 \cdot r_2 \cdots r_k}\right\} \frac{u^n}{n!},$$

yielding (8.22). Similarly, the exponential generating function of the Stirling numbers of the second kind (8.17) may be expressed as

$$\sum_{n=k}^{\infty} S(n, k) \frac{u^n}{n!} = \frac{1}{k!} \left(\sum_{r=1}^{\infty} \frac{u^r}{r!}\right)^k = \frac{1}{k!} \prod_{i=1}^{k} \left(\sum_{r_i=1}^{\infty} \frac{u^{r_i}}{r_i!}\right)$$

$$= \sum_{n=k}^{\infty} \left\{\frac{n!}{k!} \sum \frac{1}{r_1! r_2! \cdots r_k!}\right\} \frac{u^n}{n!},$$

yielding (8.23). Using the multinomial theorem instead of the Cauchy rule, the generating functions (8.18) and (8.17) may be expressed as

$$h_k(u) = \sum_{n=k}^{\infty} |s(n, k)| \frac{u^n}{n!} = \frac{1}{k!} \left(u + \frac{u^2}{2} + \cdots + \frac{u^n}{n} + \cdots\right)^k$$

$$= \sum_{n=k}^{\infty} \left\{\sum \frac{n!}{k_1! k_2! \cdots k_n!} \left(\frac{1}{1}\right)^{k_1} \left(\frac{1}{2}\right)^{k_2} \cdots \left(\frac{1}{n}\right)^{k_n}\right\} \frac{u^n}{n!},$$

and

$$f_k(u) = \sum_{n=k}^{\infty} S(n,k) \frac{u^n}{n!} = \frac{1}{k!} \left(u + \frac{u^2}{2!} + \cdots + \frac{u^n}{n!} + \cdots \right)^k$$

$$= \sum_{n=k}^{\infty} \left\{ \sum \frac{n!}{k_1! k_2! \cdots k_n!} \left(\frac{1}{1!} \right)^{k_1} \left(\frac{1}{2!} \right)^{k_2} \cdots \left(\frac{1}{n!} \right)^{k_n} \right\} \frac{u^n}{n!},$$

yielding (8.24) and (8.25), respectively.   ∎

In the next theorem, triangular recurrence relations of the Stirling numbers of the first kind and second kind are deduced.

THEOREM 8.7

  (a) *The Stirling numbers of the first kind* $s(n,k)$, $k = 0, 1, \ldots, n$, $n = 0, 1, \ldots$, *satisfy the triangular recurrence relation*

$$s(n+1, k) = s(n, k-1) - ns(n, k), \tag{8.26}$$

for $k = 1, 2, \ldots, n+1$, $n = 0, 1, \ldots$, *with initial conditions*

$$s(0,0) = 1, \quad s(n, 0) = 0, \quad n > 0, \quad s(n, k) = 0, \quad k > n.$$

  (b) *The Stirling numbers of the second kind* $S(n, k)$, $k = 0, 1, \ldots, n$, $n = 0, 1, \ldots$, *satisfy the triangular recurrence relation*

$$S(n+1, k) = S(n, k-1) + kS(n, k), \tag{8.27}$$

for $k = 1, 2, \ldots, n+1$, $n = 0, 1, \ldots$, *with initial conditions*

$$S(0,0) = 1, \quad S(n, 0) = 0, \quad n > 0, \quad S(n, k) = 0, \quad k > n.$$

PROOF   (a) Expanding both members of the recurrence relation $(t)_{n+1} = (t-n)(t)_n$ into powers of t, according to (8.2), we get the relation

$$\sum_{k=0}^{n+1} s(n+1, k) t^k = (t-n) \sum_{r=0}^{n} s(n, r) t^r$$

$$= \sum_{k=1}^{n+1} s(n, k-1) t^k - \sum_{k=0}^{n} ns(n, k) t^k,$$

which implies (8.26). The initial conditions follow directly from (8.2).

  (b) Expanding both members of the recurrence relation $t^{n+1} = t \cdot t^n$ into factorials of t, according to (8.3), we have

$$\sum_{k=0}^{n+1} S(n+1, k)(t)_k = t \sum_{r=0}^{n} S(n, r)(t)_r.$$

Since $(t)_{r+1} = (t-r)(t)_r$, whence $t(t)_r = (t)_{r+1} + r(t)_r$, we deduce the relation

$$\sum_{k=0}^{n+1} S(n+1,k)(t)_k = \sum_{r=0}^{n} S(n,r)(t)_{r+1} + \sum_{r=0}^{n} rS(n,r)(t)_r$$

$$= \sum_{k=1}^{n+1} S(n,k-1)(t)_k + \sum_{k=0}^{n} kS(n,k)(t)_k,$$

which implies (8.27). The initial conditions follow directly from (8.3). ∎

Multiplying the triangular recurrence relation (8.26) by $(-1)^{n-k+1}$ and using (8.5), we deduce the following corollary.

COROLLARY 8.3
The signless Stirling numbers of the first kind $|s(n,k)|$, $k = 0, 1, \ldots, n$, $n = 0, 1, \ldots$, satisfy the triangular recurrence relation

$$|s(n+1,k)| = |s(n,k-1)| + n|s(n,k)|, \tag{8.28}$$

for $k = 1, 2, \ldots, n+1$, $n = 0, 1, \ldots$, with initial conditions

$$|s(0,0)| = 1, \ |s(n,0)| = 0, \ n > 0, \ |s(n,k)| = 0, \ k > n.$$

The signless Stirling numbers of the first kind $|s(n,k)|$ can be tabulated by using the triangular recurrence relation (8.28) and its initial conditions. Table 8.1 gives the numbers $|s(n,k)|$, $k = 1, 2, \ldots, n$, $n = 1, 2, \ldots, 9$.

Table 8.1 **Signless Stirling Numbers of the First Kind $|s(n,k)|$**

n \ k	1	2	3	4	5	6	7	8	9
1	1								
2	1	1							
3	2	3	1						
4	6	11	6	1					
5	24	50	35	10	1				
6	120	274	225	85	15	1			
7	720	1764	1624	735	175	21	1		
8	5040	13068	13132	6769	1960	322	28	1	
9	40320	109584	188124	67284	22449	4536	546	36	1

The Stirling numbers of the second kind $S(n,k)$ can also be tabulated by using the triangular recurrence relation (8.27) and its initial conditions. Table 8.2 gives the numbers $S(n,k)$, $k = 1, 2, \ldots, n$, $n = 1, 2, \ldots, 10$.

Table 8.2   *Stirling Numbers of the Second Kind $S(n,k)$*

k \ n	1	2	3	4	5	6	7	8	9	10
1	1									
2	1	1								
3	1	3	1							
4	1	7	6	1						
5	1	15	25	10	1					
6	1	31	90	65	15	1				
7	1	63	301	350	140	21	1			
8	1	127	966	1701	1050	266	28	1		
9	1	255	3025	7770	6951	2646	462	36	1	
10	1	511	9330	34105	42525	22827	5880	750	45	1

Vertical and horizontal recurrence relations of the Stirling numbers of the first and second kind are derived in the following theorems.

THEOREM 8.8
The Stirling numbers of the first kind $s(n,k)$, $k = 0, 1, \ldots, n$, $n = 0, 1, \ldots$, with $s(0,0) = 1$, satisfy (a) the vertical recurrence relation

$$s(n+1, k+1) = \sum_{r=k}^{n} (-1)^{n-r} (n)_{n-r} s(r, k) \qquad (8.29)$$

and (b) *the horizontal recurrence relation*

$$s(n+1, k+1) = \sum_{r=k}^{n} (-1)^{r-k} \binom{r}{k} s(n, r). \qquad (8.30)$$

PROOF   (a) According to (8.2), for $n = 0, 1, \ldots$ and since $s(n+1, 0) = 0$, we have

$$\sum_{k=0}^{n} s(n+1, k+1) t^k = (t-1)(t-2) \cdots (t-n) = (t-1)_n.$$

Expanding $(t-1)_n$ into factorials of t by Vandermonde's formula,

$$(t-1)_n = \sum_{r=0}^{n} \binom{n}{r}(-1)_{n-r}(t)_r = \sum_{r=0}^{n}(-1)^{n-r}(n)_{n-r}(t)_r,$$

we find

$$\sum_{k=0}^{n} s(n+1, k+1)t^k = \sum_{r=0}^{n}(-1)^{n-r}(n)_{n-r}(t)_r.$$

Further, expanding the factorials in the right-hand side into powers of t, by using (8.2), we get the relation

$$\sum_{k=0}^{n} s(n+1, k+1)t^k = \sum_{r=0}^{n}(-1)^{n-r}(n)_{n-r}\sum_{k=0}^{r} s(r,k)t^k$$

$$= \sum_{k=0}^{n}\left\{\sum_{r=k}^{n}(-1)^{n-r}(n)_{n-r}s(r,k)\right\}t^k,$$

from which (8.29) is deduced.

(b) Differentiating the generating function (8.16),

$$\sum_{n=k}^{\infty} s(n+1, k+1)\frac{u^{n+1}}{(n+1)!} = \frac{[\log(1+u)]^{k+1}}{(k+1)!},$$

we get

$$\sum_{n=k}^{\infty} s(n+1, k+1)\frac{u^n}{n!} = \frac{(1+u)^{-1}[\log(1+u)]^k}{k!}$$

and so

$$\sum_{n=k}^{\infty} s(n+1, k+1)\frac{u^n}{n!} = \frac{e^{-\log(1+u)}[\log(1+u)]^k}{k!}$$

$$= \sum_{r=k}^{\infty}(-1)^{r-k}\frac{[\log(1+u)]^r}{(r-k)!k!}.$$

Expanding the terms of the last member of this relation into powers of u, by using (8.16), and interchanging the order of summation, we deduce the relation

$$\sum_{n=k}^{\infty} s(n+1, k+1)\frac{u^n}{n!} = \sum_{r=k}^{\infty}(-1)^{r-k}\binom{r}{k}\sum_{n=r}^{\infty} s(n,r)\frac{u^n}{n!}$$

$$= \sum_{n=k}^{\infty}\left\{\sum_{r=k}^{n}(-1)^{r-k}\binom{r}{k}s(n,r)\right\}\frac{u^n}{n!},$$

which implies (8.30). ∎

THEOREM 8.9
*The Stirling numbers of the second kind $S(n,k)$, $k = 0, 1, \ldots, n$, $n = 0, 1, \ldots,$
with $S(0,0) = 1$, satisfy* (a) *the vertical recurrence relation*

$$S(n+1, k+1) = \sum_{r=k}^{n} \binom{n}{r} S(r, k) \qquad (8.31)$$

and (b) *the horizontal recurrence relation*

$$S(n, k) = \sum_{r=k}^{n} (-1)^{r-k} (r)_{r-k} S(n+1, r+1). \qquad (8.32)$$

PROOF   (a) According to (8.3), for $n = 0, 1, \ldots$ and since $S(n+1, 0) = 0$,
we have

$$\sum_{k=0}^{n} S(n+1, k+1)(t+1)_{k+1} = (t+1)^{n+1}.$$

Further, since $(t+1)_{k+1} = (t+1)(t)_k$, we conclude that

$$\sum_{k=0}^{n} S(n+1, k+1)(t)_k = (t+1)^n.$$

Expanding the right-hand side into powers of t, by Newton's binomial formula, and
in the resulting expression, expanding the powers into factorials, by using (8.3),
we successively get

$$\sum_{k=0}^{n} S(n+1, k+1)(t)_k = \sum_{r=0}^{n} \binom{n}{r} \sum_{k=0}^{r} S(r, k)(t)_k$$

$$= \sum_{k=0}^{n} \left\{ \sum_{r=k}^{n} \binom{n}{r} S(r, k) \right\} (t)_k,$$

yielding (8.31).

(b) Multiplying the recurrence relation

$$S(n+1, r+1) = S(n, r) + (r+1)S(n, r+1)$$

by $(-1)^{r-k}(r)_{r-k}$ and summing for $r = k, k+1, \ldots, n$, since $S(n, n+1) = 0$,
we get

$$\sum_{r=k}^{n} (-1)^{r-k} (r)_{r-k} S(n+1, r+1)$$

$$= \sum_{r=k}^{n} (-1)^{r-k} (r)_{r-k} S(n, r) + \sum_{r=k}^{n-1} (-1)^{r-k} (r+1)_{r+1-k} S(n, r+1)$$

$$= \sum_{r=k}^{n} (-1)^{r-k} (r)_{r-k} S(n, r) - \sum_{j=k+1}^{n} (-1)^{j-k} (j)_{j-k} S(n, j) = S(n, k),$$

establishing (8.32). ∎

The triangular recurrence relation (8.27) can be used for the determination of the (power) generating function of the sequence of the Stirling numbers of the second kind $S(n, k)$, $n = k, k+1, \ldots$, for fixed k. This generating function (a) suitably expanded leads to an interesting expression of the numbers $S(n, k)$ and (b) transformed yields the expansion of the inverse factorials into inverse powers. Specifically, we have the following theorems.

THEOREM 8.10
The (power) generating function of the Stirling numbers of the second kind $S(n, k)$, $n = k, k+1, \ldots$, for fixed k, is given by

$$\phi_k(u) = \sum_{n=k}^{\infty} S(n, k)u^n = u^k \prod_{j=1}^{k}(1 - ju)^{-1}, \quad k = 1, 2, \ldots. \qquad (8.33)$$

PROOF Multiplying (8.27) by u^{n+1} and summing for $n = k-1, k, \ldots$, since $S(k-1, k) = 0$, we get

$$\sum_{n=k-1}^{\infty} S(n+1, k)u^{n+1} = u \sum_{n=k-1}^{\infty} S(n, k-1)u^n + ku \sum_{n=k}^{\infty} S(n, k)u^n,$$

for $k = 1, 2, \ldots$. Consequently,

$$\phi_k(u) = u\phi_{k-1}(u) + ku\phi_k(u), \quad k = 1, 2, \ldots$$

and

$$\phi_k(u) = u(1 - ku)^{-1}\phi_{k-1}(u), \quad k = 1, 2, \ldots.$$

Applying this recurrence relation repeatedly, we find

$$\phi_k(u) = u^2(1 - ku)^{-1}(1 - (k-1)u)^{-1}\phi_{k-2}(u) = \cdots$$
$$= u^k(1 - ku)^{-1}(1 - (k-1)u)^{-1} \cdots (1 - u)^{-1}\phi_0(u)$$

and since, according to the initial conditions of (8.27) $\phi_0(u) = 1$, we deduce (8.33). ∎

THEOREM 8.11
The Stirling number of the second kind $S(n, k)$, $k = 0, 1, \ldots, n$, $n = 0, 1, \ldots$, is given by the sum

$$S(n, k) = \sum 1^{r_1} 2^{r_2} \cdots k^{r_k}, \qquad (8.34)$$

where the summation is extended over all integers $r_j \geq 0$, $j = 1, 2, \ldots, k$, with $r_1 + r_2 + \cdots + r_k = n - k$, or equivalently by the sum

$$S(n, k) = \sum i_1 i_2 \cdots i_{n-k}, \tag{8.35}$$

where the summation is extended over all $(n-k)$-combinations $\{i_1, i_2, \ldots, i_{n-k}\}$ with repetition of the k positive integers $\{1, 2, \ldots, k\}$.

PROOF   Expanding each factor in (8.33) by using the geometric series, we find

$$\phi_k(u) = \sum_{n=k}^{\infty} S(n, k)u^n = u^k \prod_{j=1}^{k} \left(\sum_{r_j=0}^{\infty} j^{r_j} u^{r_j} \right)$$

$$= \sum_{n=k}^{\infty} \left\{ \sum 1^{r_1} 2^{r_2} \cdots k^{r_k} \right\} u^n,$$

where, in the inner sum, the summation is extended over all integers $r_j \geq 0$, $j = 1, 2, \ldots, k$, with $r_1 + r_2 + \cdots + r_k = n - k$. This relation implies (8.34), as well as its equivalent expression (8.35).   ∎

THEOREM 8.12

(a) *The inverse factorial $(t)_{-k}$, $k = 1, 2, \ldots$, is expanded into a series of inverse powers t^{-n}, $n = k, k + 1, \ldots$, as*

$$(t)_{-k} = \sum_{n=k}^{\infty} (-1)^{n-k} S(n, k) t^{-n}, \quad k = 1, 2, \ldots, \ t > k. \tag{8.36}$$

(b) *The inverse power t^{-k}, $k = 1, 2, \ldots$, is expanded into a series of inverse factorials $(t)_{-n}$, $n = k, k + 1, \ldots$, as*

$$t^{-k} = \sum_{n=k}^{\infty} (-1)^{n-k} s(n, k)(t)_{-n}, \quad k = 1, 2, \ldots. \tag{8.37}$$

PROOF   (a) Setting $u = -1/t$ in (8.33), we get

$$\sum_{n=k}^{\infty} (-1)^n S(n, k) t^{-n} = \frac{(-1)^k}{(t + k)_k}$$

and, since $(t)_{-k} = 1/(t + k)_k$, $k = 1, 2, \ldots$, we conclude (8.36).

(b) Multiplying both members of the expansion

$$(t)_{-n} = \sum_{r=n}^{\infty} (-1)^{r-n} S(r, n) t^{-r}$$

by $(-1)^{n-k}s(n,k)$ and summing for $n = k, k+1, \ldots$, we find

$$\sum_{n=k}^{\infty}(-1)^{n-k}s(n,k)(t)_{-n} = \sum_{n=k}^{\infty}\sum_{r=n}^{\infty}(-1)^{r-n}S(r,n)s(r,k)t^{-r}$$

$$= \sum_{r=k}^{\infty}(-1)^{r-k}\left\{\sum_{n=k}^{r}S(r,n)s(n,k)\right\}t^{-r}.$$

By the second of the orthogonality relations (8.13), it holds

$$\sum_{n=k}^{r}S(r,n)s(n,k) = \delta_{r,k}$$

and so

$$\sum_{n=k}^{\infty}(-1)^{n-k}s(n,k)(t)_{-n} = \sum_{r=k}^{\infty}(-1)^{r-k}\delta_{r,k}t^{-r},$$

which implies (8.37).   ∎

REMARK 8.4   Expansion (8.36), by replacing t by $-t$ and since $(-1)^k(-t)_{-k} = 1/(t-1)_k$, may be transformed to the expansion

$$\frac{1}{(t)_{k+1}} = \sum_{n=k}^{\infty}S(n,k)\frac{1}{t^{n+1}}, \quad k = 1,2,\ldots, \ t > k.$$

Similarly, (8.37) may be rewritten as

$$\frac{1}{t^{k+1}} = \sum_{n=k}^{\infty}s(n,k)\frac{1}{(t)_{n+1}}, \quad k = 1,2,\ldots$$

or

$$\frac{1}{t^{k+1}} = \sum_{n=k}^{\infty}|s(n,k)|\frac{1}{(t+n)_{n+1}}, \quad k = 1,2,\ldots$$

Note that, according to the last expansion, the probabilities $q(n;k)$, $n = k, k+1,$ $\ldots$, derived in the second part of Example 8.1, also sum to unity.   ∎

Example 8.7 A Markov chain
  Suppose that balls are successively drawn one after the other from an urn initially containing m white balls, according to the following scheme. After each trial, if the drawn ball is white, a black ball is placed in the urn, while if the drawn ball is black, it is placed back in the urn. Determine (a) the probability $p(k;n)$ of drawing k white balls in n trials, with $n \leq m$, and (b) the probability $q(n;k)$ that n trials are required until the k-th white balls is drawn, with $k \leq m$.

(a) The probability of drawing a white ball at any trial, given that j white balls are drawn in the previous trials, is given by $p_j = (m-j)/m$ $j = 0, 1, \ldots, n-1$. Then the probability $p(k; n)$ of drawing k white balls in n trials satisfies the triangular recurrence relation

$$p(k; n) = \frac{k}{m} p(k; n-1) + \frac{m-k+1}{m} p(k-1; n-1),$$

for $k = 1, 2, \ldots, n$, $n = 1, 2, \ldots$, with initial conditions

$$p(0; 0) = 1, \ p(0; n) = 0, \ n > 0, \ p(k; n) = 0, \ k > n.$$

Further, since $p(1; n) = m/m^n$ and $p(n; n) = (m)_n/m^n$, we may set

$$p(k; n) = c_{n,k} \frac{(m)_k}{m^n}, \ k = 0, 1, \ldots, n.$$

Then the coefficient $c_{n,k}$ satisfies the triangular recurrence relation

$$c_{n,k} = k c_{n-1,k} + c_{n-1,k-1}, \ k = 1, 2, \ldots, n, \ n = 1, 2, \ldots,$$

with initial conditions

$$c_{0,0} = 1, \ c_{n,0} = 0, \ n > 0, \ c_{n,k} = 0, \ k > n.$$

Comparing this recurrence with the triangular recurrence relation (8.27) of the Stirling numbers of the second kind, we get $c_{n,k} = S(n, k)$ and so

$$p(k; n) = \frac{S(n, k)(m)_k}{m^n}, \ k = 1, 2, \ldots, n.$$

Note that, according to (8.3), these probabilities sum to unity.

(b) The probability $q(n; k)$ that n trials are required until the k-th white ball is drawn equals the probability $p(k - 1; n - 1)$ of drawing $k - 1$ white balls in $n - 1$ trials multiplied by the probability $p_{k-1} = (m - k + 1)/m$ of drawing a white ball at the n-trial, given that the $k - 1$ balls are drawn in the previous trials. Thus

$$q(n; k) = \frac{S(n - 1, k - 1)(m)_k}{m^n}, \ n = k, k + 1, \ldots.$$

Note that these probabilities, according to the first expansion in Remark 8.4, also sum to unity. □

8.4 GENERALIZED FACTORIAL COEFFICIENTS

Consider the generalized factorial of t of order n and scale parameter s,

$$(st)_n = st(st - 1) \cdots (st - n + 1), \ n = 1, 2, \ldots, \ (st)_0 = 1, \qquad (8.38)$$

with s a real number. This can be expressed as a polynomial of factorials of t of degree n. Specifically, we successively get the expressions

$$(st)_0 = (t)_0 = 1, \ (st)_1 = s(t)_1,$$

$$(st)_2 = st[s(t-1) + (s-1)] = s^2(t)_2 + (s)_2(t)_1,$$

$$(st)_3 = s^2(t)_2[s(t-2) + 2(s-1)] + (s)_2t[s(t-1) + (s-2)]$$
$$= s^3(t)_3 + 3s(s)_2(t)_2 + (s)_3(t)_1$$

and, generally,

$$(st)_n = \sum_{k=0}^{n} C(n,k;s)(t)_k, \quad n = 0, 1, \ldots . \tag{8.39}$$

In particular, for $s = -1$ and introducing the coefficient $L(n,k) = C(n,k;-1)$, we deduce the expression

$$(-t)_n = \sum_{k=0}^{n} L(n,k)(t)_k, \quad n = 0, 1, \ldots . \tag{8.40}$$

Further, since $(-t)_n = (-1)^n(t + n - 1)_n$ and setting $|L(n,k)| = (-1)^n L(n,k)$, we get

$$(t + n - 1)_n = \sum_{k=0}^{n} |L(n,k)|(t)_k, \quad n = 0, 1, \ldots . \tag{8.41}$$

Then, the following definition is introduced.

DEFINITION 8.2   *The coefficient $C(n,k;s)$ of the k-th order factorial of t of the expansion (8.39) of the n-th order generalized factorial of t, with scale parameter s, is called generalized factorial coefficient. In particular, the coefficients $L(n,k)$ and $|L(n,k)|$ of expansions (8.40) and (8.41) are called Lah and signless or absolute Lah numbers, respectively.*

Clearly,

$$C(n,k;s) = 0, \ k > n, \ C(0,0;s) = 1.$$

Further, expansion (8.39) implies that the generalized factorial coefficients are differences of generalized factorials. Specifically, expanding the function $f(t) = (st)_n$ into factorials of t, by using Newton's formula and since $\Delta^k(st)_n = 0$, $k > n$, we get

$$(st)_n = \sum_{k=0}^{n} \left[\frac{1}{k!}\Delta^k(st)_n\right]_{t=0} \cdot (t)_k, \ n = 0, 1, \ldots$$

and so, by (8.39), we deduce that

$$C(n, k; s) = \left[\frac{1}{k!} \Delta^k (st)_n \right]_{t=0}, \quad k = 0, 1, \ldots, n, \; n = 0, 1, \ldots. \quad (8.42)$$

Also, putting $t = bu$ and $s = a/b$ in (8.39), it follows that

$$(au)_n = \sum_{k=0}^{n} C(n, k; s)(bu)_k, \quad n = 0, 1, \ldots, \; s = a/b. \quad (8.43)$$

Finally, note that the rising factorial of t of order n, $(t + n - 1)_n$, using Vandermonde's formula, may be expanded into factorials of t as

$$(t + n - 1)_n = \sum_{k=0}^{n} \binom{n}{k} (n-1)_{n-k} (t)_k = \sum_{k=0}^{n} \frac{n!}{k!} \binom{n-1}{k-1} (t)_k$$

and so, by (8.41) we get for the Lah numbers the expression

$$|L(n, k)| = (-1)^n L(n, k) = \frac{n!}{k!} \binom{n-1}{k-1}, \quad (8.44)$$

for $k = 1, 2, \ldots, n$, $n = 1, 2, \ldots$.

REMARK 8.5 (a) The generalized factorial of t of order n and increment a, denoted by $(t)_{n,a}$, is defined for t and a real numbers and n integer by

$$(t)_{n,a} = t(t - a) \cdots (t - na + a), \quad n = 1, 2, \ldots, \; (t)_{0,a} = 1$$

and for $t \neq -ra$, $r = 1, 2, \ldots, n$, by

$$(t)_{-n,a} = \frac{1}{(t + na)(t + na - a) \cdots (t + a)}, \quad n = 1, 2, \ldots.$$

Clearly, $b^{-n}(t)_{n,b} = (t/b)_n$, $a^{-k}(t)_{k,a} = (t/a)_k$ and so, replacing t by t/a in (8.39), the generalized factorial of t of order n and increment b is expressed as a polynomial of generalized factorials of t with increment a as

$$(t)_{n,b} = \sum_{k=0}^{n} b^n a^{-k} C(n, k; s)(t)_{k,a}, \quad s = a/b, \; n = 0, 1, \ldots.$$

(b) A function $f(t)$ for which the differences with increment a at zero, $[\Delta_a^k f(t)]_{t=0}$, $k = 0, 1, \ldots$, exist may be expanded, according to Newton's formula, into generalized factorials of t as

$$f(t) = \sum_{k=0}^{\infty} \left[\frac{a^{-k}}{k!} \Delta_a^k f(t) \right]_{t=0} \cdot (t)_{k,a}.$$

Note that this formula may be deduced as follows:

$$f(t) = [E_a^{t/a} f(u)]_{u=0} = [(\Delta_a + 1)^{t/a} f(u)]_{u=0}$$

$$= \sum_{k=0}^{\infty} \left[\frac{1}{k!} \Delta_a^k f(u) \right]_{u=0} \cdot (t/a)_k = \sum_{k=0}^{\infty} \left[\frac{a^{-k}}{k!} \Delta_a^k f(u) \right]_{u=0} \cdot (t)_{k,a}.$$

In the case of $f(t) = (t)_{n,b}$ and since $\Delta_a^k (t)_{n,b} = 0, k > n$, we get

$$(t)_{n,b} = \sum_{k=0}^{n} \left[\frac{a^{-k}}{k!} \Delta_a^k (t)_{n,b} \right]_{t=0} \cdot (t)_{k,a}$$

and so

$$C(n, k; s) = \left[\frac{b^{-n}}{k!} \Delta_a^k (t)_{n,b} \right]_{t=0}, \quad k = 0, 1, \dots, n, \quad n = 0, 1, \dots .$$

where $s = a/b$. ∎

The generalized factorial coefficient $C(n, k; s)$ is a polynomial in s of degree n. Specifically, the following theorem is derived.

THEOREM 8.13
The generalized factorial coefficient $C(n, k; s)$ is a polynomial in s of degree n, the coefficient of the general term of which is a product of the Stirling numbers of the first and second kind:

$$C(n, k; s) = \sum_{r=k}^{n} s(n, r) S(r, k) s^r. \tag{8.45}$$

PROOF   Expanding the generalized factorial of t of order n and scale parameter s into powers of $u = st$ by using (8.2) and, in the resulting expression, expanding the powers of t into factorials of t by using (8.3), we deduce the expression

$$(st)_n = \sum_{r=0}^{n} s(n, r) s^r t^r = \sum_{r=0}^{n} s(n, r) s^r \sum_{k=0}^{r} S(r, k)(t)_k$$

$$= \sum_{k=0}^{n} \left\{ \sum_{r=k}^{n} s(n, r) S(r, k) s^r \right\} (t)_k,$$

which, compared with (8.39), implies (8.45). ∎

REMARK 8.6   In the particular case $s = 1$ and, since by (8.39) $C(n, k; 1) = \delta_{n,k}$, it follows as a corollary of (8.45) the first of (8.13). Also, in the case $s = -1$, we have $C(n, k; -1) = L(n, k)$ and so from (8.45), using (8.5) and (8.44), we deduce the relation

$$\sum_{r=k}^{n} |s(n, r)| S(r, k) = \frac{n!}{k!} \binom{n-1}{k-1}.$$

for $k = 1, 2, \ldots, n$, $n = 1, 2, \ldots$. ∎

Note that, for fixed n, the factorial generating function of the sequence of the generalized factorial coefficients $C(n, k; s)$, $k = 0, 1, \ldots, n$, according to (8.39), is given by

$$C_n(t; s) = \sum_{k=0}^{n} C(n, k; s)(t)_k = (st)_n, \ n = 0, 1, \ldots.$$

Further, from (8.39), using Newton's general binomial theorem, we deduce the double generating function

$$g(t, u; s) = \sum_{n=0}^{\infty} \sum_{k=0}^{n} C(n, k; s)(t) \frac{u^n}{n!} = (1 + u)^{st}. \tag{8.46}$$

The exponential generating function of the sequence of the generalized factorial coefficients $C(n, k; s)$, $n = k, k + 1, \ldots$, for fixed k, is derived in the next theorem.

THEOREM 8.14
The exponential generating function of the generalized factorial coefficients $C(n, k; s)$, $n = k, k + 1, \ldots$, for fixed k, is given by

$$f_k(u; s) = \sum_{n=k}^{\infty} C(n, k; s) \frac{u^n}{n!} = \frac{[(1 + u)^s - 1]^k}{k!}, \ k = 0, 1, \ldots. \tag{8.47}$$

PROOF Interchanging the order of summation in (8.46) we get

$$g(t, u; s) = \sum_{k=0}^{\infty} \sum_{n=k}^{\infty} C(n, k; s) \frac{u^n}{n!} (t)_k = \sum_{k=0}^{\infty} f_k(u; s)(t)_k$$

and, since

$$g(t, u; s) = (1 + [(1 + u)^s - 1])^t = \sum_{k=0}^{\infty} \binom{t}{k} [(1 + u)^s - 1]^k$$

$$= \sum_{k=0}^{\infty} \frac{[(1 + u)^s - 1]^k}{k!} (t)_k,$$

we deduce (8.47). ∎

An explicit expression of the generalized factorial coefficient $C(n, k; s)$ is derived in the following theorem.

THEOREM 8.15
The generalized factorial coefficient $C(n, k; s)$, $k = 0, 1, \ldots, n$, $n = 0, 1, \ldots$, is given by the sum

$$C(n, k; s) = \frac{1}{k!} \sum_{r=0}^{k} (-1)^{k-r} \binom{k}{r} (sr)_n. \tag{8.48}$$

PROOF   Expanding the generating function (8.47) into powers of u, we get the expression

$$\sum_{n=k}^{\infty} C(n, k; s) \frac{u^n}{n!} = \frac{1}{k!} \sum_{r=0}^{k} (-1)^{k-r} \binom{k}{r} (1+u)^{sr}$$

$$= \frac{1}{k!} \sum_{r=0}^{k} (-1)^{k-r} \binom{k}{r} \sum_{n=0}^{\infty} \binom{sr}{n} u^n$$

$$= \sum_{n=0}^{\infty} \left\{ \frac{1}{k!} \sum_{r=0}^{k} (-1)^{k-r} \binom{k}{r} (sr)_n \right\} \frac{u^n}{n!},$$

which implies (8.48). ∎

REMARK 8.7   (a) Expression (8.48), of the generalized factorial coefficient, may also be deduced from (8.42), by using the expression of the k-th power of the difference operator in terms of shift operator,

$$\Delta^k = \sum_{r=0}^{k} (-1)^{k-r} \binom{k}{r} E^r.$$

Thus

$$C(n, k; s) = \left[\frac{1}{k!} \Delta^k (st)_n \right]_{t=0} = \frac{1}{k!} \sum_{r=0}^{k} (-1)^{k-r} \binom{k}{r} [E^r (st)_n]_{t=0}$$

and, since $[E^r (st)_n]_{t=0} = (sr)_n$, (8.48) is deduced. ∎

Expressions of the generalized factorial coefficient $C(n, k; s)$ as multiple sums over all compositions, as well as over all partitions of n into k parts, are given in the following theorem.

THEOREM 8.16
The generalized factorial coefficient $C(n, k; s)$, $k = 0, 1, \ldots, n$, $n = 0, 1, \ldots$, is given by

$$C(n, k; s) = \frac{n!}{k!} \sum \binom{s}{r_1} \binom{s}{r_2} \cdots \binom{s}{r_k}, \qquad (8.49)$$

where the summation is extended over all nonnegative integer solutions of the equation $r_1 + r_2 + \cdots + r_k = n$. Alternatively,

$$C(n, k; s) = \sum \frac{n!}{k_1! k_2! \cdots k_n!} \binom{s}{1}^{k_1} \binom{s}{2}^{k_2} \cdots \binom{s}{n}^{k_n}, \qquad (8.50)$$

where the summation is extended over all nonnegative integer solutions of the equations $k_1 + 2k_2 + \cdots + nk_n = n$ and $k_1 + k_2 + \cdots + k_n = k$.

PROOF   Expanding $(1 + u)^s - 1$ into powers of u and then using the Cauchy rule of multiplication of series, the exponential generating function (8.47) may be expressed as

$$\sum_{n=k}^{\infty} C(n, k; s) \frac{u^n}{n!} = \frac{1}{k!} \left[\sum_{r=1}^{\infty} \binom{s}{r} u^r \right]^k = \frac{1}{k!} \prod_{i=1}^{k} \left[\sum_{r_i=1}^{\infty} \binom{s}{r_i} u^{r_i} \right]$$

$$= \sum_{n=k}^{\infty} \left\{ \frac{n!}{k!} \sum \binom{s}{r_1} \binom{s}{r_2} \cdots \binom{s}{r_k} \right\} \frac{u^n}{n!},$$

yielding (8.49). Using the multinomial theorem instead of the Cauchy rule, we get the expansion

$$\sum_{n=k}^{\infty} C(n, k; s) \frac{u^n}{n!} = \frac{1}{k!} \left[\binom{s}{1} u + \binom{s}{2} u^2 + \cdots + \binom{s}{n} u^n + \cdots \right]^k$$

$$= \sum_{n=k}^{\infty} \left\{ \sum \frac{n!}{k_1! k_2! \cdots k_n!} \binom{s}{1}^{k_1} \binom{s}{2}^{k_2} \cdots \binom{s}{n}^{k_n} \right\} \frac{u^n}{n!},$$

which implies (8.50).   ∎

Limiting expressions as $s \to 0$ and $s \to \infty$ and an orthogonality relation for the generalized factorial coefficient are deduced in the following theorems.

THEOREM 8.17
Let $C(n, k; s)$ be the generalized factorial coefficient. Then

$$\lim_{s \to 0} s^{-k} C(n, k; s) = s(n, k), \quad \lim_{s \to \infty} s^{-n} C(n, k; s) = S(n, k), \qquad (8.51)$$

where $s(n, k)$ and $S(n, k)$ are the Stirling numbers of the first and second kind, respectively.

PROOF   By using $\lim_{s \to 0} s^{-1}[(1 + u)^s - 1] = \log(1 + u)$, we deduce for the generating function (8.47) the limiting expression

$$
\lim_{s \to 0} s^{-k} f_k(u; s) = \sum_{n=k}^{\infty} [\lim_{s \to 0} s^{-k} C(n, k; s)] \frac{u^n}{n!}
$$

$$
= \frac{\lim_{s \to 0} s^{-k} [(1 + u)^s - 1]^k}{k!} = \frac{[\log(1 + u)]^k}{k!},
$$

which, compared with the generating function (8.16) of the Stirling numbers of the first kind, implies the first of (8.51). Also, since $\lim_{s \to \infty} (1 + u/s)^s = e^u$, we deduce the limiting expression

$$
\lim_{s \to \infty} f_k(u/s; s) = \sum_{n=k}^{\infty} [\lim_{s \to \infty} s^{-n} C(n, k; s)] \frac{u^n}{n!}
$$

$$
= \frac{\lim_{s \to \infty} [(1 + u/s)^s - 1]^k}{k!} = \frac{(e^u - 1)^k}{k!},
$$

which, compared with the generating function (8.17) of the Stirling numbers of the second kind, implies the second of (8.51).   ∎

THEOREM 8.18

The generalized factorial coefficients $C(n, k; s)$, $k = 0, 1, \ldots, n$, $n = 0, 1, \ldots$, satisfy the relation

$$
\sum_{r=k}^{n} C(n, r; s_1) C(r, k; s_2) = C(n, k; s_1 s_2). \tag{8.52}
$$

In particular, they satisfy the orthogonality relation

$$
\sum_{r=k}^{n} C(n, r; s) C(r, k; s^{-1}) = \delta_{n,k}, \tag{8.53}
$$

where $\delta_{n,k} = 1$, if $k = n$ and $\delta_{n,k} = 0$, if $k \neq n$ is the Kronecker delta.

PROOF   Expanding the n-th order factorial of $s_1 s_2 t$ into factorials of $s_2 t$, using (8.39), and in the resulting expression expanding the factorials of $s_2 t$ into factorials of t, we get the relation

$$(s_1 s_2 t)_n = \sum_{r=0}^{n} C(n, r; s_1)(s_2 t)_r$$

$$= \sum_{r=0}^{n} C(n, r; s_1) \sum_{k=0}^{r} C(r, k; s_2)(t)_k$$

$$= \sum_{k=0}^{n} \left\{ \sum_{r=k}^{n} C(n, r; s_1) C(r, k; s_2) \right\} (t)_k.$$

Further, according to (8.39), we have

$$(s_1 s_2 t)_n = \sum_{k=0}^{n} C(n, k; s_1 s_2)(t)_k$$

and so

$$\sum_{k=0}^{n} \left\{ \sum_{r=k}^{n} C(n, r; s_1) C(r, k; s_2) \right\} (t)_k = \sum_{k=0}^{n} C(n, k; s_1 s_2)(t)_k.$$

The last relation implies (8.52). In the particular case of $s_1 = s$ and $s_2 = s^{-1}$, since $C(n, k; 1) = \delta_{n,k}$, (8.53) is deduced. ∎

Recurrence relations for the generalized factorial coefficients are given in the following theorems.

THEOREM 8.19
The generalized factorial coefficients $C(n, k; s)$, $k = 0, 1, \ldots, n$, $n = 0, 1, \ldots$, satisfy the triangular recurrence relation

$$C(n + 1, k; s) = (sk - n)C(n, k; s) + sC(n, k - 1; s), \qquad (8.54)$$

for $k = 1, 2, \ldots, n + 1$, $n = 0, 1, \ldots$, with initial conditions

$$C(0, 0; s) = 1, \ C(n, 0; s) = 0, \ n > 0, \ C(n, k; s) = 0, \ k > n.$$

PROOF Expanding both members of the recurrence relation $(st)_{n+1} = (st - n)(st)_n$ into factorials of t, according to (8.39), we have

$$\sum_{k=0}^{n+1} C(n + 1, k; s)(t)_k = st \sum_{r=0}^{n} C(n, r; s)(t)_r - n \sum_{k=0}^{n} C(n, k; s)(t)_k.$$

Since $(t)_{r+1} = (t-r)(t)_r$, whence $t(t)_r = (t)_{r+1} + r(t)_r$, the right-hand side of this relation may be written as

$$s \sum_{r=0}^{n} C(n,r;s)(t)_{r+1} + s \sum_{r=0}^{n} rC(n,r;s)(t)_r - n \sum_{k=0}^{n} C(n,k;s)(t)_k$$

and so

$$\sum_{k=0}^{n+1} C(n+1,k;s)(t)_k = \sum_{k=0}^{n}(sk-n)C(n,k;s)(t)_k + \sum_{k=1}^{n+1} sC(n,k-1;s)(t)_k.$$

Equating the coefficients of $(t)_k$ in this relation, we conclude (8.54). The initial conditions follow directly from (8.39). ∎

REMARK 8.8 (a) If the parameter s is a positive integer, then the numbers $C(n,k;s)$, $n = k, k+1, \ldots, sk$, $k = 0, 1, \ldots$, are positive integers. Indeed, from the triangular recurrence relation (8.54) and its initial conditions, it follows that these numbers result from successive summation of positive integers. Further, in addition to $C(n,k;s) = 0$, $k > n$, it follows from (8.48) that $C(n,k;s) = 0$, $sk < n$.

(b) Also, if s is a positive integer, then the numbers

$$|C(n,k;-s)| = (-1)^n C(n,k;-s), \ n = k, k+1, \ldots, \ k = 0, 1, \ldots,$$

are positive integers since, by (8.54), we have

$$|C(n+1,k;-s)| = (sk+n)|C(n,k;-s)| + s|C(n,k-1;-s)|, \quad (8.55)$$

for $k = 1, 2, \ldots, n+1$, $n = 0, 1, \ldots$, with

$$|C(0,0;-s)| = 1, \ |C(n,0;-s)| = 0, \ n > 0, \ |C(n,k;-s)| = 0, \ k > n;$$

this means that these numbers result from successive summation of positive integers. ∎

THEOREM 8.20
The generalized factorial coefficients $C(n,k;s)$, $k = 0, 1, \ldots, n$, $n = 0, 1, \ldots$, with $C(0,0;s) = 1$, satisfy the vertical recurrence relation

$$C(n+1,k+1;s) = \sum_{r=k}^{n} \binom{n}{r} (s)_{n-r+1} C(r,k;s). \quad (8.56)$$

PROOF (a) According to (8.39) and since $C(n+1,0;s) = 0$, $n = 0, 1, \ldots$, we have

$$\sum_{k=0}^{n} C(n+1,k+1;s)(t+1)_{k+1} = (st+s)_{n+1}.$$

Further, since $(t+1)_{k+1} = (t+1)(t)_k$, we conclude that

$$\sum_{k=0}^{n} C(n+1, k+1; s)(t)_k = s(st + s - 1)_n.$$

Expanding the right-hand side into factorials of st, by Vandermonde's formula, and in the resulting expression expanding the factorials of st into factorials of t, by using (8.39), we successively get

$$\sum_{k=0}^{n} C(n+1, k+1; s)(t)_k = \sum_{r=0}^{n} \binom{n}{r} (s)_{n-r+1} \sum_{k=0}^{r} C(r, k; s)(t)_k$$

$$= \sum_{k=0}^{n} \left\{ \sum_{r=k}^{n} \binom{n}{r} (s)_{n-r+1} C(r, k; s) \right\} (t)_k,$$

yielding (8.56). ∎

The generalized factorial coefficients $C(n, k; s)$ can be tabulated by using the triangular recurrence relation (8.54) and its initial conditions. Table 8.3 gives the numbers $C(n, k; s)$, $k = 1, 2, \ldots, n$, $n = 1, 2, \ldots, 5$.

Table 8.3 *Generalized Factorial Coefficients* $C(n, k; s)$

k / n	1	2	3	4	5
1	$(s)_1$				
2	$(s)_2$	s^2			
3	$(s)_3$	$3(s)_2 s$	s^3		
4	$(s)_4$	$7(s)_3 s + 3(s)_2 s$	$6(s)_2 s^2$	s^4	
5	$(s)_5$	$15(s)_4 s + 20(s)_3 s$	$25(s)_3 s^2 + 15(s)_2 s^2$	$10(s)_2 s^3$	s^5

The numbers $C(n, k; s)$, $k = 0, 1, \ldots, n$, $n = 0, 1, \ldots$, are defined as the coefficients of the factorials in the expansion of the generalized factorial, with scale parameter s. These numbers emerge also in the expansion of the inverse generalized factorials into inverse factorials. Specifically, we have the following theorem.

THEOREM 8.21
The inverse generalized factorial of t, with scale parameter s, $(st)_{-k}$, $k = 1, 2, \ldots$, is expanded into a series of inverse factorials $(t)_{-n}$, $n = k, k+1, \ldots$, as

$$(st)_{-k} = \sum_{n=k}^{\infty} (-1)^{n-k} C(n, k; s^{-1})(t)_{-n}, \ k = 1, 2, \ldots . \qquad (8.57)$$

PROOF   Multiplying both members of the recurrence relation

$$C(n+1, k; s^{-1}) = (s^{-1}k - n)C(n, k; s^{-1}) + s^{-1}C(n, k-1; s^{-1})$$

by $(-1)^{n-k+1}s(t)_{-n}$ and, on the left-hand side, using the recurrence relation $(t)_{-n} = (t+n+1)(t)_{-n-1}$, we get the relation

$$
\begin{aligned}
&(-1)^{n-k+1}sC(n+1, k; s^{-1})t(t)_{-n-1} \\
&+(-1)^{n-k+1}s(n+1)C(n+1, k; s^{-1})(t)_{-n-1} \\
&= (-1)^{n-k}snC(n, k; s^{-1})(t)_{-n} - (-1)^{n-k}kC(n, k; s^{-1})(t)_{-n} \\
&+(-1)^{n-k-1}C(n, k-1; s^{-1})(t)_{-n},
\end{aligned}
$$

Summing the last relation for $n = k-1, k, \ldots$, we derive for the sum

$$c_k(t; s) = \sum_{n=k}^{\infty} (-1)^{n-k} C(n, k; s^{-1})(t)_{-n}, \ k = 0, 1, \ldots ,$$

the recurrence relation

$$c_k(t; s) = (st+k)^{-1}c_{k-1}(t; s), \ k = 1, 2, \ldots , \ c_0(t; s) = 1.$$

Iterating it, we conclude that

$$c_k(t; s) = [(st+k)(st+k-1) \cdots (st+1)]^{-1} c_0(t; s) = (st)_{-k},$$

and so (8.57) is established.   ∎

REMARK 8.9   (a) Expansion (8.57) with $t = bu$ and $s = a/b$ becomes

$$(au)_{-k} = \sum_{n=k}^{\infty} (-1)^{n-k} C(n, k; s^{-1})(bu)_{-n}, \ s = a/b, \ k = 1, 2, \ldots .$$

(b) Using the relations $(t)_{-n} = a^n(u)_{-n,a}$, $(st)_{-k} = b^k(u)_{-k,b}$, with $s = a/b$ and $u = at$, (8.57) may be rewritten as

$$(u)_{-k,b} = \sum_{n=k}^{\infty} (-1)^{n-k} b^{-k} a^n C(n, k; s^{-1})(u)_{-n,a}, \ s = a/b, \ k = 1, 2, \ldots.$$

(c) Setting $u = -t$ in (8.57) and since $(-1)^k(-su)_{-k} = 1/(su - 1)_k$, $(-1)^n(-u)_{-n} = 1/(u - 1)_n$, we get

$$\frac{1}{(su - 1)_k} = \sum_{n=k}^{\infty} C(n, k; s^{-1}) \frac{1}{(u - 1)_n}, \ k = 1, 2, \ldots.$$

Dividing both members by su, we deduce the expansion

$$\frac{1}{(su)_{k+1}} = \sum_{n=k}^{\infty} s^{-1} C(n, k; s^{-1}) \frac{1}{(u)_{n+1}}, \ k = 0, 1, \ldots,$$

which, equivalently, may be written as

$$\frac{1}{(at)_{k+1}} = \sum_{n=k}^{\infty} s^{-1} C(n, k; s^{-1}) \frac{1}{(bt)_{n+1}}, \ k = 0, 1, \ldots,$$

or as

$$\frac{1}{(t)_{k+1}} = \sum_{n=k}^{\infty} s C(n, k; s) \frac{1}{(st)_{n+1}}, \ k = 0, 1, \ldots,$$

an expression useful in waiting-time probability problems. ∎

Example 8.8 Coupon collector's problem

Consider an urn containing s identical series of coupons, each consisting of m coupons bearing the numbers $1, 2, \ldots, m$. Suppose that coupons are randomly drawn, one after the other, without replacement. Calculate (a) the probability $p(k; n)$ of drawing exactly k of the numbers $\{1, 2, \ldots, m\}$ in n drawings, with $n \leq m$, and (b) the probability $q(n; k)$ that n drawings are required until the k-th different number is drawn, with $k \leq m$.

(a) Let A_j be the event that number j is not drawn in the n drawings, $j = 1, 2, \ldots, m$. Then $p(k; n)$ is the probability that exactly $m - k$ among the m events $A_1, A_2, \ldots, A_m$ occur. Further, for any r indices $\{i_1, i_2, \ldots, i_r\}$ out of the m indices $\{1, 2, \ldots, m\}$,

$$p_r = P(A_{i_1} A_{i_2} \cdots A_{i_r}) = \frac{(s(m - r))_n}{(sm)_n}, \ r = 1, 2, \ldots, m.$$

Therefore, the events $A_1, A_2, \ldots, A_m$ are exchangeable and according to Corollary 4.4, the probability $p(k; n)$ is given by

$$p(k; n) = \binom{m}{n} \sum_{r=0}^{k} (-1)^{k-r} \binom{k}{r} \frac{(sr)_n}{(sm)_n}$$

and so, by virtue of (8.48),

$$p(k; n) = \frac{C(n, k; s)(m)_k}{(sm)_n}, \ k = 1, 2, \ldots, n.$$

(b) The probability $q(n; k)$ that n drawings are required until the k-th different number is drawn equals the probability $p(k-1; n-1)$ of drawing $k-1$ of the m numbers in $n-1$ drawings multiplied by the probability $s(m-k+1)/(sm-n+1)$ that one of the $m-k+1$ not already drawn numbers is drawn at the n-th drawing. Hence

$$q(n; k) = \frac{C(n-1, k-1; s)s(m)_k}{(sm)_n}, \ n = k, k+1, \ldots, sm.$$

Note that these probabilities, according to Remark 8.9(c), sum to unity.   ꠱

8.5 NON-CENTRAL STIRLING AND RELATED NUMBERS

The Stirling numbers of the first and second kind and the generalized factorial coefficients have been generalized in several directions. In this section, the corresponding non-central numbers are briefly presented. Additional properties and applications of these numbers are given in the exercises.

Consider the non-central factorial of t of order n, $(t-r)_n$, where the non-centrality parameter r is a real number and let

$$(t-r)_n = \sum_{k=0}^{n} s(n, k; r)t^k, \ n = 0, 1, \ldots . \tag{8.58}$$

Also, let

$$(t+r)^n = \sum_{k=0}^{n} S(n, k; r)(t)_k, \ n = 0, 1, \ldots . \tag{8.59}$$

The coefficients $s(n, k; r)$ and $S(n, k; r)$ are called *non-central Stirling numbers of the first and second kind*, respectively. Clearly,

$$s(n, k; r) = S(n, k; r) = 0, \ k > n, \ s(0, 0; r) = S(0, 0; r) = 1.$$

Further, replacing t by $-t$ in expansion (8.58) and since $(t+r+n-1)_n = (-1)^n(-t-r)_n$, we deduce the expression

$$(t+r+n-1)_n = \sum_{k=0}^{n} |s(n,k;r)| t^k, \ n = 0, 1, \ldots, \quad (8.60)$$

where the coefficient

$$|s(n,k;r)| = (-1)^{n-k} s(n,k;r), \ k = 0,1,\ldots,n, \ \ n = 0,1,\ldots, \quad (8.61)$$

for $r > 0$, as sum of products of positive numbers, is positive. Specifically,

$$|s(n,k;s)| = \sum (r+i_1)(r+i_2)\cdots(r+i_{n-k}),$$

where the summation is extended over all $(n-k)$-combinations $\{i_1, i_2, \ldots, i_{n-k}\}$ of the n integers $\{0, 1, \ldots, n-1\}$. The coefficient $|s(n,k;r)|$, for $r > 0$, of expansion (8.60), of the rising non-central factorials into powers, is called *non-central signless or absolute Stirling number of the first kind*.

Expansions (8.58) and (8.60) imply that the non-central Stirling numbers of the first kind are derivatives of factorials, while expansion (8.59) entails that the non-central Stirling numbers of the second kind are differences of powers. Specifically,

$$s(n,k;r) = \left[\frac{1}{k!} D^k (t)_n \right]_{t=-r}, \ k = 0, 1, \ldots, n, \ n = 0, 1, \ldots,$$

$$|s(n,k;r)| = \left[\frac{1}{k!} D^k (t+n-1)_n \right]_{t=r}, \ k = 0, 1, \ldots, n, \ n = 0, 1, \ldots$$

and

$$S(n,k;r) = \left[\frac{1}{k!} \Delta^k t^n \right]_{t=r}, \ k = 0, 1, \ldots, n, \ n = 0, 1, \ldots.$$

Note that, for $r = 0$, these numbers reduce to the corresponding usual (central) Stirling numbers. For $r \neq 0$ the non-central Stirling numbers may be expressed in terms of the corresponding central Stirling numbers. Specifically, expanding the rising non-central factorial of t of order n, $(t+r+n-1)_n$, into powers of $u = t+r$, using (8.4), and then expanding the power of $u = t+r$ into powers of t, using Newton's binomial formula, we deduce the expression

$$|s(n,k;r)| = \sum_{j=k}^{n} \binom{j}{k} r^{j-k} |s(n,j)|.$$

Also, expanding the non-central factorial of t of order n, $(t+r+n-1)_n$, into factorials of t, using Vandermonde's formula, and then expanding the factorials of t into powers of t, using (8.2), we conclude the expression

$$|s(n,k;r)| = \sum_{j=k}^{n} \binom{n}{j} (r+n-j-1)_{n-j}|s(j,k)|.$$

Similarly

$$S(n,k;r) = \sum_{j=k}^{n} \binom{j}{k} (r)_{j-k} S(n,j)$$

and

$$S(n,k;r) = \sum_{j=k}^{n} \binom{n}{j} r^{n-j} S(j,k).$$

The non-central Stirling numbers retain the orthogonality relation of the central Stirling numbers. Specifically, for any real parameter r,

$$\sum_{j=k}^{n} s(n,j;r)S(j,k;r) = \delta_{n,k}, \quad \sum_{j=k}^{n} S(n,j;r)s(j,k;r) = \delta_{n,k},$$

where $\delta_{n,k} = 1$, if $k = n$ and $\delta_{n,k} = 0$, if $k \neq n$ is the Kronecker delta.

The exponential generating functions of $|s(n,k;r)|$, $n = k, k+1, \ldots$ and $S(n,k;r)$, $n = k, k+1, \ldots$, for fixed k and r, may be obtained as

$$g_k(u;r) = \sum_{n=k}^{\infty} |s(n,k;r)| \frac{u^n}{n!} = (1-u)^{-r} \frac{[-\log(1-u)]^k}{k!}, \quad k = 0,1,\ldots$$

and

$$f_k(u;r) = \sum_{n=k}^{\infty} S(n,k;r) \frac{u^n}{n!} = e^{ru} \frac{(e^u - 1)^k}{k!}, \quad k = 0,1,\ldots,$$

respectively.

The non-central Stirling number of the second kind, on using the expression of the k-th power of the difference operator in terms of the shift operator, is expressed in the form of a single sum of elementary terms as

$$S(n,k;r) = \frac{1}{k!} \sum_{j=0}^{k} (-1)^{k-j} \binom{k}{j} (r+j)^n. \qquad (8.62)$$

Triangular recurrence relations for the non-central Stirling numbers, analogous to those for the central Stirling numbers, can be similarly deduced:

$$s(n+1,k;r) = s(n,k-1;r) - (n+r)s(n,k;r),$$

for $k = 1, 2, \ldots, n+1$, $n = 0, 1, \ldots$, with

$$s(0,0;r) = 1, \ s(n,0;r) = (-r)_n, \ n > 0, \ s(n,k;r) = 0, \ k > n.$$

and

$$S(n+1, k; r) = S(n, k-1; r) + (k+r)S(n, k; r),$$

for $k = 1, 2, \ldots, n+1$, $n = 0, 1, \ldots$, with

$$S(0, 0; r) = 1, \ S(n, 0; r) = r^n, \ n > 0, \ S(n, k; r) = 0, \ k > n.$$

Using (8.61), the first recurrence relation yields for the non-central signless Stirling numbers of the first kind the triangular recurrence relation

$$|s(n+1, k; r)| = |s(n, k-1; r)| + (n+r)|s(n, k; r)|,$$

for $k = 1, 2, \ldots, n+1$, $n = 0, 1, \ldots$, with

$$|s(0, 0; r)| = 1, \ |s(n, 0; r)| = (r+n-1)_n, \ n > 0, \ |s(n, k; r)| = 0, \ k > n.$$

The triangular recurrence relation for the non-central Stirling numbers of the second kind can be used for the derivation of the (power) generating function

$$\phi_k(u; r) = \sum_{n=k}^{\infty} S(n, k; r)u^n = u^k \prod_{j=0}^{k}(1 - ru - ju)^{-1},$$

for $k = 1, 2, \ldots$. Setting $u = 1/t$, we get

$$\sum_{n=k}^{\infty} S(n, k; r)\frac{1}{t^{n+1}} = \frac{1}{(t-r)_{k+1}}, \ k = 1, 2, \ldots .$$

Using the orthogonality relation of the non-central Stirling numbers, this expansion can be inverted as

$$\sum_{n=k}^{\infty}(-1)^{n-k}s(n, k; r)\frac{1}{(t-r)_{n+1}} = \frac{1}{t^{k+1}}, \ k = 1, 2, \ldots .$$

Equivalently, this expression may be written as

$$\sum_{n=k}^{\infty}|s(n, k; r)|\frac{1}{(t+r+n)_{n+1}} = \frac{1}{t^{k+1}}, \ k = 1, 2, \ldots .$$

Consider the non-central generalized factorial of t of order n, scale parameter s and non-centrality parameter r, $(st+r)_n$, and let

$$(st + r)_n = \sum_{k=0}^{n} C(n, k; s, r)(t)_k, \ n = 0, 1, \ldots . \tag{8.63}$$

The coefficients $C(n, k; s, r)$ are called *non-central generalized factorial co-efficients* or *Gould-Hopper numbers*. Clearly,

$$C(n, k; s, r) = 0, \ k > n, \ C(0, 0; s, r) = 1.$$

Further, expansion (8.63) implies that the non-central generalized factorial coefficients are differences of non-central generalized factorials. Specifically,

$$C(n, k; s, r) = \left[\frac{1}{k!} \Delta^k (st + r)_n \right]_{t=0}, \ k = 0, 1, \dots, n, \ n = 0, 1, \dots.$$

Note that, for $r = 0$, these numbers reduce to the corresponding central numbers. For $r \neq 0$, the non-central generalized factorial coefficients may be expressed in terms of the corresponding central generalized factorial coefficients. Specifically,

$$C(n, k; s, r) = \sum_{j=k}^{n} \binom{j}{k} (r/s)_{j-k} C(n, j; s)$$

and

$$C(n, k; s, r) = \sum_{j=k}^{n} \binom{n}{j} (r)_{n-j} C(j, k; s).$$

The non-central generalized factorial coefficient $C(n, k; s, \rho s - r)$ is a polynomial in s of degree n, the coefficient of general term of which is a product of the non-central Stirling numbers of the first and second kind,

$$C(n, k; s, \rho s - r) = \sum_{j=k}^{n} s(n, j; r) S(j, k; \rho) s^j.$$

The exponential generating function of the sequence $C(n, k; s, r)$, $n = k$, $k + 1, \dots$, for fixed k and r, may be obtained as

$$f_k(u; s, r) = \sum_{n=k}^{\infty} C(n, k; s, r) \frac{u^n}{n!}$$

$$= (1 + u)^r \frac{[(1 + u)^s - 1]^k}{k!}, \ k = 0, 1, \dots.$$

The non-central generalized factorial coefficients, on using the expression of the k-th power of the difference operator in terms of the shift operator, is expressed in the form of a single sum of elementary terms as

$$C(n, k; s, r) = \frac{1}{k!} \sum_{j=0}^{k} (-1)^{k-j} \binom{k}{j} (sj + r)_n.$$

A triangular recurrence relation for the non-central generalized factorial coefficients, analogous to that for the central generalized factorial coefficients, can be similarly deduced:

$$C(n+1,k;s,r) = (sk+r-n)C(n,k;s,r) + sC(n,k-1;s,r),$$

for $k = 1, 2, \ldots, n+1$, $n = 0, 1, \ldots$, with

$$C(0,0;s,r) = 1, \ C(n,0;s,r) = (r)_n, \ n > 0, \ C(n,k;s,r) = 0, \ k > n.$$

This triangular recurrence relation can be used to show that

$$\frac{1}{(t)_{k+1}} = \sum_{n=k}^{\infty} sC(n,k;s,r)\frac{1}{(st+r)_{n+1}}.$$

8.6 BIBLIOGRAPHIC NOTES

The Stirling numbers were so named by N. Nielsen (1906) in honor of James Stirling, who introduced them in his *Methodus Differentialis* (1730) without using any notation for them. The notation adopted in this book is due to J. Riordan (1958). Recurrence relations and certain number theoretic properties of the Stirling numbers of the first kind were derived by L. Lagrange (1770). P. S. Laplace (1812) and A. Cayley (1887) provided several approximations of the Stirling numbers of the second kind. J. A. Grunert (1822, 1843), A. Cauchy (1833), O. Schlömilch (1852, 1895), G. Boole (1860), L. Schlafli (1867), J. Blissard (1867) and J. Worpitzky (1883) explored further the Stirling numbers of both kinds.

The work of N. E. Nörlund (1924) inspired several publications on certain generalizations of the Stirling numbers and their connection with the Bernoulli and the generalized Bernoulli numbers. Also, the books by N. Nielsen (1906, 1923), E. Netto (1927), J. F. Steffensen (1927) and L. M. Milne-Thomson (1933) are worth mentioning.

A thorough presentation of the Stirling numbers and their most important properties provided by Ch. Jordan (1933) in an excellent paper, which was included as Chapter 4 in his classical book on the calculus of finite differences (1939a), revived the interest in the Stirling numbers. Since then, a large number of publications on these numbers appeared in the literature. The more recent book of L. Comtet (1974) devotes a chapter to these numbers and provides a very rich bibliography. The expression of the n-th power of the operator $\Theta = tD$ in terms of powers of the operator D was obtained by J. A. Grunert (1843). The derivation of this expression in Example 8.4 is due to Ch. Jordan (1933, 1939a), who also derived the expression of the n-th power of the operator $\Psi = t\Delta$ in terms of powers of the operator Δ given in Example 8.5. H. W. Gould (1964) expressed the operator $(a^t D)^n$ in terms of powers of the operator D with the Stirling numbers of the first kind as coefficients. L. Comtet (1973) expressed the more general operator $(\lambda(t)D)^n$ in terms of powers of the operator D with coefficients Bell partition polynomials. A generalization to another direction was discussed by L. Carlitz (1932).

O. Schlömilch (1852) used the same symbol, C_k^n, to denote the Stirling numbers of both kinds calling them *factorial coefficients*. In this unified notation, $C_k^n = |s(n,k)|$ and $C_k^{-n} = S(k,n)$. N. E. Nörlund (1924) introduced the generalized Bernoulli numbers $B_n^{(r)}$, $n = 0, 1, \ldots$, with generating function the r-th power of the generating function of usual Bernoulli numbers. Then $|s(n,k)| = \binom{n-1}{k-1} B_{n-k}^{(n)}$ and $S(n,k) = \binom{n}{k} B_{n-k}^{(-k)}$. (see Exercise 19). Properties of the generalized Bernoulli numbers were studied by L. Carlitz (1960). F. N. David and D. E. Barton (1962) devoted a chapter to these numbers.

The first short table of the Stirling numbers of the second kind, up to $n = 9$, was published by James Stirling (1730). Extensive tables of the Stirling numbers of both kinds were constructed by H. Gupta (1950), R. A. Fisher and F. Yates (1953), F. N. David, M. G. Kendall and D. E. Barton (1966) and M. Abramowitz and I. A. Stegun (1965). A variety of asymptotic expressions for the Stirling numbers exist in the literature. The approximate expressions given in Exercise 7 are due to Ch. Jordan (1933, 1939a). Several references on other approximations are given in the review paper by Ch. A. Charalambides and J. Singh (1988).

$\cdot$ The coefficients $L(n,k)$, $k = 0, 1, \ldots, n$, $n = 0, 1, \ldots$ of the expansion of the rising factorials into falling factorials, introduced by I. Lah (1955), were called *Lah numbers* by J. Riordan (1958). Extending these numbers, Ch. A. Charalambides (1976, 1977a, 1979a) systematically studied the coefficients $C(n, k; s)$ of the expansion of the generalized factorials into falling factorials. These numbers were noted before by Ch. Jordan (1933) and appeared as coefficients of a generalized Hermite polynomial in E. T. Bell (1934b), and in several other forms in H. W. Gould (1958), R. Shumway and J. Gurland (1960), L. Bernstein (1965), L. Carlitz (1965), W. Feller (1968), L. Comtet (1973). Later, L. Carlitz (1979) studied these numbers under the name *degenerate Stirling numbers*.

The non-central Stirling numbers of the second kind appeared in N. Nielsen (1906) as differences of the powers at an arbitrary point. J. Riordan (1937) used them as connection constants of power moments about an arbitrary point and factorial moments (see Example 8.3). Also, they appeared as coefficients in a modification of the classical occupancy problem discussed by D. E. Barton and F. N. David (1959). Recently, these numbers were studied by L. Carlitz (1980a,b) as weighted Stirling numbers, by M.

Koutras (1982) as non-central Stirling numbers, by A. Z. Broder (1984) as r-Stirling numbers and also by R. Shanmugan (1984). The differences of the generalized factorials at an arbitrary point (Gould-Hopper numbers) were studied by Ch. A. Charalambides and M. Koutras (1983).

The associated Stirling numbers, introduced by J. Riordan (1958), are closely related to the numbers of Ch. Jordan (1933, 1939a) and M. Ward (1934), which are the coefficients of the representation of $s(n, n - k)$ and $S(n, n - k)$ as sums of binomials of n. These numbers were further discussed by L. Carlitz (1971). Recurrence relations and other properties of the r-associated Stirling numbers and generalized factorial coefficients were discussed by J. Riordan (1958), L. Comtet (1974) and Ch. A. Charalambides (1974).

8.7 EXERCISES

1. *Additional vertical recurrence relations for the Stirling numbers.* Show that (a) the Stirling numbers of the first kind $s(n, k)$, $k = 0, 1, \ldots, n$, $n = 0, 1, \ldots$, with $s(0, 0) = 1$, satisfy the recurrence relation

$$s(n, k) = \sum_{r=k}^{n} n^{r-k} s(n + 1, r + 1), \ k = 0, 1, \ldots, n, \ n = 0, 1, \ldots,$$

and (b) the Stirling numbers of the second kind $S(n, k)$, $k = 0, 1, \ldots, n$, $n = 0, 1, \ldots$, with $S(0, 0) = 1$, satisfy the vertical recurrence relation

$$S(n, k) = \sum_{r=k}^{n} k^{n-r} S(r - 1, k - 1), \ k = 1, 2, \ldots, n, \ n = 1, 2, \ldots.$$

2. Show that

$$\sum_{r=k}^{n} S(n, r) s(r + 1, k + 1) = (-1)^{n-k} \binom{n}{k}$$

and

$$\sum_{r=k}^{n} S(n + 1, r + 1) s(r, k) = \binom{n}{k}.$$

3*. Show that

$$S(n, n - k) = \sum_{r=0}^{k} \binom{k - n}{k + r} \binom{k + n}{k - r} |s(k + r, r)|.$$

4. Show that

$$\binom{k+r}{k} s(n, k+r) = \sum_{j=k}^{n-r} \binom{n}{j} s(j, k) s(n-j, r)$$

and conclude its inverse relation

$$\binom{n}{i} s(n-i, r) = \sum_{k=i}^{n-r} \binom{k+r}{k} S(k, i) s(n, k+r).$$

5. Show that

$$\binom{k+r}{k} S(n, k+r) = \sum_{j=k}^{n-r} \binom{n}{j} S(j, k) S(n-j, r)$$

and conclude its inverse relation

$$\binom{n}{i} S(n-i, r) = \sum_{k=i}^{n-r} \binom{k+r}{k} s(k, i) S(n, k+r).$$

6*. Let $C(n, k; s)$ be the generalized factorial coefficient. Show that

$$(-1)^k C(n, n-k; s^{-1}) = \sum_{r=0}^{k} \binom{k-n}{k+r} \binom{k+n}{k-r} C(k+r, r; s).$$

7. (*Continuation*). Show that

$$\binom{k+r}{k} C(n, k+r; s) = \sum_{j=k}^{n-r} \binom{n}{j} C(j, k; s) C(n-j, r; s)$$

and

$$\binom{n}{i} C(n-i, r; s) = \sum_{k=i}^{n-r} \binom{k+r}{k} C(k, i; s^{-1}) C(n, k+r; s).$$

8*. Show that the signless Stirling number of the first kind is given by

$$|s(n+1, k+1)|$$
$$= \frac{n!}{k!} \sum \frac{(-1)^{r_2+r_4+\cdots+r_{2m}} k!}{1^{r_1} r_1! 2^{r_2} r_2! \cdots k^{r_k} r_k!} (\zeta_n(1))^{r_1} (\zeta_n(2))^{r_2} \cdots (\zeta_n(k))^{r_k},$$

where $\zeta_n(s) = \sum_{j=1}^{n} 1/j^s$, $m = [n/2]$ and the summation is extended over all nonnegative integer solutions of the equation $r_1 + 2r_2 + \cdots + kr_k = k$.

9*. *Asymptotic expressions for the Stirling numbers.* For fixed k and $n \to \infty$, show that

$$|s(n+1, k+1)| \cong n![\log(n+1) + C]^k/k!$$

and

$$S(n, k) \cong k^n/k!,$$

where $C = 0.57721$ is the Euler's constant. Also, for fixed k, s and $n \to \infty$ derive for the generalized factorial coefficients $C(n, k; s)$ the asymptotic expression

$$C(n, k; s) \cong (sk)_n/k!.$$

10. *Associated Stirling numbers of the first kind.* Using the triangular recurrence relation

$$s(n+1, k) = s(n, k-1) - ns(n, k), \quad k = 1, 2, \ldots, n+1, \quad n = 0, 1, \ldots,$$

$$s(0, 0) = 1, \quad s(n, 0) = 0, \quad n > 0, \quad s(n, k) = 0, \quad k > n,$$

show that

$$s(n, n) = 1, \quad s(n, n-1) = -\binom{n}{2}, \quad s(n, n-2) = 2\binom{n}{3} + 3\binom{n}{4}$$

and, generally, setting

$$s(n, n-k) = \sum_{j=0}^{k} s_2(k+j, j)\binom{n}{k+j},$$

derive the bivariate generating function

$$g_2(t, u) = \sum_{n=0}^{\infty} \sum_{k=0}^{[n/2]} s_2(n, k) t^k \frac{u^n}{n!} = (1+u)^t e^{-ut}$$

and deduce the generating function

$$f_{k,2}(u) = \sum_{n=2k}^{\infty} s_2(n, k)\frac{u^n}{n!} = \frac{[\log(1+u) - u]^k}{k!}, \quad k = 0, 1, \ldots.$$

Further, show that

$$s_2(n+1, k) = -ns_2(n, k) - ns_2(n-1, k-1),$$

for $n = 2k, 2k+1, \ldots$, $k = 1, 2, \ldots$, with

$$s_2(0, 0) = 1, \quad s_2(n, 0) = 0, \quad n > 0, \quad s_2(n, k) = 0, \quad 2k > n,$$

and

$$s_2(n,k) = \sum_{j=0}^{k} (-1)^j \binom{n}{j} s(n-j, k-j),$$

for $n = 2k, 2k+1, \ldots, k = 0, 1, \ldots$.

11. (*Continuation*). Consider, more generally, the numbers $s_r(n,k)$, $n = rk, rk+1, \ldots, k = 0, 1, \ldots, r = 2, 3, \ldots$, defined by their exponential generating function

$$f_{k,r}(u) = \sum_{n=rk}^{\infty} s_r(n,k) \frac{u^n}{n!} = \frac{1}{k!} \left[\log(1+u) - \sum_{j=1}^{r-1} (-1)^{j-1} \frac{u^j}{j} \right]^k.$$

Derive the recurrence relation

$$s_r(n+1, k) = (-1)^{r-1} (n)_{r-1} s_r(n-r+1, k-1) - n s_r(n,k),$$

for $n = rk, rk+1, \ldots, k = 1, 2, \ldots, r = 2, 3, \ldots$, with

$$s_r(0,0) = 1, \ s_r(n,0) = 0, \ n > 0, \ s_r(n,k) = 0, \ rk > n.$$

Also, show that

$$s_{r+1}(n,k) = \sum_{j=0}^{k} (-1)^{rj} \frac{(n)_{rj}}{j! r^j} s_r(n-rj, k-j), \ r = 1, 2, \ldots$$

and

$$s_r(n,k) = \sum_{j=0}^{k} (-1)^{rj+1} \frac{(n)_{rj}}{j! r^j} s_{r+1}(n-rj, k-j), \ r = 1, 2, \ldots.$$

12. *Associated Stirling numbers of the second kind.* Using the triangular recurrence relation

$$S(n+1, k) = S(n, k-1) + kS(n,k), \ k = 1, 2, \ldots, n+1, \ n = 0, 1, \ldots,$$

$$S(0,0) = 1, \ S(n,0) = 0, \ n > 0, \ S(n,k) = 0, \ k > n,$$

show that

$$S(n,n) = 1, \ S(n, n-1) = \binom{n}{2}, \ S(n, n-2) = \binom{n}{3} + 3\binom{n}{4}$$

and, generally, setting

$$S(n, n-k) = \sum_{j=0}^{k} S_2(k+j, j) \binom{n}{k+j},$$

derive the bivariate generating function

$$g_2(t, u) = \sum_{n=0}^{\infty} \sum_{k=0}^{[n/2]} S_2(n, k) t^k \frac{u^n}{n!} = \exp[t(e^u - 1 - u)]$$

and deduce the generating function

$$f_{k,2}(u) = \sum_{n=2k}^{\infty} S_2(n, k) \frac{u^n}{n!} = \frac{(e^u - 1 - u)^k}{k!}, \ k = 0, 1, \dots .$$

Also, show that

$$S_2(n + 1, k) = kS_2(n, k) + nS_2(n - 1, k - 1),$$

for $n = 2k, 2k + 1, \dots, k = 1, 2, \dots$, with

$$S_2(0, 0) = 1, \ S_2(n, 0) = 0, \ n > 0, \ S_2(n, k) = 0, \ 2k > n,$$

and

$$S_2(n, k) = \sum_{j=0}^{k} (-1)^j \binom{n}{j} S(n - j, k - j),$$

for $n = 2k, 2k + 1, \dots, k = 0, 1, \dots .$

13. (*Continuation*). Consider, more generally, the numbers $S_r(n, k)$, $n = rk, rk + 1, \dots, k = 0, 1, \dots, r = 1, 2, \dots$, defined by their exponential generating function

$$f_{k,r}(u) = \sum_{n=rk}^{\infty} S_r(n, k) \frac{u^n}{n!} = \frac{1}{k!} \left(e^u - \sum_{j=0}^{r-1} \frac{u^j}{j!} \right)^k .$$

Derive the recurrence relation

$$S_r(n + 1, k) = kS_r(n, k) + \binom{n}{r - 1} S_r(n - r + 1, k - 1),$$

for $n = rk, rk + 1, \dots, k = 1, 2, \dots, r = 1, 2, \dots$, with

$$S_r(0, 0) = 1, \ S_r(n, 0) = 0, \ n > 0, \ S_r(n, k) = 0, \ rk > n.$$

Further, show that

$$S_{r+1}(n, k) = \sum_{j=0}^{k} (-1)^j \frac{(n)_{rj}}{j!(r!)^j} S_r(n - rj, k - j), \ r = 1, 2, \dots$$

and

$$S_r(n, k) = \sum_{j=0}^{k} \frac{(n)_{rj}}{j!(r!)^j} S_{r+1}(n - rj, k - j), \ r = 1, 2, \ldots .$$

14. *Associated generalized factorial coefficients.* Using the triangular recurrence relation

$$C(n + 1, k; s) = (sk - n)C(n, k; s) + sC(n, k - 1; s),$$

for $k = 1, 2, \ldots, n + 1$, $n = 0, 1, \ldots$, with initial conditions

$$C(0, 0; s) = 1, \ C(n, 0; s) = 0, \ n > 0, \ C(n, k; s) = 0, \ k > n,$$

show that

$$C(n, n; s) = s^n, \ C(n, n - 1; s) = s^{n-2}(s)_2 \binom{n}{2},$$

$$C(n, n - 2; s) = s^{n-3}(s)_3 \binom{n}{3} + 3s^{n-4}[(s)_2]^2 \binom{n}{4}$$

and, generally, setting

$$C(n, n - k; s) = \sum_{j=0}^{k} C_2(k + j, j; s)s^{n-k-j} \binom{n}{k+j},$$

derive the bivariate generating function

$$g_2(t, u; s) = \sum_{n=0}^{\infty} \sum_{k=0}^{[n/2]} C_2(n, k; s)t^k \frac{u^n}{n!} = \exp\{t[(1 + u)^s - 1 - su]\},$$

and conclude the generating function

$$f_{k,2}(u; s) = \sum_{n=2k}^{\infty} C_2(n, k; s)\frac{u^n}{n!} = \frac{[(1 + u)^s - 1 - su]^k}{k!}, \ k = 0, 1, \ldots .$$

Also, show that

$$C_2(n + 1, k; s) = (sk - n)C_2(n, k; s) + n(s)_2 C_2(n - 1, k - 1; s),$$

for $n = 2k, 2k + 1, \ldots$, $k = 1, 2, \ldots$, with

$$C_2(0, 0; s) = 1, \ C_2(n, 0; s) = 0, \ n > 0, \ C_2(n, k; s) = 0, \ 2k > n,$$

and

$$C_2(n, k; s) = \sum_{j=0}^{k} (-1)^j \binom{n}{j} s^j C(n - j, k - j; s),$$

for $n = 2k, 2k + 1, \ldots, k = 0, 1, \ldots$.

15. (*Continuation*). Consider, more generally, the numbers $C_r(n, k; s)$, $n = rk, rk + 1, \ldots, k = 0, 1, \ldots, r = 1, 2, \ldots$, defined by their exponential generating function

$$f_{k,r}(u; s) = \sum_{n=rk}^{\infty} C_r(n, k; s) \frac{u^n}{n!} = \frac{1}{k!} \left[(1 + u)^s - \sum_{j=0}^{r-1} \binom{s}{j} u^j \right]^k.$$

Derive the recurrence relation

$$C_r(n + 1, k; s) = (sk - n) C_r(n, k; s)$$
$$+ \binom{n}{r - 1} (s)_r C_r(n - r + 1, k - 1; s),$$

for $n = rk, rk + 1, \ldots, k = 1, 2, \ldots, r = 1, 2, \ldots$, with

$$C_r(0, 0; s) = 1, \ C_r(n, 0; s) = 0, \ n > 0, \ C_r(n, k; s) = 0, \ rk > n.$$

Also, show that

$$C_{r+1}(n, k; s) = \sum_{j=0}^{k} (-1)^j \frac{(n)_{rj}}{j!} \binom{s}{r}^j C_r(n - rj, k - j; s), \ r = 1, 2, \ldots$$

and

$$C_r(n, k; s) = \sum_{j=0}^{k} \frac{(n)_{rj}}{j!} \binom{s}{r}^j C_{r+1}(n - rj, k - j; s), \ r = 1, 2, \ldots.$$

16. *Cauchy numbers.* The sequence of the Cauchy numbers C_n, $n = 0, 1, \ldots$, has generating function

$$f(t) = \sum_{n=0}^{\infty} C_n \frac{t^n}{n!} = \frac{t}{\log(1 + t)}.$$

Show that

$$C_n = \sum_{k=1}^{n} (-1)^{k-1} \frac{(n)_k}{k + 1} C_{n-k}, \ n = 1, 2, \ldots, \ C_0 = 1$$

and

$$C_n = \sum_{k=0}^{n} \frac{1}{k + 1} s(n, k), \ \sum_{n=0}^{r} S(r, n) C_n = \frac{1}{r + 1},$$

where $s(n, k)$ and $S(r, n)$ are the Stirling numbers of the first and second kind, respectively.

17. Bernoulli numbers. The sequence of the Bernoulli numbers B_n, $n = 0, 1, \ldots$, has generating function

$$g(t) = \sum_{n=0}^{\infty} B_n \frac{t^n}{n!} = \frac{t}{e^t - 1}.$$

Show that

$$B_n = \sum_{k=0}^{n} \binom{n}{k} B_k, \ n = 1, 2, \ldots, \ B_0 = 1$$

and

$$B_n = \sum_{k=0}^{n} \frac{(-1)^k k!}{k+1} S(n, k), \ \sum_{n=0}^{r} s(r, n) B_n = \frac{(-1)^r r!}{r+1},$$

where $s(n, k)$ and $S(r, n)$ are the Stirling numbers of the first and second kind, respectively.

18. (Continuation). (a) Show that $B_{2r+1} = 0, r = 1, 2, \ldots$, and conclude that

$$1 + \sum_{r=1}^{\infty} (-1)^r 2^{2r} B_{2r} \frac{t^{2r}}{(2r)!} = t \cot t.$$

(b) Using the expansion

$$1 + 2 \sum_{n=0}^{\infty} \frac{t^2}{t^2 - n^2 \pi^2} = t \cot t,$$

show that

$$B_{2r} = (-1)^{r+1} 2(2r)!(2\pi)^{-2r} \zeta(2r), \ \zeta(s) = \sum_{n=1}^{\infty} n^{-s}$$

and conclude that $|B_{2r}| = (-1)^{r+1} B_{2r}, \ r = 1, 2, \ldots$, and

$$\zeta(2r) = \frac{(2\pi)^{2r}}{2(2r)!} |B_{2r}|, \ r = 1, 2, \ldots.$$

19. Nörlund-Bernoulli numbers. The sequence of numbers $B_n^{(r)}$, $n = 0, 1, \ldots$, where r is a real number, with generating function

$$g_r(t) = \sum_{n=0}^{\infty} B_n^{(r)} \frac{t^n}{n!} = \left(\frac{t}{e^t - 1} \right)^r,$$

has been defined by Nörlund and, in the particular case of $r = 1$, reduces to the sequence of Bernoulli numbers. Show that

$$B_n^{(r+1)} = \left(1 - \frac{n}{r}\right) B_n^{(r)} - n B_{n-1}^{(r)}, \ n = 1, 2, \dots, \ B_0^{(r)} = 1$$

and conclude that

$$B_n^{(r)} = s(r, r - n) / \binom{r-1}{n}, \ n = 0, 1, \dots, r - 1, \ r = 1, 2, \dots,$$

$$B_n^{(-r)} = S(n + r, r) / \binom{n+r}{n}, \ n = 0, 1, \dots, \ r = 1, 2, \dots,$$

where $s(n, k)$ and $S(n, k)$ are the Stirling numbers of the first and second kind, respectively.

20. Consider the sequence $B(n; s)$, $n = 0, 1, \dots$, where s is a real number, with generating function

$$g(t; s) = \sum_{n=0}^{\infty} B(n; s) \frac{t^n}{n!} = \frac{t}{(1+t)^s - 1}.$$

Show that

$$B(n; s) = \sum_{k=0}^{n} \binom{n}{k} (s)_{n-k} B(k; s), \ n = 1, 2, \dots, \ B(0; s) = 1/s$$

and

$$B_n = \sum_{k=0}^{n} \frac{(1/s)_{k+1}}{k+1} C(n, k; s),$$

where $C(n, k; s)$ is the generalized factorial coefficient. Further, show that

$$\lim_{s \to 0} s B(n; s) = C_n, \ \lim_{s \to \infty} s^{-n+1} B(n; s) = B_n,$$

where C_n and B_n are the Cauchy and Bernoulli numbers, respectively.

21*. *Carlitz-Riordan numbers of the first kind.* Consider the expansion of central factorial of t of order n,

$$t^{[n]} = t \left(t + \frac{n}{2} - 1\right) \left(t + \frac{n}{2} - 2\right) \cdots \left(t - \frac{n}{2} + 2\right) \left(t - \frac{n}{2} + 1\right)$$

$$= t \left(t + \frac{n}{2} - 1\right)_{n-1}, \ n = 1, 2, \dots,$$

with $t^{[0]} = 1$, into powers of t:

$$t^{[n]} = \sum_{k=0}^{n} r(n, k) t^k, \ n = 0, 1, \dots.$$

Show that

$$g(t, u) = \sum_{n=0}^{\infty} \sum_{k=0}^{n} r(n, k) t^k \frac{u^n}{n!} = \left((u/2) + \sqrt{1 + (u/2)^2} \right)^{2t},$$

and conclude that

$$f_k(u) = \sum_{n=k}^{\infty} r(n, k) \frac{u^n}{n!} = \frac{\left[2 \log \left((u/2) + \sqrt{1 + (u/2)^2} \right) \right]^k}{k!}, \quad k = 0, 1, \ldots .$$

Derive the recurrence relation

$$r(n + 2, k) = r(n, k - 2) - (n/2)^2 r(n, k),$$

for $k = 2, 3, \ldots, n + 2$, $n = 0, 1, \ldots$, with

$$r(0, 0) = 1, \quad r(n, k) = 0, \quad k > n, \quad r(2n, 2k + 1) = 0, \quad r(2n + 1, 2k) = 0.$$

22*. *Carlitz-Riordan numbers of the second kind.* Consider the expansion of the n-th order power of t into central factorials of it:

$$t^n = \sum_{k=0}^{n} R(n, k) t^{[k]}, \quad n = 0, 1, \ldots .$$

Show that

$$\sum_{j=k}^{n} r(n, j) R(j, k) = \delta_{n,k}, \quad \sum_{j=k}^{n} R(n, j) r(j, k) = \delta_{n,k}$$

and

$$f_k(u) = \sum_{n=k}^{\infty} R(n, k) \frac{u^n}{n!} = \frac{\left(e^{u/2} - e^{-u/2} \right)^k}{k!}, \quad k = 0, 1, \ldots .$$

Derive the explicit expression

$$R(n, k) = \frac{1}{k!} \sum_{j=0}^{k} (-1)^j \binom{k}{j} \left(\frac{k}{2} - j \right)^n.$$

23*. (*Continuation*). Show that

$$R(n + 2, k) = R(n, k - 2) - (k/2)^2 R(n, k),$$

for $k = 2, 3, \ldots, n + 2$, $n = 0, 1, \ldots$, with

$$R(0, 0) = 1, \quad R(n, k) = 0, \quad k > n, \quad R(2n, 2k + 1) = 0, \quad R(2n + 1, 2k) = 0$$

and deduce the generating function

$$\phi_k(u) = \sum_{n=k}^{\infty} R(n,k)u^n = \begin{cases} u^{2s} \displaystyle\prod_{j=1}^{s}(1 - j^2 u^2)^{-1}, & k = 2s \\ u^{2s+1} \displaystyle\prod_{j=1}^{s}(1 - (j + 1/2)^2 u^2)^{-1}, & k = 2s + 1. \end{cases}$$

24. Consider the sequence of non-central signless Stirling numbers of the first kind $|s(n,k;r)|$, $k = 0, 1, \dots, n$, $n = 0, 1, \dots$, for fixed r, which are defined by (see Section 8.5)

$$(t + r + n - 1)_n = \sum_{k=0}^{n} |s(n,k;r)| t^k, \ n = 0, 1, \dots .$$

Show that

$$g(t, u; r) = \sum_{n=0}^{\infty} \sum_{k=0}^{n} |s(n,k;r)| t^k \frac{u^n}{n!} = (1 - u)^{-t-r}$$

and conclude that

$$f_k(u; r) = \sum_{n=k}^{\infty} |s(n,k;r)| \frac{u^n}{n!} = (1 - u)^{-r} \frac{[-\log(1 - u)]^k}{k!}, \ k = 0, 1, \dots .$$

25. (*Continuation*). Show that the non-central signless Stirling numbers of the first kind $|s(n,k;r)|$, $k = 0, 1, \dots, n$, $n = 0, 1, \dots$, satisfy the triangular recurrence relation

$$|s(n + 1, k; r)| = |s(n, k - 1; r)| + (n + r)|s(n, k; r)|,$$

for $k = 1, 2, \dots, n + 1$, $n = 0, 1, \dots$, with initial conditions

$$|s(0, 0; r)| = 1, \ |s(n, 0; r)| = (r + n - 1)_n, \ n > 0, \ |s(n, k; r)| = 0, \ k > n.$$

Also, show that

$$|s(n, k; r)| = \sum_{j=k}^{n} \binom{n}{j} (m + n - j - 1)_{n-j} |s(j, k; r - m)|$$

and conclude that

$$|s(n, k; r)| = \sum_{j=k}^{n} (n)_{n-j} |s(j, k; r - 1)|$$

and

$$|s(n,k;r)| = \sum_{j=k}^{n} \binom{n}{j} (r+n-j-1)_{n-j}|s(j,k)|.$$

26. Suppose that, balls are successively drawn one after the other from an urn initially containing w white and b black balls, according to the following scheme. After each trial the drawn ball is placed back in the urn along with s black balls. Show that (a) the probability $p(k;n,r)$ of drawing k white balls in n trials is given by

$$p(k;n,r) = \frac{|s(n,k;r)|\theta^k}{(\theta+r+n-1)_n}, \quad k = 0,1,\ldots,n$$

and (b) the probability $q(n;k,r)$ that n trials are required until the k-th white ball is drawn is given by

$$q(n;k,r) = \frac{|s(n-1,k-1;r)|\theta^k}{(\theta+r+n-1)_n}, \quad n = k,k+1,\ldots,$$

where $\theta = w/s$ and $r = b/s$.

27. Consider the sequence of non-central Stirling numbers of the second kind $S(n,k;r)$, $k = 0,1,\ldots,n$, $n = 0,1,\ldots$, for fixed r, which are defined by (see Section 8.5)

$$(t+r)^n = \sum_{k=0}^{n} S(n,k;r)(t)_k, \quad n = 0,1,\ldots.$$

Show that

$$f(t,u;r) = \sum_{n=0}^{\infty}\sum_{k=0}^{n} S(n,k;r)(t)_k \frac{u^n}{n!} = e^{(t+r)u}$$

and conclude that

$$f_k(u;r) = \sum_{n=k}^{\infty} S(n,k;r)\frac{u^n}{n!} = e^{ru}\frac{(e^u-1)^k}{k!}, \quad k = 0,1,\ldots.$$

Further, derive the explicit expression

$$S(n,k;r) = \frac{1}{k!}\sum_{j=0}^{k}(-1)^{k-j}\binom{k}{j}(r+j)^n.$$

28. (*Continuation*). Show that the non-central Stirling numbers of the second kind $S(n,k;r)$, $k = 0,1,\ldots,n$, $n = 0,1,\ldots$, satisfy the triangular recurrence relation

$$S(n+1,k;r) = S(n,k-1;r) + (k+r)S(n,k;r),$$

for $k = 1, 2, \ldots, n+1$, $n = 0, 1, \ldots$, with initial conditions

$$S(0, 0; r) = 1, \ S(n, 0; r) = r^n, \ n > 0, \ S(n, k; r) = 0, \ k > n.$$

Also, show that

$$S(n, k; r) = \sum_{j=k}^{n} \binom{n}{j} m^{n-j} S(j, k; r - m)$$

and conclude that

$$S(n, k; r) = \sum_{j=k}^{n} \binom{n}{j} S(j, k; r - 1)$$

and

$$S(n, k; r) = \sum_{j=k}^{n} \binom{n}{j} r^{n-j} S(j, k).$$

29. (*Continuation*). Show that

$$\sum_{j=k}^{n} s(n, j; r_1) S(j, k; r_2) = \binom{n}{k} (r_2 - r_1)_{n-k},$$

$$\sum_{j=k}^{n} S(n, j; r_1) s(j, k; r_2) = \binom{n}{k} (r_1 - r_2)^{n-k},$$

and conclude that

$$\sum_{j=k}^{n} s(n, j; r) S(j, k; r) = \delta_{n,k}, \ \sum_{j=k}^{n} S(n, j; r) s(j, k; r) = \delta_{n,k}.$$

30. (*Continuation*). Show that

$$\binom{k+m}{k} s(n, k+m; r_1 + r_2) = \sum_{j=k}^{n-m} \binom{n}{j} s(j, k; r_1) s(n - j, m; r_2)$$

and conclude its inverse relation

$$\binom{n}{i} s(n - i, m; r_2) = \sum_{k=i}^{n-m} \binom{k+m}{k} S(k, i; r_1) s(n, k+m; r_1 + r_2).$$

31. (*Continuation*). Show that

$$\binom{k+m}{k} S(n, k+m; r_1 + r_2) = \sum_{j=k}^{n-m} \binom{n}{j} S(j, k; r_1) S(n - j, m; r_2)$$

and conclude its inverse relation

$$\binom{n}{i} S(n-i,m;r_2) = \sum_{k=i}^{n-m} \binom{k+m}{k} s(k,i;r_1)S(n,k+m;r_1+r_2).$$

32. (*Continuation*). Show that

$$\phi_k(u;r) = \sum_{n=k}^{\infty} S(n,k;r)u^n = u^k \prod_{j=0}^{k}(1-ru-ju)^{-1}, \; k=1,2,\dots$$

and conclude the inverse relations

$$(t+r)_{-k} = \sum_{n=k}^{\infty}(-1)^{n-k}S(n-1,k-1;r+1)t^{-n}, \; k=1,2,\dots,$$

$$(t-r)^{-k} = \sum_{n=k}^{\infty}(-1)^{n-k}s(n-1,k-1;r+1)(t)_{-n}, \; k=1,2,\dots.$$

33. Suppose that, balls are successively drawn one after the other from an urn initially containing m white and r black balls, according to the following scheme. After each trial, if the drawn ball is white, a black ball is placed in the urn, while if the drawn ball is black, it is placed back in the urn. Show that (a) the probability $p(k;n,r)$ of drawing k white balls in n trials, with $n \leq m$, is given by

$$p(k;n,r) = \frac{S(n,k;r)(m)_k}{(m+r)^n}, \; k=0,1,\dots,n$$

and (b) the probability $q(n;k,r)$ that n trials are required until the k-th white ball is drawn, with $k \leq m$, is given by

$$q(n;k,r) = \frac{S(n-1,k-1;r)(m)_k}{(m+r)^n}, \; n=k,k+1,\dots.$$

34. Consider the sequence of the non-central generalized factorial coefficients $C(n,k;s,r)$, $k=0,1,\dots,n$, $n=0,1,\dots$, for fixed s and r, which are defined by (see Section 8.5)

$$(st+r)_n = \sum_{k=0}^{n} C(n,k;s,r)(t)_k, \; n=0,1,\dots.$$

Show that

$$f(t,u;s,r) = \sum_{n=0}^{\infty}\sum_{k=0}^{n} C(n,k;s,r)(t)_k \frac{u^n}{n!} = (1+u)^{st+r}$$

and conclude that

$$f_k(u; s, r) = \sum_{n=k}^{\infty} C(n, k; s, r) \frac{u^n}{n!} = (1 + u)^r \frac{[(1 + u)^s - 1]^k}{k!}, \quad k = 0, 1, \ldots.$$

Using a suitable generating function, show that

$$C(n, k; s, r) = \sum_{j=k}^{n} \binom{j}{k} (r/s)_{j-k} C(n, j; s),$$

and

$$C(n, k; s, r) = \sum_{j=k}^{n} \binom{n}{j} (r)_{n-j} C(j, k; s).$$

35. (*Continuation*). Show that the non-central generalized factorial coefficient $C(n, k; s, r)$, $k = 0, 1, \ldots, n$, $n = 0, 1, \ldots$, is given by the sum

$$C(n, k; s, r) = \frac{1}{k!} \sum_{j=0}^{k} (-1)^{k-j} \binom{k}{r} (sj + r)_n.$$

36. (*Continuation*). Show that the non-central generalized factorial coefficients $C(n, k; s, r)$, $k = 0, 1, \ldots, n$, $n = 0, 1, \ldots$, satisfy the triangular recurrence relation

$$C(n + 1, k; s, r) = (sk + r - n)C(n, k; s, r) + sC(n, k - 1; s, r),$$

for $k = 1, 2, \ldots, n + 1$, $n = 0, 1, \ldots$, with initial conditions

$$C(0, 0; s, r) = 1, \; C(n, 0; s, r) = (r)_n, \; n > 0, \; C(n, k; s, r) = 0, \; k > n.$$

37. (*Continuation*). Show that

$$\lim_{s \to 0} s^{-k} C(n, k; s, r) = s(n, k; -r)$$

and, for $\lim_{s \to \infty} r/s = \rho$, that

$$\lim_{s \to \infty} s^{-n} C(n, k; s, r) = S(n, k; \rho).$$

38. (*Continuation*). *Non-central Lah numbers.* Consider the expansion

$$(-(t - r))_n = \sum_{k=0}^{n} L(n, k; r)(t)_k, \quad n = 0, 1, \ldots.$$

The coefficient $L(n, k; r)$ is called *non-central Lah number*; it is a particular case, $s = -1$, of the non-central generalized factorial coefficient. Show that

$$L(n, k; r) = (-1)^n \frac{n!}{k!} \binom{n - r - 1}{k - r - 1}.$$

39. (*Continuation*). Show that

$$C(n, k; s, \rho s - r) = \sum_{j=k}^{n} s(n, j; r) S(j, k; \rho) s^j$$

and conclude that

$$\sum_{j=k}^{n} s(n, j; r) S(j, k; \rho) = \binom{n}{k} (\rho - r)_{n-k},$$

$$\sum_{j=k}^{n} (-1)^{n-j} s(n, j; r) S(j, k; \rho) = \binom{n}{k} (\rho - r + n - 1)_{n-k}.$$

40. (*Continuation*). Show that

$$\sum_{j=k}^{n} C(n, j; s_1, r_1) C(j, k; s_2, r_2) = C(n, k; s_1 s_2, r_1 + r_2 s_1)$$

and conclude that

$$\sum_{j=k}^{n} C(n, j; s, r_1) C(j, k; s^{-1}, r_2) = \binom{n}{k} (r_1 + r_2 s)_{n-k},$$

and

$$\sum_{j=k}^{n} C(n, j; s, r) C(j, k; s^{-1}, -rs^{-1}) = \delta_{n,k}.$$

41. (*Continuation*). Let

$$\phi_k(u; s, r) = \sum_{n=k}^{\infty} C(n - 1, k - 1; s, r) \frac{1}{(u)_n}, \quad k = 1, 2, \dots.$$

Show that

$$\phi_k(u; s, r) = \frac{1}{(u - r)/s - (k - 1)} \phi_{k-1}(u; s, r), \quad k = 2, 3, \dots,$$

with $\phi_1(u; s, r) = 1/(u - r)$, and conclude that

$$\phi_k(u; s, r) = \frac{1}{s((u - r)/s)_k}$$

and

$$\frac{1}{(t)_k} = \sum_{n=k}^{\infty} sC(n - 1, k - 1; s, r)\frac{1}{(st + r)_n}.$$

42. (*Continuation*). Show that

$$\binom{k + m}{k} C(n, k + m; s, r_1 + r_2) =$$

$$\sum_{j=k}^{n-m} \binom{n}{j} C(j, k; s, r_1)C(n - j, m; s, r_2)$$

and conclude its inverse relation

$$\binom{n}{i} C(n - i, m; s, r_2) =$$

$$\sum_{k=i}^{n-m} \binom{k + m}{k} C(k, i; s^{-1}, -r_1 s^{-1})C(n, k + m; s, r_1 + r_2).$$

43. (*Continuation*). Show that

$$(t^{b+1} D)^n f(t) = \sum_{k=0}^{n}(-1)^{n-k} b^n a^{-k} C(n, k; s, r) t^{bn-ak}(Dt^{a+1})^k f(t),$$

where $s = a/b$, $r = (a + 1)/b$, and conclude that

$$(t^{b+1} D)^n f(t) = \sum_{k=0}^{n}(-1)^n b^n C(n, k; s) t^{bn+k} D^k f(t), \ s = -1/b$$

and

$$(tD)^n f(t) = \sum_{k=0}^{n} S(n, k) t^k D^k f(t).$$

44. Consider an urn containing s identical series of coupons, each consisting of m coupons bearing the numbers $1, 2, \ldots, m$ and r additional coupons, all bearing the number $m + 1$. Suppose that coupons are drawn one after the other, at random, without replacement. Show that (a) the probability

$p(k; n)$ of drawing exactly k of the m numbers $\{1, 2, \ldots, m\}$ in n drawings, with $n \leq m$, is given by

$$p(k; n) = \frac{C(n, k; s, r)(m)_k}{(sm + r)_n}, \quad k = 0, 1, \ldots, n$$

and (b) the probability $q(n; k)$ that n drawings are required until the k-th different number, among the m numbers $\{1, 2, \ldots, m\}$, is drawn, with $k \leq m$, is given by

$$q(n; k) = \frac{C(n - 1, k - 1; s, r)s(m)_k}{(sm + r)_n}, \quad n = k, k + 1, \ldots, sm + r.$$

45. (*Continuation*). Suppose that from the urn coupons are drawn one after the other, at random, by returning in the urn, after each drawing, the chosen coupon together with another coupon bearing the same number. Show that (a) the probability $p(k; n)$ of drawing exactly k of the m numbers $\{1, 2, \ldots, m\}$ in n drawings is given by

$$p(k, n) = \frac{|C(n, k; -s, -r)|(m + k - 1)_k}{(sm + r + n - 1)_n}, \quad k = 0, 1, \ldots, n$$

where $|C(n, k; -s, -r)| = (-1)^n C(n, k; -s, -r)$. Further show that (b) the probability $q(n; k)$ that n drawings are required until the k-th different number, among the m numbers $\{1, 2, \ldots, m\}$, is drawn is given by

$$q(n; k) = \frac{|C(n - 1, k - 1; -s, -r)|s(m + k - 1)_r}{(sm + r + n - 1)_n}, \quad n = k, k + 1, \ldots.$$

Chapter 9

DISTRIBUTIONS AND OCCUPANCY

9.1 INTRODUCTION

A considerable number of combinatorial configurations, such as permutations, combinations and partitions of a finite set under various conditions, can be described by the model of distribution (allocation) of balls (objects) into urns (cells, boxes), introduced by P. MacMahon. It is an advantageous approach, since urn models can be easily visualized and are very flexible. Further, these models admit many equivalent interpretations. In general, the balls are of the type $(r_1, r_2, \ldots, r_m)$ with $r_1 + r_2 + \cdots + r_m = n$, that is, r_i of the n balls are of the i-th kind, $i = 1, 2, \ldots, m$. The urns are of the type $(s_1, s_2, \ldots, s_\nu)$ with $s_1 + s_2 + \cdots + s_\nu = k$, that is, s_j of the k urns are of the j-th kind, $j = 1, 2, \ldots, \nu$. Moreover, the urns may be of limited or unlimited capacity and the balls in each urn may be ordered or unordered. The enumeration of the assignments of the balls to the urns is a distribution problem, while the enumeration of the balls in specified or arbitrary urns is an occupancy problem.

This general consideration reveals a host of enumeration problems. Particular cases of such problems have been examined in Chapter 2 as applications of enumeration of certain permutations and combinations. In the present chapter, more general distribution and occupancy problems are studied by using the inclusion and exclusion principle and generating functions. Specifically, the classical occupancy problem and some of its modifications are examined in length. Then the distributions of balls into urns, when the ordering of the balls in the urns counts, are enumerated. Further, the problem of enumeration of the distributions of balls of a general specification into distinguishable urns is treated by using the inclusion and exclusion principle. As applications, a variety of committee problems is discussed. Finally, the use of generating functions in coping with the same general problem is demonstrated.

9.2 CLASSICAL OCCUPANCY AND MODIFICATIONS

Let us first consider n distinguishable balls and k distinguishable urns and assume that the capacity of each urn is unlimited. The urns as distinguishable, without any loss of generality, may be numbered from 1 to k. Then, the number of distributions of n distinguishable balls into k distinguishable urns equals

$$U(k,n) = k^n,$$

the number of n-permutations of the set $\{1, 2, \ldots, k\}$, with repetition (see Theorem 2.3). Further, the number of distributions of n distinguishable balls into k distinguishable urns with r_j balls in the j-th urn, $j = 1, 2, \ldots, k$, $r_1 + r_2 + \cdots + r_k = n$, equals

$$C(n, r_1, r_2, \ldots, r_{k-1}) = \binom{n}{r_1, r_2, \ldots, r_{k-1}} = \frac{n!}{r_1! r_2! \cdots r_{k-1}! r_k!},$$

the number of divisions of the set of n balls $\{1, 2, \ldots, n\}$ into k subsets containing $r_1, r_2, \ldots, r_k$ balls, respectively (see Theorem 2.8).

In the classical occupancy problem, the number of distributions with a given number of occupied urns is of interest. The next theorem is concerned with this number.

THEOREM 9.1
The number of distributions of n distinguishable balls into k distinguishable urns so that r urns are occupied is given by

$$N(n, k, r) = (k)_r S(n, r), \tag{9.1}$$

where

$$S(n, r) = \frac{1}{r!} \sum_{j=0}^{r} (-1)^j \binom{r}{j} (r - j)^n,$$

is the Stirling number of the second kind.

PROOF   Consider the set Ω of distributions of n distinguishable balls into k distinguishable urns and let A_i be the subset of these distributions in which the i-th urn remains empty, $i = 1, 2, \ldots, k$. Then, for any selection of s indices $\{i_1, i_2, \ldots, i_s\}$, out of the k indices $\{1, 2, \ldots, k\}$,

$$\nu_s = N(A_{i_1} A_{i_2} \cdots A_{i_s}) = (k - s)^n, \quad s = 1, 2, \ldots, k,$$

which is the number of distributions of n distinguishable balls into the remaining (after excluding the s specified urns) $k - s$ distinguishable urns. This expression implies the exchangeability of the sets $A_1, A_2, \ldots, A_k$. Further, the number $N(n, k, r)$ of distributions of n distinguishable balls into k distinguishable urns, which leave $k - r$ urns empty (and consequently r urns occupied), equals the number $N_{k,k-r}$ of elements of Ω that are contained in $k - r$ among the k sets $A_1, A_2, \ldots, A_k$. Thus, according to Corollary 4.4, we deduce the expression

$$N(n, k, r) = \binom{k}{r} \sum_{j=0}^{r} (-1)^j \binom{r}{j} (r - j)^n,$$

which, upon using the explicit expression of the Stirling numbers of the second kind (see Section 8.3), implies (9.1).   ∎

COROLLARY 9.1

The number of distributions of n distinguishable balls into k distinguishable urns so that, out of s specified urns, r urns are occupied is given by

$$N(n, k, s, r) = (s)_r S(n, r; k - s) \tag{9.2}$$

where

$$S(n, r; k - s) = \frac{1}{r!} \sum_{j=0}^{r} (-1)^j \binom{r}{j} (k - s + r - j)^n,$$

is the non-central Stirling number of the second kind.

PROOF   Note that i balls can be selected from n distinguishable balls in $\binom{n}{i}$ ways, $i = 0, 1, \ldots, n$. Further, for each selection, the i balls, according to Theorem 9.1, can be distributed into the s specified urns, so that r among them are occupied in

$$N(i, s, r) = \binom{s}{r} \sum_{j=0}^{r} (-1)^j \binom{r}{j} (r - j)^i$$

ways, while the remaining $n - i$ balls can be distributed into the remaining $k - s$ urns, without any restriction, in $(k - s)^{n-i}$ ways, $i = 0, 1, \ldots, n$. So, according to the multiplication principle, there are

$$\binom{n}{i} N(i, s, r)(k - s)^{n-i} = \binom{s}{r} \sum_{j=0}^{r} (-1)^j \binom{r}{j} \binom{n}{i} (r - j)^i (k - s)^{n-i}$$

different such distributions. Thus, summing for $i = 0, 1, \ldots, n$, according to the addition principle, the number of distributions of n distinguishable balls into k

distinguishable urns so that r among s specified urns are occupied is obtained as

$$N(n,k,s,r) = \binom{s}{r} \sum_{j=0}^{r} (-1)^j \binom{r}{j} \sum_{i=0}^{n} \binom{n}{i} (r-j)^i (k-s)^{n-i}$$

$$= \binom{s}{r} \sum_{j=0}^{r} (-1)^j \binom{r}{j} (k-s+r-j)^n.$$

The last expression, using the explicit expression of the non-central Stirling numbers of the second kind (see Section 8.5), implies (9.2). ∎

COROLLARY 9.2

The number of distributions of n distinguishable balls into k indistinguishable urns so that r urns are occupied is given by

$$S(n,r) = \frac{1}{r!} \sum_{j=0}^{r} (-1)^j \binom{r}{j} (r-j)^n, \tag{9.3}$$

the Stirling number of the second kind, while, without any restriction, it is given by

$$B(n,k) = \sum_{r=1}^{k} S(n,r), \tag{9.4}$$

the Bell number.

PROOF   Note first that r urns can be selected from k indistinguishable urns in only one way. Further, let $R(n,r)$ and $N(n,r)$ be the number of distributions of n distinguishable balls into r indistinguishable urns and into r distinguishable urns, respectively, so that no urn remains empty. Consider a distribution of n distinguishable balls into r indistinguishable urns so that no urn remains empty. If the r indistinguishable urns are transformed to distinguishable, by numbering them, and permuted in all possible ways, $r!$ distributions of n distinguishable balls into r distinguishable urns, so that no urn remains empty, are constructed. Consequently $r!R(n,r) = N(n,r)$ and since, by Theorem 9.1, $N(n,r) = r!S(n,r)$, it follows that $R(n,r) = S(n,r)$. Thus, the number $R(n,k,r)$ of distributions of n distinguishable balls into k indistinguishable urns so that r urns are occupied is given by (9.3).

As regards the evaluation of the number $B(n,k)$ of distributions of n distinguishable balls into k indistinguishable urns, without any restriction, note that, in such a distribution, 1 or 2 or , ..., or k urns are occupied. Since the number of distributions of n distinguishable balls into k indistinguishable urns with r occupied urns equals $S(n,r)$, summing for all values $r = 1, 2, \ldots, k$, according to the addition principle, we deduce (9.4). ∎

Consider now n indistinguishable (like) balls and k distinguishable urns. Again, the urns as distinguishable may be numbered from 1 to k. The number of distributions of n indistinguishable balls into k distinguishable urns equals the number of nonnegative integer solutions of the linear equation $r_1 + r_2 + \cdots + r_k = n$, where r_j is the number of balls in the j-th urn, $j = 1, 2, \ldots, k$. Thus, according to Theorem 2.12, the number of distributions of n indistinguishable balls into k distinguishable urns equals

$$\binom{k+n-1}{n},$$

the number of n-combinations of the k urns $\{1, 2, \ldots, k\}$ with repetition. Further, the number of distributions of n indistinguishable balls into k distinguishable urns so that no urn remains empty equals the number of positive integer solutions of the linear equation $r_1 + r_2 + \cdots + r_k = n$, which, according to Corollary 2.5, equals

$$\binom{n-1}{k-1},$$

the number of n-combinations of the k urns $\{1, 2, \ldots, k\}$, with repetition and the restriction that each urn is included at least once. As a consequence of these remarks, the following corollary, analogous to Theorem 9.1, is deduced.

COROLLARY 9.3

The number of distributions of n indistinguishable balls into k distinguishable urns so that r urns are occupied is given by

$$\binom{k}{r}\binom{n-1}{r-1}. \tag{9.5}$$

PROOF   Note that r urns can be selected from k distinguishable urns in $\binom{k}{r}$ ways and n indistinguishable balls can be distributed into these r urns, so that no urn remains empty, in $\binom{n-1}{r-1}$ ways. Thus, according to the multiplication principle, the required number of distributions is deduced as (9.5).   ∎

Let us, more generally, consider n indistinguishable (like) balls and k distinguishable urns, each divided into s distinguishable cells (compartments). Further, assume that each cell is of capacity limited to one ball. Then, the number of distributions of n indistinguishable balls into the k distinguishable urns (sk distinguishable cells) equals

$$\binom{sk}{n},$$

while the number of distributions of n indistinguishable balls into the k distinguishable urns with r_j balls in the j-th urn, $j = 1, 2, \ldots, k$, $r_1 + r_2 + \cdots + r_k = n$, equals

$$\binom{s}{r_1}\binom{s}{r_2}\cdots\binom{s}{r_k}.$$

The next theorem is analogous to Theorem 9.1.

THEOREM 9.2

The number of distributions of n indistinguishable balls into k distinguishable urns, each with s distinguishable cells of capacity limited to one ball, so that r urns are occupied, is given by

$$R(n, k, s, r) = \frac{(k)_r}{n!}C(n, r; s) \tag{9.6}$$

where

$$C(n, r; s) = \frac{1}{r!}\sum_{j=0}^{r}(-1)^{r-j}\binom{r}{j}(sj)_n$$

is the generalized factorial coefficient.

PROOF   Consider the set Ω of distributions of n indistinguishable (like) balls into k distinguishable urns, each with s distinguishable cells of capacity limited to one ball. Let A_i be the subset of these distributions in which the i-th urn remains empty, $i = 1, 2, \ldots, k$. Then, for any selection of j indices $\{i_1, i_2, \ldots, i_j\}$, out of the k indices $\{1, 2, \ldots, k\}$,

$$\nu_j = N(A_{i_1}A_{i_2}\cdots A_{i_j}) = \binom{s(k-j)}{n}, \quad j = 1, 2, \ldots, k,$$

which is the number of distributions of n indistinguishable balls into the remaining (after excluding the j specified urns) $k - j$ distinguishable urns ($s(k - j)$ distinguishable cells each of capacity limited to one ball). This expression implies the exchangeability of the sets $A_1, A_2, \ldots, A_k$. Further, the number $R(n, k, s, r)$ of distributions of n indistinguishable balls into k distinguishable urns that leave $k - r$ urns empty (and consequently r urns occupied) equals the number $N_{k,k-r}$ of elements of Ω that are contained in $k - r$ among the k sets $A_1, A_2, \ldots, A_k$. Thus, according to Corollary 4.4, we get the expression

$$R(n, k, s, r) = \binom{k}{r}\sum_{j=0}^{r}(-1)^j\binom{r}{j}\binom{s(r-j)}{n},$$

which, using the explicit expression of the generalized factorial coefficient (see Section 8.4), implies (9.6).   ∎

REMARK 9.1   The number of distributions of n distinguishable balls into k distinguishable urns, each with s distinguishable cells of capacity limited to one ball, equals

$$(sk)_n,$$

while, with the restriction that r urns are occupied, is given by

$$n!R(n,k,s,r) = (k)_r C(n,r;s), \qquad (9.7)$$

where $C(n,r;s)$ is generalized factorial coefficient.   ∎

Assume now that, in the preceding model, the cells are of unlimited capacity. Then, the number of distributions of n indistinguishable balls into the k distinguishable urns (sk distinguishable cells), equals

$$\binom{sk+n-1}{n} = (-1)^n \binom{-sk}{n},$$

while the number of distributions of n indistinguishable balls into the k distinguishable urns, each with s distinguishable cells, with r_j balls in the j-th urn, $j = 1, 2, \ldots, k$, $r_1 + r_2 + \cdots + r_k = n$, is given by

$$\binom{s+r_1-1}{r_1}\binom{s+r_2-1}{r_2}\cdots\binom{s+r_k-1}{r_k} = (-1)^n \binom{-s}{r_1}\binom{-s}{r_2}\cdots\binom{-s}{r_k}$$

The next theorem is analogous to Theorem 9.1 and 9.2.

THEOREM 9.3
The number of distributions of n indistinguishable balls into k distinguishable urns, each with s distinguishable cells of unlimited capacity, so that r urns are occupied, is given by

$$T(n,k,s,r) = \frac{(k)_r}{n!}|C(n,r;-s)|, \qquad (9.8)$$

where

$$|C(n,r;-s)| = (-1)^n C(n,r;-s) = \frac{1}{r!}\sum_{j=0}^{r}(-1)^{r-j}\binom{r}{j}(sj+n-1)_n$$

is the generalized factorial coefficient.

PROOF   Consider the set Ω of distributions of n indistinguishable (like) balls into k distinguishable urns, each with s distinguishable cells of unlimited capacity. Let A_i be the subset of these distributions, in which the i-th urn remains empty,

$i = 1, 2, \ldots, k$. Then, for any selection of j indices $\{i_1, i_2, \ldots, i_j\}$, out of the k indices $\{1, 2, \ldots, k\}$,

$$\nu_j = N(A_{i_1} A_{i_2} \cdots A_{i_j}) = \binom{s(k - j) + n - 1}{n}, \quad j = 1, 2, \ldots, k,$$

which is the number of distributions of n indistinguishable balls into the remaining (after excluding the j specified urns) $k - j$ distinguishable urns ($s(k - j)$ distinguishable cells, each of unlimited capacity). This expression implies the exchangeability of the sets $A_1, A_2, \ldots, A_k$. Further, the number $T(n, k, s, r)$ of distributions of n indistinguishable balls into k distinguishable urns, which leave $k - r$ urns empty (and consequently r urns occupied), equals the number $N_{k,k-r}$ of elements of Ω that are contained in $k - r$ among the k sets $A_1, A_2, \ldots, A_k$. Thus, according to Corollary 4.4, we deduce the expression

$$T(n, k, s, r) = \binom{k}{r} \sum_{j=0}^{r} (-1)^j \binom{r}{j} \binom{s(r - j) + n - 1}{n},$$

which, upon introducing the explicit expression of the generalized factorial coefficient, implies (9.8).   ∎

REMARK 9.2   The enumeration of distributions of indistinguishable balls into indistinguishable urns with or without restrictions cannot be handled by using the inclusion and exclusion principle. Further, it cannot be connected with the enumeration of distributions of indistinguishable balls into distinguishable urns, as the enumeration of distinguishable balls into indistinguishable urns is connected with the enumeration of distributions of distinguishable balls into distinguishable urns. This problem will be separately examined in the next chapter.   ∎

Example 9.1 The Morse code
   A letter in the Morse code is formed by a succession of dashes and dots with repetitions allowed. The two symbols, dash and dot, may be considered as two distinguishable urns and the n different positions (first, second, $\ldots$, n-th) of a letter as n distinguishable balls. Thus, a formation of a letter of n symbols corresponds to a distribution of n distinguishable balls into $k = 2$ distinguishable urns. Consequently, there are 2^n different letters of n symbols. Further, the number of letters of n symbols that include both the dash and dot symbols, according to Theorem 9.1, equals

$$\sum_{j=0}^{2} (-1)^j \binom{2}{j} (2 - j)^n = 2^n - 2.$$

In the more general k-symbol code, there are k^n different letters of n symbols. Further, the number of letters of n symbols that include r out of s specified symbols, according to Corollary 9.1, is given by

$$N(n, k, s, r) = (s)_r S(n, r; k - s),$$

where $S(n, r; k - s)$ is the non-central Stirling number of the second kind. ⬚

Example 9.2 Decomposition of a product of primes into factors
Consider a product $p_n = a_1 a_2 \cdots a_n$, where a_i, $i = 1, 2, \ldots, n$, are different prime numbers. Find the number of decompositions of the product p_n, of n different prime numbers, into k factors.

The n different prime numbers $\{a_1, a_2, \ldots, a_n\}$ may be considered as n distinguishable balls and the k factors, which are not ordered, as k indistinguishable urns. Thus, a decomposition of a product of n different prime numbers into k factors corresponds to a distribution of n distinguishable balls into k indistinguishable urns with no urn empty and vice versa. For instance, the different decompositions of the product $p_3 = a_1 a_2 a_3$ of $n = 3$ different prime numbers into $k = 2$ factors, and their corresponding distributions of three distinguishable balls into two indistinguishable urns with no urn empty, are the following:

$$(a_1)(a_2 a_3), \quad (a_2)(a_1 a_3), \quad (a_3)(a_1 a_2).$$

Hence, the number of decompositions of a product of n different prime numbers into k factors equals the number of distributions of n distinguishable balls into k indistinguishable urns with no urn empty. According to Corollary 9.2, this is given by $S(n, k)$, the Stirling number of the second kind. ⬚

Example 9.3 Selection of a central committee
For the selection of the n members of the central committee of a federation of k associations, each association submits a list of s candidates. Find the number of different selections of the n members of the central committee in which r associations are represented in the committee.

The n members of the central committee may be considered as n indistinguishable balls and the k associations as k distinguishable urns, each with s distinguishable cells of capacity limited to one ball. Thus, a selection of the n members of the central committee from the k lists, each with s candidates, corresponds to a distribution of n indistinguishable balls into k distinguishable urns, each with s distinguishable cells of capacity limited to one ball. Consequently, the number of selections of the n members of the central committee from the k lists, each with s candidates, so that r associations are represented in the committee, according to Theorem 9.2, is given by

$$R(n, k, s, r) = \frac{(k)_r}{n!} C(n, r; s),$$

where $C(n, r; s)$ is the generalized factorial coefficient. ⬚

9.3   ORDERED DISTRIBUTIONS AND OCCUPANCY

The case with ordered balls in each urn is of particular interest. The urns with ordered balls are simply called *ordered urns*. The enumeration of the distributions of balls of any specification into distinguishable ordered urns, with or without restrictions, is closely related to the enumeration of the distributions of indistinguishable (like) balls into distinguishable (non-ordered) urns. Specifically, we have the next theorem.

THEOREM 9.4

The number of distributions of n balls of the type $(r_1, r_2, \ldots, r_m)$, where $r_1 + r_2 + \cdots + r_m = n$, into k distinguishable ordered urns without any restriction is given by

$$\frac{n!}{r_1! r_2! \cdots r_m!} \binom{k + n - 1}{n} = \frac{(k + n - 1)_n}{r_1! r_2! \cdots r_m!}, \tag{9.9}$$

and, with the restriction that r urns are occupied, is given by

$$\frac{n!}{r_1! r_2! \cdots r_m!} \binom{k}{r} \binom{n - 1}{r - 1}. \tag{9.10}$$

PROOF   Note that any distribution of n balls of the type $(r_1, r_2, \ldots, r_m)$ into k distinguishable ordered urns, with or without restrictions, may be carried out in two consecutive stages. Initially, the n balls are distributed into the k distinguishable urns, without taking into account which elements and in what order are placed into each urn. The number $Q(n, k)$ of these distributions equals the number of the distributions of n indistinguishable balls into k distinguishable urns. This number, in the absence of any restriction, is given by

$$\binom{k + n - 1}{n},$$

and, with the restriction that r urns are occupied, according to Corollary 9.3, is given by

$$\binom{k}{r} \binom{n - 1}{r - 1}.$$

Further, for each such distribution, by taking into account that the n balls are of the type $(r_1, r_2, \ldots, r_m)$ and permuting them, all possible orderings in the urns are accomplished. Since the number of permutations of the n balls of the type

$(r_1, r_2, \ldots, r_m)$, according to Theorem 2.4, is given by

$$\frac{n!}{r_1! r_2! \cdots r_m!},$$

applying the multiplication principle, (9.9) and (9.10) are deduced. ∎

The following corollary of Theorem 9.4 is concerned with the particular case of n distinguishable balls.

COROLLARY 9.4

The number of distributions of n distinguishable balls, into k distinguishable ordered urns, without any restriction, is given by

$$n! \binom{k + n - 1}{n} = (k + n - 1)_n, \tag{9.11}$$

and, with the restriction that r urns are occupied, is given by

$$n! \binom{k}{r} \binom{n - 1}{r - 1}. \tag{9.12}$$

Example 9.4 Priority lists
Suppose that n persons have applied for certain vacant positions in k organizations of the public sector. Each candidate is placed on only one of k different priority lists and each organization is filling its vacant positions from the corresponding priority list. Find the number of ways of placing the n candidates on the k priority lists.

The n candidates may be considered as n distinguishable balls and the k different priority lists as k distinguishable ordered urns. Then, the number of ways of placing the n candidates on the k priority lists without any restriction, according to (9.11), is given by

$$n! \binom{k + n - 1}{n} = (k + n - 1)_n,$$

and, with the restriction that each of r organizations should fill at least one position, according to (9.12), is given by

$$n! \binom{k}{r} \binom{n - 1}{r - 1}. \quad \square$$

9.4   BALLS OF GENERAL SPECIFICATION AND DISTINGUISHABLE URNS

In the introduction to this chapter, the balls of a general specification were classified according to the kind to which they belong and the notation $(r_1, r_2, \ldots, r_m)$, with $r_1 + r_2 + \cdots + r_m = n$, where r_i is the number of balls of the i-th kind, $i = 1, 2, \ldots, m$, was used. These balls may also be classified according to their multiplicity and the notation $[m_1, m_2, \ldots, m_n]$, with $m_1 + 2m_2 + \cdots + nm_n = n$, where $m_i \geq 0$ is the number of distinguishable kinds of balls, each including i like balls, $i = 1, 2, \ldots, n$, may be used. In this case, the partition notation $(1^{m_1} 2^{m_2} \cdots n^{m_n})$ may also be used.

THEOREM 9.5

The number of distributions of n balls of the specification $[m_1, m_2, \ldots, m_n]$, with $m_i \geq 0$, $i = 1, 2, \ldots, n$ and $m_1 + 2m_2 + \cdots + nm_n = n$, into k distinguishable urns is given by

$$U(m_1, m_2, \ldots, m_n; k) = \binom{k}{1}^{m_1} \binom{k+1}{2}^{m_2} \cdots \binom{k+n-1}{n}^{m_n}. \quad (9.13)$$

PROOF   Note first that the number of distributions of i like balls into k distinguishable urns equals $\binom{k+i-1}{i}$, $i = 1, 2, \ldots, n$. Further, by the multiplication principle, the number of distributions of $m_i \geq 0$ distinguishable kinds of balls, each including i like balls, into k distinguishable urns is given by $\binom{k+i-1}{i}^{m_i}$, $i = 1, 2, \ldots, n$. Applying, once more, the multiplication principle, we conclude that the number of distributions of n balls of the specification $[m_1, m_2, \ldots, m_n]$, with $m_i \geq 0$, $i = 1, 2, \ldots, n$ and $m_1 + 2m_2 + \cdots + nm_n = n$, into k distinguishable urns is given by the expression (9.13).   ∎

THEOREM 9.6

The number of distributions of n balls of the specification $[m_1, m_2, \ldots, m_n]$, with $m_i \geq 0$, $i = 1, 2, \ldots, n$ and $m_1 + 2m_2 + \cdots + nm_n = n$, into k distinguishable urns so that r urns are occupied is given by

$$R(m_1, m_2, \ldots, m_n; k, r) = \binom{k}{r} \sum_{j=0}^{r} (-1)^j \binom{r}{j} U(m_1, m_2, \ldots, m_n; r - j). \quad (9.14)$$

PROOF   Consider the set Ω of distributions of n balls of the specification $[m_1, m_2, \ldots, m_n]$ into k distinguishable urns and let A_i be the subset of these distributions in which the i-th cell remains empty, $i = 1, 2, \ldots, k$. Then, for any selection of s indices $\{i_1, i_2, \ldots, i_s\}$ out of the k indices $\{1, 2, \ldots, k\}$, according to Theorem 9.5,

$$\nu_s = N(A_{i_1} A_{i_2} \cdots A_{i_s}) = U(m_1, m_2, \ldots, m_n; k - s), \quad s = 1, 2, \ldots, k.$$

This expression implies the exchangeability of the sets $A_1, A_2, \ldots, A_k$. Further, the number $R(m_1, m_2, \ldots, m_n; k, r)$ of distributions of n balls of the specification $[m_1, m_2, \ldots, m_n]$ into k distinguishable urns, which leave $k - r$ urns empty (and consequently r urns occupied), equals the number $N_{k,k-r}$ of elements of Ω that are contained in $k - r$ among the k sets $A_1, A_2, \ldots, A_k$. Thus, according to Corollary 4.4, (9.14) is established. ∎

Some particular cases of (9.14), of special interest, have been already examined in Section 9.2. The case of n distinguishable balls corresponds to $m_1 = n$, $m_i = 0$, $i = 2, 3, \ldots, n$, while the case of n indistinguishable (like) balls corresponds to $m_i = 0$, $i = 1, 2, \ldots, n-1$, $m_n = 1$. Another particular case that corresponds to $m_s = m$, $m_i = 0$, $i = 1, 2, \ldots, s-1, s+1, \ldots, n$ is explicitly presented in the next corollary.

COROLLARY 9.5

The number of distributions of sm balls, which belong in m distinguishable kinds with s like balls of each kind, into k distinguishable urns so that r urns are occupied is given by

$$P(m, s, k, r) = \binom{k}{r} \sum_{j=0}^{r} (-1)^{r-j} \binom{r}{j} \binom{j + s - 1}{s}^m. \tag{9.15}$$

Let us now consider the case of the existence of restrictions in the capacity of the urns. Specifically, the placement of a ball into an urn excludes the possibility of placing any other like ball into it. Thus, the balls that are allowed to be placed in any urn are all distinguishable.

THEOREM 9.7

The number of distributions of n balls of the specification $[m_1, m_2, \ldots, m_n]$, with $m_i \geq 0$, $i = 1, 2, \ldots, n$ and $m_1 + 2m_2 + \cdots + nm_n = n$, into k distinguishable urns, each of which may accommodate at most one ball from each kind, is given by

$$V(m_1, m_2, \ldots, m_n; k) = \binom{k}{1}^{m_1} \binom{k}{2}^{m_2} \cdots \binom{k}{n}^{m_n}. \tag{9.16}$$

PROOF   Note first that, since each urn may accommodate at most one ball from each kind, the number of distributions of i like balls into k distinguishable urns equals $\binom{k}{i}$, $i = 1, 2, \ldots, n$. Further, by the multiplication principle, the number of distributions of $m_i \geq 0$ distinguishable kinds of balls, each including i like balls, into k distinguishable urns is given by $\binom{k}{i}^{m_i}$, $i = 1, 2, \ldots, n$. Applying, once more, the multiplication principle, we conclude that the number of distributions of n balls of the specification $[m_1, m_2, \ldots, m_n]$, with $m_i \geq 0$, $i = 1, 2, \ldots, n$ and $m_1 + 2m_2 + \cdots + nm_n = n$, into k distinguishable urns, each of which may accommodate at most one ball from each kind, is given by (9.16). ∎

Using the inclusion and exclusion principle and Theorem 9.7, the next theorem (analogous to Theorem 9.6) is deduced.

THEOREM 9.8

The number of distributions of n balls of the specification $[m_1, m_2, \ldots, m_n]$, with $m_i \geq 0$, $i = 1, 2, \ldots, n$ and $m_1 + 2m_2 + \cdots + nm_n = n$, into k distinguishable urns, each of which may accommodate at most one ball from each kind, so that r urns are occupied, is given by

$$Q(m_1, m_2, \ldots, m_n; k, r) = \binom{k}{r} \sum_{j=0}^{r} (-1)^j \binom{r}{j} V(m_1, m_2, \ldots, m_n; r - j).$$
$$(9.17)$$

COROLLARY 9.6

The number of distributions of sm balls, which belong in m distinguishable kinds with s like balls of each kind, into k distinguishable urns, each of which may accommodate at most one ball from each kind, so that r urns are occupied, is given by

$$Q(m, s, k, r) = \binom{k}{r} \sum_{j=0}^{r} (-1)^{r-j} \binom{r}{j} \binom{j}{s}^{m}.$$
$$(9.18)$$

Example 9.5 Formation of committees

Assume that, for the $sm = n$ positions of m committees, each with s positions, there are k candidates. Find the number of ways the n members of the committees can be selected so that r of the k candidates will participate in at least one committee.

The $sm = n$ positions of the m committees, each with s positions, may be considered as m distinguishable kinds of balls, each with s balls, and the k candidates as k distinguishable urns. Thus, the placement of a ball of the i-th kind into the j-th urn corresponds to the selection of the j-th candidate for a position of the i-th committee. We assume that no candidate can be assigned to more than one position of the same committee, but can participate in more than one committee. This assumption implies that each urn may accommodate at most one ball from each kind. Consequently, according to Corollary 9.6, the number of ways the n members of the committees can be selected so that r of the k candidates will participate in at least one committee is given by

$$Q(m, s, k, r) = \binom{k}{r} \sum_{j=0}^{r} (-1)^{r-j} \binom{r}{j} \binom{j}{s}^{m}.$$

In the particular case when each of the k candidates must participate in at least one committee, the number of ways the n members of the committees can be selected is given by

$$Q(m, s, k) = \sum_{j=0}^{k} (-1)^{k-j} \binom{k}{j} \binom{j}{s}^{m}. \qquad \square$$

Example 9.6 Formation of committees from grouped candidates

Assume that each of k different political groups submits a full priority list of n candidates to the president's office for the $sm = n$ positions of m committees, each with s positions. Find the number of ways the n members of the committees can be selected so that r of the k political groups will be represented in at least one committee.

As in the preceding example, the $sm = n$ positions of the m committees, each with s positions, may be considered as m distinguishable kinds of balls, each with s balls, and the k political groups as k distinguishable urns. In this case, there is no restriction on the number of balls of the same kind that each urn may accommodate. Consequently, according to Corollary 9.5, the number of ways the n members of the committees can be selected so that r of the k political groups will be represented in at least one committee is given by

$$P(m, s, k, r) = \binom{k}{r} \sum_{j=0}^{r} (-1)^{r-j} \binom{r}{j} \binom{j+s-1}{s}^{m}.$$

In the particular case when each of the k political groups must be represented in at least one committee, the number of ways the n members of the committees can be selected is given by

$$P(m, s, k) = \sum_{j=0}^{k} (-1)^{k-j} \binom{k}{j} \binom{j+s-1}{s}^{m}. \qquad \square$$

9.5 GENERATING FUNCTIONS

The use of generating functions in the study of distributions and occupancy leads to a more extensive development of this subject. In this section, we present the generating functions of the cases of distinguishable or indistinguishable balls and distinguishable urns, as well as the case of balls of a general specification and distinguishable urns.

Consider first the case of distinguishable balls and urns. Let us attach the occupancy indicator x_j to the j-th urn, $j = 1, 2, \ldots, k$, so that the homogeneous product of order n, $x_1^{r_1} x_2^{r_2} \cdots x_k^{r_k}$, with $r_1 + r_2 + \cdots + r_k = n$, uniquely determines the distribution of n distinguishable balls into k distinguishable urns, with r_j balls in the j-th urn, $j = 1, 2, \ldots, k$. Then the indicator (generating function) of the distributions of a single ball into k distinguishable urns is

$$g_1(x_1, x_2, \ldots, x_k) = x_1 + x_2 + \cdots + x_k$$

and hence the generating function of the distributions of n distinguishable balls into k distinguishable urns, with respect to the numbers of balls in the urns, is

$$g_n(x_1, x_2, \ldots, x_k) = (x_1 + x_2 + \cdots + x_k)^n. \tag{9.19}$$

Note that (9.19) expanded into powers of $x_1, x_2, \ldots, x_k$ by using the multinomial theorem,

$$\sum \frac{n!}{r_1! r_2! \cdots r_k!} x_1^{r_1} x_2^{r_2} \cdots x_k^{r_k} = (x_1 + x_2 + \cdots + x_k)^n,$$

gives the number of distributions of n distinguishable balls into k distinguishable urns with r_j balls in the j-th urn, $j = 1, 2, \ldots, k$, $r_1 + r_2 + \cdots + r_k = n$.

The exponential generating function of the sequence $g_n(x_1, x_2, \ldots, x_k)$, $n = 0, 1, \ldots$, of generating functions (9.19) is

$$G(t; x_1, x_2, \ldots, x_k) = \sum_{n=0}^{\infty} g_n(x_1, x_2, \ldots, x_k) \frac{t^n}{n!}$$
$$= \exp\{(x_1 + x_2 + \cdots + x_k)t\}$$

Therefore, the enumerator for occupancy of k distinguishable urns is given by

$$G(t; x_1, x_2, \ldots, x_k) = E(t; x_1) E(t; x_2) \cdots E(t; x_k), \tag{9.20}$$

where the enumerator for occupancy of the j-th urn is given by

$$E(t; x_j) = 1 + x_j t + \frac{x_j^2 t^2}{2!} + \cdots = e^{x_j t}, \quad j = 1, 2, \ldots, k, \tag{9.21}$$

exactly as in the case of permutations with repetition (see Section 6.3). The occupancy for each urn is specified independently of the other urns. Thus, if the j-th urn must contain at least r and at most s balls, the enumerator for its occupancy is

$$E_{r,s}(t; s_j) = \frac{x_j^r t^r}{r!} + \frac{x_j^{r+1} t^{r+1}}{(r+1)!} + \cdots + \frac{x_j^s t^s}{s!} \tag{9.22}$$

and (9.20) is accordingly modified.

In the case of n indistinguishable balls, there exists only one distribution of them into k distinguishable urns so that the j-th urn contains r_j balls, for $j = 1, 2, \ldots, k$ and $r_1 + r_2 + \cdots + r_k = n$. Hence, the generating function of the distributions of n indistinguishable balls into k distinguishable urns, with respect to the numbers of balls in the urns, is given by

$$h_n(x_1, x_2, \ldots, x_k) = \sum x_1^{r_1} x_2^{r_2} \cdots x_k^{r_k}, \tag{9.23}$$

where the summation is extended over all $r_j \geq 0$, $j = 1, 2, \ldots, k$ with $r_1 + r_2 + \cdots + r_k = n$. The function $h_n(x_1, x_2, \ldots, x_k)$ is called *homogeneous product sum* of weight k. Note that the corresponding generating function of the distributions of n distinguishable balls is given by

$$g_n(x_1, x_2, \ldots, x_k) = [h_1(x_1, x_2, \ldots, x_k)]^n,$$

where $h_1(x_1, x_2, \ldots, x_k) = x_1 + x_2 + \cdots + x_k$.

The ordinary generating function of the sequence $h_n \equiv h_n(x_1, x_2, \ldots, x_k)$, $n = 0, 1, \ldots$, of generating functions (9.23), since

$$\sum_{n=0}^{\infty} \sum x_1^{r_1} x_2^{r_2} \cdots x_k^{r_k} t^n = \sum_{r_1=0}^{\infty} (x_1 t)^{r_1} \sum_{r_2=0}^{\infty} (x_2 t)^{r_2} \cdots \sum_{r_k=0}^{\infty} (x_k t)^{r_k},$$

is deduced as

$$H(t; x_1, x_2, \ldots, x_k) = \sum_{n=0}^{\infty} h_n(x_1, x_2, \ldots, x_k) t^n$$

$$= \frac{1}{(1 - x_1 t)(1 - x_2 t) \cdots (1 - x_k t)}. \tag{9.24}$$

Therefore, the enumerator for occupancy of k distinguishable urns is given by

$$H(t; x_1, x_2, \ldots, x_k) = F(t; x_1) F(t; x_2) \cdots F(t; x_k), \tag{9.25}$$

where the enumerator for occupancy of the j-th urn is given by

$$F(t; x_j) = 1 + x_j t + x_j^2 t^2 + \cdots = (1 - x_j t)^{-1}, \quad j = 1, 2, \ldots, k, \tag{9.26}$$

exactly as in the case of combinations with repetition (see Section 6.3). The occupancy for each urn is specified independently of the other urns. Thus, if the j-th urn must contain at least r and at most s balls, the enumerator for its occupancy is

$$F_{r,s}(t; x_j) = x_j^r t^r + x_j^{r+1} t^{r+1} + \cdots + x_j^s t^s \qquad (9.27)$$

and (9.25) is accordingly modified.

Let us finally consider the general case of n balls of the specification $[m_1, m_2, \ldots, m_n]$, with $m_1 + 2m_2 + \cdots + nm_n = n$, where $m_i \geq 0$ is the number of balls of multiplicity i (that is, the number of distinguishable kinds of balls, each including i like balls), $i = 1, 2, \ldots, n$. The generating function of the distributions of i indistinguishable (like) balls into k distinguishable urns, with respect to the numbers of balls in the urns, according to (9.23), is given by $h_i(x_1, x_2, \ldots, x_k)$, the homogeneous product sum of weight i. Since the $m_i \geq 0$ distinguishable kinds of balls, each of which includes i like balls, are independently distributed, the generating function of the distributions of these balls is given by $[h_i(x_1, x_2, \ldots, x_k)]^{m_i}$, $i = 1, 2, \ldots, n$. Therefore, the generating function of the distributions of n balls of the specification $[m_1, m_2, \ldots, m_n]$, $m_i \geq 0$, $i = 1, 2, \ldots, n$, $m_1 + 2m_2 + \cdots + nm_n = n$, into k distinguishable urns, with respect to the numbers of balls in the urns, is given by

$$h_{m_1, m_2, \ldots, m_n}(x_1, x_2, \ldots, x_k) = \prod_{i=1}^{n} [h_i(x_1, x_2, \ldots, x_k)]^{m_i}. \qquad (9.28)$$

Note that (9.28), for $x_j = 1$, $j = 1, 2, \ldots, k$, yields the total number of distributions of the n balls of the specification $[m_1, m_2, \ldots, m_n]$ into k distinguishable urns. The fact that this number agrees with that of Theorem 9.5 follows from the expression $h_i(1, 1, \ldots, 1) = \binom{k+i-1}{i}$, $i = 1, 2, \ldots, n$.

The enumerator for occupancy of the j-th cell by n balls of the specification $[m_1, m_2, \ldots, m_n]$, with $m_i \geq 0$, $i = 1, 2, \ldots, n$, $m_1 + 2m_2 + \cdots + nm_n = n$ and $m_1 + m_2 + \cdots + m_n = m$, where m is the total number of distinguishable kinds of balls, is given by

$$F(t_1, t_2, \ldots, t_m; x_j) = \frac{1}{(1 - x_j t_1)(1 - x_j t_2) \cdots (1 - x_j t_m)},$$

where the term $(x_j t_1)^{\nu_1} (x_j t_2)^{\nu_2} \cdots (x_j t_m)^{\nu_m}$ indicates the occupancy of the j-th urn by $\nu = \nu_1 + \nu_2 + \cdots + \nu_m$ balls, among which ν_l are of the l-th kind, $l = 1, 2, \ldots, m$. The occupancy of each urn is specified independently of the other urns and so the enumerator for occupancy of the k urns is given by

$$H(t_1, t_2, \ldots, t_m; x_1, x_2, \ldots, x_k)$$
$$= F(t_1, t_2, \ldots, t_m; x_1) F(t_1, t_2, \ldots, t_m; x_2) \cdots F(t_1, t_2, \ldots, t_m; x_k).$$

Two illustrative examples follow.

Example 9.7
The generating function $G(t)$, of the number of distributions of n distinguishable balls into k distinguishable urns, without any restriction, is deduced from (9.20) and (9.21) by setting $x_j = 1, j = 1, 2, \ldots, k$. Then,

$$E(t; 1) = e^t, \quad G(t) \equiv G(t; 1, 1, \ldots, 1) = e^{kt} = \sum_{n=0}^{\infty} k^n \frac{t^n}{n!},$$

and hence the number of distributions of n distinguishable balls into k distinguishable urns equals

$$U(k, n) = k^n.$$

In the case of the restriction that each urn should contain at least one ball, it follows from (9.20) and (9.22) that

$$E_1(t; 1) = t + \frac{t^2}{2!} + \cdots = e^t - 1, \quad G_1(t) \equiv G_1(t; 1, 1, \ldots, 1) = (e^t - 1)^k.$$

Expanding the generating function $G_1(t)$ into powers of t by using Newton's binomial theorem, we find

$$G_1(t) = \sum_{j=0}^{k} (-1)^j \binom{k}{j} e^{(k-j)t} = \sum_{j=0}^{k} (-1)^j \binom{k}{j} \sum_{n=0}^{\infty} (k-j)^n \frac{t^n}{n!}$$

$$= \sum_{n=0}^{\infty} \left\{ \sum_{j=0}^{k} (-1)^j \binom{k}{j} (k-j)^n \right\} \frac{t^n}{n!}.$$

Therefore, the number of distributions of n distinguishable balls into k distinguishable urns so that all k urns are occupied is given by

$$N(n, k) = \sum_{j=0}^{k} (-1)^j \binom{k}{j} (k-j)^n = k! S(n, k),$$

in agreement with (9.1) in the particular case of $r = k$. $\quad\Box$

Example 9.8
The generating function, $H(t)$, of the number of distributions of n indistinguishable balls into k distinguishable urns, without any restriction, is deduced from (9.25) and (9.26) by setting $x_j = 1, j = 1, 2, \ldots, k$. Then

$$F(t; 1) = (1 - t)^{-1},$$

$$H(t) \equiv H(t; 1, 1, \ldots, 1) = (1 - t)^{-k} = \sum_{n=0}^{\infty} (-1)^n \binom{-k}{n} t^n,$$

and hence the number of distributions of n indistinguishable balls into k distinguishable urns equals

$$(-1)^n \binom{-k}{n} = \binom{k+n-1}{n},$$

in agreement with the results stated in Section 9.2. In the case of the restriction that each urn should contain at least one ball, it follows from (9.25) and (9.27) that

$$F_1(t; 1) = t + t^2 + \cdots = t(1-t)^{-1}, \quad H_1(t) \equiv H_1(t; 1, 1, \ldots, 1) = t^k(1-t)^{-k}.$$

Expanding the generating function $H_1(t)$ into powers of t by using Newton's negative binomial theorem, we find

$$G_1(t) = \sum_{j=0}^{\infty} (-1)^j \binom{-k}{j} t^{k+j} = \sum_{j=0}^{\infty} \binom{k+j-1}{k-1} t^{k+j} = \sum_{n=k}^{\infty} \binom{n-1}{k-1} t^n.$$

Therefore, the number of distributions of n indistinguishable balls into k distinguishable urns, so that all k urns are occupied, is given by

$$\binom{n-1}{k-1}.$$

Further applications may be found in the exercises. $\square$

9.6   BIBLIOGRAPHIC NOTES

P. A. MacMahon (1915, 1916) in his treatise on combinatorial analysis introduced and extensively studied, mostly through generating functions, the theory of distribution of balls (objects) into urns (cells, boxes). Theorem 9.5, which is concerned with the distributions of balls of a general specification into distinguishable urns, was derived by the aid of generating functions. Also, Section 9.5 on the study of distributions through enumerating generating function is based on the books of P. A. MacMahon (1915, 1916) and J. Riordan (1958). W. Feller (1968) treated the classical occupancy problem and the coupon collector's problem by using the inclusion and exclusion principle. N. L. Johnson and S. Kotz (1977) devoted a chapter to the occupancy theory containing a wealth of information and an extensive list of references.

9.7   EXERCISES

1. Let $N_j(n, k, r)$ be the number of distributions of n distinguishable balls into k distinguishable urns, so that r urns are occupied, each by j balls. Show that

$$N_j(n, k, r) = \binom{k}{r} \sum_{i=r}^{k} (-1)^{i-r} \binom{k-r}{i-r} \frac{n!}{(j!)^i(n-ij)!} (k-i)^{n-ij}.$$

2. (*Continuation*). (a) Show that

$$\sum_{n=0}^{\infty} N_j(n, k, r) \frac{t^n}{n!} = \binom{k}{r} \left(\frac{t^j}{j!}\right)^r \left(e^t - \frac{t^j}{j!}\right)^{k-r}$$

and hence that

$$\sum_{n=0}^{\infty} \sum_{r=0}^{k} N_j(n, k, r) x^r \frac{t^n}{n!} = \left(e^t + (x-1)\frac{t^j}{j!}\right)^k.$$

(b) If $\mu_{(i)} \equiv \mu_{(i)}(n, k, j)$, $i = 0, 1, \dots, k$, are the factorial moments of the sequence $a_r \equiv N_j(n, k, r)$, $r = 0, 1, \dots, k$, show that

$$\mu_{(i)}(n, k, j) = (k)_i \frac{n!}{(j!)^i(n-ij)!} (k-i)^{n-ij}, \quad i = 0, 1, \dots, k.$$

3. Let $B(n, k)$ be the number of distributions of n distinguishable balls into k indistinguishable urns. Show that

$$B(n, k) = \frac{1}{k!} \sum_{j=0}^{k} \binom{k}{j} j^n D_{k-j},$$

where

$$D_k = k! \sum_{i=0}^{k} \frac{(-1)^i}{i!}$$

is the k-subfactorial.

4. Let $Q_j(n, k, r)$ be the number of distributions of n indistinguishable balls into k distinguishable urns, so that r urns are occupied, each by j balls. Show that

$$Q_j(n, k, r) = \binom{k}{r} \sum_{i=r}^{k} (-1)^{i-r} \binom{k-r}{i-r} \binom{k+n-i(j+1)-1}{n-ij}.$$

5. (*Continuation*). (a) Show that

$$\sum_{n=0}^{\infty} Q_j(n,k,r)t^n = \binom{k}{r} t^{jr}[(1-t)^{-1} - t^j]^{k-r}$$

and hence that

$$\sum_{n=0}^{\infty}\sum_{r=0}^{k} Q_j(n,k,r)x^r t^n = [(1-t)^{-1} + (x-1)t^j]^k.$$

(b) If $b_{(i)} \equiv b_{(i)}(n,k,j)$, $i = 0,1,\ldots,k$, are the binomial moments of the sequence $a_r \equiv Q_j(n,k,r)$, $r = 0,1,\ldots,k$, show that

$$b_{(i)}(n,k,j) \equiv \binom{k}{i}\binom{k+n-i(j+1)-1}{n-ij}, \quad i = 0,1,\ldots,k.$$

6. Show that the number of distributions of n indistinguishable balls into k distinguishable urns, each of capacity limited to s balls, is

$$L(n,k,s) = \sum_{j=0}^{k}(-1)^j \binom{k}{j}\binom{k+n-j(s+1)-1}{k-1},$$

and, with the restriction that r urns remain empty, equals

$$L(n,k,s,r) = \binom{k}{r}\sum_{i=r}^{k}(-1)^{i-r}\binom{k-r}{i-r}L(n,k-i,s).$$

7. (*Continuation*) Let $L_j(n,k,r)$ be the number of distributions of n indistinguishable balls into k distinguishable urns, each of capacity limited to s balls so that each of r urns is occupied by j balls. Show that

$$L_j(n,k,s,r) = \binom{k}{r}\sum_{i=r}^{k}(-1)^{i-r}\binom{k-r}{i-r}L(n-ij,k-i,s)$$

and, for $j = s$, conclude that

$$L_s(n,k,s,r) = \binom{k}{r}L(n-sr,k-r,s).$$

8. (*Continuation*). (a) Show that

$$\sum_{n=0}^{s(k-r)+jr} L_j(n,k,s,r)t^n = \binom{k}{r}t^{jr}[(1-t^{s+1})(1-t)^{-1} - t^j]^{k-r}$$

and hence that

$$\sum_{n=0}^{sk} \sum_{r=0}^{k} Q_j(n, k, r) x^r t^n = [(1 - t^{s+1})(1 - t)^{-1} + (x - 1)t^j]^k.$$

(b) If $b_{(i)} \equiv b_{(i)}(n, k, s, j)$, $i = 0, 1, \ldots, k$, are the binomial moments of the sequence $a_r \equiv L_j(n, k, s, r)$, $r = 0, 1, \ldots, k$, show that

$$b_{(i)}(n, k, s, j) = \binom{k}{i} L(n - ij, k - i, s), \quad i = 0, 1, \ldots, k.$$

9. (*Continuation*). Let $M_j(n, k, s)$ be the number of distributions of n indistinguishable balls into k distinguishable urns, each of capacity limited to s balls, so that each urn is occupied by at least j balls. Show that

$$M_j(n, k, s) = L(n - kj, k, s - j)$$
$$= \sum_{i=0}^{k} (-1) \binom{k}{i} \binom{n - kj + k - i(s - j + 1) - 1}{k - 1}.$$

10. Let n like balls be distributed into k distinguishable urns. (a) If $h_n(x)$ is the enumerator for the occupancy of a single urn, show that

$$H(t; x) = \sum_{n=0}^{\infty} h_n(x) t^n = (1 - xt)^{-1} (1 - x)^{-k+1}$$

and

$$h_n(x) = \sum_{r=0}^{n} \binom{n + k - r - 2}{n - r} x^r = \sum_{s=0}^{n} \binom{n + k - 1}{n - s} (x - 1)^s.$$

(b) If $B(n, k, r)$ is the number of distributions of n like balls into k distinguishable urns so that a specified urn contains r balls, $r = 0, 1, \ldots, n$ and $b_{(s)} \equiv b_{(s)}(n, k)$, $s = 0, 1, \ldots, n$, are the binomial moments of the sequence $a_r \equiv B(n, k, r)$, $r = 0, 1, \ldots, n$, conclude that

$$B(n, k, r) = \binom{n + k - r - 2}{n - r}, \quad b_{(s)}(n, k) = \binom{n + k - 1}{n - s}.$$

11. (*Continuation*). If $h_n(x, y)$ is the enumerator for the occupancy of a pair of urns, show that

$$H(t; x, y) = \sum_{n=0}^{\infty} h_n(x, y) t^n = (1 - xt)^{-1} (1 - yt)^{-1} (1 - x)^{-k+2}$$

and hence conclude that

$$h_n(x,y) = \sum_{s=0}^{n} \sum_{r=0}^{n-s} \binom{n+k-r-s-3}{n-r-s} x^r y^s$$

$$= \sum_{j=0}^{n} \sum_{i=0}^{n-j} \binom{n+k-1}{n-i-j} (x-1)^i (y-1)^j.$$

12. Let n like balls be distributed into k distinguishable urns, each with s distinguishable cells of capacity limited to one ball. Show that (a) the enumerator for occupancy of the j-th urn is

$$F_s(t; x_j) = \sum_{r=0}^{s} \binom{s}{r} x_j^r t^r = (1 + x_j t)^s, \quad j = 1, 2, \dots, k$$

and (b) the generating function of the distributions with r_j balls in the j-th urn, $j = 1, 2, \dots, k$, is

$$h_{n,s}(x_1, x_2, \dots, x_k) = \sum \binom{x}{r_1} \binom{s}{r_2} \cdots \binom{s}{r_k} x_1^{r_1} x_2^{r_2} \cdots x_k^{r_k},$$

where the summation is extended over all $0 \le r_j \le s$, $j = 1, 2, \dots, k$, with $r_1 + r_2 + \cdots + r_k = n$.

13. (*Continuation*). In the case of cells of unlimited capacity, show that (a) the enumerator for occupancy of the j-th urn is

$$G_s(t; s_j) = \sum_{r=0}^{\infty} \binom{s+r-1}{r} x_j^r t^r$$

$$= \sum_{r=0}^{\infty} (-1)^r \binom{-s}{r} x_j^r t^r = (1 - x_j t)^{-s}, \quad j = 1, 2, \dots, k$$

and (b) the generating function of the distributions with r_j balls in the j-th urn, $j = 1, 2, \dots, k$, is

$$g_{n,s}(x_1, x_2, \dots, x_k)$$
$$= \sum \binom{s+r_1-1}{r_1} \binom{s+r_2-1}{r_2} \cdots \binom{s+r_k-1}{r_k} x_1^{r_1} x_2^{r_2} \cdots x_k^{r_k},$$

where the summation is extended over all $r_j \ge 0$, $j = 1, 2, \dots, k$, with $r_1 + r_2 + \cdots + r_k = n$.

14. Let us consider k distinguishable urns, each with s distinguishable cells of capacity limited to one ball. Show that the number of distributions of n like balls into the k urns so that each of r urns contains j balls equals

$$R_j(n, k, s, r) = \binom{k}{r} \sum_{i=r}^{k} (-1)^{i-r} \binom{k-r}{i-r} \binom{s}{j}^i \binom{s(k-i)}{n-ji}$$

and conclude that the number of distributions of n like balls into the k urns so that r urns are completely occupied is given by

$$R(n, k, s, r) = \binom{k}{r} \sum_{i=r}^{k} (-1)^{i-r} \binom{k-r}{i-r} \binom{s(k-i)}{sk-n}.$$

15. (*Continuation*). (a) Show that

$$\sum_{n=0}^{s(k-r)+jr} R_j(n, k, s, r) t^n = \binom{k}{r} \left[\binom{s}{j} t^j \right]^r \left[(1+t)^s - \binom{s}{j} t^j \right]^{k-r}$$

and hence that

$$\sum_{n=0}^{sk} \sum_{r=0}^{k} R_j(n, k, r) x^r t^n = \left[(1+t)^s + (x-1) \binom{s}{j} t^j \right]^k.$$

16. Let us consider k distinguishable urns, each with s distinguishable cells of unlimited capacity. Let $T_j(n, k, s, r)$ be the number of distributions of n like balls into the k urns so that each of r urns contains j balls. Show that

$$T_j(n, k, s, r)$$
$$= \binom{k}{r} \sum_{i=r}^{k} (-1)^{i-r} \binom{k-r}{i-r} \binom{s+j-1}{j}^i \binom{s(k-i)+n-ij-1}{n-ji}.$$

17. (*Continuation*). (a) Show that

$$\sum_{n=0}^{\infty} T_j(n, k, s, r) t^n$$
$$= \binom{k}{r} \left[\binom{s+j-1}{j} t^j \right]^r \left[(1-t)^{-s} - \binom{s+j-1}{j} t^j \right]^{k-r}$$

and hence that

$$\sum_{n=0}^{\infty} \sum_{r=0}^{k} T_j(n, k, r) x^r t^n = \left[(1-t)^{-s} + (x-1) \binom{s+j-1}{j} t^j \right]^k.$$

18. Let us consider kr distinguishable cells, each of capacity limited to one ball, arranged in k columns and r rows. Show that the number of distributions of n like balls into the kr cells so that j from s predetermined columns of cells are occupied (each by at least one ball) equals

$$R(n,k,s,r,j) = \binom{s}{j} \sum_{i=0}^{j} (-1)^i \binom{j}{i} \binom{r(k-s+j-i)}{n}.$$

19. (*Continuation*). Show that the number of distributions of n like balls into the kr cells so that no row and no column of cells remain empty equals

$$A(n,k,r) = \sum_{i=0}^{r} \sum_{j=0}^{k} (-1)^{k+r-j-i} \binom{k}{j} \binom{r}{i} \binom{ij}{n}$$

and conclude that the number of distributions of n distinguishable balls into the kr cells so that no row and no column of cells remain empty equals

$$n!A(n,k,r) = \sum_{i=0}^{r} \sum_{j=0}^{k} (-1)^{k+r-j-i} \binom{k}{j} \binom{r}{i} (ij)_n.$$

20. Let us consider kr distinguishable cells of unlimited capacity, arranged in k columns and r rows. Show that the number of distributions of n like balls into the kr cells so that j from s predetermined columns of cells are occupied (each by at least one ball) equals

$$T(n,k,s,r,j) = \binom{s}{j} \sum_{i=0}^{j} (-1)^i \binom{j}{i} \binom{r(k-s+j-i)+n-1}{n}.$$

21. (*Continuation*). Show that the number of distributions of n like balls into the kr cells so that no row and no column of cells remain empty equals

$$B(n,k,r) = \sum_{i=0}^{r} \sum_{j=0}^{k} (-1)^{k+r-j-i} \binom{k}{j} \binom{r}{i} \binom{ij+n-i}{n}.$$

22. Let $M(n,k,s)$ be the number of distributions of n distinguishable balls into k distinguishable urns, each of capacity limited to s balls. Show that

$$M_k(t) = \sum_{n=0}^{sk} M(n,k,s)\frac{t^n}{n!} = \left(1 + t + \frac{t^2}{2!} + \cdots + \frac{t^s}{s!}\right)^k$$

and

$$M(n+1,k,s) = k\left\{M(n,k,s) - \binom{n}{s}M(n-s,k-1,s)\right\}, \quad n \geq s,$$

$$M(n,k,s) = k^n, \quad n \leq s.$$

Further, derive the vertical recurrence relation

$$M(n,k,s) = \sum_{j=0}^{m}\binom{n}{j}M(n-j,k-1,s), \quad m = \min\{n,s\}.$$

23. Let $S_r(n,k)$ be the number of distributions of n distinguishable balls into k distinguishable urns so that each urn contains at least r balls. Show that

$$S_{k,r}(t) = \sum_{n=rk}^{\infty} S_r(n,k)\frac{t^n}{n!} = \left(e^t - 1 - t - \cdots - \frac{t^{r-1}}{(r-1)!}\right)^k$$

and

$$S_r(n+1,k) = k\left\{S_r(n,k) + \binom{n}{r-1}S_r(n-r+1,k-1)\right\}, \quad n \geq rk,$$

$$S_r(n,k) = 0, \quad n < rk, \quad S_r(rk,k) = \frac{(rk)!}{(r!)^k}.$$

Further, derive the recurrence relation

$$S_{r+1}(n,k) = \sum_{j=0}^{k}(-1)^j\binom{k}{j}\frac{(n)_{rj}}{(r!)^j}S_r(n-rj,k-j).$$

24. Let us consider k distinguishable urns, each with s distinguishable cells of capacity limited to one ball. Let $G(n,k,s,r)$ be the number of distributions of n like balls into the k urns so that each urn contains at most r balls. Show that

$$G_k(t) = \sum_{n=0}^{rk}G(n,k,s,r)t^n = \left[1 + \binom{s}{1}t + \binom{s}{2}t^2 + \cdots + \binom{s}{r}t^r\right]^k$$

and

$$(n+1)G(n+1,k,s,r) = (sk-n)G(n,k,s,r)$$
$$-sk\binom{s-1}{r}G(n-r,k-1,s,r), \quad n \geq r,$$

$$G(n, k, s, r) = \binom{sk}{n}, \quad n \le r.$$

Further, derive the vertical recurrence relation

$$G(n, k, s, r) = \sum_{j=0}^{m} \binom{s}{j} G(n - j, k - 1, s, r), \quad m = \min\{n, s, r\}.$$

25. (*Continuation*). Let $C_r(n, k, s)$ be the number of distributions of n like balls into the k urns so that each urn contains at least r balls. Show that

$$G_{k,r}(t) = \sum_{n=rk}^{sk} C_r(n, k, s) t^n = \left[(1 + t)^s - 1 - \binom{s}{1} t - \cdots - \binom{s}{r-1} t^{r-1} \right]^k$$

and

$$(n + 1) C_r(n + 1, k, s) = (sk - n) C_r(n, k, s)$$
$$+ sk \binom{s-1}{r-1} C_r(n - r + 1, k - 1, s), \quad rk < n < sk,$$

$$C_r(n, k, s) = 0, \quad n < rk, \quad C_r(rk, k, s) = \binom{s}{r}^k,$$

$$C_r(sk, k, s) = 1, \quad C_r(n, k, s) = 1, \quad n > sk.$$

Further, derive the recurrence relation

$$C_{r+1}(n, k, s) = \sum_{j=0}^{k} (-1)^j \binom{k}{j} \binom{s}{r}^j C_r(n - rj, k - j, s).$$

26. Let us consider k distinguishable urns, each with s distinguishable cells of unlimited capacity. Let $H(n, k, s, r)$ be the number of distributions of n like balls into the k urns so that each urn contains at most r balls. Show that

$$H_k(t) = \sum_{n=0}^{rk} H(n, k, s, r) t^n$$

$$= \left[1 + \binom{s}{1} t + \binom{s+1}{2} t^2 + \cdots + \binom{s+r-1}{r} t^r \right]^k$$

and

$$(n + 1) H(n + 1, k, s, r) = (sk + n) H(n, k, s, r)$$
$$- sk \binom{s+r}{r} H(n - r, k - 1, s, r), \quad n \ge r,$$

$$H(n,k,s,r) = \binom{sk+n-1}{n}, \quad n \le r.$$

Further, derive the vertical recurrence relation

$$H(n,k,s,r) = \sum_{j=0}^{m} \binom{s+j-1}{j} H(n-j,k-1,s,r), \quad m = \min\{n,r\}.$$

27. (*Continuation*). Let $D_r(n,k,s)$ be the number of distributions of n like balls into the k urns so that each urn contains at least r balls. Show that

$$D_{k,r}(t) = \sum_{n=k}^{\infty} D_r(n,k,s)t^n$$

$$= \left[(1-t)^{-s} - 1 - \binom{s}{1}t - \cdots - \binom{s+r-2}{r-1}t^{r-1} \right]^k$$

and

$$(n+1)D_r(n+1,k,s) = (sk+n)D_r(n,k,s)$$
$$+ sk\binom{s+r-1}{r-1}D_r(n-r+1,k-1,s), \quad n > rk,$$

$$D_r(n,k,s) = 0, \quad n < rk, \quad D_r(rk,k,s) = \binom{s+r-1}{r}^k.$$

Further, derive the recurrence relation

$$D_{r+1}(n,k,s) = \sum_{j=0}^{k}(-1)^j \binom{k}{j}\binom{s+r-1}{r}D_r(n-rj,k-j,s).$$

28. Let us consider $n+r$ balls among which n are distinguishable and the other r are indistinguishable. Let $U(n,r,k)$ be the number of distributions of these balls into k distinguishable urns without any restriction and $W(n,r,k)$ the corresponding number with the restriction that no urn remains empty. Show that

$$U(n,r,k) = \binom{k+r-1}{r}k^n$$

and

$$W(n,r,k) = \sum_{j=0}^{k}(-1)^j \binom{k}{j}\binom{k-j+r-1}{r}(k-j)^n.$$

29. (*Continuation*). Show combinatorially that

$$W(n,r,k) = \sum_{j=1}^{m} \binom{r+j-1}{k-1}(k)_j S(n,j), \quad m = \min\{n,k\},$$

where

$$S(n,j) = \frac{1}{j!}\sum_{i=0}^{j}(-1)^{j-1}\binom{j}{i}i^n$$

is the Stirling number of the second kind. Also, show that

$$W(n,r,k) = k\{W(n-1,r,k) + W(n-1,r,k-1)\},$$

$$W(n,0,k) = k^n, \quad W(0,r,k) = \binom{r+k-1}{k}, \quad W(n,r,k) = 0, \quad k > n+r$$

30. (*Continuation*). Show that

$$\sum_{r=0}^{\infty}\sum_{n=0}^{\infty} U(n,r,k)\frac{t^n}{n!}x^r = (1-x)^{-k}e^{kt}$$

and

$$\sum_{r=0}^{\infty}\sum_{n=0}^{\infty} W(n,r,k)\frac{t^n}{n!}x^r = [(1-x)^{-1}e^t - 1]^k = (1-x)^{-k}[(e^t - 1) + x]^k.$$

Further, conclude that

$$W(n,r,k) = \sum_{j=1}^{m} \binom{r+j-1}{k-1}(k)_j S(n,j), \quad m = \min\{n,k\}.$$

Chapter 10

PARTITIONS OF INTEGERS

10.1 INTRODUCTION

The theory of partitions of integers, established by Euler in the 18th century and enhanced by Hardy, Ramanujan and Rademacher, belongs to both combinatorics and number theory. As regards the combinatorial aspect of partitions, note first that the positive integer solutions of the linear equation $x_1 + x_2 + \cdots + c_k = n$ with n a positive integer, enumerated in Chapter 2, are ordered solutions. Further, the enumeration of the corresponding non-ordered solutions, the partitions of the integer n, is of great interest. Note also that, the enumeration of distributions of indistinguishable balls into indistinguishable urns is merely a partition enumeration problem.

In this chapter, after the introduction of the notion of a partition of a positive integer n, recurrence relations and generating functions for the total number of partitions of n and the numbers of partitions of n into k and into at most k parts are derived. Then, a universal generating function for the number of partitions of n into parts of specified or unspecified number, whose number of parts of any specific size belongs to a subset of non-negative integers, is obtained. As applications of this generating function several interesting sequences of partition numbers are presented. Furthermore, relations connecting various partition numbers are deduced by using their generating functions. Also, after introducing the Ferrers diagram of a partition and the notion of a conjugate (and a self-conjugate) partition, additional interrelations among certain partition numbers are derived. Euler's pentagonal theorem on the difference of the number of partitions of n into an even number of unequal parts and the number of partitions of n into an odd number of unequal parts is obtained. The last section is devoted to the derivation of combinatorial identities; the Euler and Gauss-Jacobi identities are deduced. As a complement to this subject, a collection of exercises on q-numbers, q-factorials and q-binomial coefficients, as well as on q-Stirling numbers of the first and second kind, is provided.

10.2   RECURRENCE RELATIONS AND GENERATING FUNCTIONS

DEFINITION 10.1   *A partition of a positive integer n is a non-ordered collection of positive integers whose sum equals n. These integers (terms of the sum) are called summands or parts of the partition.*

In the case of a partition of n into a specified number of parts, say k, then the term *partitions of n into k parts* is used. Since the order of parts in a partition does not count, they are registered in decreasing order of magnitude. Thus, a partition of n into k parts is a solution in positive integers of the linear equation:

$$x_1 + x_2 + \cdots + x_k = n, \quad x_1 \geq x_2 \geq \cdots \geq x_k \geq 1, \tag{10.1}$$

In a partition of n, let $y_i \geq 0$ be the number of parts that are equal to i, for $i = 1, 2, \ldots, n$. Then

$$y_1 + 2y_2 + \cdots + ny_n = n, \quad y_i \geq 0, \quad i = 1, 2, \ldots, n \tag{10.2}$$

In the case of a partition of n into k parts, in addition,

$$y_1 + y_2 + \cdots + y_n = k. \tag{10.3}$$

If $\{y_{i_1}, y_{i_2}, \ldots, y_{i_{r-1}}, y_{i_r}\}$, with $i_1 < i_2 < \cdots < i_{r-1} < i_r$, is the set of the positive y_i in (10.2), the corresponding partition of n is denoted by

$$i_r^{y_{i_r}} i_{r-1}^{y_{i_{r-1}}} \cdots i_1^{y_{i_1}},$$

omitting the exponents that are equal to 1.

REMARK 10.1   It is important to point out the clear distinction between a partition of a positive integer and a partition of a non-empty set. Thus, a partition of the positive integer n into k parts is a solution of the equation

$$x_1 + x_2 + \cdots + x_k = n, \quad x_1 \geq x_2 \geq \cdots \geq x_k \geq 1$$

in positive integers, say $\{r_1, r_2, \ldots, r_k\}$, $r_1 \geq r_2 \geq \cdots \geq r_k \geq 1$, which corresponds to a distribution of n indistinguishable balls (the units) into k indistinguishable urns containing $r_1, r_2, \ldots, r_k$ balls. In contrast, a partition of a non-empty set B, with $N(B) = n$ elements, into k subsets is a solution of the set-theoretic equation

$$X_1 + X_2 + \cdots + X_k = B, \ N(X_1) \geq N(X_2) \geq \cdots \geq N(X_k) \geq 1$$

in non-empty disjoint sets, say $\{A_1, A_2, \ldots, A_k\}$, $N(A_1) \geq N(A_2) \geq \cdots \geq N(A_k) \geq 1$, $A_i \cap A_j = \emptyset$, $i \neq j$, with $N(A_i) = r_i$, $i = 1, 2, \ldots, k$, which corresponds to a distribution of n distinguishable balls (the elements of B) into k indistinguishable urns containing $r_1, r_2, \ldots, r_k$ balls. ∎

Example 10.1

Let us determine the partitions of the integers 1, 2, 3, 4, 5 and 6. Clearly, number 1 has only one partition with one part equal to 1, number 2 has two partitions: 2, 1^2 and number 3 has three partitions: 3, 21, 1^3. Further, since $4 = 3 + 1 = 2 + 2 = 2 + 1 + 1 = 1 + 1 + 1 + 1$, the partitions of number 4 are: 4, 31, 2^2, 21^2, 1^4. Also, since $5 = 4 + 1 = 3 + 2 = 3 + 1 + 1 = 2 + 2 + 1 = 2 + 1 + 1 + 1 = 1 + 1 + 1 + 1 + 1$, the partitions of number 5 are: 5, 41, 32, 31^2, $2^2 1$, 21^3, 1^5. Finally, since $6 = 5 + 1 = 4 + 2 = 3 + 3 = 4 + 1 + 1 = 3 + 2 + 1 = 2 + 2 + 2 = 3 + 1 + 1 + 1 = 2 + 2 + 1 + 1 = 2 + 1 + 1 + 1 + 1 = 1 + 1 + 1 + 1 + 1 + 1$, the partitions of number 6 are: 6, 51, 42, 3^2, 41^2, 321, 2^3, 31^3, $2^2 1^2$, 21^4, 1^6. ▯

Let $p(n)$ be the number of partitions of n and $p(n, k)$ be the number of partitions of n into k parts. Then

$$p(n) = \sum_{k=1}^{n} p(n, k), \quad n = 1, 2, \ldots . \tag{10.4}$$

If $P(n, k)$ is the number of partitions of n into at most k parts, then

$$P(n, k) = \sum_{r=1}^{k} p(n, r), \tag{10.5}$$

$$p(n, k) = P(n, k) - P(n, k - 1) \tag{10.6}$$

and

$$P(n, k) = p(n), k \geq n. \tag{10.7}$$

A recurrence relation for the calculation of the number of partitions of n into k parts is given in the following theorem.

THEOREM 10.1
The number $p(n, k)$ of partitions of n into k parts satisfies the recurrence relation

$$p(n, k) = \sum_{r=1}^{m} p(n - k, r), \quad k = 2, 3, \ldots, n - 1, \quad n = 2, 3, \ldots, \tag{10.8}$$

with $m = \min\{k, n - k\}$ and initial conditions

$$p(n, 1) = p(n, n) = 1, \quad n = 1, 2, \ldots, \quad p(n, k) = 0, \quad k > n.$$

PROOF   Setting in the linear equation (10.1) $z_i = x_i - 1, i = 1, 2, \ldots, k$, we deduce the linear equation

$$z_1 + z_2 + \cdots + z_k = n - k, \quad z_1 \geq z_2 \geq \cdots \geq z_k \geq 0. \qquad (10.9)$$

Since the transformation is one to one, the number $p(n, k)$ of solutions of (10.1) equals the number of solutions of (10.9). If $\mathcal{P}$ is the set of solutions of (10.9) and $\mathcal{P}_r \subseteq \mathcal{P}$ is the subset of solutions of (10.9) for which $z_r \geq 1$ and $z_i = 0$, $i = r + 1, r + 2, \ldots, k, r = 1, 2, \ldots, m, m = \min\{k, n - k\}$, then $\mathcal{P}_i \cap \mathcal{P}_j = \emptyset$, $i \neq j$ and $\mathcal{P} = \mathcal{P}_1 + \mathcal{P}_2 + \cdots + \mathcal{P}_m$. Therefore, according to the addition principle,

$$p(n, k) = N(\mathcal{P}) = \sum_{r=1}^{m} N(\mathcal{P}_r)$$

and, since $N(\mathcal{P}_r)$ equals the number $p(n - k, r)$ of solutions of the linear equation

$$z_1 + z_2 + \cdots + z_r = n - k, \quad z_1 \geq z_2 \geq \cdots \geq z_r \geq 1,$$

(10.8) is established. Its initial conditions are obvious.   ∎

Using recurrence relation (10.8) and its initial conditions, Table 10.1 of the numbers $p(n, k)$ is constructed. Summing the elements of each row we get, according to (10.4), the numbers $p(n)$.

Table 10.1   The Numbers $p(n, k)$ and $p(n)$ A008284

n \ k	1	2	3	4	5	6	7	8	9	10	$p(n)$
1	1										1
2	1	1									2
3	1	1	1								3
4	1	2	1	1							5
5	1	2	2	1	1						7
6	1	3	3	2	1	1					11
7	1	3	4	3	2	1	1				15
8	1	4	5	5	3	2	1	1			22
9	1	4	7	6	5	3	2	1	1		30
10	1	5	8	9	7	5	3	2	1	1	42

REMARK 10.2   From recurrence relation (10.8) the following relations are readily deduced: for $k \geq n/2$, whence $n - k \leq n/2$, and by the expression (10.4),

$$p(n, k) = \sum_{r=1}^{n-k} p(n - k, r) = p(n - k), \qquad (10.10)$$

while, for $k < n/2$, whence $n - k > n/2$,

$$p(n,k) = \sum_{r=1}^{k} p(n-k,r) = \sum_{r=1}^{k-1} p(n-k,r) + p(n-k,k)$$
$$= p(n-1,k-1) + p(n-k,k). \qquad (10.11)$$

Note that this recurrence relation may also be derived independently of the recurrence relation (10.8) (see Exercise 1).   ∎

From Theorem 10.1, using (10.5), (10.6) and (10.7), the following corollary is deduced.

COROLLARY 10.1
The number $P(n,k)$ of partitions of n into at most k parts satisfies the recurrence relation

$$P(n,k) = P(n,k-1) + P(n-k,k), \qquad (10.12)$$

for $k = 2, 3, \ldots, n-1$, $n = 2, 3, \ldots$, with initial conditions

$$P(n,1) = 1, \quad P(0,k) = 1.$$

Using recurrence relation (10.12) and its initial conditions, Table 10.2 of the numbers $P(n,k)$ is constructed.

Table 10.2   The Numbers $P(n,k)$

k n	1	2	3	4	5	6	7	8	9	10
1	1	1	1	1	1	1	1	1	1	1
2	1	2	2	2	2	2	2	2	2	2
3	1	2	3	3	3	3	3	3	3	3
4	1	3	4	5	5	5	5	5	5	5
5	1	3	5	6	7	7	7	7	7	7
6	1	4	7	9	10	11	11	11	11	11
7	1	4	8	11	13	14	15	15	15	15
8	1	5	10	15	18	20	21	22	22	22
9	1	5	12	18	23	26	28	29	30	30
10	1	6	14	23	30	35	38	40	41	42

Bivariate and vertical generating functions for the numbers $p(n,k)$, $k = 1, 2, \ldots, n$, $n = 1, 2, \ldots$, are derived in the next theorem.

THEOREM 10.2

(a) *The bivariate generating function of the numbers $p(n, k)$ of partitions of n into k parts, $k = 1, 2, \ldots, n = 1, 2, \ldots$, is given by*

$$G(t, u) = \sum_{n=0}^{\infty} \sum_{k=0}^{n} p(n, k) u^k t^n = \prod_{i=1}^{\infty} (1 - ut^i)^{-1}, \qquad (10.13)$$

where $p(0, 0) = 1$.

(b) *The vertical generating function of the numbers $p(n, k)$ of partitions of n into k parts, $n = k, k + 1, \ldots$, for fixed k, is given by*

$$G_k(t) = \sum_{n=k}^{\infty} p(n, k) t^n = t^k \prod_{i=1}^{k} (1 - t^i)^{-1}. \qquad (10.14)$$

PROOF (a) Using the fact that the number $p(n, k)$ of partitions of n into k parts equals the number of nonnegative integer solutions of (10.2) and (10.3), it follows that

$$G(t, u) = \sum_{n=0}^{\infty} \sum_{k=0}^{n} \left(\sum u^{y_1 + y_2 + \cdots + y_n} t^{y_1 + 2y_2 + \cdots + ny_n} \right),$$

where in the inner sum the summation is extended over all nonnegative integer solutions of (10.2) and (10.3). Since this inner sum is summed over all $k = 0, 1, \ldots, n$ and $n = 0, 1, \ldots$, it follows

$$G(t, u) = \prod_{i=1}^{\infty} \left(\sum_{y_i=0}^{\infty} u^{y_i} t^{iy_i} \right) = \prod_{i=1}^{\infty} (1 - ut^i)^{-1},$$

establishing (10.13).

(b) Interchanging the order of summation in (10.13), we get

$$G(t, u) = \sum_{k=0}^{\infty} G_k(t) u^k$$

and since

$$G(t, ut) = (1 - ut) G(t, u),$$

we deduce the recurrence relation

$$t^k G_k(t) = G_k(t) - t G_{k-1}(t), \quad k = 1, 2, \ldots.$$

Consequently,

$$G_k(t) = t(1 - t^k)^{-1} G_{k-1}(t), \quad k = 1, 2, \ldots, \quad G_0(t) = 1$$

and

$$G_k(t) = t^k \prod_{i=1}^{k}(1 - t^i)^{-1},$$

which completes the proof of the theorem. ∎

The bivariate generating function (10.13) of $p(n,k)$ for $u = 1$ yields, by virtue of (10.4), the generating function of $p(n)$. This is given in the following corollary of Theorem 10.2.

COROLLARY 10.2
The generating function of the number $p(n)$ of partitions of n, $n = 1, 2, \ldots$, is given by

$$G(t) = \sum_{n=0}^{\infty} p(n)t^n = \prod_{i=1}^{\infty}(1 - t^i)^{-1}, \tag{10.15}$$

where $p(0) = 1$.

Bivariate and vertical generating functions for the numbers $P(n,k)$, $k = 0, 1, \ldots$, $n = 0, 1, \ldots$, are derived in the following corollary of Theorem 10.2.

COROLLARY 10.3
(a) *The generating function of the numbers $P(n,k)$ of partitions of n into at most k parts, $k = 0, 1, \ldots$, $n = 0, 1, \ldots$, is given by*

$$F(t,u) = \sum_{n=0}^{\infty}\sum_{k=0}^{\infty} P(n,k)u^k t^n = \prod_{i=0}^{\infty}(1 - ut^i)^{-1}. \tag{10.16}$$

(b) *The vertical generating function of the numbers $P(n,k)$, $n = 0, 1, \ldots$, for fixed k, is given by*

$$F_k(t) = \sum_{n=0}^{\infty} P(n,k)t^n = \prod_{i=1}^{k}(1 - t^i)^{-1}. \tag{10.17}$$

PROOF (a) The bivariate generating function $F(t,u)$ of the numbers $P(n,k)$, $k = 0, 1, \ldots$, $n = 0, 1, \ldots$, upon introducing (10.5), may be expressed as

$$F(t,u) = \sum_{n=0}^{\infty}\sum_{k=0}^{\infty}\sum_{r=0}^{k} p(n,r)u^k t^n = \sum_{n=0}^{\infty}\sum_{r=0}^{\infty}\left(\sum_{k=r}^{\infty} u^{k-r}\right) p(n,r)u^r t^n$$

$$= (1 - u)^{-1}\sum_{n=0}^{\infty}\sum_{r=0}^{\infty} p(n,r)u^r t^n,$$

which, by virtue of (10.13), implies (10.16).

(b) Expanding into powers of u both members of the relation

$$F(t, ut) = (1 - u)F(t, u),$$

according to

$$F(t, u) = \sum_{k=0}^{\infty} F_k(t)u^k,$$

it follows that

$$t^k F_k(t) = F_k(t) - F_{k-1}(t), \quad k = 1, 2, \ldots .$$

Thus

$$F_k(t) = (1 - t^k)^{-1} F_{k-1}(t), \quad k = 1, 2, \ldots, \quad F_0(t) = 1$$

and

$$F_k(t) = \prod_{i=1}^{k} (1 - t^i)^{-1},$$

which completes the proof of the corollary.   ∎

10.3   A UNIVERSAL GENERATING FUNCTION

The generating functions (10.13), (10.15) and (10.16) constitute particular cases of the generating function of the number of partitions of n into parts of specified or unspecified number, whose number of parts of any specific size belongs to a subset of nonnegative integers. Specifically, let $\mathbf{A} = (a_{i,j})$, $i = 1, 2, \ldots$, $j = 0, 1, \ldots$, be an infinite matrix with elements $a_{i,j} = 0$ or 1. The matrix $\mathbf{A}$ determines a sequence Y_i, $i = 1, 2, \ldots$ of subsets of nonnegative integers and vice versa as follows: for a specific $i = 1, 2, \ldots$, $Y_i = \{j : a_{i,j} = 1\}$ is the set of indices j of the elements of the i-th row of the matrix $\mathbf{A}$ that are equal to 1. Let us denote by $p(n; \mathbf{A})$ the number of partitions of n and by $p(n, k; \mathbf{A})$ the number of partitions of n into k parts whose number y_i of parts that are equal to i belongs to Y_i, $i = 1, 2, \ldots$. Also, let us denote by $P(n, k; \mathbf{A})$ the number of partitions of n into at most k parts whose number y_i of parts that are equal to i belongs to Y_i, $i = 1, 2, \ldots$. Note that, in particular for $a_{i,j} = 1$, $i = 1, 2, \ldots$, $j = 0, 1, \ldots$, whence $Y_i = \{0, 1, \ldots\}$, $i = 1, 2, \ldots$, $p(n; \mathbf{A}) \equiv p(n)$, $p(n, k; \mathbf{A}) \equiv p(n, k)$ and $P(n, k; \mathbf{A}) \equiv P(n, k)$. A bivariate generating function for $p(n, k; \mathbf{A})$, $k = 0, 1, \ldots, n$, $n = 0, 1, \ldots$ is derived in the next theorem.

THEOREM 10.3

Let $A = (a_{i,j})$, $i = 1, 2, \ldots$, $j = 0, 1, \ldots$ be an infinite matrix with elements $a_{i,j} = 0$ or 1 and $p(n, k; A)$ be the number of partitions of n into k parts whose number y_i of parts that are equal to i belongs to the set $Y_i = \{j : a_{i,j} = 1\}$, $i = 1, 2, \ldots$. Then

$$G(t, u; A) = \sum_{n=0}^{\infty} \sum_{k=0}^{n} p(n, k; A) u^k t^n = \prod_{i=1}^{\infty} \left(\sum_{j=0}^{\infty} a_{i,j} u^j t^{ij} \right). \quad (10.18)$$

PROOF   The number $p(n, k; A)$ equals the number of nonnegative integer solutions of (10.2) and (10.3) with

$$y_i \in Y_i = \{j : a_{i,j} = 1\}, \quad i = 1, 2, \ldots . \quad (10.19)$$

Consequently,

$$G(t, u; A) = \sum_{n=0}^{\infty} \sum_{k=0}^{n} \left(\sum u^{y_1 + y_2 + \cdots + y_n} t^{y_1 + 2y_2 + \cdots + ny_n} \right),$$

where, the summation in the inner sum is extended over all nonnegative integer solutions of (10.2) and (10.3) that satisfy (10.19). Introducing the elements $a_{i,j}$ of the matrix A into this sum, we get the expression

$$G(t, u; A) = \sum_{n=0}^{\infty} \sum_{k=0}^{n} \left(\sum a_{1,y_1} a_{2,y_2} \cdots a_{n,y_n} u^{y_1 + y_2 + \cdots + y_n} t^{y_1 + 2y_2 + \cdots + ny_n} \right),$$

where, the summation in the inner sum is extended over all nonnegative integer solutions of (10.2) and (10.3). Note that this transformation of the inner sum incorporates condition (10.19), adding zero terms and thus saves from the concern of selecting the solutions of (10.2) and (10.3) that satisfy (10.19). Since this inner sum is summed over all $k = 0, 1, \ldots, n$ and $n = 0, 1, \ldots$, it follows

$$G(t, u; A) = \prod_{i=1}^{\infty} \left(\sum_{y_i=0}^{\infty} a_{i,y_i} u^{y_i} t^{iy_i} \right) = \prod_{i=1}^{\infty} \left(\sum_{j=0}^{\infty} a_{i,j} u^j t^{ij} \right),$$

which is the required expression of the generating function.   ∎

COROLLARY 10.4

Let $A = (a_{i,j})$, $i = 1, 2, \ldots$, $j = 0, 1, \ldots$, be an infinite matrix with elements $a_{i,j} = 0$ or 1. Also let $p(n; A)$ be the number of partitions of n and $P(n, k; A)$ be the number of partitions of n into at most k parts whose number y_i of parts that *are equal to i belongs to the set $Y_i = \{j : a_{i,j} = 1\}$, $i = 1, 2, \ldots$. Then*

$$G(t; \boldsymbol{A}) = \sum_{n=0}^{\infty} p(n; \boldsymbol{A}) t^n = \prod_{i=1}^{\infty} \left(\sum_{j=0}^{\infty} a_{i,j} t^{ij} \right) \qquad (10.20)$$

and

$$F(t, u; \boldsymbol{A}) = \sum_{n=0}^{\infty} \sum_{k=0}^{\infty} P(n, k; \boldsymbol{A}) u^k t^n = (1 - u)^{-1} \prod_{i=1}^{\infty} \left(\sum_{j=0}^{\infty} a_{i,j} u^j t^{ij} \right). \qquad (10.21)$$

PROOF Expressing the number $p(n; \boldsymbol{A})$ as a sum of the numbers $p(n, k; \boldsymbol{A})$, $k = 0, 1, \ldots, n$, we get

$$G(t; \boldsymbol{A}) = \sum_{n=0}^{\infty} p(n; \boldsymbol{A}) t^n = \sum_{n=0}^{\infty} \sum_{k=0}^{n} p(n, k; \boldsymbol{A}) t^n$$

and so, from (10.18) with $u = 1$, (10.20) is deduced. Similarly we get

$$F(t, u; \boldsymbol{A}) = \sum_{n=0}^{\infty} \sum_{k=0}^{\infty} P(n, k; \boldsymbol{A}) u^k t^n = \sum_{n=0}^{\infty} \sum_{k=0}^{\infty} \sum_{r=0}^{k} p(n, r; \boldsymbol{A}) u^k t^n$$

$$= \sum_{n=0}^{\infty} \sum_{r=0}^{\infty} \left(\sum_{k=r}^{\infty} u^{k-r} \right) p(n, r; \boldsymbol{A}) u^r t^n$$

$$= (1 - u)^{-1} \sum_{n=0}^{\infty} \sum_{r=0}^{n} p(n, r; \boldsymbol{A}) u^r t^n$$

and so, by virtue of (10.18), (10.21) is deduced. ∎

Several particular cases of the generating functions (10.18) and (10.20) are presented in the following examples.

Example 10.2 Partitions into unequal parts
Let $q(n)$ be the number of partitions of n into unequal (different) parts and $q(n, k)$ be the number of partitions of n into k unequal (different) parts. Determine the generating functions

$$H(t) = \sum_{n=0}^{\infty} q(n) t^n, \quad H(t, u) = \sum_{n=0}^{\infty} \sum_{k=0}^{n} q(n, k) u^k t^n, \quad H_k(t) = \sum_{n=0}^{\infty} q(n, k) t^n.$$

Note that, in this case, the number y_i of parts that are equal to i may take the values 0 or 1, $i = 1, 2, \ldots$. Consequently, $q(n) = p(n; \boldsymbol{A})$, $q(n, k) = p(n, k; \boldsymbol{A})$, with $a_{i,j} = 1, j = 0, 1$ and $a_{i,j} = 0, j = 2, 3, \ldots, i = 1, 2, \ldots$. Thus, by (10.18) and (10.20),

$$H(t, u) = \sum_{n=0}^{\infty} \sum_{k=0}^{n} q(n, k) u^k t^n = \prod_{i=1}^{\infty} (1 + u t^i)$$

and

$$H(t) = \sum_{n=0}^{\infty} q(n) t^n = \prod_{i=1}^{\infty} (1 + t^i).$$

For the calculation of the third generating function, note that

$$H(t, u) = (1 + ut) H(t, ut)$$

and since

$$H(t, u) = \sum_{k=0}^{\infty} H_k(t) u^k,$$

it follows that

$$H_k(t) = t^k H_k(t) + t^k H_{k-1}(t), \quad k = 1, 2, \ldots .$$

Therefore

$$H_k(t) = t^k (1 - t^k)^{-1} H_{k-1}(t), \quad k = 1, 2, \ldots, \quad H_0(t) = 1$$

and

$$H_k(t) = \sum_{n=k(k+1)/2}^{\infty} q(n, k) t^n = \prod_{j=1}^{k} t^j (1 - t^j)^{-1} = t^{1+2+\cdots+k} \prod_{j=1}^{k} (1 - t^j)^{-1}.$$

Hence

$$H_k(t) = t^{k(k+1)/2} \prod_{j=1}^{k} (1 - t^j)^{-1}. \qquad \square$$

Example 10.3 Partitions into even and odd parts

(a) Let $p_0(n)$ be the number of partitions of n into even parts and $p_0(n, k)$ be the number of partitions of n into k even parts. Determine the generating functions

$$G_0(t) = \sum_{n=0}^{\infty} p_0(n) t^n, \quad G_0(t, u) = \sum_{n=0}^{\infty} \sum_{k=0}^{n} p_0(n, k) u^k t^n.$$

In this case, the number y_{2i} of parts that are equal to $2i$ belongs to the set $Y_{2i} = \{0, 1, \ldots\}$, while the number y_{2i-1} of parts that are equal to $2i - 1$ belongs to the set $Y_{2i-1} = \{0\}, i = 1, 2, \ldots$. Consequently, $p_0(n) = p(n; \mathbf{A}), p_0(n, k) =$

$p(n, k; \boldsymbol{A})$, with $a_{i,0} = 1$, $a_{2i,j} = 1$ and $a_{2i-1,j} = 0$, $i = 1, 2, \ldots, j = 1, 2, \ldots$. Thus, by (10.18) and (10.20),

$$G_0(t, u) = \sum_{n=0}^{\infty} \sum_{k=0}^{n} p_0(n, k) u^k t^n = \prod_{i=1}^{\infty} (1 - ut^{2i})^{-1}$$

and

$$G_0(t) = \sum_{n=0}^{\infty} p_0(n) t^n = \prod_{i=1}^{\infty} (1 - t^{2i})^{-1}.$$

(b) Let $p_1(n)$ be the number of partitions of n into odd parts and $p_1(n, k)$ be the number of partitions of n into k odd parts. Then, proceeding as in the previous case, we conclude that

$$G_1(t, u) = \sum_{n=0}^{\infty} \sum_{k=0}^{n} p_1(n, k) u^k t^n = \prod_{i=1}^{\infty} (1 - ut^{2i-1})^{-1}$$

and

$$G_1(t) = \sum_{n=0}^{\infty} p_1(n) t^n = \prod_{i=1}^{\infty} (1 - t^{2i-1})^{-1}.$$

(c) Let $q_0(n)$ be the number of partitions of n into unequal even parts and $q_1(n)$ the number of partitions of n into unequal odd parts. Determine the generating functions

$$H_0(t) = \sum_{n=0}^{\infty} q_0(n) t^n, \quad H_1(t) = \sum_{n=0}^{\infty} q_1(n) t^n.$$

In the first case $y_{2i} = 0$ or 1, $y_{2i-1} = 0$, $i = 1, 2, \ldots$ and so $q_0(n) = p(n; \boldsymbol{A})$, with $a_{2i,j} = 1$, $j = 0, 1$, $a_{2i,j} = 0$, $j = 2, 3, \ldots$, $i = 1, 2, \ldots$ and $a_{2i-1,0} = 1$, $a_{2i-1,j} = 0$, $j = 1, 2, \ldots$, $i = 1, 2, \ldots$. In the second case $y_{2i-1} = 0$ or 1, $y_{2i} = 0$, $i = 1, 2, \ldots$ and so $q_1(n) = p(n; \boldsymbol{A})$, with $a_{2i-1,j} = 1$, $j = 0, 1$, $a_{2i-1,j} = 0$, $j = 2, 3, \ldots$, $i = 1, 2, \ldots$ and $a_{2i,0} = 1$, $a_{2i,j} = 0$, $j = 1, 2, \ldots$, $i = 1, 2, \ldots$. Thus, by virtue of (10.20), it follows that

$$H_0(t) = \sum_{n=0}^{\infty} q_0(n) t^n = \prod_{i=1}^{\infty} (1 + t^{2i})$$

and

$$H_1(t) = \sum_{n=0}^{\infty} q_1(n) t^n = \prod_{i=1}^{\infty} (1 + t^{2i-1}). \qquad \Box$$

Example 10.4 *Partitions into parts of restricted size*

Let $R(n, k)$ be the number of partitions of n with no part greater than k. Determine the generating

$$R_k(t) = \sum_{n=0}^{\infty} R(n, k) t^n.$$

Note that $R(n, k) = p(n; \boldsymbol{A})$, with $a_{i,0} = 1$, $i = 1, 2, \ldots$, $a_{i,j} = 1$, $i = 1, 2, \ldots, k$, $a_{i,j} = 0$, $i = k+1, k+2, \ldots, j = 1, 2, \ldots$. So, according to (10.20),

$$R_k(t) = \sum_{n=0}^{\infty} R(n, k)t^n = \prod_{i=1}^{k} (1 - t^i)^{-1}.$$

Note that this generating function equals the generating function (10.17), $R_k(t) = F_k(t)$. Consequently, the number $R(n, k)$, of partitions of n with no part greater than k, equals the number $P(n, k)$, of partitions of n into at most k parts, $R(n, k) = P(n, k)$.   ☐

Example 10.5 Partitions into an even and an odd number of parts

(a) Let $P_0(n)$ be the number of partitions of n into an even number of parts and $P_1(n)$ be the number of partitions of n into an odd number of parts. Determine the generating functions

$$A_0(t) = \sum_{n=0}^{\infty} P_0(n)t^n, \quad A_1(t) = \sum_{n=0}^{\infty} P_1(n)t^n.$$

The numbers $P_0(n)$ and $P_1(n)$ may be expressed in terms of the number $p(n, k)$ of partitions of n into k parts, $k = 1, 2, \ldots, n$, as

$$P_0(n) = \sum_{r=0}^{[n/2]} p(n, 2r) = \frac{1}{2} \sum_{k=0}^{n} \{p(n, k) + (-1)^k p(n, k)\}$$

and

$$P_1(n) = \sum_{r=0}^{[(n-1)/2]} p(n, 2r + 1) = \frac{1}{2} \sum_{k=0}^{n} \{p(n, k) - (-1)^k p(n, k)\},$$

respectively. Therefore, using (10.13) with $u = \pm 1$, it follows that

$$A_0(t) = \sum_{n=0}^{\infty} P_0(n)t^n = \frac{1}{2} \left\{ \prod_{i=1}^{\infty} (1 - t^i)^{-1} + \prod_{i=1}^{\infty} (1 + t^i)^{-1} \right\}$$

and

$$A_1(t) = \sum_{n=0}^{\infty} P_1(n)t^n = \frac{1}{2} \left\{ \prod_{i=1}^{\infty} (1 - t^i)^{-1} - \prod_{i=1}^{\infty} (1 + t^i)^{-1} \right\}.$$

(b) Let $Q_0(n)$ be the number of partitions of n into an even number of unequal parts and $Q_1(n)$ be the number of partitions of n into an odd number of unequal parts. Determine the generating functions

$$B_0(t) = \sum_{n=0}^{\infty} Q_0(n)t^n, \quad B_1(t) = \sum_{n=0}^{\infty} Q_1(n)t^n.$$

The numbers $Q_0(n)$ and $Q_1(n)$ may be expressed in terms of the number $q(n, k)$ of partitions of n into k unequal parts, $k = 1, 2, \ldots, n$, as

$$Q_0(n) = \sum_{r=0}^{[n/2]} q(n, 2r) = \frac{1}{2} \sum_{k=0}^{n} \{q(n, k) + (-1)^k q(n, k)\}$$

and

$$Q_1(n) = \sum_{r=0}^{[(n-1)/2]} q(n, 2r + 1) = \frac{1}{2} \sum_{k=0}^{n} \{q(n, k) - (-1)^k q(n, k)\},$$

respectively. Therefore, using the generating function $H(t, u)$ of Example 10.2 with $u = \pm 1$, it follows that

$$B_0(t) = \sum_{n=0}^{\infty} Q_0(n) t^n = \frac{1}{2} \left\{ \prod_{i=1}^{\infty} (1 + t^i) + \prod_{i=1}^{\infty} (1 - t^i) \right\}$$

and

$$B_1(t) = \sum_{n=0}^{\infty} Q_1(n) t^n = \frac{1}{2} \left\{ \prod_{i=1}^{\infty} (1 + t^i) - \prod_{i=1}^{\infty} (1 - t^i) \right\}. \qquad \square$$

Example 10.6 Partitions into specified parts

In the partitions of a positive integer n, without any restriction,

$$x_1 + x_2 + \cdots + x_k = n, \quad x_1 \geq x_2 \geq \cdots \geq x_k \geq 1, \quad k = 1, 2, \ldots, n,$$

the parts x_i, $i = 1, 2, \ldots, n$ belong to the set $\{1, 2, \ldots, n\}$. Let us now consider the partitions of n whose parts x_i, $i = 1, 2, \ldots, n$ belong to a subset $\{i_1, i_2, \ldots, i_r\}$ of $\{1, 2, \ldots, n\}$. In this case, if y_{i_s} is the number of parts that are equal to i_s, $s = 1, 2, \ldots, r$, then

$$i_1 y_{i_1} + i_2 y_{i_2} + \cdots + i_r y_{i_r} = n, \quad y_{i_s} \geq 0, \quad s = 1, 2, \ldots, r.$$

The number $D(n; i_1, i_2, \ldots, i_r)$ of partitions of n whose parts belong to the subset $\{i_1, i_2, \ldots, i_r\}$ of $\{1, 2, \ldots, n\}$ equals the number of nonnegative integer solutions of this equation. Determine the generating function

$$D_{i_1, i_2 \ldots, i_r}(t) = \sum_{n=0}^{\infty} D(n; i_1, i_2, \ldots, i_r) t^n.$$

Note that $D(n; i_1, i_2, \ldots, i_r) = p(n; \boldsymbol{A})$, where the elements of the matrix $\boldsymbol{A} = (a_{i,j})$, $i = 1, 2, \ldots, j = 0, 1, \ldots$, are given by $a_{i,0} = 1$, $i = 1, 2, \ldots,$

$a_{i,j} = 1$, $i \in \{i_1, i_2, \ldots, i_r\}$, $a_{i,j} = 0$, $i \notin \{i_1, i_2, \ldots, i_r\}$, $j = 1, 2, \ldots$. Hence, by (10.20),

$$D_{i_1, i_2, \ldots, i_r}(t) = \sum_{n=0}^{\infty} D(n; i_1, i_2, \ldots, i_r) t^n = \prod_{s=1}^{r} (1 - t^{i_s})^{-1}.$$

The *money-changing problem* is a characteristic example requiring the calculation of partitions into specified parts. Specifically, let us consider a cashier that has an unrestricted number of coins of 5, 10, 25 and 50 cents. Calculate the number of ways of forming a given amount of, say, $ 1. This number equals $D(100; 5, 10, 25, 50) = D(20; 1, 2, 5, 10)$, the number of nonnegative integer solutions of the linear equation $5y_5 + 10y_{10} + 25y_{25} + 50y_{50} = 100$ or, equivalently, the equation $y_1 + 2y_2 + 5y_5 + 10y_{10} = 20$, which equals the coefficient of t^{20} in the expansion of the generating function

$$D_{1,2,5,10}(t) = [(1 - t)(1 - t^2)(1 - t^5)(1 - t^{10})]^{-1}. \quad \square$$

10.4 INTERRELATIONS AMONG PARTITION NUMBERS

The generating functions of the numbers of partitions that satisfy certain conditions may be used to deduce relations between these numbers. Such relations are derived in the following theorems.

THEOREM 10.4
The number $q(n)$ of partitions of n into unequal parts equals the number $p_1(n)$ of partitions of n into odd parts:

$$q(n) = p_1(n). \tag{10.22}$$

PROOF The generating function of the number $q(n)$ of partitions of n into unequal parts is given by (see Example 10.2)

$$H(t) = \sum_{n=0}^{\infty} q(n) t^n = \prod_{i=1}^{\infty} (1 + t^i)$$

and the generating function of the number $p_1(n)$ of partitions of n into odd parts is given by (see Example 10.3)

$$G_1(t) = \sum_{n=0}^{\infty} p_1(n) t^n = \prod_{i=1}^{\infty} (1 - t^{2i-1})^{-1}.$$

Using the identity

$$(1 + t^i)(1 - t^i) = (1 - t^{2i}), \quad i = 1, 2, \ldots,$$

the generating function $H(t)$ may be written in the form

$$H(t) = \prod_{i=1}^{\infty}(1 - t^{2i})(1 - t^i)^{-1} = \prod_{i=1}^{\infty}(1 - t^{2i})\prod_{j=1}^{\infty}(1 - t^j)^{-1}.$$

If, in the last product the terms with j even are separated from those with j odd, then it is written as

$$\prod_{j=1}^{\infty}(1 - t^j)^{-1} = \prod_{i=1}^{\infty}(1 - t^{2i})^{-1}\prod_{i=1}^{\infty}(1 - t^{2i-1})^{-1}$$

and so

$$H(t) = \prod_{i=1}^{\infty}(1 - t^{2i-1})^{-1} = G_1(t).$$

The last relation implies (10.22).   ∎

THEOREM 10.5
The number $P(n,k)$ of partitions of n into at most k parts, which equals the number $R(n,k)$ of partitions of n with no part greater than k, is equal to the number $p(n + k, k)$ of partitions of $n + k$ into k parts:

$$P(n, k) = R(n, k) = p(n + k, k). \tag{10.23}$$

PROOF   The generating function of the numbers $P(n, k) = R(n, k)$, $n = 0, 1, \ldots$ is (see Corollary 10.3 and Example 10.4)

$$F_k(t) = \sum_{n=0}^{\infty} P(n, k)t^n = \sum_{n=0}^{\infty} R(n, k)t^n = \prod_{i=1}^{k}(1 - t^i)^{-1},$$

while the generating function of the numbers $p(r, k)$, $r = k, k + 1, \ldots$, is (see Theorem 10.2)

$$G_k(t) = \sum_{r=k}^{\infty} p(r, k)t^r = t^k \prod_{i=1}^{k}(1 - t^i)^{-1}.$$

Consequently, $F_k(t) = t^{-k}G_k(t)$ and

$$\sum_{n=0}^{\infty} P(n, k)t^n = \sum_{r=k}^{\infty} p(r, k)t^{r-k} = \sum_{n=0}^{\infty} p(n + k, k)t^n.$$

The last relation implies (10.23).   ∎

THEOREM 10.6
The number $q(n, k)$ of partitions of n into k unequal parts equals the number $P(n - k(k + 1)/2, k)$ of partitions of $n - k(k + 1)/2$ into at most k parts, which is equal to the number $p(n - k(k - 1)/2, k)$ of partitions of $n - k(k - 1)/2$ into k parts:

$$q(n, k) = P(n - k(k + 1)/2, k) = p(n - k(k - 1)/2, k). \qquad (10.24)$$

PROOF Since (see Example 10.2 and Corollary 10.3)

$$H_k(t) = \sum_{n=k(k+1)/2}^{\infty} q(n, k)t^n = t^{k(k+1)/2} \prod_{i=1}^{k} (1 - t^i)^{-1}$$

and

$$F_k(t) = \sum_{r=0}^{\infty} P(r, k)t^r = \prod_{i=1}^{k} (1 - t^i)^{-1},$$

it follows that $H_k(t) = t^{k(k+1)/2} F_k(t)$ and so

$$\sum_{n=k(k+1)/2}^{\infty} q(n, k)t^n = \sum_{r=0}^{\infty} P(r, k)t^{r+k(k+1)/2}$$

$$= \sum_{n=k(k+1)/2}^{\infty} P(n - k(k + 1)/2, k)t^n.$$

The last relation implies the first part of (10.24). Its second part is deduced from (10.23) by noting that $n - k(k + 1)/2 + k = n - k(k - 1)/2$. ∎

REMARK 10.3 The following combinatorial derivation of (10.24) possesses its own merits. The number $q(n, k)$ of partitions of n into k unequal parts equals the number of integer solutions of the linear equation

$$x_1 + x_2 + \cdots + x_k = n, \quad x_1 > x_2 > \cdots > x_k \geq 1. \qquad (10.25)$$

(a) Putting $z_i = x_i - (k - i + 1), i = 1, 2, \ldots, k$, equation (10.25) is transformed to the equation

$$z_1 + z_2 + \cdots + z_k = n - k(k + 1)/2, \quad z_1 \geq z_2 \geq \cdots \geq z_{k-1} \geq z_k \geq 0,$$

whose number of integer solutions equals the number $P(n - k(k + 1)/2, k)$ of partitions of $n - k(k + 1)/2$ into at most k parts.
(b) Setting $w_i = x_i - (k - i), i = 1, 2, \ldots, k$, equation (10.25) is transformed to the equation

$$w_1 + w_2 + \cdots + w_k = n - k(k - 1)/2, \quad w_1 \geq w_2 \geq \cdots \geq w_{k-1} \geq w_k \geq 1,$$

whose number of integer solutions equals the number $p(n - k(k-1)/2, k)$ of partitions of $n - k(k-1)/2$ into k parts. ∎

A simple and instructive representation of the partition of a positive integer n is its Ferrers diagram, which is defined as follows.

Ferrers diagram of a partition of a positive integer n into parts x_i, $i = 1, 2, \ldots, k, \ldots,$

$$x_1 + x_2 + \cdots + x_k + \cdots = n, \quad x_1 \geq x_2 \geq \cdots \geq x_k \geq \cdots \geq 1, \quad (10.26)$$

is called the diagram of (equidistant) points whose i-th horizontal row of points (numbering from the bottom up) has x_i points, $i = 1, 2, \ldots, k, \ldots$. To each partition there corresponds its conjugate partition, defined as follows.

Conjugate partition of the partition (10.26) is called the partition of n into parts y_j, $j = 1, 2, \ldots, r, \ldots,$

$$y_1 + y_2 + \cdots + y_r + \cdots = n, \quad y_1 \geq y_2 \geq \cdots \geq y_r \geq \cdots \geq 1, \quad (10.27)$$

whose Ferrers diagram the number y_j of points of the j-th horizontal row equals the number of points of the j-th column (numbering from left to right) of the Ferrers diagram of the partition (10.26).

Self-conjugate partition is called a partition that coincides with its conjugate partition.

FIGURE 10.1
Ferrers diagram

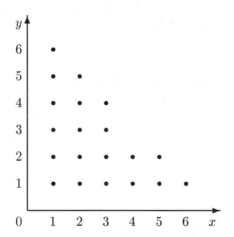

Example 10.7

Figure 10.1 gives the Ferrers diagram of the partition $653^2 21$ of 20. Clearly, its conjugate partition is $6542^2 1$. ▯

Some interesting interrelations among partition numbers are derived by considering the Ferrers diagrams of partitions and their conjugate partitions. Thus we have:

THEOREM 10.7
The number $p(n, k)$ of partitions of n into k parts equals the number of partitions of n into parts of which the maximum is k.

PROOF   Consider a partition of n into k parts and its Ferrers diagram. Associate to it its conjugate partition, which is also a partition of n into parts of which the maximum is k. Since this correspondence is one to one, the required conclusion is deduced.   ∎

THEOREM 10.8
The number of self-conjugate partitions of n equals the number $q_1(n)$ of partitions of n into unequal odd parts.

PROOF   Consider a self-conjugate partition of n into parts x_i, $i = 1, 2, \ldots$,

$$x_1 + x_2 + \cdots + x_k + \cdots = n, \quad x_1 \geq x_2 \geq \cdots \geq x_k \geq \cdots \geq 1, \quad (10.28)$$

and its Ferrers diagram. Note that this diagram is symmetric with respect to the line $x = y$. This symmetry implies that the pairs of straight-line sections defined by the points $\{(i, i), (x_i, i)\}$ and $\{(i, i), (i, y_i)\}$, $i = 1, 2, \ldots, k \ldots$, since $x_i = y_i$, $i = 1, 2, \ldots, k, \ldots$, form isosceles angles. The number z_i of points of the diagram that lie on the i-th isosceles angle is odd,

$$z_i = 2(x_i - 1) + 1, \quad i = 1, 2, \ldots, k, \ldots, \quad (10.29)$$

and since $x_1 \geq x_2 \geq \cdots \geq x_k \geq \cdots \geq 1$, we have in addition

$$z_1 > z_2 > \cdots > z_k > \cdots \geq 1.$$

Let us now correspond to the self-conjugate partition of n (10.28) the partition of n into unequal odd parts,

$$z_1 + z_2 + \cdots + z_k + \cdots = n, \quad z_1 > z_2 > \cdots > z_k > \cdots \geq 1. \quad (10.30)$$

This correspondence of (10.28) to (10.30), according to (10.29), is one-to-one and thus the number of self-conjugate partitions of n equals the number of partitions of n into unequal odd parts.   ∎

REMARK 10.4   The one-to-one correspondence between the self-conjugate partitions of n and the partitions of n into unequal odd parts is further clarified by considering the following specific cases.

FIGURE 10.2

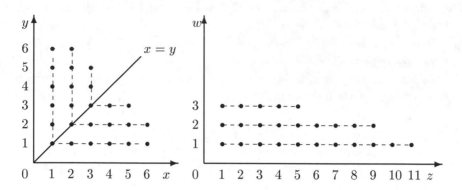

(a) Consider the self-conjugate partition $6^2 53^2 2$ of 25 into the parts $x_1 = 6$, $x_2 = 6$, $x_3 = 5$, $x_4 = 3$, $x_5 = 3$, $x_6 = 2$. Figure 10.2 (left) represents its Ferrers diagram in which the three pairs of straight-line sections:

$$(\{(1,1),(6,1)\}, \{(1,1),(1,6)\}), \quad (\{(2,2),(6,2)\}, \{(2,2),(2,6)\}),$$

$$(\{(3,3),(5,3)\}, \{(3,3),(3,5)\})$$

form isosceles angles. The number z_i, of points of this diagram that lie on the i-th isosceles angle, is $z_i = 2(x_i - 1) + 1$, $i = 1, 2, 3$: $z_1 = 11$, $z_2 = 9$, $z_3 = 5$. Hence, to the self-conjugate partition $6 + 6 + 5 + 3 + 3 + 2 = 25$ there corresponds the partition $11 + 9 + 5 = 25$ into unequal odd parts the Ferrers diagram of which is represented in Figure 10.2 (right).

FIGURE 10.3

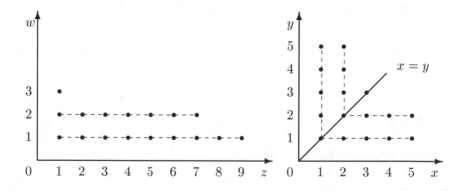

(b) Consider the partition 971 of 17 into unequal odd parts: $z_1 = 9$, $z_2 = 7$, $z_3 = 1$. Figure 10.3 (left) represents its Ferrers diagram. We have $x_i = y_i =$

$i + (z_i - 1)/2$, $i = 1, 2, 3$: $x_1 = y_1 = 5$, $x_2 = y_2 = 5$, $x_3 = y_3 = 3$ and so the Ferrers diagram of the corresponding self-conjugate partition forms three isosceles angles:

$$(\{(1,1), (5,1)\}, \{(1,1), (1,5)\}), \quad (\{(2,2), (5,2)\}, \{(2,2), (2,5)\}),$$

$$(\{(3,3), (3,3)\}, \{(3,3), (3,3)\})$$

Note that the third angle degenerates to the point $(3, 3)$ of the straight line $x = y$. Figure 10.3 (right) represents this diagram. From this diagram it follows that the self-conjugate partition that corresponds to the partition 971 is $5^2 32^2$. ∎

THEOREM 10.9 Euler's pentagonal theorem

Let $Q_0(n)$ be the number of partitions of n into an even number of unequal parts and $Q_1(n)$ be the number of partitions of n into an odd number of unequal parts. Then, the difference $E(n) = Q_0(n) - Q_1(n)$ is given by

$$E(n) = \begin{cases} (-1)^k, & n = (3k^2 \pm k)/2 \\ 0, & otherwise. \end{cases} \tag{10.31}$$

PROOF   Let us, initially, consider the Ferrers diagram F of a partition of n in any (even or odd) number of unequal parts and let:

(i) A be the straight-line section that starts from the easternmost point (of the first horizontal line) of the diagram F, forms an angle of $3\pi/4$ with the horizontal axis and thus, passing through only the outer points of F, ends in the northernmost of them.

(ii) B be the straight-line section that starts from the northernmost point (of the first column) of the diagram F, is parallel to the horizontal axis and thus, passing through only the outer points of F that have the same ordinate, ends in the easternmost of them. It should be noted that B, as well as A, may contain only one point of F (degenerate to a point straight-line sections).

Further, let $N(A) = r$ and $N(B) = s$ be the number of points of the diagram F that lie on A and B respectively.

Now, we associate with the partition of n into unequal parts with Ferrers diagram F, the partition of n into unequal parts with Ferrers diagram G that is defined as follows:

(a) If $r < s$, the Ferrers diagram G is deduced from F by removing the points that lie on A and adjoining them to the first columns, one to each one of them (Figure 10.4), except when A and B have a common point and $r = s - 1$. In the latter case the corresponding partition of n into unequal parts has $k = r$ parts and $x_k = s = k + 1$ so that

$$n = 2k + (2k - 1) + \cdots + (k + 2) + (k + 1) = (3k^2 + k)/2 \tag{10.32}$$

and to its Ferrers diagram F we do not associate any diagram G (Figure 10.5).

FIGURE 10.4

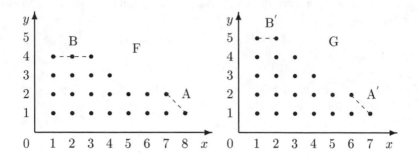

FIGURE 10.5

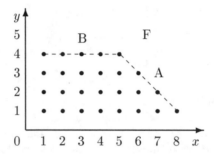

(b) If $r \geq s$, the Ferrers diagram G is deduced from F by removing the points that lie on B and adjoining them to the first rows, one to each one of them (Figure 10.6), except when A and B have a common point and $r = s$. In the latter case the corresponding partition of n into unequal parts has $k = r$ parts and $x_k = s = k$ so that

$$n = (2k - 1) + (2k - 2) + \cdots + (k + 1) + k = (3k^2 - k)/2 \quad (10.33)$$

and to its Ferrers diagram F we do not associate any diagram G (Figure 10.7).

The transition from F to G changes (increases or decreases) by one the number of parts of the corresponding partition, without altering the inequality of the parts. Thus, if F belongs to the set of Ferrers diagrams $\mathcal{F}_0$, of the partitions of n into an even number of unequal parts, then G belongs to the set of Ferrers diagrams $\mathcal{F}_1$, of the partitions of n into an odd number of unequal parts, while, if F belongs to $\mathcal{F}_1$, then G belongs to $\mathcal{F}_0$ and vice versa. Consequently, if $n \neq (3k^2 - k)/2$ and $n \neq (3k^2 + k)/2$, then the correspondence is one to one and so $E(n) = Q_0(n) - Q_1(n) = 0$. If $n = (3k^2 \pm k)/2$ and k is even, the Ferrers diagram of the partition (10.32) or (10.33) belongs to $\mathcal{F}_0$ and has no image in $\mathcal{F}_1$, while if $n = (3k^2 \pm k)/2$ and k is odd the Ferrers diagram of the partition (10.32) or (10.33) belongs to $\mathcal{F}_1$

FIGURE 10.6

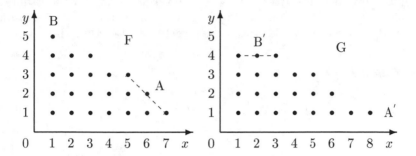

FIGURE 10.7

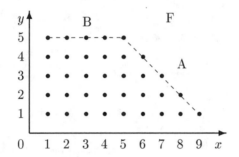

and has no archetype (pre-image) in $\mathcal{F}_0$. Hence $E(n) = Q_0(n) - Q_1(n) = (-1)^k$, for $n = (3k^2 \pm k)/2$, and the proof of the theorem is completed. ∎

10.5 COMBINATORIAL IDENTITIES

Some interesting combinatorial identities may be deduced from the relation

$$G(t, u; \boldsymbol{A}) = \sum_{k=0}^{\infty} G_k(t; \boldsymbol{A}) u^k,$$

which connects the generating functions

$$G(t, u; \boldsymbol{A}) = \sum_{n=0}^{\infty} \sum_{k=0}^{n} p(n, k; \boldsymbol{A}) u^k t^n, \quad G_k(t; \boldsymbol{A}) = \sum_{n=k}^{\infty} p(n, k; \boldsymbol{A}) t^n,$$

of the numbers $p(n, k; \boldsymbol{A})$ of partitions of n into k parts, whose number of parts equal to i belongs to the set $\{j : a_{i,j} = 1\}$, $i = 1, 2, \ldots$, where

$\mathbf{A} = (a_{i,j})$, $i = 1, 2, \ldots$, $j = 0, 1, \ldots$, with $a_{i,j} = 0$ or 1. In the literature, in expressing these identities instead of the variables t and u the variables q and x, respectively, are used. Further, such an identity is referred to as q-*identity*. Two typical q-identities are given in the following theorem.

THEOREM 10.10

$$\prod_{i=1}^{\infty}(1 - xq^i)^{-1} = \sum_{k=0}^{\infty} x^k q^k \prod_{j=1}^{k}(1 - q^j)^{-1}, \qquad (10.34)$$

$$\prod_{i=1}^{\infty}(1 + xq^i) = \sum_{k=0}^{\infty} x^k q^{k(k+1)/2} \prod_{j=1}^{k}(1 - q^j)^{-1}. \qquad (10.35)$$

PROOF   Introducing into the relation

$$G(t, u) = \sum_{k=0}^{\infty} G_k(t) u^k,$$

expressions (10.13) and (10.14) of the generating functions,

$$G(t, u) = \sum_{n=0}^{\infty} \sum_{k=0}^{n} p(n, k) u^k t^n = \prod_{i=1}^{\infty}(1 - ut^i)^{-1},$$

$$G_k(t) = \sum_{n=k}^{\infty} p(n, k) t^n = \prod_{j=1}^{k}(1 - t^j)^{-1},$$

of the numbers $p(n, k)$ of partitions of n into k parts, and putting $t = q$ and $u = x$, (10.34) is readily deduced. Similarly, introducing into the relation

$$H(t, u) = \sum_{k=0}^{\infty} H_k(t) u^k,$$

the expressions of the generating functions (see Example 10.2),

$$H(t, u) = \sum_{n=0}^{\infty} \sum_{k=0}^{n} q(n, k) u^k t^n = \prod_{i=1}^{\infty}(1 + ut^i),$$

$$H_k(t) = \sum_{n=k(k+1)/2}^{\infty} q(n, k) t^n = t^{k(k+1)/2} \prod_{j=1}^{k}(1 - t^j)^{-1},$$

of the numbers $q(n, k)$ of partitions of n into k unequal (different) parts, and putting $t = q$ and $u = x$, (10.35) is deduced.   ∎

Combinatorial identities may also be deduced from the relation

$$G(t; \boldsymbol{A}) = \sum_{n=0}^{\infty} p(n; \boldsymbol{A})t^n$$

when the number $p(n; \boldsymbol{A})$, of partitions of n whose number of parts equal to i belongs to the set $\{j : a_{i,j} = 1\}$, $i = 1, 2, \ldots$, where $\boldsymbol{A} = (a_{i,j})$, $i = 1, 2, \ldots$, $j = 0, 1, \ldots$, with $a_{i,j} = 0$ or 1, can be evaluated independently of the generating function $G(t; \boldsymbol{A})$. A characteristic case of such q-identity is *Euler's identity* that is deduced in the following corollary of Euler's pentagonal theorem.

COROLLARY 10.5

$$\prod_{i=1}^{\infty}(1 - q^i) = 1 + \sum_{k=1}^{\infty}(-1)^k \{q^{k(3k-1)/2} + q^{k(3k+1)/2}\}. \qquad (10.36)$$

PROOF   The generating functions of the numbers $Q_0(n)$ and $Q_1(n)$ of partitions of n into an even and odd number of unequal parts, respectively, are given by (see Example 10.5)

$$B_0(t) = \sum_{n=0}^{\infty} Q_0(n)t^n = \frac{1}{2} \left\{ \prod_{i=1}^{\infty}(1 + t^i) + \prod_{i=1}^{\infty}(1 - t^i) \right\}$$

and

$$B_1(t) = \sum_{n=0}^{\infty} Q_1(n)t^n = \frac{1}{2} \left\{ \prod_{i=1}^{\infty}(1 + t^i) - \prod_{i=1}^{\infty}(1 - t^i) \right\}.$$

Hence

$$\sum_{n=0}^{\infty} \{Q_0(n) - Q_1(n)\}t^n = \prod_{i=1}^{\infty}(1 - t^i).$$

Further, the difference $E(n) = Q_0(n) - Q_1(n)$ has been evaluated in Theorem 10.9 as

$$E(n) = \begin{cases} (-1)^k, & n = (3k^2 \pm k)/2 \\ 0, & \text{otherwise} \end{cases}$$

and so

$$\sum_{n=0}^{\infty} E(n)t^n = 1 + \sum_{k=1}^{\infty}(-1)^k \{t^{k(3k-1)/2} + t^{k(3k+1)/2}\}.$$

Consequently,

$$\prod_{i=1}^{\infty}(1 - t^i) = 1 + \sum_{k=1}^{\infty}(-1)^k \{t^{k(3k-1)/2} + t^{k(3k+1)/2}\}$$

and, putting $t = q$, (10.36) is deduced.   ∎

COROLLARY 10.6
The number $p(n)$ of partitions of n satisfies the recurrence relation

$$p(n) = \sum_{k=1}^{n_1}(-1)^{k-1}p(n - k(3k - 1)/2) + \sum_{k=1}^{n_2}(-1)^{k-1}p(n - k(3k + 1)/2),$$

(10.37)

where n_i is the largest positive integer satisfying for $i = 1$ the inequality $n_1(3n_1 - 1) \leq 2n$ and for $i = 2$ the inequality $n_2(3n_2 + 1) \leq 2n$.

PROOF   Expression (10.15) may be rewritten as

$$\left\{\prod_{i=1}^{\infty}(1 - t^i)\right\} \cdot \left\{\sum_{n=0}^{\infty}p(n)t^n\right\} = 1.$$

Replacing the first factor of the left-hand side by expression (10.36), it follows that

$$\left\{1 - \sum_{k=1}^{\infty}(-1)^{k-1}\{t^{k(3k-1)/2} + t^{k(3k+1)/2}\}\right\} \cdot \left\{\sum_{n=0}^{\infty}p(n)t^n\right\} = 1$$

and, since $p(0) = 1$,

$$\sum_{n=1}^{\infty}p(n)t^n = \left\{\sum_{n=0}^{\infty}p(n)t^n\right\} \cdot \left\{\sum_{k=1}^{\infty}(-1)^{k-1}\{t^{k(3k-1)/2} + t^{k(3k+1)/2}\}\right\}$$

$$= \sum_{n=1}^{\infty}\left\{\sum_{k=1}^{n_1}(-1)^{k-1}p(n - k(3k - 1)/2)\right.$$

$$\left. + \sum_{k=1}^{n_2}(-1)^{k-1}p(n - k(3k + 1)/2)\right\}t^n.$$

Equating the coefficients of t^n in both members of the last identity, (10.37) is deduced.   ∎

In the next theorem, using (10.34) and (10.35), the Gauss-Jacobi identity is derived.

THEOREM 10.11 Gauss-Jacobi identity

$$\prod_{i=1}^{\infty}(1 - q^{2i})(1 + xq^{2i-1})(1 + x^{-1}q^{2i-1}) = \sum_{k=-\infty}^{+\infty}x^k q^{k^2}.$$

(10.38)

PROOF   Replacing q by q^2 in (10.35) and, in the resulting expression, replacing xq by x, it follows that

$$\prod_{i=1}^{\infty}(1 + xq^{2i-1}) = \sum_{n=0}^{\infty} x^n q^{n^2} \prod_{j=1}^{n}(1 - q^{2j})^{-1}.$$

Multiplying both members by $\prod_{i=1}^{\infty}(1 - q^{2i})$ and using the relation

$$\prod_{i=1}^{\infty}(1 - q^{2i}) \prod_{j=1}^{n}(1 - q^{2j})^{-1} = \prod_{j=1}^{\infty}(1 - q^{2n+2j}),$$

we get

$$\prod_{i=1}^{\infty}(1 - q^{2i})(1 + xq^{2i-1}) = \sum_{n=0}^{\infty} x^n q^{n^2} \prod_{j=1}^{\infty}(1 - q^{2n+2j})$$

$$= \sum_{n=-\infty}^{\infty} x^n q^{n^2} \prod_{j=1}^{\infty}(1 - q^{2n+2j}),$$

where the last equation holds since the additional terms are zero because, for any negative integer n, the factor $j = -n$ in the product is zero. Also, replacing q by q^2 and x by $-q^{2n}$ in (10.35), we get

$$\prod_{j=1}^{\infty}(1 - q^{2n+2j}) = \sum_{r=0}^{\infty} x^r q^{r(r+1)+2rn} \prod_{i=1}^{r}(1 - q^{2i})^{-1}$$

and, introducing it into the previous expression, we sequentially deduce

$$\prod_{i=1}^{\infty}(1 - q^{2i})(1 + xq^{2i-1}) = \sum_{n=-\infty}^{+\infty} x^n q^{n^2} \sum_{r=0}^{\infty} x^r q^{r^2+r+2rn} \prod_{i=1}^{r}(1 - q^{2i})^{-1}$$

$$= \sum_{r=0}^{\infty}(-1)^r(-x^{-1}q)^r \prod_{i=1}^{r}(1 - q^{2i})^{-1} \sum_{n=-\infty}^{+\infty} x^{n+r} q^{(n+r)^2}$$

$$= \sum_{r=0}^{\infty}(-1)^r(-x^{-1}q)^r \prod_{i=1}^{r}(1 - q^{2i})^{-1} \sum_{k=-\infty}^{+\infty} x^k q^{k^2}.$$

Further, replacing q by q^2 in (10.34) and, in the resulting expression, replacing xq by $-x^{-1}$, it follows that

$$\left[\prod_{i=1}^{\infty}(1 + x^{-1}q^{2i-1})\right]^{-1} = \sum_{r=0}^{\infty}(-1)^r(-x^{-1}q)^r \prod_{i=1}^{r}(1 - q^{2i})^{-1}.$$

Introducing it into the previous expression, we deduce (10.38).   ∎

10.6 BIBLIOGRAPHIC NOTES

The roots of partitions of integers, according to L. E. Dickson (1920), may be traced in a letter from Gottfried Wilhelm Leibnitz to John Bernoulli in 1669. Their systematic study and development started in 1746 with Leonhard Euler and his two-volume work *Introductio in Analysin Infinitorum*. The proof of Euler's pentagonal theorem given here is due to Fabian Franklin (1881). The contribution of G. H. Hardy and Ramanujan (1917, 1918) and Rademacher (1937a,b, 1940, 1943) enhanced this subject. In his two-volume treatise, P. A. MacMahon (1915, 1916) discussed at length combinatorial aspects of partitions. For further reading on partitions of integers, refer to the books of J. Riordan (1958), L. Comtet (1974) and G. E. Andrews (1976).

10.7 EXERCISES

1. Show combinatorially that the number $p(n, k)$ of partitions of n into k parts satisfies the recurrence relation

$$p(n, k) = p(n - 1, k - 1) + p(n - k, k), \quad k < n/2,$$

with initial conditions

$$p(n, 1) = p(n, n) = 1, \quad n = 1, 2, \ldots .$$

2. Show combinatorially that the number $P(n, k)$ of partitions of n into at most k parts satisfies the recurrence relation

$$P(n, k) = P(n, k - 1) + P(n - k, k), \quad k = 2, 3, \ldots, n, \quad n = 2, 3, \ldots,$$

with initial conditions

$$P(n, 1) = 1, \quad n = 1, 2, \ldots, \quad P(0, k) = 1, \quad k = 0, 1, \ldots .$$

3. *Partitions into parts of restricted number and restricted size.* Let $P(n, k, r)$ be the number of partitions of n into at most k parts, none of which is greater than r. Show that

$$F_r(t, u) = \sum_{n=0}^{\infty} \sum_{k=0}^{\infty} P(n, k, r) u^k t^n = \prod_{i=1}^{r} (1 - ut^i)^{-1}$$

and

$$F_{r,k}(t) = \sum_{n=0}^{\infty} P(n,k,r)t^n = \prod_{i=1}^{r}(1 - t^{r+i})(1 - t^i)^{-1}.$$

4. Let $r(n,k)$ be the number of partitions of n whose smallest part equals k. Show that

$$\Psi_k(t) = \sum_{n=0}^{\infty} r(n,k)t^n = t^k \prod_{i=k}^{\infty}(1 - ut^i)^{-1}$$

and

$$r(n,k) = r(n-1, k-1) - r(n-k, k-1),$$

for $k = 2, 3, \ldots, [(n+1)/2]$, $n = 3, 4, \ldots$, with

$$r(n,1) = p(n-1), n = 2, 3, \ldots, r(n,n) = 1, n = 1, 2, \ldots, r(n,k) = 0, n < k.$$

5. Show combinatorially that the number $q(n,k)$ of partitions of n into k unequal parts satisfies the recurrence relation

$$q(n,k) = q(n-k, k) + q(n-k, k-1), \quad k = 2, 3, \ldots, [n/2], \quad n = 4, 5, \ldots,$$

with initial conditions

$$q(n,1) = 1, \, n = 1, 2, \ldots, \, q(n,k) = 0, \, n < k(k+1)/2, \, q(k(k+1)/2, k) = 1.$$

6. Let $Q(n,k)$ be the number of partitions of n into at most k unequal parts. Show that

$$F(t,u) = \sum_{n=0}^{\infty}\sum_{k=0}^{\infty} Q(n,k)u^k t^n = (1-u)^{-1}\prod_{i=1}^{\infty}(1 + ut^i)$$

and, using the identity

$$(1-u)F(t,u) = (1 - u^2 t^2)F(t, ut),$$

conclude that the generating function

$$F_k(t) = \sum_{n=0}^{\infty} Q(n,k)t^n$$

satisfies the recurrence relation

$$(1 - t^k)F_k(t) = F_{k-1}(t) - t^k F_{k-2}(t), \quad k = 2, 3, \ldots,$$

with $F_0(t) = 2$, $F_1(t) = (1-t)^{-1}$.

7. *Partitions into even and odd unequal parts.* Let $q_0(n, k)$ be the number of partitions of n into k even unequal parts and $q_1(n, k)$ be the number of partitions of n into k odd unequal parts. Show that

$$H_0(t, u) = \sum_{n=0}^{\infty} \sum_{k=0}^{n} q_0(n, k) u^k t^n = \prod_{i=1}^{\infty} (1 + ut^{2i}),$$

$$H_1(t, u) = \sum_{n=0}^{\infty} \sum_{k=0}^{n} q_1(n, k) u^k t^n = \prod_{i=1}^{\infty} (1 + ut^{2i-1})$$

and

$$H_{0,k}(t) = \sum_{n=0}^{\infty} q_0(n, k) t^n = t^{k(k+1)} \prod_{i=1}^{k} (1 - t^{2i})^{-1},$$

$$H_{1,k}(t) = \sum_{n=0}^{\infty} q_1(n, k) t^n = t^{k^2} \prod_{i=1}^{k} (1 - t^{2i})^{-1}.$$

8. *Partitions into parts of restricted size.* Let $p(n, k, r)$ be the number of partitions of n into k parts, none of which is greater than r. Show that

$$G_r(t, u) = \sum_{n=0}^{\infty} \sum_{k=0}^{n} p(n, k, r) u^k t^n = \prod_{i=1}^{r} (1 - ut^i)^{-1}$$

and conclude that

$$(1 - ut) G_r(t, u) = (1 - ut^{r+1}) G_r(t, ut).$$

Using this recurrence relation show that

$$G_{r,k}(t) = \sum_{n=k}^{\infty} p(n, k, r) t^n = t^k \prod_{i=1}^{k} (1 - t^{r+i-1})(1 - t^i)^{-1}.$$

9. Let

$$\sum_{r=0}^{\infty} G_r(t, u) g_r(t) u^r = G(t, u),$$

where $G_r(t, u)$ is the generating function of the number $p(n, k, r)$ of partitions of n into k parts, none of which is greater than r and $G(t, u)$ is the generating function of the number $p(n, k)$ of partitions of n into k parts. Using the relations

$$(1 - ut) G(t, u) = G(t, ut),$$

$$(1 - ut) \{ G_r(t, u) + ut^{r+1} G_{r+1}(t, u) \} = G_r(t, ut),$$

show that

$$g_r(t) = t^{r^2} \prod_{i=1}^{r} (1 - t^i)^{-1}$$

and conclude the identity

$$\sum_{r=0}^{\infty} t^{r^2} \prod_{i=1}^{r} (1 - t^i)^{-2} = \prod_{i=1}^{\infty} (1 - t^i)^{-1}.$$

10. Show that

$$\prod_{i=0}^{k-1} (1 + t^{2^i}) = \sum_{n=0}^{2^k - 1} t^n$$

and thus conclude that

$$\prod_{i=0}^{\infty} (1 + t^{2^i}) = (1 - t)^{-1}, \quad |t| < 1.$$

Using the last identity, show that for every nonnegative integer n there exists a unique finite sequence $n_i \in \{0, 1\}$, $i = 0, 1, \dots, r$, such that

$$n = \sum_{i=0}^{r} n_i 2^i,$$

where the positive integer r is determined by the inequalities $2^r \le n < 2^{r+1}$.

11. (*Continuation*). Show that the number $R_0(n)$ of partitions of n into an even number of parts, with values (sizes) in the set $\{1, 2, 2^2, \dots\}$, equals the number $R_1(n)$ of partitions of n into an odd number of such parts, for $n = 2, 3, \dots$.

12. Show that

$$\prod_{i=0}^{k-1} \left(\sum_{r_i=0}^{i} t^{r_i i!} \right) = \sum_{n=0}^{k!-1} t^n$$

and thus conclude that

$$\prod_{i=0}^{\infty} \left(\sum_{r_i=0}^{i} t^{r_i i!} \right) = (1 - t)^{-1}, \quad |t| < 1.$$

Using the last identity, show that for every nonnegative integer n there exists a unique finite sequence $n_i \in \{0, 1, \dots, i\}$, $i = 1, 2, \dots, r$, such that

$$n = \sum_{i=1}^{r} n_i i!,$$

where the positive integer r is determined by the inequalities $r! \leq n < (r+1)!$.

13. *Partitions into unequal parts of restricted size.* Let $q(n,k,r)$ be the number of partitions of n into k unequal parts, none of which is greater than r. Show that

$$H_r(t,u) = \sum_{n=0}^{\infty} \sum_{k=0}^{n} q(n,k,r) u^k t^n = \prod_{i=1}^{r} (1 + ut^i)$$

and conclude that

$$(1 + ut) H_r(t, ut) = (1 + ut^{r+1}) H_r(t, u).$$

Using this recurrence relation, show that

$$H_{r,k}(t) = \sum_{n=k(k+1)/2}^{\infty} p(n,k,r) t^n = t^{k(k+1)/2} \prod_{i=1}^{k} (1 - t^{r-i+1})(1 - t^i)^{-1},$$

for $k < r$, and

$$H_{r,r}(t) = t^{r(r+1)/2}.$$

14. *Sum of multinomial coefficients.* The multinomial coefficient

$$\binom{n}{k_1, k_2, \ldots, k_{r-1}} = \frac{n!}{k_1! k_2! \cdots k_{r-1}! k_r!}, \quad k_r = n - (k_1 + k_2 + \cdots + k_{r-1})$$

is a symmetric function of the variables $k_1, k_2, \ldots, k_{r-1}, k_r$:

$$\binom{n}{k_1, k_2, \ldots, k_{r-1}} = \binom{n}{k_1, k_2, \ldots, k_{r-2}, k_r} = \cdots = \binom{n}{k_2, k_3, \ldots, k_{r-1}, k_r}.$$

Thus, for the representative of this class of multinomial coefficients, with n and $k_1, k_2, \ldots, k_r$ fixed, we may assume that $k_1 \geq k_2 \geq \cdots \geq k_r$.

(a) Show that the number of different classes of multinomial coefficients, for given positive integers n and r, equals the number $P(n,r)$ of partitions of n into at most r parts, which, when $r \geq n$, equals the number $p(n)$ of partitions of n.

(b) Let

$$M(n) = \sum \frac{n!}{k_1! k_2! \cdots k_r!},$$

where the summation is extended over all nonnegative integer solutions of the linear equation $k_1 + k_2 + \cdots + k_r = n$, $k_1 \geq k_2 \geq \cdots \geq k_r \geq 0$. This is the sum of the multinomial coefficients, representatives of the $p(n)$ classes. Show that

$$\sum_{n=0}^{\infty} M(n) \frac{t^n}{n!} = \prod_{i=1}^{\infty} \left(1 - \frac{t^i}{i!}\right)^{-1}.$$

15*. *Perfect partitions.* A partition of a positive integer n that contains just one partition for every positive integer r less than n is called *perfect partition of n*. Thus, an integer solution of the equation $y_1 + 2y_2 + \cdots + ny_n = n$, $y_i \geq 0$, $i = 1, 2, \ldots, n$ is a perfect partition of n, if for every $r = 1, 2, \ldots, n - 1$, the equation $z_1 + 2z_2 + \cdots + rz_r = r$, $0 \leq z_i \leq y_i$, $i = 1, 2, \ldots, r$, has a unique solution. For example, the partitions 31^2, $2^2 1$, 1^5 are perfect partitions of $n = 5$. Show that the number of perfect partitions of n equals the number of ordered factorizations of $n+1$ without unit factors.

16. *Compositions of integers.* A composition of a positive integer n is an ordered collection of positive integers whose sum equals n. Thus a composition of n into k parts (summands) is a solution $(r_1, r_2, \ldots, r_k)$ in positive integers of the linear equation $x_1 + x_2 + \cdots + x_k = n$. Show that the number $c(n, k)$ of compositions of n into k parts equals

$$c(n, k) = \binom{n-1}{k-1}, \quad k \leq n$$

and, consequently,

$$G(t, u) = \sum_{n=1}^{\infty} \sum_{k=1}^{n} c(n, k) u^k t^n = \frac{ut}{1 - (u+1)t}.$$

17. *(Continuation).* Let $c(n, k; S)$ be the number of compositions of n into k parts of sizes x_i, $i = 1, 2, \ldots, k$ belonging to the set $S = \{s_1, s_2, \ldots, s_r, \ldots\}$, with $1 \leq s_1 < s_2 < \cdots < s_r < \cdots$. Show that

$$G(t, u; S) = 1 + \sum_{n=1}^{\infty} \sum_{k=1}^{n} c(n, k; S) u^k t^n = \left(1 - u \sum_{i=1}^{\infty} t^{s_i} \right)^{-1}.$$

18. *(Continuation).* Let $c_2(n)$ be the number of compositions of n with no part greater than 2. Show that

$$1 + \sum_{n=1}^{\infty} c_2(n) t^n = (1 - t - t^2)^{-1}$$

and thus conclude that $c_2(n) = f_n$, $n = 1, 2, \ldots$, where f_n is the *Fibonacci number*.

19. *(Continuation).* Let $c_r(n)$ be the number of compositions of n with no part greater than r. Setting $c_r(0) = 1$, show that

$$C_r(t) = \sum_{n=0}^{\infty} c_r(n) t^n = (1 - t - t^2 - \cdots - t^r)^{-1},$$

and
$$c_r(n) = c_r(n-1) + c_r(n-2) + \cdots + c_r(n-r), \; n \geq r,$$
$$c_r(n) = 2^{n-1}, \; n < r.$$

Thus, conclude that $Q_r(n) = 2c_r(n)$ is the number of n-permutations of 2 with repetition that include no r consecutive objects alike.

20. Let $p(n,k)$ be the number of partitions of n into k parts and $q(n,k)$ the number of partitions of n into k unequal parts. Show that

$$k!q(n,k) \leq \binom{n-1}{k-1} \leq k!p(n,k)$$

and, using the relation $p(n,k) = q(n+r,k)$, $r = k(k-1)/2$, conclude that

$$k!p(n,k) \leq \binom{n-r-1}{k-1}, \quad r = k(k-1)/2.$$

21. Using the Gauss-Jacobi identity, show that

$$\prod_{i=1}^{\infty}(1 - q^{2ri})(1 + q^{2ri-r+s})(1 + q^{2ri-r-s}) = \sum_{k=-\infty}^{+\infty} q^{rk^2+sk}, \quad r,s \geq 0$$

and

$$\prod_{i=1}^{\infty}(1 - q^{2ri})(1 - q^{2ri-r+s})(1 - q^{2ri-r-s}) = \sum_{k=-\infty}^{+\infty} (-1)^k q^{rk^2+sk}, \quad r,s \geq 0.$$

Putting $r = 3/2$, $s = 1/2$, conclude Euler's identity:

$$\prod_{i=1}^{\infty}(1 - q^i) = 1 + \sum_{k=1}^{\infty}(-1)^k \{q^{k(3k-1)/2} + q^{k(3k+1)/2}\}.$$

Further, setting $r = 1/2$, $s = \varepsilon + 1/2$ and taking the limit as $\varepsilon \to 0$, conclude the Jacobi identity:

$$\prod_{i=1}^{\infty}(1 - q^i) = 1 + \sum_{k=1}^{\infty}(-1)^k (2k+1) q^{k(k+1)/2}.$$

22*. Let
$$g_i(t) = \sum_{k=0}^{\infty} u_i(k)t^k, \quad i = 1,2,\ldots$$

and
$$a(n) = \sum_{\mathcal{P}_n} u_1(k_1)u_2(k_2)\cdots u_n(k_n), \quad n = 0,1,\ldots,$$

where the summation is extended over the set $\mathcal{P}_n$ of partitions of n, that is over all nonnegative integer solutions $(k_1, k_2, \ldots, k_n)$ of the equation $k_1 + 2k_2 + \cdots + nk_n = n$. Show that

$$G(t) = \sum_{n=0}^{\infty} a(n)t^n = \prod_{i=1}^{\infty} g_i(t^i)$$

and conclude that

$$\sum_{n=0}^{\infty} p(n)t^n = \prod_{i=1}^{\infty} (1 - t^i)^{-1}, \quad p(n) = \sum_{\mathcal{P}_n} 1.$$

23*. (*Continuation*). Let

$$A(k_1, k_2, \ldots, k_n; u_1, u_2, \ldots, u_n) = u_1 k_1 + u_2 k_2 + \cdots + u_n k_n.$$

Show that

$$\sum_{n=0}^{\infty} \left\{ \sum_{\mathcal{P}_n} A(k_1, k_2, \ldots, k_n; u_1, u_2, \ldots, u_n) \right\} t^n$$

$$= \prod_{i=1}^{\infty} (1 - t^i)^{-1} \sum_{j=1}^{\infty} u_j t^j (1 - t^j)^{-1}$$

and conclude that

$$\sum_{\mathcal{P}_n} A(k_1, k_2, \ldots, k_n; u_1, u_2, \ldots, u_n) = \sum_{k=1}^{n} p(n - k) \sum_{d|k} u_d.$$

In particular, for $u_j = j$, $j = 1, 2, \ldots$, conclude that

$$np(n) = \sum_{k=1}^{n} p(n - k)\sigma(k), \quad \sigma(k) = \sum_{d|k} d.$$

24. *Partitions of ordered pairs of integers.* A *partition* of an ordered pair (m, n) of nonnegative integers, different $(0, 0)$, is a non-ordered collection of nonnegative integers (x_i, y_i), $i = 1, 2, \ldots$, different $(0, 0)$, whose sum equals (m, n). In a partition of a bipartite number (m, n), let $z_{i,j}$ be the number of ordered pairs that are equal to (i, j), $i = 0, 1, \ldots, m$, $j = 0, 1, \ldots, n$, $(i, j) \neq (0, 0)$. Then

$$\sum_{i=1}^{m} i \sum_{j=0}^{n} z_{i,j} = m, \quad \sum_{j=1}^{n} j \sum_{i=0}^{m} z_{i,j} = n.$$

In the case of a partition of (m, n) into k parts, in addition,

$$\sum_{j=0}^{n} \sum_{\substack{i=0 \\ (i,j) \neq (0,0)}}^{m} z_{i,j} = k.$$

Let $p_{m,n}$ be the number of partitions of the bipartite (m, n) and $p_{m,n,k}$ the number of partitions of the bipartite number (m, n) into k parts. Prove that

$$F(t, u, w) = \sum_{n=0}^{\infty} \sum_{m=0}^{\infty} \sum_{k=0}^{m+n} p_{m,n,k} t^m u^n w^k = \prod_{j=0}^{\infty} \prod_{\substack{i=0 \\ (i,j) \neq (0,0)}}^{\infty} (1 - w t^i u^j)^{-1},$$

where $p_{0,0,0} = 1$, and conclude that

$$F(t, u) = \sum_{n=0}^{\infty} \sum_{m=0}^{\infty} p_{m,n} t^m u^n = \prod_{j=0}^{\infty} \prod_{\substack{i=0 \\ (i,j) \neq (0,0)}}^{\infty} (1 - t^i u^j)^{-1},$$

where $p_{0,0} = 1$.

25. *q-number, q-factorial and q-binomial coefficient.* Let $0 < q < 1$, x be a real number and k a positive integer. The number $[x]_q = (1 - q^x)/(1 - q)$ is called *q-real number* and, in particular, $[k]_q = (1 - q^k)/(1 - q)$ is called *q-positive integer*. The factorial of the q-number $[x]_q$ of order k, which is defined by

$$[x]_{k,q} = [x]_q [x-1]_q \cdots [x-k+1]_q = \frac{(1 - q^x)(1 - q^{x-1}) \cdots (1 - q^{x-k+1})}{(1 - q)^k},$$

is called *q-factorial of x of order k.* In particular, $[k]_q! = [1]_q [2]_q \cdots [k]_q = (1 - q)(1 - q^2) \cdots (1 - q^k)/(1 - q)^k$. The *q-binomial coefficient* is defined by

$$\begin{bmatrix} x \\ k \end{bmatrix}_q = \frac{[x]_{k,q}}{[k]_q!} = \frac{(1 - q^x)(1 - q^{x-1}) \cdots (1 - q^{x-k+1})}{(1 - q)(1 - q^2) \cdots (1 - q^k)}.$$

For x a real number and k a positive integer, derive the recurrence relations

$$\begin{bmatrix} x \\ k \end{bmatrix}_q = \begin{bmatrix} x-1 \\ k-1 \end{bmatrix}_q + q^k \begin{bmatrix} x-1 \\ k \end{bmatrix}_q$$

and

$$\begin{bmatrix} x \\ k \end{bmatrix}_q = \frac{[x]_q}{[k]_q} \begin{bmatrix} x-1 \\ k-1 \end{bmatrix}_q.$$

Also, show that

$$\begin{bmatrix} -x \\ k \end{bmatrix}_q = (-1)^k q^{-kx-k(k-1)/2} \begin{bmatrix} x+k-1 \\ k \end{bmatrix}_q,$$

$$\begin{bmatrix} x \\ k \end{bmatrix}_{1/q} = q^{-k(x-k)} \begin{bmatrix} x \\ k \end{bmatrix}_q$$

and

$$\lim_{q \to 1} \begin{bmatrix} x \\ k \end{bmatrix}_q = \binom{x}{k}.$$

26. (*Continuation*). For n, k and m positive integers, show that (a)

$$\begin{bmatrix} n \\ k \end{bmatrix}_q = \begin{bmatrix} n \\ n-k \end{bmatrix}_q$$

and (b)

$$\begin{bmatrix} n \\ m \end{bmatrix}_q \begin{bmatrix} m \\ k \end{bmatrix}_q = \begin{bmatrix} n \\ k \end{bmatrix}_q \begin{bmatrix} n-k \\ m-k \end{bmatrix}_q = \begin{bmatrix} n \\ m-k \end{bmatrix}_q \begin{bmatrix} n-m+k \\ k \end{bmatrix}_q,$$

$$\begin{bmatrix} n \\ m \end{bmatrix}_q \begin{bmatrix} n-m \\ k \end{bmatrix}_q = \begin{bmatrix} n \\ k \end{bmatrix}_q \begin{bmatrix} n-k \\ m \end{bmatrix}_q = \begin{bmatrix} n \\ m+k \end{bmatrix}_q \begin{bmatrix} m+k \\ m \end{bmatrix}_q.$$

27. (*Continuation*). For n, k and m positive integers, show that

$$\sum_{r=m}^{n} q^{-rk} \begin{bmatrix} r-1 \\ k-1 \end{bmatrix}_q = q^{-nk} \begin{bmatrix} n \\ k \end{bmatrix}_q - q^{(m-1)k} \begin{bmatrix} m-1 \\ k \end{bmatrix}_q$$

and conclude that

$$\sum_{r=k}^{n} q^{-rk} \begin{bmatrix} r-1 \\ k-1 \end{bmatrix}_q = q^{-nk} \begin{bmatrix} n \\ k \end{bmatrix}_q.$$

28. (*Continuation*). For n, k and m positive integers, show that

$$\sum_{r=k}^{m} (-1)^{r-k} q^{r(r+1)/2} \begin{bmatrix} n+1 \\ r+1 \end{bmatrix}_q = q^{k(k+1)/2} \begin{bmatrix} n \\ k \end{bmatrix}_q$$

$$+ (-1)^{m-k} q^{(m+1)(m+2)/2} \begin{bmatrix} n \\ m+1 \end{bmatrix}_q$$

and conclude that

$$\sum_{r=k}^{n} (-1)^{r-k} q^{r(r+1)/2} \begin{bmatrix} n+1 \\ r+1 \end{bmatrix}_q = q^{k(k+1)/2} \begin{bmatrix} n \\ k \end{bmatrix}_q.$$

29. *q-binomial and negative binomial formulae.* For n a positive integer and x a real number, prove that

$$\prod_{i=1}^{n}(1+xq^{i-1}) = \sum_{k=0}^{n} q^{k(k-1)/2} \begin{bmatrix} n \\ k \end{bmatrix}_q x^k, \quad 0 < q < 1$$

and

$$\prod_{i=1}^{n}(1-xq^{i-1})^{-1} = \sum_{k=0}^{\infty} \begin{bmatrix} n+k-1 \\ k \end{bmatrix}_q x^k, \quad 0 < q < 1.$$

Also, prove that

$$(q^x)^n = \sum_{k=0}^{n}(-1)^k q^{k(k-1)/2}(1-q)^k \begin{bmatrix} n \\ k \end{bmatrix}_q [x]_{k,q}.$$

30. (*Continuation*). For n and k positive integers, show that

$$\begin{bmatrix} r+s \\ n \end{bmatrix}_q = \sum_{k=0}^{n} q^{(n-k)(r-k)} \begin{bmatrix} r \\ k \end{bmatrix}_q \begin{bmatrix} s \\ n-k \end{bmatrix}_q$$

and

$$\begin{bmatrix} r+s+n-1 \\ n \end{bmatrix}_q = \sum_{k=0}^{n} q^{r(n-k)} \begin{bmatrix} r+k-1 \\ k \end{bmatrix}_q \begin{bmatrix} s+n-k-1 \\ n-k \end{bmatrix}_q.$$

More generally, for x and y real numbers and n a positive integer, show that

$$\begin{bmatrix} x+y \\ n \end{bmatrix}_q = \sum_{k=0}^{n} q^{(n-k)(x-k)} \begin{bmatrix} x \\ k \end{bmatrix}_q \begin{bmatrix} y \\ n-k \end{bmatrix}_q$$

or, equivalently, that

$$[x+y]_{n,q} = \sum_{k=0}^{n} q^{(n-k)(x-k)} \begin{bmatrix} n \\ k \end{bmatrix}_q [x]_{k,q}[y]_{n-k,q}.$$

31. (*Continuation*). For n and k positive integers, show that

$$\sum_{r=k}^{n}(-1)^{n-r} q^{(n-r)(n-r-1)/2} \begin{bmatrix} n \\ r \end{bmatrix}_q \begin{bmatrix} r \\ k \end{bmatrix}_q = \delta_{n,k}$$

and

$$\sum_{r=k}^{n}(-1)^{r-k} q^{(r-k)(r-k-1)/2} \begin{bmatrix} n \\ r \end{bmatrix}_q \begin{bmatrix} r \\ k \end{bmatrix}_q = \delta_{n,k}.$$

32*. *q-Stirling numbers of the first kind.* Consider the n-th order factorial of the q-number $[t]_q = (1 - q^t)/(1 - q)$, $0 < q < 1$ (n-th order q-factorial of t):

$$[t]_{n,q} = [t]_q[t-1]_q \cdots [t-n+1]_q$$
$$= q^{-n(n-1)/2}[t]_q([t]_q - [1]_q) \cdots ([t]_q - [n-1]_q).$$

This is a polynomial of the q-number $[t]_q$ of degree n:

$$[t]_{n,q} = q^{-n(n-1)/2} \sum_{k=0}^{n} s(n,k|q)[t]_q^k, \quad n = 0, 1, \ldots.$$

The coefficient $s(n, k|q)$ is called *q-Stirling number of the first kind.*
(a) Show that

$$s(n, k|q) = (-1)^{n-k} \sum [i_1]_q [i_2]_q \cdots [i_{n-k}]_q,$$

where the summation is extended over all $(n - k)$-combinations $\{i_1, i_2, \ldots, i_{n-k}\}$ of the $n - 1$ positive integers $\{1, 2, \ldots, n - 1\}$.
  (b) Derive the triangular recurrence relation

$$s(n + 1, k|q) = s(n, k - 1|q) - \{n\}_q s(n, k|q),$$

for $k = 1, 2, \ldots, n + 1$, $n = 0, 1, \ldots$, with

$$s(0, 0|q) = 1, \quad s(n, k|q) = 0, \quad k > n,$$

and the vertical recurrence relation

$$s(n + 1, k + 1|q) = \sum_{r=k}^{n} (-1)^{n-r}[n]_{n-r,q} s(r, k|q).$$

33*. *q-Stirling numbers of the second kind.* Consider the expansion of the n-th power of the q-number $[t]_q$ into factorials of the same number:

$$[t]_q^n = \sum_{k=0}^{n} q^{k(k-1)/2} S(n, k|q)[t]_{k,q}, \quad n = 0, 1, \ldots.$$

The coefficient $S(n, k|q)$ is called *q-Stirling number of the second kind.*
(a) Derive the triangular recurrence relation

$$S(n + 1, k|q) = S(n, k - 1|q) + [k]_q S(n, k|q),$$

for $k = 1, 2, \ldots, n + 1$, $n = 0, 1, \ldots$, with

$$S(0, 0|q) = 1, \quad S(n, k|q) = 0, \quad k > n,$$

and the generating function

$$\phi_k(u|q) = \sum_{n=k}^{\infty} S(n,k|q)u^n = u^k \prod_{j=1}^{k}(1 - [j]_q u)^{-1}, \quad k = 1, 2, \ldots$$

and conclude the expansion

$$\frac{1}{[t]_{k+1,q}} = q^{k(k+1)/2} \sum_{n=k}^{\infty} S(n,k|q)\frac{1}{[t]_q^{n+1}}$$

and its inverse

$$\frac{1}{[t]_q^{k+1}} = \sum_{n=k}^{\infty} q^{-n(n+1)/2} s(n,k|q)\frac{1}{[t]_{n+1,q}}.$$

(b) Further, deduce the vertical recurrence relation

$$S(n,k|q) = \sum_{r=k}^{n}[k]_q^{n-r} S(r-1,k-1|q),$$

and the expressions

$$S(n,k|q) = \sum [1]_q^{r_1}[2]_q^{r_2}\cdots[k]_q^{r_k} = \sum [i_1]_q[i_2]_q\cdots[i_{n-k}]_q,$$

where the summation in the first sum is extended over all integers $r_j \geq 0$, $j = 1, 2, \ldots, k$, with $r_1 + r_2 + \cdots + r_k = n - k$, while, in the second sum, is extended over all $(n-k)$-combinations $\{i_1, i_2, \ldots, i_{n-k}\}$, with repetition, of the k positive integers $\{1, 2, \ldots, k\}$.

(c) Derive the orthogonal relations

$$\sum_{r=k}^{n} s(n,r|q)S(r,k|q) = \delta_{n,k}, \quad \sum_{r=k}^{n} S(n,r|q)s(r,k|q) = \delta_{n,k}.$$

34*. (*Continuation*). Show that the q-Stirling number of the second kind $S(n,k|q)$, $k = 0, 1, \ldots, n$, $n = 0, 1, \ldots$, is given by the sum

$$S(n,k|q) = \frac{q^{-k(k-1)/2}}{[k]_q!} \sum_{r=0}^{k}(-1)^{k-r} q^{(k-r)(k-r-1)/2} \begin{bmatrix} k \\ r \end{bmatrix}_q [r]_q^n.$$

Alternatively,

$$S(n,k|q) = \frac{1}{(1-q)^{n-k}} \sum_{r=k}^{n}(-1)^{k-r} \binom{n}{r} \begin{bmatrix} r \\ k \end{bmatrix}_q.$$

Inverting the last expression, deduce for the q-Stirling number of the first kind $s(n, k|q)$, $k = 0, 1, \ldots, n$, $n = 0, 1, \ldots$ the explicit expression

$$s(n, k|q) = \frac{1}{(1-q)^{n-k}} \sum_{r=k}^{n} (-1)^{r-k} q^{(n-r)(n-r-1)/2} \begin{bmatrix} n \\ r \end{bmatrix}_q \binom{r}{k}.$$

35*. *q-Lah numbers.* Consider the expansion

$$[-t]_{n,q^{-1}} = q^{n(n-1)/2} \sum_{k=0}^{n} q^{k(k-1)/2} L(n, k|q) [t]_{k,q}.$$

Since $[-t]_{n,q^{-1}} = [-1]_q^n [t+n-1]_{n,q}$ and setting $|L(n, k|q)| = [-1]_q^n L(n, k|q)$, it can be written as

$$[t + n - 1]_{n,q} = q^{n(n-1)/2} \sum_{k=0}^{n} q^{k(k-1)/2} |L(n, k|q)| [t]_{k,q}.$$

The coefficients $L(n, k|q)$ and $|L(n, k|q)|$ are called *q-Lah* and *signless q-Lah numbers*, respectively. Show that

$$|L(n, k|q)| = q^{-n(n-1)/2+k(k-1)/2} \frac{[n]_q!}{[k]_q!} \begin{bmatrix} n-1 \\ k-1 \end{bmatrix}_q.$$

Chapter 11

PARTITION POLYNOMIALS

11.1 INTRODUCTION

The partition polynomials, introduced by E. T. Bell (1927), are multi-variable polynomials that are defined by a sum extended over all partitions of their index. Such a sum had been previously used by F. Faa di Bruno (1855) to express the derivatives of a composite function in terms of the derivatives of the component functions. The later partition polynomials have found many applications in combinatorics, probability theory and statistics, as well as in number theory. Since they are quite general polynomials, they include as particular cases the exponential polynomials and their inverses, the logarithmic polynomials, as well as the potential polynomials, owing their particular names to the form of their generating functions.

In this chapter, the partition polynomials are presented along with some of their combinatorial and probabilistic applications. Specifically, the exponential and partial Bell partition polynomials are examined first. Generating functions and recurrence relations are derived. As an example, the general arithmetical function, that led E. T. Bell to introduce the exponential partition polynomials is presented. Then, the general partition polynomials are examined; the Faa di Bruno formula is expressed in terms of these polynomials. Their use in expressing the probability function of a compound discrete distribution in terms of the component probability functions is discussed. The logarithmic and potential polynomials, constituting particular cases of the partition polynomials, are further explored. The Lagrange formula for the inversion of a power series is derived by the aid of a generalization of Faa di Bruno formula. J. Touchard (1939), in studying the problem of enumeration of permutations with a given number of partially ordered cycles introduced the partition polynomials that were later called Touchard polynomials. These polynomials are briefly presented at the end of this chapter.

11.2   EXPONENTIAL BELL PARTITION POLYNOMIALS

DEFINITION 11.1   *The polynomial $B_n \equiv B_n(x_1, x_2, \ldots, x_n)$ in the variables $x_1, x_2, \ldots, x_n$ defined by the sum*

$$B_n = \sum \frac{n!}{k_1!(1!)^{k_1} k_2!(2!)^{k_2} \cdots k_n!(n!)^{k_n}} x_1^{k_1} x_2^{k_2} \cdots x_n^{k_n}, \qquad (11.1)$$

where the summation is extended over all partitions of n, that is, over all nonnegative integer solutions $(k_1, k_2, \ldots, k_n)$ of the equation $k_1 + 2k_2 + \cdots + nk_n = n$, is called exponential Bell partition polynomial.

DEFINITION 11.2   *The polynomial $B_{n,k} \equiv B_{n,k}(x_1, x_2 \ldots, x_n)$ in the variables $x_1, x_2, \ldots, x_n$ of degree k defined by the sum*

$$B_{n,k} = \sum \frac{n!}{k_1!(1!)^{k_1} k_2!(2!)^{k_2} \cdots k_n!(n!)^{k_n}} x_1^{k_1} x_2^{k_2} \cdots x_n^{k_n}, \qquad (11.2)$$

where the summation is extended over all partitions of n into k parts, that is, over all nonnegative integer solutions $(k_1, k_2, \ldots, k_n)$ of the equations $k_1 + 2k_2 + \cdots + nk_n = n$, $k_1 + k_2 + \cdots + k_n = k$, is called (exponential) partial Bell partition polynomial.

Definitions 11.1 and 11.2 directly imply the following relations:

$$B_n(ax_1, a^2 x_2, \ldots, a^n x_n) = a^n B_n(x_1, x_2, \ldots, x_n), \qquad (11.3)$$

$$B_{n,k}(abx_1, a^2 bx_2, \ldots, a^n bx_n) = a^n b^k B_{n,k}(x_1, x_2, \ldots, x_n), \qquad (11.4)$$

$$B_n(x_1, x_2, \ldots, x_n) = \sum_{k=0}^{n} B_{n,k}(x_1, x_2, \ldots, x_n), \quad n = 0, 1, \ldots. \qquad (11.5)$$

REMARK 11.1   The number of monomials contained in the exponential Bell partition polynomial (11.1) equals the number $p(n)$ of partitions of n and in the partial Bell partition polynomial (11.2) equals the number $p(n, k)$ of partitions of n into k parts. These remarks justify the term *partition* that is included in the name of polynomials (11.1) and (11.2). The term *partial* that is used for polynomial (11.2) in contrast to polynomial (11.1) is justified by the fact, that in (11.2), the summation is extended over a subset of the partitions of n and, specifically, over the partitions of n into k parts. The exponential form of their generating functions (see Theorem 11.1 that follows) justifies the term *exponential* that is included in the name of polynomial (11.1) and, occasionally, in the name of polynomial (11.2).

The coefficients of polynomials (11.1) and (11.2) are positive integers. Specifically, the coefficient of the general term of the exponential Bell partition polynomial (11.1) equals the number of partitions of a finite set of n elements into subsets, among which $k_i \geq 0$ subsets include i elements each, for $i = 1, 2, \ldots, n$. Also, the coefficient of the general term of the partial Bell partition polynomial (11.2) equals the number of partitions of a finite set of n elements into k subsets, among which $k_i \geq 0$ subsets include i elements each, for $i = 1, 2, \ldots, n$. Thus, polynomials (11.1) and (11.2) are the multivariable generating functions of the numbers of partitions of a finite set of n elements into any number of subsets and into k subsets, respectively, with respect to the numbers $k_i \geq 0$ of subsets including i elements each, $i = 1, 2, \ldots, n$. ∎

Bivariate and vertical generating functions for the partial Bell partition polynomials are derived in the following theorem.

THEOREM 11.1

(a) *The bivariate generating function of the partial Bell partition polynomials $B_{n,k}(x_1, x_2, \ldots, x_n)$, $k = 0, 1, \ldots, n$, $n = 0, 1, \ldots$, with $B_{0,0} \equiv 1$, is given by*

$$B(t, u) = \sum_{n=0}^{\infty} \sum_{k=0}^{n} B_{n,k}(x_1, x_2, \ldots, x_n) u^k \frac{t^n}{n!} = \exp\{u[g(t) - x_0]\}. \quad (11.6)$$

(b) *The vertical generating function of the partial Bell partition polynomials $B_{n,k}(x_1, x_2, \ldots, x_n)$, $n = k, k+1, \ldots$, for fixed k, is given by*

$$B_k(t) = \sum_{n=k}^{\infty} B_{n,k}(x_1, x_2, \ldots, x_n) \frac{t^n}{n!} = \frac{[g(t) - x_0]^k}{k!}, \quad k = 0, 1, \ldots, \quad (11.7)$$

where

$$g(t) = \sum_{r=0}^{\infty} x_r \frac{t^r}{r!}. \quad (11.8)$$

PROOF   (a) Multiplying (11.2) by $u^k t^n / n!$ and summing the resulting expression for $k = 0, 1, \ldots, n$ and $n = 0, 1, \ldots$, it follows that

$$B(t, u) = \sum_{n=0}^{\infty} \sum_{k=0}^{n} B_{n,k}(x_1, x_2, \ldots, x_n) u^t \frac{t^n}{n!}$$

$$= \sum_{n=0}^{\infty} \sum_{k=0}^{n} \sum \frac{n!}{k_1! k_2! \cdots k_n!} \left(\frac{x_1 u t}{1!}\right)^{k_1} \left(\frac{x_2 u t^2}{2!}\right)^{k_2} \cdots \left(\frac{x_n u t^n}{n!}\right)^{k_n},$$

where the summation in the inner sum is extended over all nonnegative integer solutions $(k_1, k_2, \ldots, k_n)$ of $k_1 + 2k_2 + \cdots + nk_n = n$ and $k_1 + k_2 + \cdots + k_n = k$. Since this inner sum is summed over all $k = 0, 1, \ldots, n$ and $n = 0, 1, \ldots$, it follows that the summation is extended over all $k_r = 0, 1, \ldots$, for $r = 1, 2, \ldots$, and so

$$B(t, u) = \prod_{r=1}^{\infty} \sum_{k_r=0}^{\infty} \frac{1}{k_r!} \left(\frac{x_r u t^r}{r!} \right)^{k_r} = \prod_{r=1}^{\infty} \exp \left(\frac{x_r u t^r}{r!} \right)$$

$$= \exp \left(u \sum_{r=1}^{\infty} x_r \frac{t^r}{r!} \right).$$

Using (11.8), the last expression implies (11.6).

(b) Interchanging the order of summation, (11.6) may be written as

$$B(t, u) = \sum_{k=0}^{\infty} \sum_{n=k}^{\infty} B_{n,k}(x_1, x_2, \ldots, x_n) u^k \frac{t^n}{n!} = \sum_{k=0}^{\infty} B_k(t) u^k$$

and, since $B(t, u) = \exp\{u[g(t) - x_0]\}$, it follows that

$$\sum_{k=0}^{\infty} B_k(t) u^k = \sum_{k=0}^{\infty} \frac{[g(t) - x_0]^k}{k!} u^k.$$

Equating the coefficients of u^k in both sides of the last expression, (11.7) is deduced. ∎

The generating function of the exponential Bell partition polynomials, by virtue of (11.5), may be deduced from (11.6) by setting $u = 1$.

COROLLARY 11.1

The generating function of the exponential Bell partition polynomials $B_n \equiv B_n(x_1, x_2, \ldots, x_n)$, $n = 0, 1, \ldots$, with $B_0 \equiv 1$, is given by

$$B(t) = \sum_{n=0}^{\infty} B_n(x_1, x_2, \ldots, x_n) \frac{t^n}{n!} = \exp[g(t) - x_0], \qquad (11.9)$$

where $g(t) = \sum_{r=0}^{\infty} x_r t^r / r!$.

Recurrence relations for the exponential and the partial Bell partition polynomials are derived in the following theorem.

THEOREM 11.2

(a) *The exponential Bell partition polynomials $B_n \equiv B_n(x_1, x_2, \ldots, x_n)$, $n =$*

$0, 1, \ldots,$ with $B_0 \equiv 1,$ *satisfy the recurrence relation*

$$B_{n+1} = \sum_{r=0}^{n} \binom{n}{r} x_{r+1} B_{n-r}, \quad n = 0, 1, \ldots. \tag{11.10}$$

(b) *The partial Bell partition polynomials* $B_{n,k} \equiv B_{n,k}(x_1, x_2, \ldots, x_n),$ $k = 0, 1, \ldots, n,$ $n = 0, 1, \ldots,$ *with* $B_{0,0} \equiv 1,$ *satisfy the recurrence relations*

$$B_{n+1,k+1} = \sum_{r=0}^{n-k} \binom{n}{r} x_{r+1} B_{n-r,k} \tag{11.11}$$

and

$$B_{n+1,k+1} = \frac{1}{k+1} \sum_{r=0}^{n-k} \binom{n+1}{r+1} x_{r+1} B_{n-r,k}, \tag{11.12}$$

for $k = 0, 1, \ldots, n,$ $n = 0, 1, \ldots.$

PROOF (a) Differentiating the generating function (11.9) and expanding the resulting expression,

$$\frac{d}{dt} B(t) = B(t) \frac{d}{dt} g(t),$$

into powers of t it follows that

$$\sum_{n=0}^{\infty} B_{n+1} \frac{t^n}{n!} = \left[\sum_{r=0}^{\infty} x_{r+1} \frac{t^r}{r!} \right] \cdot \left[\sum_{n=0}^{\infty} B_n \frac{t^n}{n!} \right]$$

and so

$$\sum_{n=0}^{\infty} B_{n+1} \frac{t^n}{n!} = \sum_{n=0}^{\infty} \left[\sum_{r=0}^{n} \binom{n}{r} x_{r+1} B_{n-r} \right] \frac{t^n}{n!}.$$

Equating the coefficients of $t^n/n!$ in both sides of the last expression, (11.10) is deduced.

(b) Differentiating the generating function (11.7) and expanding the resulting expression,

$$\frac{d}{dt} B_{k+1}(t) = B_k(t) \frac{d}{dt} g(t),$$

into powers of t, it follows that

$$\sum_{n=k}^{\infty} B_{n+1,k+1} \frac{t^n}{n!} = \left[\sum_{r=0}^{\infty} x_{r+1} \frac{t^r}{r!} \right] \cdot \left[\sum_{n=k}^{\infty} B_{n,k} \frac{t^n}{n!} \right]$$

$$= \sum_{n=k}^{\infty} \left[\sum_{r=0}^{n-k} \binom{n}{r} x_{r+1} B_{n-r,k} \right] \frac{t^n}{n!}.$$

Equating the coefficients of $t^n/n!$ in both sides of the last expression, (11.11) is deduced. Further, the generating function (11.7) may be written in the form

$$(k+1)B_{k+1}(t) = [g(t) - x_0]B_k(t),$$

which, expanded into powers of t, yields

$$\sum_{n=k}^{\infty}(k+1)B_{n+1,k+1}\frac{t^{n+1}}{(n+1)!} = \left[\sum_{r=0}^{\infty}x_{r+1}\frac{t^{r+1}}{(r+1)!}\right] \cdot \left[\sum_{n=k}^{\infty}B_{n,k}\frac{t^n}{n!}\right]$$

$$= \sum_{n=k}^{\infty}\left[\sum_{r=0}^{n-k}\binom{n+1}{r+1}x_{r+1}B_{n-r,k}\right]\frac{t^{n+1}}{(n+1)!}.$$

Equating the coefficients of $t^{n+1}/(n+1)!$ in both sides of the last expression, (11.12) is deduced. ∎

Using recurrence relation (11.11), a table of partial Bell partition polynomials $B_{n,k} \equiv B_{n,k}(x_1, x_2 \ldots, x_n)$ can be constructed. Table 11.1 gives the polynomial $B_{n,k}$ for $k = 1, 2, \ldots, n$, $n = 1, 2, \ldots, 10$. From this table, using the relation $B_n = \sum_{k=1}^{n} B_{n,k}$, the exponential Bell partition polynomials may be deduced.

Example 11.1 An arithmetical function

The Bell partition polynomials were introduced by E. T. Bell (1927) as new arithmetical functions, unifying a diversity of such functions. Let S be any set of positive integers and

$$x_n = a\sum_{d}du^{n/d}, \quad n = 1, 2, \ldots,$$

where the summation is extended over all divisors d of n that belong to S. Consider the polynomials $B_n(x_1, 1!x_2, \ldots, (n-1)!x_n)$, $n = 1, 2, \ldots$. Show that

$$\sum_{n=0}^{\infty}B_n(x_1, 1!x_2, \ldots, (n-1)!x_n)\frac{t^n}{n!} = \prod_{k \in S}(1 - ut^k)^{-a}.$$

The generating function of the sequence x_n, $n = 1, 2, \ldots$,

$$h(t) = \sum_{n=1}^{\infty}x_n t^n = a\sum_{n=1}^{\infty}\sum_{d}du^{n/d}t^n,$$

on replacing the variables (d, n) by the variables (k, r), with $k = d$, $r = n/d$, and summing first for $r = 1, 2, \ldots$ and then for $k \in S$, may be obtained as

$$h(t) = a\sum_{k \in S}k\sum_{r=1}^{\infty}(ut^k)^r = a\sum_{k \in S}kut^k(1 - ut^k)^{-1}.$$

Table 11.1 Partial Bell Partition Polynomials $B_{n,k}$

$B_{1,1} = x_1$; $B_{2,1} = x_2$; $B_{2,2} = x_1^2$; $B_{3,1} = x_3$; $B_{3,2} = 3x_1x_2$; $B_{3,3} = x_1^3$;

$B_{4,1} = x_4$; $B_{4,2} = 4x_1x_3 + 3x_2^2$; $B_{4,3} = 6x_1^2x_2$; $B_{4,4} = x_1^4$;

$B_{5,1} = x_5$; $B_{5,2} = 5x_1x_4 + 10x_2x_3$; $B_{5,3} = 10x_1^2x_3 + 15x_1x_2^2$;

$B_{5,4} = 10x_1^3x_2$; $B_{5,5} = x_1^5$;

$B_{6,1} = x_6$; $B_{6,2} = 6x_1x_5 + 15x_2x_4 + 10x_3^2$;

$B_{6,3} = 15x_1^2x_4 + 60x_1x_2x_3 + 15x_2^3$;

$B_{6,4} = 20x_1^3x_3 + 45x_1^2x_2^2$; $B_{6,5} = 15x_1^4x_2$; $B_{6,6} = x_1^6$;

$B_{7,1} = x_7$; $B_{7,2} = 7x_1x_6 + 21x_2x_5 + 35x_3x_4$;

$B_{7,3} = 21x_1^2x_5 + 105x_1x_2x_4 + 70x_1x_3^2 + 105x_2^2x_3$;

$B_{7,4} = 35x_1x_4 + 105x_1x_3^2 + 210x_1^2x_2x_3$; $B_{7,5} = 35x_1^4x_3 + 105x_1^3x_2^2$;

$B_{7,6} = 21x_1^5x_2$; $B_{7,7} = x_1^7$;

$B_{8,1} = x_8$; $B_{8,2} = 8x_1x_7 + 28x_2x_6 + 56x_3x_5 + 35x_4^2$;

$B_{8,3} = 28x_1^2x_6 + 168x_1x_2x_5 + 280x_1x_3x_4 + 210x_2^2x_4 + 280x_2x_3^2$;

$B_{8,4} = 56x_1^3x_5 + 420x_1^2x_2x_4 + 280x_1^2x_3^2 + 840x_1x_2^2x_3 + 105x_2^4$;

$B_{8,5} = 70x_1^4x_4 + 560x_1^3x_2x_3 + 420x_1^2x_2^3$; $B_{8,6} = 56x_1^5x_3 + 210x_1^4x_2^2$;

$B_{8,7} = 28x_1^6x_2$; $B_{8,8} = x_1^8$;.

$B_{9,1} = x_9$; $B_{9,2} = 9x_1x_8 + 36x_2x_7 + 84x_3x_6 + 126x_4x_5$;

$B_{9,3} = 36x_1^2x_7 + 252x_1x_2x_6 + 504x_1x_3x_5 + 378x_2^2x_5 + 315x_1x_4^2$
$\qquad + 1260x_2x_3x_4 + 280x_3^3$;

$B_{9,4} = 84x_1^3x_6 + 756x_1^2x_2x_5 + 1260x_1^2x_3x_4 + 1890x_1x_2^2x_4$
$\qquad + 2520x_1x_2x_3^2 + 1260x_2^3x_3$;

$B_{9,5} = 126x_1^4x_5 + 1260x_1^3x_2x_4 + 840x_1^3x_3^2 + 3780x_1^2x_2^2x_3 + 945x_1x_2^4$;

$B_{9,6} = 126x_1^5x_4 + 1260x_1^4x_2x_3 + 1260x_1^3x_2^3$;

$B_{9,7} = 84x_1^6x_3 + 378x_1^5x_2^2$; $B_{9,8} = 36x_1^7x_2$; $B_{9,9} = x_1^9$;

$B_{10,1} = x_1^{10}$; $B_{10,2} = 10x_1x_9 + 45x_2x_8 + 120x_3x_7 + 210x_4x_6 + 126x_5^2$;

$B_{10,3} = 45x_1^2x_8 + 360x_1x_2x_7 + 840x_1x_3x_6 + 630x_2^2x_6 + 1260x_1x_4x_5$
$\qquad + 2520x_2x_3x_5 + 1575x_2x_4^2 + 2100x_3^2x_4$;

$B_{10,4} = 120x_1^3x_7 + 1260x_1^2x_2x_6 + 2520x_1^2x_3x_5 + 3780x_1x_2^2x_5 + 1575x_1^2x_4^2$
$\qquad + 12600x_1x_2x_3x_4 + 3150x_2^3x_4 + 2800x_1x_3^3 + 6300x_2^2x_3^2$;

$B_{10,5} = 210x_1^4x_6 + 2520x_1^3x_2x_5 + 4200x_1^3x_3x_4 + 9450x_1^2x_2^2x_4$
$\qquad + 12600x_1^2x_2x_3^2 + 12600x_1x_2^3x_3 + 945x_2^5$;

$B_{10,6} = 252x_1^5x_5 + 3150x_1^4x_2x_4 + 2100x_1^4x_3^2 + 12600x_1^3x_2^2x_3 + 4725x_1^2x_2^4$;

$B_{10,7} = 210x_1^6x_4 + 2520x_1^5x_2x_3 + 3150x_1^4x_2^3$;

$B_{10,8} = 120x_1^7x_3 + 630x_1^6x_2^2$; $B_{10,9} = 45x_1^8x_2$; $B_{10,10} = x_1^{10}$.

Differentiating the logarithm of the function

$$G(t) = \prod_{k \in S} (1 - ut^k)^{-a},$$

we get

$$\frac{G'(t)}{G(t)} = a \sum_{k \in S} kut^{k-1}(1 - ut^k)^{-1}$$

and so

$$\frac{G'(t)}{G(t)} = \frac{h(t)}{t}.$$

Integrating the last equation in the interval $(0, t)$ and using the initial condition $G(0) = 1$ we find the relation

$$G(t) = exp\left(\sum_{r=1}^{\infty} x_r \frac{t^r}{r}\right),$$

from which, by virtue of (11.9), the required expression is deduced. ☐

Example 11.2 Dinner choices in a Chinese restaurant

A Chinese restaurant offers x_1 different dinner choices for one person eating alone, x_2 different choices for two persons eating together, x_3 still different choices for three persons eating together, etc. Suppose that a group of n persons arrive at this restaurant to have dinner. In how many different ways may they choose their dinner?

The set (group) of the n persons $\{p_1, p_2, \ldots, p_n\}$, according to Theorem 2.10, may be partitioned into subsets (subgroups), among which $k_i \geq 0$ include i persons each, $i = 1, 2, \ldots, n$, in

$$\frac{n!}{k_1!(1!)^{k_1} k_2!(2!)^{k_2} \cdots k_n!(n!)^{k_n}}, \quad k_1 + 2k_2 + \cdots + nk_n = n$$

different ways. Further, there are x_i different choices for each subset (subgroup) of i persons eating together, $i = 1, 2, \ldots, n$. Thus, for fixed $k_i \geq 0, i = 1, 2, \ldots, n$, with $k_1 + 2k_2 + \cdots + nk_n = n$, there are

$$\frac{n!}{k_1!(1!)^{k_1} k_2!(2!)^{k_2} \cdots k_n!(n!)^{k_n}} x_1^{k_1} x_2^{k_2} \cdots x_n^{k_n}$$

different ways for the n persons to choose their dinner. Summing these ways for all $k_i \geq 0, i = 1, 2, \ldots, n$, with $k_1 + 2k_2 + \cdots + nk_n = n$, using (11.1), it follows that there are $B_n(x_1, x_2, \ldots, x_n)$ different ways n persons may choose their dinner. ☐

11.3 GENERAL PARTITION POLYNOMIALS

DEFINITION 11.3 *The polynomial $P_n \equiv P_n(x_1, x_2, \ldots, x_n; c_1, c_2, \ldots, c_n)$ in the variables $x_1, x_2, \ldots, x_n$ and parameters $c_1, c_2, \ldots, c_n$ defined by the sum*

$$P_n = \sum \frac{n!}{k_1!(1!)^{k_1} k_2!(2!)^{k_2} \cdots k_n!(n!)^{k_n}} c_k x_1^{k_1} x_2^{k_2} \cdots x_n^{k_n}, \quad (11.13)$$

where $k = k_1 + k_2 + \cdots + k_n$ and the summation is extended over all partitions of n, that is, over all nonnegative integer solutions $(k_1, k_2, \ldots, k_n)$ of the equation $k_1 + 2k_2 + \cdots + nk_n = n$, is called partition polynomial.

The partition polynomial $P_n(x_1.x_2, \ldots, x_n; c_1, c_2, \ldots, c_n)$, using the symbolic calculus, may be expressed by the exponential Bell partition polynomial $B_n(x_1, x_2, \ldots, x_n)$ as

$$P_n(x_1, x_2, \ldots, x_n; c_1, c_2, \ldots, c_n) = B_n(cx_1, cx_2, \ldots, cx_n), \ c^k \equiv c_k, \quad (11.14)$$

since

$$\sum \frac{n!}{k_1!(1!)^{k_1} k_2!(2!)^{k_2} \cdots k_n!(n!)^{k_n}} c_k x_1^{k_1} x_2^{k_2} \cdots x_n^{k_n}$$

$$= \sum \frac{n!}{k_1!(1!)^{k_1} k_2!(2!)^{k_2} \cdots k_n!(n!)^{k_n}} (cx_1)^{k_1} (cx_2)^{k_2} \cdots (cx_n)^{k_n},$$

with $c^k \equiv c_k$. This symbolic expression of the partition polynomial will be used in the sequel. Further, the partition polynomial $B_n(cx_1, cx_2, \ldots, cx_n)$, $c^k \equiv c_k$, $k = 1, 2, \ldots, n$, can be expressed as a linear combination of the partial Bell partition polynomials $B_{n,k}(x_1, x_2, \ldots, x_n)$, $k = 1, 2, \ldots, n$ with coefficients c_k, $k = 1, 2, \ldots, n$. Specifically, from Definitions 11.3 and 11.2 and putting $B_0 = c_0$, it follows directly that

$$B_n(cx_1, cx_2, \ldots, cx_n) = \sum_{k=0}^{n} c_k B_{n,k}(x_1, x_2, \ldots, x_n), \quad (11.15)$$

for $n = 0, 1, \ldots$.

A generating function and a recurrence relation for the partition polynomials may be deduced (at least symbolically) from the generating function and the recurrence relation for the exponential Bell partition polynomials by using the symbolic expression (11.14). Specifically, the following theorem is deduced from (11.9) and (11.10).

THEOREM 11.3

(a) *The generating function of the sequence of the partition polynomials* $B_n(cx_1, cx_2, \ldots, cx_n)$, $c^k \equiv c_k$, $n = 0, 1, \ldots$, *with* $B_0 \equiv c_0$, *is given by*

$$P(t) = \sum_{n=0}^{\infty} B_n(cx_1, cx_2, \ldots, cx_n) \frac{t^n}{n!} = \sum_{k=0}^{\infty} c_k \frac{[g(t) - x_0]^k}{k!}, \quad (11.16)$$

where $g(t) = \sum_{r=0}^{\infty} x_r t^r / r!$, *or symbolically by*

$$\exp(tB) = \exp[c(e^{tx} - x_0)], \quad B^n \equiv B_n, \quad c^k \equiv c_k, \quad x^r \equiv x_r.$$

(b) *The partition polynomials* $B_n(cx_1, cx_2, \ldots, cx_n)$, $c^k \equiv c_k$, $n = 0, 1, \ldots$ *with* $B_0 \equiv c_0$, *satisfy the recurrence relation*

$$B_{n+1}(cx_1, cx_2, \ldots, cx_{n+1})$$
$$= \sum_{r=0}^{n} \binom{n}{r} x_{r+1} c B_{n-r}(cx_1, cx_2, \ldots, cx_{n-r}), \quad c^k \equiv c_k. \quad (11.17)$$

In the next theorem, the derivative of order n of a composite function is expressed as a partition polynomial of the derivatives up to order n of the component functions.

THEOREM 11.4 Faà di Bruno formula
Let $f(u)$ and $g(t)$ be two functions of real variables for which all the derivatives,

$$g_r = \left[\frac{d^r g(t)}{dt^r} \right]_{t=a}, \quad r = 0, 1, \ldots, \quad f_k = \left[\frac{d^k f(u)}{du^k} \right]_{u=g(a)}, \quad k = 0, 1, \ldots,$$

exist. Then the derivatives of the composite function $h(t) = f(g(t))$,

$$h_n = \left[\frac{d^n h(t)}{dt^n} \right]_{t=a}, \quad n = 0, 1, \ldots,$$

are given by

$$h_n = \sum_{k=0}^{n} f_k B_{n,k}(g_1, g_2, \ldots, g_n)$$
$$= B_n(fg_1, fg_2, \ldots, fg_n), \quad f^k \equiv f_k. \quad (11.18)$$

PROOF   By successive differentiation of the composite function $h(t) = f(g(t))$, it follows that

$$h_0 = f_0, \ h_1 = f_1 g_1, \ h_2 = f_1 g_2 + f_2 g_1^2, \ h_3 = f_1 g_3 + 3 f_2 g_2 g_1 + f_3 g_1^3$$

and, in general,

$$h_n = \sum_{k=0}^{n} f_k h_{n,k}, \quad n = 0, 1, \dots,$$ (11.19)

where $h_{0,0} = 1$ and the polynomials $h_{n,k} = h_{n,k}(g_1, g_2, \dots, g_n)$ of the derivatives $g_1, g_2, \dots, g_n$, $k = 1, 2, \dots, n$, $n = 1, 2, \dots$, do not depend on the derivatives $f_1, f_2, \dots, f_n$. Hence they may be determined by an arbitrary choice of the function f. The exponential function

$$f(u) = e^{cu},$$

with c an arbitrary parameter, which is the most convenient, entails

$$f_k = \left[\frac{d^k f(u)}{du^k} \right]_{u=g(a)} = \left[\frac{d^k e^{cu}}{du^k} \right]_{u=g(a)} = c^k e^{cg(a)},$$

and

$$h_n = \left[\frac{d^n h(t)}{dt^n} \right]_{t=a} = \left[\frac{d^n}{dt^n} e^{cg(t)} \right]_{t=a}.$$

Thus, according to (11.19), it follows that

$$e^{-cg(a)} \left[\frac{d^n}{dt^n} e^{cg(t)} \right]_{t=a} = \sum_{k=0}^{n} h_{n,k} c^k, \quad n = 0, 1, \dots.$$

Further,

$$e^{-cg(a)} \left[\frac{d^n}{dt^n} e^{cg(t)} \right]_{t=a} = \left[\frac{d^n}{ds^n} e^{c[g(s+a)-g(a)]} \right]_{s=0},$$

with

$$g(s+a) - g(a) = \sum_{r=1}^{\infty} \left[\frac{d^r g(t)}{dt^r} \right]_{t=a} \cdot \frac{s^r}{r!} = \sum_{r=1}^{\infty} g_r \frac{s^r}{r!}$$

and so

$$\sum_{k=0}^{n} h_{n,k} c^k = \left[\frac{d^n}{ds^n} \exp \left\{ c \sum_{r=1}^{\infty} g_r \frac{s^r}{r!} \right\} \right]_{s=0}, \quad n = 0, 1, \dots.$$

Multiplying this expression by $s^n/n!$ and summing for $n = 0, 1, \dots$ we get the generating function

$$\sum_{n=0}^{\infty} \sum_{k=0}^{n} h_{n,k} c^k \frac{s^n}{n!} = \sum_{n=0}^{\infty} \left[\frac{d^n}{ds^n} \exp \left\{ c \sum_{r=1}^{\infty} g_r \frac{s^r}{r!} \right\} \right]_{s=0} \cdot \frac{s^n}{n!}$$

$$= \exp \left\{ c \sum_{r=1}^{\infty} g_r \frac{s^r}{r!} \right\},$$

which, compared to (11.6), implies

$$h_{n,k} = h_{n,k}(g_1, g_2, \dots, g_n) = B_{n,k}(g_1, g_2, \dots, g_n), \quad n = 0, 1, \dots$$

and, according to (11.19), the first part of (11.18) is deduced. The second part of (11.18) is a direct consequence of (11.15). ∎

Example 11.3 Compound discrete distributions

Consider a discrete distribution assuming nonnegative integer values, with probability function p_k, $k = 0, 1, \dots$ and probability generating function

$$f(u) = \sum_{k=0}^{\infty} p_k u^k.$$

Further, consider another discrete distribution assuming nonnegative integer values, with probability function q_r, $r = 0, 1, \dots$ and probability generating function

$$g(t) = \sum_{r=0}^{\infty} q_r t^r.$$

Then, the distribution with probability generating function the composite function $h(t) = f(g(t))$ is called *compound distribution*. Its probability function P_n, $n = 0, 1, \dots$, may be expressed in terms of p_k, $k = 0, 1, \dots$ and q_r, $r = 0, 1, \dots$. Indeed, since

$$P_n = \frac{1}{n!} \left[\frac{d^n h(t)}{dt^n} \right]_{t=0}, \quad n = 0, 1, \dots,$$

using Theorem 11.4, with

$$g_r = \left[\frac{d^r g(t)}{dt^r} \right]_{t=0} = r! q_r, \quad r = 1, 2, \dots, \quad g_0 = g(0) = q_0,$$

and

$$f_k = \left[\frac{d^k f(u)}{du^k} \right]_{u=g(0)=q_0} = \sum_{j=k}^{\infty} (j)_k p_j q_0^{j-k}, \quad k = 0, 1, \dots,$$

we deduce the expression

$$P_n = \frac{1}{n!} \sum_{k=0}^{n} f_k B_{n,k}(q_1, 2! q_2, \dots, n! q_n)$$

$$= \frac{1}{n!} B_n(f q_1, f 2! q_2, \dots, f n! q_n), \quad f^k \equiv f_k,$$

for $n = 0, 1, \dots$. The factorial moments of the compound distribution,

$$M_{(n)} = \sum_{j=n}^{\infty} (j)_n P_j, \quad n = 1, 2, \dots,$$

may be expressed in terms of the factorial moments

$$\mu_{(r)} = \sum_{j=r}^{\infty} (j)_r q_j, \quad r = 1, 2, \ldots, \quad m_{(k)} = \sum_{j=k}^{\infty} (j)_k p_j, \quad k = 1, 2, \ldots,$$

on the distributions p_k, $k = 0, 1, \ldots$ and q_r, $r = 0, 1, \ldots$. Indeed, since

$$M_{(n)} = \left[\frac{d^n h(t)}{dt^n} \right]_{t=1}, \quad n = 1, 2, \ldots,$$

using Theorem 11.4, with

$$g_r = \left[\frac{d^r g(t)}{dt^r} \right]_{t=1} = \mu_{(r)}, \quad r = 1, 2, \ldots, \quad g_0 = g(1) = 1,$$

and

$$f_k = \left[\frac{d^k f(u)}{du^k} \right]_{u=g(1)=1} = m_{(k)}, \quad k = 1, 2, \ldots,$$

we deduce the expression

$$M_{(n)} = \sum_{k=0}^{n} m_{(k)} B_{n,k}(\mu_{(1)}, \mu_{(2)}, \ldots, \mu_{(n)})$$

$$= B_n(m\mu_{(1)}, m\mu_{(2)}, \ldots, m\mu_{(n)}), \quad m^k \equiv m_{(k)},$$

for $n = 1, 2, \ldots$. ∎

REMARK 11.2 The discrete distribution with probability function

$$Q_j = \frac{a_j \theta^j}{f(\theta)}, \quad j = 0, 1, \ldots, \quad 0 < \theta < \rho,$$

where the coefficients a_j, $j = 0, 1, \ldots$ are independent of θ and

$$f(\theta) = \sum_{j=0}^{\infty} a_j \theta^j, \quad 0 < \theta < \rho,$$

is called *power series distribution* with parameter θ and series function $f(\theta)$. In the preceding example, if $q_0 = 0$, then

$$f_k = \left[\frac{d^k f(u)}{du^k} \right]_{u=q_0=0} = k! p_k, \quad k = 0, 1, \ldots,$$

while if $q_0 > 0$, then

$$f_k = q_0^{-k} f(q_0) \sum_{j=k}^{\infty} (j)_k \frac{p_j q_0^j}{f(q_0)} = q_0^{-k} f(q_0) m_{(k)}(q_0), \quad k = 0, 1, \ldots,$$

where $m_{(k)}(q_0)$ is the factorial moment of order k of the power series distribution with parameter q_0 and series function $f(q_0) = \sum_{j=0}^{\infty} p_j q_0^j$. ∎

11.4 LOGARITHMIC PARTITION POLYNOMIALS

The logarithmic partition polynomials, which are the inverse of the exponential Bell partition polynomials, are defined as follows.

DEFINITION 11.4 *The polynomial $L_n \equiv L_n(x_1, x_2, \ldots, x_n)$ in the variables $x_1, x_2, \ldots, x_n$ defined by the sum*

$$L_n = \sum \frac{n!(-1)^{k-1}(k-1)!}{k_1!(1!)^{k_1} k_2!(2!)^{k_2} \cdots k_n!(n!)^{k_n}} x_1^{k_1} x_2^{k_2} \cdots x_n^{k_n}, \qquad (11.20)$$

where $k = k_1 + k_2 + \cdots + k_n$ and the summation is extended over all partitions of n, that is, over all nonnegative integer solutions $(k_1, k_2, \ldots, k_n)$ of the equation $k_1 + 2k_2 + \cdots + nk_n = n$, is called logarithmic partition polynomial.

The logarithmic partition polynomial $L_n(x_1, x_2, \ldots, x_n)$ constitutes a particular case of the partition polynomial $B_n(cx_1, cx_2, \ldots, cx_n)$, $c^k \equiv c_k$. Specifically, for $n = 0, 1, \ldots,$

$$L_n(x_1, x_2, \ldots, x_n) = B_n(cx_1, cx_2, \ldots, cx_n), \quad c^k \equiv c_k, \qquad (11.21)$$

with $c_0 = 0$, $c_k = (-1)^{k-1}(k-1)!$, $k = 1, 2, \ldots, n$. Note that $B_0 = c_0$ and so $L_0 = 0$. Further, Definitions 11.4 and 11.2 imply, for $n = 1, 2, \ldots$, the relation

$$L_n(x_1, x_2, \ldots, x_n) = \sum_{k=1}^{n} (-1)^{k-1}(k-1)! B_{n,k}(x_1, x_2, \ldots, x_n). \quad (11.22)$$

The generating function of the logarithmic partition polynomials may be deduced from the generating function (11.16) of the partition polynomials by setting $c_k = (-1)^{k-1}(k-1)!$, $k = 1, 2, \ldots$, and using (11.21) and the expansion of the logarithm, $\log(1 + t) = \sum_{k=1}^{\infty} (-1)^{k-1} t^k / k$. Thus the following theorem is deduced.

THEOREM 11.5
The generating function of the logarithmic partition polynomials $L_n(x_1, x_2, \ldots, x_n)$, $n = 1, 2, \ldots$, with $L_0 \equiv 0$, is given by

$$L(t) = \sum_{n=1}^{\infty} L_n(x_1, x_2, \ldots, x_n) \frac{t^n}{n!} = \log\{1 + [g(t) - x_0]\}, \qquad (11.23)$$

where $g(t) = \sum_{r=0}^{\infty} x_r t^r / r!$.

REMARK 11.3 The recurrence relation for the partition polynomials, (11.17), in the particular case of $c_0 = 0$, $c_k = (-1)^{k-1}(k-1)!$, $k = 1, 2, \ldots$, does not yield a corresponding recurrence relation for logarithmic partition polynomials. This is because, according to the symbolic calculus and (11.15),

$$cB_n(cx_1, cx_2, \ldots, cx_n) = \sum_{k=0}^{n} c_{k+1} B_{n,k}(x_1, x_2, \ldots, x_n),$$

a relation, which for $c_0 = 0$, $c_k = (-1)^{k-1}(k-1)!$, $k = 1, 2, \ldots$, yields the expression

$$cB_n(cx_1, cx_2, \ldots, cx_n) = \sum_{k=0}^{n} (-1)^k k! B_{n,k}(x_1, x_2, \ldots, x_n)$$

$$= \sum_{k=0}^{n} (-1)_k B_{n,k}(x_1, x_2, \ldots, x_n).$$

Consequently,

$$cB_n(cx_1, cx_2, \ldots, cx_n) = C_{n,-1}(x_1, x_2, \ldots, x_n), \quad c^k \equiv c_k,$$

for $c_0 = 0$, $c_k = (-1)^{k-1}(k-1)!$, $k = 1, 2, \ldots, n$, where

$$C_{n,s}(x_1, x_2, \ldots, x_n) = \sum_{k=0}^{n} (s)_k B_{n,k}(x_1, x_2, \ldots, x_n), \quad n = 0, 1, \ldots,$$

with s a real number, is the potential partition polynomial examined in the next section. Thus, recurrence relation (11.17) for $c_0 = 0$, $c_k = (-1)^{k-1}(k-1)!$, $k = 1, 2, \ldots$, upon using (11.21), yields

$$L_{n+1}(x_1, x_2, \ldots, x_{n+1}) = \sum_{r=0}^{n} \binom{n}{r} x_{r+1} C_{n-r,-1}(x_1, x_2, \ldots, x_{n-r}),$$

for $n = 0, 1, \ldots$, with $C_{0,-1} = 1$. $\blacksquare$

A recurrence relation for the logarithmic partition polynomials may be obtained by differentiating generating function (11.23). Specifically, the following theorem is derived.

THEOREM 11.6
The logarithmic partition polynomials $L_n \equiv L_n(x_1, x_2, \ldots, x_n)$, $n = 1, 2, \ldots$, with $L_1 = x_1$, satisfy the recurrence relation

$$L_{n+1} = x_{n+1} - \sum_{r=1}^{n} \binom{n}{r} x_r L_{n-r+1}, \quad n = 1, 2, \ldots . \tag{11.24}$$

PROOF   Differentiating generating function (11.23) and multiplying the resulting expression by $1 + [g(t) - x_0]$, we get the relation

$$\frac{d}{dt}L(t) = \frac{d}{dt}g(t) - [g(t) - x_0]\frac{d}{dt}L(t).$$

Expanding it into powers of t, it follows that

$$\sum_{n=0}^{\infty} L_{n+1}\frac{t^n}{n!} = \sum_{n=0}^{\infty} x_{n+1}\frac{t^n}{n!} - \left[\sum_{r=1}^{\infty} x_r\frac{t^r}{r!}\right] \cdot \left[\sum_{n=0}^{\infty} L_{n+1}\frac{t^n}{n!}\right]$$

and so

$$\sum_{n=0}^{\infty} L_{n+1}\frac{t^n}{n!} = \sum_{n=0}^{\infty} x_{n+1}\frac{t^n}{n!} - \sum_{n=1}^{\infty}\left[\sum_{r=1}^{n}\binom{n}{r}x_r L_{n-r+1}\right]\frac{t^n}{n!}.$$

Equating the coefficients of $t^n/n!$ in both sides of the last expression, (11.24) is deduced.   ■

Example 11.4 Connection of cumulants and moments
  The cumulant generating function

$$K(t) = \sum_{n=1}^{\infty} \kappa_n \frac{t^n}{n!}$$

is usually defined through the moment generating function

$$M(t) = 1 + \sum_{n=1}^{\infty} \mu'_n \frac{t^n}{n!}$$

by $K(t) = \log M(t)$. Thus

$$\sum_{n=1}^{\infty} \kappa_n \frac{t^n}{n!} = \log\left[1 + \sum_{n=1}^{\infty} \mu'_n \frac{t^n}{n!}\right].$$

Comparing this relation with (11.23), it follows that

$$\kappa_n = L_n(\mu'_1, \mu'_2, \ldots, \mu'_n), \quad n = 1, 2, \ldots.$$

The logarithmic partition polynomials, according to (11.9) and (11.23), are the inverse of the exponential Bell partition polynomials and so

$$\mu'_n = B_n(\kappa_1, \kappa_2, \ldots, \kappa_n), \quad n = 1, 2, \ldots. \quad \square$$

Example 11.5 Elementary and power sum symmetric functions

A polynomial $P(x_1, x_2, \dots, x_n)$ in n variables is called *symmetric function* if, for any permutation $(j_1, j_2, \dots, j_n)$ of the n indices $\{1, 2, \dots, n\}$, it holds that $P(x_{j_1}, x_{j_2}, \dots, x_{j_n}) = P(x_1, x_2, \dots, x_n)$.

The *elementary* symmetric function $a_k = a_k(x_1, x_2, \dots, x_n)$ is defined by the sum

$$a_k(x_1, x_2, \dots, x_n) = \sum x_{i_1} x_{i_2} \cdots x_{i_k}, \quad k = 1, 2, \dots, n,$$

where the summation is extended over all k-combinations $\{i_1, i_2, \dots, i_k\}$ of the n indices $\{1, 2, \dots, n\}$. Therefore, with $a_0 = 1$,

$$\sum_{k=0}^{n} a_k(x_1, x_2, \dots, x_n) t^k = \prod_{i=1}^{n} (1 + x_i t). \tag{11.25}$$

The *power sum* symmetric function $s_k = s_k(x_1, x_2, \dots, x_n)$ is defined by the sum

$$s_k(x_1, x_2, \dots, x_n) = \sum_{i=1}^{n} x_i^k, \quad k = 1, 2, \dots.$$

Its generating function is readily deduced as

$$\sum_{k=1}^{\infty} s_k(x_1, x_2, \dots, x_n) \frac{t^k}{k} = \sum_{k=1}^{\infty} \sum_{i=1}^{n} x_i^k \frac{t^k}{k} = \sum_{i=1}^{n} \log(1 - x_i t)$$

$$= -\log \prod_{i=1}^{n} (1 - x_i t). \tag{11.26}$$

Express the elementary symmetric functions $a_k = a_k(x_1, x_2, \dots, x_n)$, $k = 1, 2, \dots, n$ in terms of the power sum symmetric functions $s_k = s_k(x_1, x_2, \dots, x_n)$, $k = 1, 2, \dots$, and vice versa.

Note first that, in (11.25) and (11.26), n may be regarded as being as large as may be necessary. Then,

$$1 + \sum_{k=1}^{\infty} (-1)^k a_k t^k = \exp\left(-\sum_{r=1}^{\infty} s_r \frac{t^r}{r}\right)$$

and

$$\sum_{k=1}^{\infty} (-1)^{k-1} s_k \frac{t^k}{k} = \log\left(1 + \sum_{r=1}^{\infty} a_r t^r\right).$$

Therefore, according to (11.9) and (11.23), it follows that

$$a_k = (-1)^k B_k(-s_1, -s_2, \dots, -(k-1)! s_k)/k!$$

and

$$s_k = (-1)^{k-1} L_k(a_1, 2! a_2, \dots, k! a_k)/(k-1)!. \qquad \square$$

11.5   POTENTIAL PARTITION POLYNOMIALS

DEFINITION 11.5   *The polynomial $C_{n,s} \equiv C_{n,s}(x_1, x_2, \dots, x_n)$ in the variables $x_1, x_2, \dots, x_n$ defined for a real (or complex) number s by the sum*

$$C_{n,s} = \sum \frac{n!}{k_1!(1!)^{k_1} k_2!(2!)^{k_2} \cdots k_n!(n!)^{k_n}} (s)_k x_1^{k_1} x_2^{k_2} \cdots x_n^{k_n}, \quad (11.27)$$

where $k = k_1 + k_2 + \cdots + k_n$ and the summation is extended over all partitions of n, that is, over all nonnegative integer solutions $(k_1, k_2, \dots, k_n)$ of the equation $k_1 + 2k_2 + \cdots + nk_n = n$, is called potential partition polynomial.

The potential partition polynomial $C_{n,s}(x_1, x_2, \dots, x_n)$ constitutes a particular case of the partition polynomial $B_n(cx_1, cx_2, \dots, cx_n)$, $c^k \equiv c_k$. Specifically, for $n = 0, 1, \dots$,

$$C_{n,s}(x_1, x_2, \dots, x_n) = B_n(cx_1, cx_2, \dots, cx_n), \quad c^k \equiv c_k, \quad (11.28)$$

with $c_k = (s)_k$, $k = 0, 1, \dots, n$. Further, Definitions 11.5 and 11.2 imply the relation

$$C_{n,s}(x_1, x_2, \dots, x_n) = \sum_{k=0}^{n} (s)_k B_{n,k}(x_1, x_2, \dots, x_n), \quad n = 0, 1, \dots.$$
$$(11.29)$$

The generating function of the potential partition polynomials may be deduced from the generating function of the partition polynomials, (11.16), by setting $c_k = (s)_k$, $k = 0, 1, \dots$, and using (11.28) and the expansion

$$(1 + t)^s = \sum_{k=0}^{\infty} \binom{s}{k} t^k.$$

Thus the following theorem is deduced.

THEOREM 11.7
The generating function of the potential partition polynomials $C_{n,s}(x_1, x_2, \dots, x_n)$, $n = 0, 1, \dots$, is given by

$$C_s(t) = \sum_{n=0}^{\infty} C_{n,s}(x_1, x_2, \dots, x_n) \frac{t^n}{n!} = [1 + (g(t) - x_0)]^s, \quad (11.30)$$

where $g(t) = \sum_{r=0}^{\infty} x_r t^r / r!$.

REMARK 11.4   The recurrence relation for the partition polynomials, (11.17), in the particular case of $c_k = (s)_k$, $k = 1, 2, \ldots$, and since

$$cB_n(cx_1, cx_2, \ldots, cx_n) = \sum_{k=0}^{n} (s)_{k+1} B_{n,k}(x_1, x_2, \ldots, x_n)$$

$$= s \sum_{k=0}^{n} (s-1)_k B_{n,k}(x_1, x_2, \ldots, x_n)$$

$$= s C_{n,s-1}(x_1, x_2, \ldots, x_n),$$

yields for the potential partition polynomials $C_{n,s} \equiv C_{n,s}(x_1, x_2, \ldots, x_n)$, $n = 0, 1, \ldots$, the recurrence relation

$$C_{n+1,s} = s \sum_{k=0}^{n} \binom{n}{k} x_{k+1} C_{n-k,s-1}. \tag{11.31}$$

Note that this recurrence relation expresses the potential partition polynomial $C_{n+1,s}$, with parameter s, in terms of the potential partition polynomials $C_{k,s-1}$, $k = 0, 1, \ldots, n$, with parameter $s - 1$. ∎

A recurrence relation for the potential polynomials, in which s remains constant, is obtained in the next theorem.

THEOREM 11.8
The potential partition polynomials $C_{n,s} \equiv C_{n,s}(x_1, x_2, \ldots, x_n)$, $n = 0, 1, \ldots$, with $C_{0,s} = 1$, satisfy the recurrence relation

$$C_{n+1,s} = \sum_{r=0}^{n} \left\{ s \binom{n}{r} - \binom{n}{r+1} \right\} x_{r+1} C_{n-r,s}, \quad n = 0, 1, \ldots. \tag{11.32}$$

PROOF   Differentiating generating function (11.30) and multiplying the resulting expression by $1 + [g(t) - x_0]$, we get the relation

$$\frac{dC_s(t)}{dt} = sC_s(t)\frac{dg(t)}{dt} - [g(t) - x_0]\frac{dC_s(t)}{dt}.$$

Expanding it into powers of t, it follows that

$$\sum_{n=0}^{\infty} C_{n+1,s}\frac{t^n}{n!} = s\left[\sum_{r=0}^{\infty} x_{r+1}\frac{t^r}{r!}\right] \cdot \left[\sum_{n=0}^{\infty} C_{n,s}\frac{t^n}{n!}\right]$$

$$- \left[\sum_{r=0}^{\infty} x_{r+1}\frac{t^{r+1}}{(r+1)!}\right] \cdot \left[\sum_{n=1}^{\infty} C_{n,s}\frac{t^{n-1}}{(n-1)!}\right]$$

and so

$$\sum_{n=0}^{\infty} C_{n+1,s} \frac{t^n}{n!} = \sum_{n=0}^{\infty} \left[s \sum_{r=0}^{n} \binom{n}{r} x_{r+1} C_{n-r,s} \right] \frac{t^n}{n!}$$

$$- \sum_{n=1}^{\infty} \left[\sum_{r=0}^{n-1} \binom{n}{r+1} x_{r+1} C_{n-r,s} \right] \frac{t^n}{n!}.$$

Equating the coefficients of $t^n/n!$ in both sides of the last expression, (11.32) is deduced. ∎

The potential partition polynomial $C_{n,s}(x_1, x_2, \ldots, x_n)$, $n = 0, 1, \ldots$, in the case of a positive integer s, may be expressed as a partial Bell partition polynomial. Specifically, we have the following theorem.

THEOREM 11.9
If s is a positive integer, then for $n = 0, 1, \ldots$,

$$C_{n,s}(x_1, x_2, \ldots, x_n) = \frac{n! s!}{(n+s)!} B_{n+s,s}(1, 2x_1, 3x_2, \ldots, (n+1)x_n).$$
$$(11.33)$$

PROOF   For $y_1 = 1$, $y_r = r x_{r-1}$, $r = 2, 3, \ldots$,

$$\left(1 + \sum_{r=1}^{\infty} x_r \frac{t^r}{r!} \right)^s = t^{-s} \left(\sum_{r=1}^{\infty} y_r \frac{t^r}{r!} \right)^s.$$

If s is a positive integer, expanding the left-hand side according to (11.30) and the right-hand side according to (11.7), we get the relation

$$\sum_{n=0}^{\infty} C_{n,s}(x_1, x_2, \ldots, x_n) \frac{t^n}{n!} = \sum_{j=s}^{\infty} s! B_{j,s}(y_1, y_2, \ldots, y_{j-s+1}) \frac{t^{j-s}}{j!}$$

$$= \sum_{n=0}^{\infty} \frac{n! s!}{(n+s)!} B_{n+s,s}(x_1, x_2, \ldots, x_{n+1}) \frac{t^n}{n!},$$

which implies (11.33). ∎

An interesting property of the potential partition polynomial is given in the following theorem.

THEOREM 11.10
Let $C_{n,s} \equiv C_{n,s}(x_1, x_2, \ldots, x_n)$, $n = 0, 1, \ldots$, be the potential partition poly-

nomial. Then, for any real (or complex) number s,

$$C_{n,s} = \sum_{r=0}^{n} \binom{s}{r} \binom{n-s}{n-r} C_{n,r}. \qquad (11.34)$$

PROOF   Generating function (11.30) may be expressed as

$$C_s(t) = [1 + (g(t) - x_0)]^s = \sum_{k=0}^{\infty} \binom{s}{k} [g(t) - x_0]^k. \qquad (11.35)$$

Note that the function

$$h_k(t) = [g(t) - x_0]^k = t^k \left(\sum_{j=1}^{\infty} x_j \frac{t^{j-1}}{j!} \right)^k$$

has the factor t^k and, thus, for the determination of $C_{n,s}$, which is the coefficients of $t^n/n!$, the first n terms of the sum in the right-hand side of (11.35) are sufficient. Consequently, the potential partition polynomial $C_{n,s}$ is the coefficient of $t^n/n!$ in the expansion

$$\sum_{k=0}^{n} \binom{s}{k} [g(t) - x_0]^k = \sum_{k=0}^{n} \sum_{r=0}^{k} (-1)^{k-r} \binom{s}{k} \binom{k}{r} C_r(t)$$

$$= \sum_{k=0}^{n} \sum_{r=0}^{k} (-1)^{k-r} \binom{s}{k} \binom{k}{r} \sum_{n=0}^{\infty} C_{n,r} \frac{t^n}{n!}$$

$$= \sum_{n=0}^{\infty} \left\{ \sum_{k=0}^{n} \sum_{r=0}^{k} (-1)^{k-r} \binom{s}{k} \binom{k}{r} C_{n,r} \right\} \frac{t^n}{n!}.$$

Hence we get

$$C_{n,s} = \sum_{k=0}^{n} \sum_{r=0}^{k} (-1)^{k-r} \binom{s}{k} \binom{k}{r} C_{n,r}$$

$$= \sum_{r=0}^{n} \left\{ \sum_{k=r}^{n} (-1)^{k-r} \binom{s}{k} \binom{k}{r} \right\} C_{n,r}$$

and, since, by the horizontal recurrence relation of binomial coefficients (2.9),

$$\sum_{k=r}^{n} (-1)^{k-r} \binom{s}{k} \binom{k}{r} = \binom{s}{r} \sum_{k=r}^{n} (-1)^{k-r} \binom{s-r}{k-r}$$

$$= (-1)^{n-r} \binom{s}{r} \binom{s-r-1}{n-r} = \binom{s}{r} \binom{n-s}{n-r},$$

(11.34) is deduced.   ∎

REMARK 11.5   The following formulation of Theorem 11.10 is of interest. For any real (or complex) number s and for every function $h(t)$, with $h(t_0) = 1$, for which the derivatives at t_0 exist,

$$\left[\frac{d^n(h(t))^s}{dt^n}\right]_{t=t_0} = \sum_{r=0}^{n} \binom{s}{r}\binom{n-s}{n-r}\left[\frac{d^n(h(t))^r}{dt^n}\right]_{t=t_0}. \tag{11.36}$$

Indeed (11.30), with $h(t) = 1 + g(t-t_0) - g(0)$ and $g(t) = \sum_{r=0}^{\infty} x_r t^r / r!$, yields

$$\left[\frac{d^n(h(t))^s}{dt^n}\right]_{t=t_0} = \left[\frac{d^n(1+\{g(t)-x_0\})^s}{dt^n}\right]_{t=0} = C_{n,s}(x_1, x_2, \dots, x_n)$$

and so (11.36) is readily deduced from (11.34).   ∎

Example 11.6 Homogeneous product sum symmetric function
The *homogeneous product sum* symmetric function $h_k = h_k(x_1, x_2, \dots, x_n)$ is defined by the sum

$$h_k(x_1, x_2, \dots, x_n) = \sum x_1^{r_1} x_2^{r_2} \dots x_n^{r_n}, \quad k = 1, 2, \dots,$$

where the summation is extended over all $r_i = 0, 1, \dots, k$, $i = 1, 2, \dots, n$, such that $r_1 + r_2 + \cdots + r_n = k$. Therefore, with $h_0 = 1$,

$$\sum_{k=0}^{\infty} h_k(x_1, x_2, \dots, x_n)t^k = \sum_{k=0}^{\infty} \sum \prod_{i=1}^{n}(x_i t)^{r_i} = \prod_{i=1}^{n}\sum_{r_i=0}^{\infty}(x_i t)^{r_i}$$

and so

$$\sum_{k=0}^{\infty} h_k(x_1, x_2, \dots, x_n)t^k = \prod_{i=1}^{n}(1-x_i t)^{-1}. \tag{11.37}$$

Express the homogeneous product sum symmetric functions $h_k = h_k(x_1, x_2, \dots, x_n)$, $k = 1, 2, \dots$ in terms of (a) the elementary symmetric functions $a_k = a_k(x_1, x_2, \dots, x_n)$, $k = 1, 2, \dots$ and (b) the power sum symmetric functions $s_k = s_k(x_1, x_2, \dots, x_n)$, and vice versa (see Example 11.5).

Generating functions (11.25) and (11.26) are connected with (11.37) by the relations

$$1 + \sum_{k=1}^{\infty}(-1)^k h_k t^k = \left(1 + \sum_{r=1}^{\infty} a_r t^r\right)^{-1},$$

$$1 + \sum_{k=1}^{\infty}(-1)^k a_k t^k = \left(1 + \sum_{r=1}^{\infty} h_r t^r\right)^{-1},$$

and

$$1 + \sum_{k=1}^{\infty} h_k t^k = \exp\left(\sum_{r=1}^{\infty} s_r \frac{t^r}{r}\right),$$

$$1 + \sum_{k=1}^{\infty} s_k \frac{t^k}{k} = \log\left(1 + \sum_{r=1}^{\infty} h_r t^r\right).$$

Therefore, according to (11.9), (11.23) and (11.30), it follows that

$$h_k = (-1)^k C_{k,-1}(a_1, 2!a_2, \dots, k!a_k)/k!,$$

$$a_k = (-1)^k C_{k,-1}(h_1, 2!h_2, \dots, k!h_k)/k!$$

and

$$h_k = B_k(s_1, s_2, \dots, (k-1)!s_k)/k!,$$

$$s_k = L_k(h_1, 2!h_2, \dots, k!h_k)/(k-1)!. \qquad \Box$$

11.6 INVERSION OF POWER SERIES

Consider the power series

$$\phi(t) = \sum_{n=1}^{\infty} \phi_n \frac{t^n}{n!}, \quad \phi_1 \neq 0, \tag{11.38}$$

which has no constant term, and let

$$\phi^{-1}(u) = \sum_{n=1}^{\infty} \psi_n \frac{u^n}{n!} \tag{11.39}$$

be its inverse series, $\phi^{-1}(\phi(t)) = \phi(\phi^{-1}(t)) = t$. Since

$$\left[\frac{d\phi^{-1}(u)}{du}\right]_{u=\phi(t)} \cdot \left[\frac{d\phi(t)}{dt}\right] = 1,$$

it follows that $\psi_1 = \phi_1^{-1}$. The coefficients ψ_n, $n = 2, 3, \dots$, may be expressed as potential polynomials of the coefficients ϕ_n, $n = 2, 3, \dots$. Interesting on its own a preliminary result, is derived in the following lemma, which constitutes a generalization of Faà di Bruno formula.

LEMMA 11.1

  Let $f(u)$ and $g(t)$ be two functions of real variables for which all the derivatives,

$$g_n = \left[\frac{d^n g(t)}{dt^n}\right]_{t=0}, \quad f_n = \left[\frac{d^n f(u)}{du^n}\right]_{u=g(0)}, \quad n = 1, 2, \ldots,$$

exist. Further, let $h(t) = f(g(t))$ and

$$h_n = \left[\frac{d^n h(t)}{dt^n}\right]_{t=0}, \quad n = 1, 2, \ldots.$$

Then, for $r = 1, 2, \ldots, n$ and $n = 1, 2, \ldots$,

$$B_{n,r}(h_1, h_2, \ldots, h_n) = \sum_{k=r}^{n} B_{k,r}(f_1, f_2, \ldots, f_k) B_{n,k}(g_1, g_2, \ldots, g_n). \tag{11.40}$$

PROOF   According to (11.7),

$$\sum_{n=r}^{\infty} B_{n,r}(h_1, h_2, \ldots, h_n)\frac{t^n}{n!} = \frac{[h(t) - h(0)]^r}{r!}, \quad r = 1, 2, \ldots.$$

Further,

$$\sum_{k=r}^{\infty} B_{k,r}(f_1, f_2, \ldots, f_k)\frac{(u - u_0)^k}{k!} = \frac{[f(u) - f(u_0)]^r}{r!}, \quad r = 1, 2, \ldots,$$

where

$$f_n = \left[\frac{d^n f(u)}{du^n}\right]_{u=u_0}, \quad n = 1, 2, \ldots.$$

Consequently, with $u = g(t)$, $u_0 = g(0)$ and, since

$$\sum_{n=k}^{\infty} B_{n,k}(g_1, g_2, \ldots, g_n)\frac{t^n}{n!} = \frac{[g(t) - g(0)]^k}{k!}, \quad k = 1, 2, \ldots,$$

it follows that

$$\frac{[f(g(t)) - f(g(0))]^r}{r!} = \sum_{k=r}^{\infty} B_{k,r}(f_1, f_2, \ldots, f_k)\frac{[g(t) - g(0)]^k}{k!}$$

$$= \sum_{k=r}^{\infty} B_{k,r}(f_1, f_2, \ldots, f_k)\sum_{n=k}^{\infty} B_{n,k}(g_1, g_2, \ldots, g_n)\frac{t^n}{n!}$$

$$= \sum_{n=r}^{\infty}\left\{\sum_{k=r}^{n} B_{k,r}(f_1, f_2, \ldots, f_k) B_{n,k}(g_1, g_2, \ldots, g_n)\right\}\frac{t^n}{n!}.$$

Thus

$$\sum_{n=r}^{\infty} B_{n,r}(h_1, h_2, \ldots, h_n) \frac{t^n}{n!}$$

$$= \sum_{n=r}^{\infty} \left\{ \sum_{k=r}^{n} B_{k,r}(f_1, f_2, \ldots, f_k) B_{n,k}(g_1, g_2, \ldots, g_n \right\} \frac{t^n}{n!}$$

and, equating the coefficients of $t^n/n!$ in both members of this relation, (11.40) is deduced. ∎

Note that expression (11.40) for $r = 1$, since $B_{n,1}(h_1, h_2, \ldots, h_n) = h_n$, $B_{k,1}(f_1, f_2, \ldots, f_k) = f_k$, reduces to Faa di Bruno formula (11.18).

THEOREM 11.11 Inversion formula of Lagrange
Let $\phi(t)$ be power series (11.38), which has no constant term, and $\phi^{-1}(u)$ its inverse power series (11.39). Then for $k = 1, 2, \ldots, n$ and $n = 1, 2, \ldots,$

$$\left[\frac{d^n}{du^n} (\phi^{-1}(u))^k \right]_{u=0} = k(n-1)_{k-1} \left[\frac{d^{n-k}}{dt^{n-k}} \left(\frac{\phi(t)}{t} \right)^{-n} \right]_{t=0}. \quad (11.41)$$

PROOF Applying Lemma 11.1 with $f(u) = \phi(u)$, $g(t) = \phi^{-1}(t)$, whence $h(t) = f(g(t)) = t$, and since

$$B_{n,r}(h_1, h_2, \ldots, h_n) = \frac{1}{r!} \left[\frac{d^n t^r}{dt^n} \right]_{t=0} = \delta_{n,r},$$

$$B_{k,r}(\phi_1, \phi_2, \ldots, \phi_k) = \frac{1}{r!} \left[\frac{d^k}{dt^k} (\phi(t))^r \right]_{t=0},$$

$$B_{n,k}(\psi_1, \psi_2, \ldots, \psi_n) = \frac{1}{k!} \left[\frac{d^n}{dt^n} (\phi^{-1}(t))^k \right]_{t=0}, \quad \psi_r = \left[\frac{d^r \phi^{-1}(t)}{dt^r} \right]_{t=0},$$

it follows that

$$\sum_{k=r}^{n} \frac{1}{k!} \left[\frac{d^n}{dt^n} (\phi^{-1}(t))^k \right]_{t=0} \cdot \frac{1}{r!} \left[\frac{d^k}{dt^k} (\phi(t))^r \right]_{t=0} = \delta_{n,r}.$$

Thus, for (11.41) to hold, it suffices to prove that

$$c_{n,r} \equiv \sum_{k=r}^{n} \frac{k}{n} \binom{n}{k} \left[\frac{d^{n-k}}{dt^{n-k}} \left(\frac{\phi(t)}{t} \right)^{-n} \right]_{t=0} \cdot \frac{1}{r!} \left[\frac{d^k}{dt^k} (\phi(t))^r \right]_{t=0} = \delta_{n,r}.$$

Using the relation

$$k \left[\frac{d^k}{dt^k} (\phi(t))^r \right]_{t=0} = \left[\frac{d^k}{dt^k} \left\{ t \frac{d}{dt} (\phi(t))^r \right\} \right]_{t=0}$$

$$= r \left[\frac{d^k}{dt^k} \left\{ t(\phi(t))^{r-1} \frac{d\phi(t)}{dt} \right\} \right]_{t=0},$$

$c_{n,r}$ may be written as

$$c_{n,r} = \frac{1}{n(r-1)!} \sum_{k=r}^{n} \binom{n}{k} \left[\frac{d^{n-k}}{dt^{n-k}} \left(\frac{\phi(t)}{t} \right)^{-n} \right]_{t=0}$$

$$\times \left[\frac{d^k}{dt^k} \left\{ t(\phi(t))^{r-1} \frac{d\phi(t)}{dt} \right\} \right]_{t=0}$$

and so, according to the formula of the n-th derivative of the product of two functions,

$$c_{n,r} = \frac{1}{n(r-1)!} \left[\frac{d^n}{dt^n} \left\{ \left(\frac{\phi(t)}{t} \right)^{-n} t(\phi(t))^{r-1} \frac{d\phi(t)}{dt} \right\} \right]_{t=0}$$

$$= \frac{1}{n(r-1)!} \left[\frac{d^n}{dt^n} \left\{ t^{n+1} (\phi(t))^{-n+r-1} \frac{d\phi(t)}{dt} \right\} \right]_{t=0}.$$

In particular, for $r = n$,

$$c_{n,n} = \frac{1}{n!} \left[\frac{d^n}{dt^n} \left\{ t^{n+1} (\phi(t))^{-1} \frac{d\phi(t)}{dt} \right\} \right]_{t=0}$$

$$= \frac{1}{n!} \left[\frac{d^n}{dt^n} \left\{ t^{n+1} \frac{d \log \phi(t)}{dt} \right\} \right]_{t=0}$$

and so, setting $x_0 = \phi_1$, $x_i = \phi_1^{-1} \phi_{i+1}/(i+1)$, $i = 1, 2, \ldots$, and using (11.23), it follows that

$$c_{n,n} = \frac{1}{n!} \left[\frac{d^n}{dt^n} \left\{ t^{n+1} \frac{d \log(x_0 t)}{dt} + t^{n+1} \frac{d}{dt} \log \left(1 + \sum_{i=1}^{\infty} x_i \frac{t^i}{i!} \right) \right\} \right]_{t=0}$$

$$= \frac{1}{n!} \left[\frac{d^n}{dt^n} \left\{ t^n + t^{n+1} \frac{d}{dt} \sum_{j=1}^{\infty} L_j(x_1, x_2, \ldots, x_j) \frac{t^j}{j!} \right\} \right]_{t=0}$$

$$= \frac{1}{n!} \left[\frac{d^n}{dt^n} \left\{ t^n + \sum_{j=1}^{\infty} L_j(x_1, x_2, \ldots, x_j) \frac{t^{n+j}}{(j-1)!} \right\} \right]_{t=0} = 1.$$

Also, for $r < n$,

$$c_{n,r} = -\frac{1}{n(n-r)(r-1)!} \left[\frac{d^n}{dt^n} \left\{ t^{n+1} \frac{d}{dt} (\phi(t))^{-n+r} \right\} \right]_{t=0}$$

and so, setting $x_0 = \phi_1$, $x_i = \phi_1^{-1}\phi_{i+1}/(i+1)$, $i = 1, 2, \ldots$, $d_{n,r} = -n(n-r)(r-1)! x_0^{-n+r} c_{n,r}$ and using (11.30), it follows that

$$
\begin{aligned}
d_{n,r} &= \left[\frac{d^n}{dt^n} \left\{ t^{n+1} \frac{d}{dt} \left(1 + \sum_{i=1}^{\infty} x_i \frac{t^i}{i!} \right)^{-n+r} \right\} \right]_{t=0} \\
&= \left[\frac{d^n}{dt^n} \left\{ t^{n+1} \frac{d}{dt} \sum_{j=0}^{\infty} C_{j,-n+r}(x_1, x_2, \ldots, x_j) \frac{t^{j-n+r}}{j!} \right\} \right]_{t=0} \\
&= \left[\frac{d^n}{dt^n} \left\{ \sum_{j=0}^{\infty} (j-n+r) C_{j,-n+r}(x_1.x_2, \ldots, x_j) \frac{t^{j+r}}{j!} \right\} \right]_{t=0} = 0.
\end{aligned}
$$

Therefore $c_{n,r} = \delta_{n,r}$ and the proof of (11.41) is completed. ∎

COROLLARY 11.2

Let $\phi(t)$ be power series (11.38), which has no constant term, and $\phi^{-1}(u)$ its inverse power series (11.39). Then the n-order coefficient $\psi_{n,k}$, $n = k, k+1, \ldots$ of the k-th power of $\phi^{-1}(u)$,

$$
\sum_{n=k}^{\infty} \psi_{n,k} \frac{u^n}{n!} = [\phi^{-1}(u)]^k, \quad k = 1, 2, \ldots,
$$

is given by

$$
\psi_{n,k} = \frac{k(n-1)_{k-1}}{\phi_1^n} C_{n-k,-n} \left(\frac{\phi_2}{2\phi_1}, \frac{\phi_3}{3\phi_1}, \ldots, \frac{\phi_{n-k+1}}{(n-k+1)\phi_1} \right). \quad (11.42)
$$

In particular (for $k = 1$),

$$
\psi_n = \frac{1}{\psi_1^n} C_{n-1,-n} \left(\frac{\phi_2}{2\phi_1}, \frac{\phi_3}{3\phi_1}, \ldots, \frac{\phi_n}{n\phi_1} \right). \quad (11.43)
$$

PROOF   According to (11.38),

$$
\left(\frac{\phi(t)}{t} \right)^{-n} = \left(\sum_{n=1}^{\infty} \phi_n \frac{t^{n-1}}{n!} \right)^{-n} = \frac{1}{\phi_1^n} \left(1 + \sum_{r=1}^{\infty} \frac{\phi_{r+1}}{(r+1)\phi_1} \cdot \frac{t^r}{r!} \right)^{-n}
$$

and so, using (11.30),

$$
\left(\frac{\phi(t)}{t} \right)^{-n} = \frac{1}{\phi_1^n} \sum_{j=0}^{\infty} C_{j,-n} \left(\frac{\phi_2}{2\phi_1}, \ldots, \frac{\phi_{j+1}}{(j+1)\phi_1} \right) \frac{t^j}{j!}.
$$

Hence

$$
\left[\frac{d^{n-k}}{dt^{n-k}}\left(\frac{\phi(t)}{t}\right)^{-n}\right]_{u=0} = \frac{1}{\phi_1^n}C_{n-k,-n}\left(\frac{\phi_2}{2\phi_1}, \frac{\phi_3}{3\phi_1}, \ldots, \frac{\phi_{n-k+1}}{(n-k+1)\phi_1}\right)
$$

and by (11.41) we get

$$
\psi_{n,k} = \left[\frac{d^n}{du^n}(\phi^{-1}(u))^k\right]_{u=0}
$$
$$
= \frac{k(n-1)_{k-1}}{\phi_1^n}C_{n-k,-n}\left(\frac{\phi_2}{2\phi_1}, \frac{\phi_3}{3\phi_1}, \ldots, \frac{\phi_{n-k+1}}{(n-k+1)\phi_1}\right),
$$

which is the required expression (11.42).   ∎

COROLLARY 11.3
If

$$
\phi(t) = t\left(1 + \sum_{r=1}^{\infty} x_r \frac{t^{sr}}{r!}\right),
$$

with s a positive integer, then

$$
\phi^{-1}(u) = u\left(1 + \sum_{n=1}^{\infty} y_n \frac{u^{sn}}{n!}\right),
$$

where

$$
y_n = C_{n,-sn-1}(x_1, x_2, \ldots, x_n)/(sn+1)
$$
$$
= \sum_{k=1}^{n}(-1)^k(sn+k)_{k-1}B_{n,k}(x_1, x_2, \ldots, x_n). \tag{11.44}
$$

PROOF   The inversion formula of Lagrange (11.41), in the particular case of $k = 1$ and for $sn + 1$ instead of n, reduces to

$$
y_n = \frac{n!}{(sn+1)!}\left[\frac{d^{sn+1}}{du^{sn+1}}\phi^{-1}(u)\right]_{u=0}
$$
$$
= \frac{n!}{(sn+1)!}\left[\frac{d^{sn}}{dt^{sn}}\left(\frac{\phi(t)}{t}\right)^{-sn-1}\right]_{t=0}
$$
$$
= \frac{n!}{(sn+1)!}\left[\frac{d^{sn}}{dt^{sn}}\left(1 + \sum_{r=1}^{\infty} x_r \frac{t^{sr}}{r!}\right)^{-sn-1}\right]_{t=0}.
$$

Thus, using (11.30), it follows that

$$y_n = \frac{n!}{(sn+1)!} \left[\frac{d^{sn}}{dt^{sn}} \sum_{j=0}^{\infty} C_{j,-sn-1}(x_1, x_2, \ldots, x_n) \frac{t^{sj}}{j!} \right]_{t=0},$$

yielding the first part of (11.44). Further, from (11.29) and since

$$\frac{(-sn-1)_k}{sn+1} = \frac{(-sn-1)(-sn-2)\cdots(-sn-k)}{sn+1}$$

$$= (-1)^k (sn+k)_{k-1},$$

the second part of (11.44) is readily deduced. ∎

In the following theorem two useful forms of Lagrange formula are derived.

THEOREM 11.12
Let $f(t) = \sum_{k=0}^{\infty} f_k t^k / k!$ and $u = \phi(t) = \sum_{k=0}^{\infty} a_k t^k / k!$, with $a_0 \neq 0$. Then

$$f(t) = f(0) + \sum_{n=1}^{\infty} \left[\frac{d^{n-1}}{dt^{n-1}} \left(g^n(t) \frac{df(t)}{dt} \right) \right]_{t=0} \cdot \frac{u^n}{n!} \qquad (11.45)$$

and

$$f(t) \left(1 - u \frac{dg(t)}{dt} \right)^{-1} = \sum_{n=0}^{\infty} \left[\frac{d^n}{dt^n} \left(g^n(t) f(t) \right) \right]_{t=0} \cdot \frac{u^n}{n!}, \qquad (11.46)$$

with $u = t/g(t)$.

PROOF   Introducing into the expansion of $f(t)$ the inverse series $t = \phi^{-1}(u)$ and expanding the resulting expression into powers of u, we have

$$f(t) = f(0) + \sum_{k=1}^{\infty} \frac{1}{k!} \left[\frac{d^k f(t)}{dt^k} \right]_{t=0} \cdot (\phi^{-1}(u))^k$$

$$= f(0) + \sum_{k=1}^{\infty} \frac{1}{k!} \left[\frac{d^k f(t)}{dt^k} \right]_{t=0} \sum_{n=k}^{\infty} \left[\frac{d^n}{du^n} (\phi^{-1}(u))^k \right]_{u=0} \cdot \frac{u^n}{n!}.$$

Using (11.41) and putting $u = \phi(t) = t/g(t)$, we get

$$f(t) = f(0) + \sum_{k=1}^{\infty} \sum_{n=k}^{\infty} \binom{n-1}{k-1} \left[\frac{d^k f(t)}{dt^k} \right]_{t=0} \cdot \left[\frac{d^{n-k} g^n(t)}{dt^{n-k}} \right]_{t=0} \cdot \frac{u^n}{n!}$$

$$= f(0) + \sum_{n=1}^{\infty} \left\{ \sum_{k=1}^{n} \binom{n-1}{k-1} \left[\frac{d^{n-k} g^n(t)}{dt^{n-k}} \right]_{t=0} \cdot \left[\frac{d^{k-1}}{dt^{k-1}} \frac{df(t)}{dt} \right]_{t=0} \right\} \cdot \frac{u^n}{n!}$$

and, according to the formula of the $(n-1)$-th order derivative of the product of two functions, (11.45) is deduced.

Differentiating formula (11.45) with respect to u, we get

$$\frac{df(t)}{du} = \sum_{n=0}^{\infty} \left[\frac{d^n}{dt^n} \left(g^{n+1}(t) \frac{df(t)}{dt} \right) \right]_{t=0} \cdot \frac{u^n}{n!}.$$

Further,

$$\frac{df(t)}{du} = \frac{df(t)}{dt} \bigg/ \frac{du}{dt}$$

and, from $ug(t) = t$, by differentiation with respect to t, it follows that

$$g(t)\frac{du}{dt} + u\frac{dg(t)}{dt} = 1.$$

Therefore

$$\frac{df(t)}{du} = g(t)\frac{df(t)}{dt} \left(1 - u\frac{dg(t)}{dt} \right)^{-1}$$

and

$$g(t)\frac{df(t)}{dt} \left(1 - u\frac{dg(t)}{dt} \right)^{-1} = \sum_{n=0}^{\infty} \left[\frac{d^n}{dt^n} \left(g^{n+1}(t) \frac{df(t)}{dt} \right) \right]_{t=0} \cdot \frac{u^n}{n!}.$$

Replacing $g(t)df(t)/dt$ by $f(t)$, (11.46) is deduced.   ∎

Example 11.7

Determine the inverse power series of (a) $\phi(t) = te^{-t}$ and (b) $\psi(t) = t(1+t)^{-r}$.

(a) By (11.41), with $k = 1$, it follows that

$$\left[\frac{d^n}{du^n} \phi^{-1}(u) \right]_{u=0} = \left[\frac{d^{n-1}}{dt^{n-1}} \left(\frac{\phi(t)}{t} \right)^{-n} \right]_{t=0} = \left[\frac{d^{n-1}}{dt^{n-1}} e^{nt} \right]_{t=0} = n^{n-1}$$

and so the inverse of $\phi(t) = te^{-t}$ is the power series

$$\phi^{-1}(u) = \sum_{n=1}^{\infty} n^{n-1} \frac{u^n}{n!}.$$

(b) Again by (11.41), with $k = 1$, it follows that

$$\left[\frac{d^n}{du^n} \psi^{-1}(u) \right]_{u=0} = \left[\frac{d^{n-1}}{dt^{n-1}} \left(\frac{\psi(t)}{t} \right)^{-n} \right]_{t=0}$$

$$= \left[\frac{d^{n-1}}{dt^{n-1}} (1+t)^{rn} \right]_{t=0} = (rn)_{n-1}$$

and so the inverse of $\psi(t) = t(1+t)^{-r}$ is the power series

$$\psi^{-1}(u) = \sum_{n=1}^{\infty} \frac{1}{n} \binom{rn}{n-1} u^n. \qquad \square$$

Example 11.8

Determine the factorial moments of the sequence of probabilities

$$p_k = e^{-\lambda k} \frac{(\lambda k)^{k-1}}{k!}, \quad k = 1, 2, \dots, \ 0 < \lambda < \infty.$$

The generating function of this sequence of probabilities may be obtained as

$$P(z) = \sum_{k=1}^{\infty} p_k z^k = \frac{1}{\lambda} \sum_{k=1}^{\infty} k^{k-1} \frac{(z\lambda e^{-\lambda})^k}{k!} = \frac{\phi^{-1}(z\lambda e^{-\lambda})}{\lambda},$$

where

$$\phi^{-1}(u) = \sum_{k=1}^{\infty} k^{k-1} \frac{u^k}{k!}$$

is the inverse power series of $\phi(t) = te^{-t}$ (see Example 11.7). The generating function of the factorial moments $\mu_{(n)}$, $n = 1, 2, \dots$,

$$B(z) = 1 + \sum_{n=1}^{\infty} \mu_{(n)} \frac{z^n}{n!},$$

is connected with the generating function $P(z)$ by $B(z) = P(1+z)$ and so

$$B(z) = \frac{\phi^{-1}((1+z)\lambda e^{-\lambda})}{\lambda}.$$

Since $t = \phi^{-1}(u)$ is the inverse of $u = \phi(t) = te^{-t}$, it follows that $(1+z)\lambda e^{-\lambda} = te^{-t}$ and $z = (t/\lambda)e^{-\lambda[(t/\lambda)-1]} - 1$. Putting $w = (t/\lambda) - 1$ and $f(w) = (1+w)e^{-\lambda w} - 1$, we have $z = f(w)$ and

$$B(z) = \frac{\phi^{-1}((1+z)\lambda e^{-\lambda})}{\lambda} = 1 + f^{-1}(z),$$

where $w = f^{-1}(z)$ is the inverse of $z = f(w) = (1+w)e^{-\lambda w} - 1$. Therefore

$$f^{-1}(z) = 1 + \sum_{n=1}^{\infty} \mu_{(n)} \frac{z^n}{n!}$$

and

$$f(w) = (1 + w) \sum_{n=0}^{\infty} (-1)^n \lambda^n \frac{w^n}{n!} - 1$$

$$= \sum_{n=1}^{\infty} (-1)^n \lambda^n \frac{w^n}{n!} + \sum_{n=0}^{\infty} (-1)^n \lambda^n \frac{w^{m+1}}{n!}$$

$$= \sum_{n=1}^{\infty} (-1)^{n-1} (n - \lambda) \lambda^{n-1} \frac{w^n}{n!}.$$

The coefficients $\mu_{(n)}$, $n = 1, 2, \ldots$, of $w = f^{-1}(z)$, using (11.43) and (11.29), are obtained as

$$\mu_{(n)} = \sum_{k=0}^{n-1} \frac{(-1)^{n-1}(-n)_k \lambda^{n-1}}{(1-\lambda)^{n+k}} B_{n-1,k}(1 - \lambda/2, -\lambda/3, \ldots, 1 - \lambda/n).$$

In particular, for $n = 1, 2$,

$$\mu = \mu_{(1)} = (1 - \lambda)^{-1}, \quad \mu_{(2)} = \lambda(2 - \lambda)(1 - \lambda)^{-3}$$

and

$$\sigma^2 = \mu_{(2)} + \mu_{(1)} - \mu_{(1)}^2 = \lambda(1 - \lambda)^{-3}. \quad \square$$

11.7 TOUCHARD POLYNOMIALS

DEFINITION 11.6 *The polynomial $T_{n,k} \equiv T_{n,k}(x_1, x_2, \ldots, x_n; y_1, y_2, \ldots, y_n)$ in the variables $x_1, x_2, \ldots, x_n$ and $y_1, y_2, \ldots, y_n$ defined by the sum*

$$T_{n,k} = \sum \frac{n!}{k_1! k_2! \cdots k_n! r_1! r_2! \cdots r_n!} \left(\frac{x_1}{1!}\right)^{k_1} \cdots \left(\frac{x_n}{n!}\right)^{k_n} \left(\frac{y_1}{1!}\right)^{r_1} \cdots \left(\frac{y_n}{n!}\right)^{r_n},$$

$$(11.47)$$

where the summation is extended over all nonnegative integer solutions $(k_1, k_2, \ldots, k_n, r_1, r_2, \ldots, r_n)$ of the equations

$$\sum_{i=1}^{n} i(k_i + r_i) = n, \quad \sum_{i=1}^{n} k_i = k, \quad (11.48)$$

is called Touchard polynomial.

Note that

$$T_{n,0}(x_1, x_2, \ldots, x_n; y_1, y_2, \ldots, y_n) = B_n(y_1, y_2, \ldots, y_n)$$

and

$$T_{n,k}(x_1, x_2, \ldots, x_n; 0, 0, \ldots, 0) = B_{n,k}(x_1, x_2, \ldots, x_n),$$

where $B_n(y_1, y_2, \ldots, y_n)$ and $B_{n,k}(x_1, x_2, \ldots, x_n)$ are the exponential and the partial Bell partition polynomials, respectively. Also

$$T_{n,k}(abx_1, a^2bx_2, \ldots, a^nbx_n; ay_1, a^2y_2, \ldots, a^ny_n)$$
$$= a^n b^k T_{n,k}(x_1, x_2, \ldots, x_n; y_1, y_2, \ldots, y_n).$$

Further, summing the Touchard polynomials for $k = 0, 1, \ldots, n$ and setting $s_i = k_i + r_i$, $i = 1, 2, \ldots, n$, it follows that

$$\sum_{k=0}^{n} T_{n,k}(x_1, x_2, \ldots, x_n; y_1, y_2, \ldots, y_n)$$

$$= \sum \frac{n!}{s_1! s_2! \cdots s_n!} \prod_{i=1}^{n} \sum_{k_i=0}^{s_i} \binom{s_i}{k_i} \left(\frac{x_i}{i!}\right)^{k_i} \left(\frac{y_i}{i!}\right)^{s_i - k_i}$$

$$= \sum \frac{n!}{s_1! s_2! \cdots s_n!} \left(\frac{x_1 + y_1}{1!}\right)^{s_1} \left(\frac{x_2 + y_2}{2!}\right)^{s_2} \cdots \left(\frac{x_n + y_n}{n!}\right)^{s_n},$$

where the summation is extended over all partitions of n, that is, over all nonnegative integer solutions $(s_1, s_2, \ldots, s_n)$ of the equation $s_1 + 2s_2 + \cdots + ns_n = n$. Hence

$$\sum_{k=0}^{n} T_{n,k}(x_1, x_2, \ldots, x_n; y_1, y_2, \ldots, y_n)$$
$$= B_n(x_1 + y_1, x_2 + y_2, \ldots, x_n + y_n), \quad n = 0, 1, \ldots.$$

This relation may also be deduced from the bivariate generating function $T(t, u)$ of the Touchard polynomials (see Theorem 11.13 that follows) by putting $u = 1$ and comparing the resulting generating function with the generating function of the exponential Bell partition polynomials (11.9).

REMARK 11.6 The coefficients of the Touchard polynomials (11.47) are positive integers. Specifically,

$$B(n, k; k_1, k_2, \ldots, k_n; r_1, r_2, \ldots, r_n)$$

$$= \frac{n!}{k_1!(1!)^{k_1} k_2!(2!)^{k_2} \cdots k_n!(n!)^{k_n} r_1!(1!)^{r_1} r_2!(2!)^{r_2} \cdots r_n!(n!)^{r_n}},$$

with $\sum_{i=1}^{n} i(k_i + r_i) = n$ and $\sum_{i=1}^{n} k_i = k$, equals the number of partitions $\{A_1, A_2, \ldots, A_k, B_1, B_2, \ldots, B_r, \ldots\}$ of a finite set W_n of n elements, where

$\{A_1, A_2, \ldots, A_k\}$ is a partition of a set $U \subseteq W_n$ into k subsets, among which $k_i \geq 0$ subsets include i elements each, $i = 1, 2, \ldots, n$, and $\{B_1, B_2, \ldots, B_r, \ldots\}$ is a partition of the set $W_n - U$ into subsets, among which $r_i \geq 0$ subsets include i elements each, $i = 1, 2, \ldots, n$. ∎

Bivariate and vertical generating functions for the Touchard polynomials are derived in the following theorem.

THEOREM 11.13

(a) *The bivariate generating function of the Touchard polynomials, $T_{n,k} \equiv T_{n,k}(x_1, x_2, \ldots, x_n; y_1, y_2, \ldots, y_n)$, $k = 0, 1, \ldots, n$, $n = 0, 1, \ldots$, with $T_{0,0} \equiv 1$, is given by*

$$T(t, u) = \sum_{n=0}^{\infty} \sum_{k=0}^{n} T_{n,k}(x_1, x_2, \ldots, x_n; y_1, y_2, \ldots, y_n) u^k \frac{t^n}{n!}$$
$$= \exp\{u[g(t) - x_0] + [h(t) - y_0]\}. \tag{11.49}$$

(b) *The vertical generating function of the Touchard polynomials $T_{n,k} \equiv T_{n,k}(x_1, x_2, \ldots, x_n; y_1, y_2, \ldots, y_n)$, $n = k, k+1, \ldots$, for fixed k, is given by*

$$T_k(t) = \sum_{n=k}^{\infty} T_{n,k}(x_1, x_2, \ldots, x_n; y_1, y_2, \ldots, y_n) \frac{t^n}{n!}$$
$$= \frac{[g(t) - x_0]^k}{k!} \exp[h(t) - y_0], \quad k = 0, 1, \ldots, \tag{11.50}$$

where

$$g(t) = \sum_{j=0}^{\infty} x_j \frac{t^j}{j!}, \quad h(t) = \sum_{j=0}^{\infty} y_j \frac{t^j}{j!}. \tag{11.51}$$

PROOF (a) Multiplying (11.47) by $u^k t^n / n!$ and summing the resulting expression for $k = 0, 1, \ldots, n$ and $n = 0, 1, \ldots$, it follows that

$$T(t, u) = \sum_{n=0}^{\infty} \sum_{k=0}^{n} T_{n,k}(x_1, x_2, \ldots, x_n; y_1, y_2, \ldots, y_n) u^k \frac{t^n}{n!}$$
$$= \sum_{n=0}^{\infty} \sum_{k=0}^{n} \sum \frac{n!}{k_1! k_2! \cdots k_n! r_1! r_2! \cdots r_n!}$$
$$\times \left(\frac{x_1 ut}{1!}\right)^{k_1} \cdots \left(\frac{x_n ut^n}{n!}\right)^{k_n} \left(\frac{y_1 t}{1!}\right)^{r_1} \cdots \left(\frac{y_n t^n}{n!}\right)^{r_n},$$

where the summation in the inner sum is extended over all nonnegative integer solutions $(k_1, k_2, \ldots, k_n, r_1, r_2, \ldots, r_n)$ of $\sum_{i=1}^{n} i(k_i + r_i) = n$ and $\sum_{i=1}^{n} k_i = k$.

Since this inner sum is summed over all $k = 0, 1, \ldots, n$ and $n = 0, 1, \ldots$, it follows that the summation is extended over all $k_i = 0, 1, \ldots, r_i = 0, 1, \ldots$, for $i = 1, 2, \ldots$. Therefore,

$$
T(t, u) = \prod_{i=1}^{\infty} \left\{ \sum_{k_i=0}^{\infty} \frac{1}{k_i!} \left(\frac{x_i u t^i}{i!} \right)^{k_i} \sum_{r_i=0}^{\infty} \frac{1}{r_i!} \left(\frac{y_i t^i}{i!} \right)^{r_i} \right\}
$$

$$
= \prod_{i=1}^{\infty} \exp \left(\frac{x_i u t^i}{i!} + \frac{y_i t^i}{i!} \right) = \exp \left(u \sum_{i=1}^{\infty} x_i \frac{t^i}{i!} + \sum_{i=1}^{\infty} y_i \frac{t^i}{i!} \right).
$$

Using (11.51), the last expression implies (11.49).

(b) Interchanging the order of summation (11.49) may be expressed as

$$
T(t, u) = \sum_{k=0}^{\infty} \sum_{n=k}^{\infty} T_{n,k}(x_1, x_2, \ldots, x_n; y_1, y_2, \ldots, y_n) u^k \frac{t^n}{n!}
$$

$$
= \sum_{k=0}^{\infty} T_k(t) u^k
$$

and, since $T(t, u) = \exp\{u[g(t) - x_0] + [h(t) - y_0]\}$, it follows that

$$
\sum_{k=0}^{\infty} T_k(t) u^k = \exp[h(t) - y_0] \sum_{k=0}^{\infty} \frac{[g(t) - x_0]^k}{k!} u^k.
$$

Equating the coefficients of u^k in both sides of the last expression, (11.50) is deduced. ∎

The Touchard polynomial is expressed as a finite sum of Bell partition polynomials in the following theorem.

THEOREM 11.14

The Touchard polynomial $T_{n,k}(x_1, x_2, \ldots, x_n; y_1, y_2 \ldots, y_n)$ is expressed in terms of the partial Bell polynomials $B_{r,k}(x_1, x_2, \ldots, x_r)$ and the exponential Bell polynomials $B_{n-r}(y_1, y_2, \ldots, y_{n-r})$, for $r = k, k+1, \ldots, n$, as

$$
T_{n,k}(x_1, x_2, \ldots, x_n; y_1, y_2, \ldots, y_n)
$$

$$
= \sum_{r=k}^{n} \binom{n}{r} B_{r,k}(x_1, x_2, \ldots, x_r) B_{n-r}(y_1, y_2, \ldots, y_{n-r}). \tag{11.52}
$$

PROOF Expanding the right-hand side of (11.50) into powers of t, using (11.7) and (11.9), we get the relation

$$T_k(t) = \left[\sum_{r=k}^{\infty} B_{r,k}(x_1, x_2, \dots, x_r)\frac{t^n}{n!}\right] \cdot \left[\sum_{n=0}^{\infty} B_n(y_1, y_2, \dots, y_n)\frac{t^n}{n!}\right]$$

$$= \sum_{n=k}^{\infty} \left[\sum_{r=k}^{n} \binom{n}{r} B_{r,k}(x_1, x_2, \dots, x_r)B_{n-r}(y_1, y_2, \dots, y_{n-r})\right]\frac{t^n}{n!},$$

which, by virtue of

$$T_k(t) = \sum_{n=k}^{\infty} T_{n,k}(x_1, x_2, \dots, x_n; y_1, y_2, \dots, y_n)\frac{t^n}{n!},$$

implies (11.52).   ∎

In the next theorem a recurrence relation for the Touchard polynomials is deduced.

THEOREM 11.15
The Touchard polynomials $T_{n,k} \equiv T_{n,k}(x_1, x_2, \dots, x_n ; y_1, y_2, \dots, y_n)$, $k = 0, 1, \dots, n$, $n = 0, 1, \dots$, with $T_{0,0} \equiv 1$, satisfy the recurrence relation

$$T_{n+1,k+1} = \sum_{r=0}^{n-k} \binom{n}{r} x_{r+1}T_{n-r,k} + \sum_{r=0}^{n-k-1} \binom{n}{r} y_{r+1}T_{n-r,k+1}. \quad (11.53)$$

PROOF   Differentiating the generating function (11.50) and expanding the resulting expression,

$$\frac{d}{dt}T_{k+1}(t) = T_k(t)\frac{d}{dt}g(t) + T_{k+1}(t)\frac{d}{dt}h(t),$$

into powers of t, it follows that

$$\sum_{n=k}^{\infty} T_{n+1,k+1}\frac{t^n}{n!}$$

$$= \left[\sum_{r=0}^{\infty} x_{r+1}\frac{t^r}{r!}\right]\left[\sum_{n=k}^{\infty} T_{n,k}\frac{t^n}{n!}\right] + \left[\sum_{r=0}^{\infty} y_{r+1}\frac{t^r}{r!}\right]\left[\sum_{n=k}^{\infty} T_{n,k+1}\frac{t^n}{n!}\right]$$

$$= \sum_{n=k}^{\infty} \left[\sum_{r=0}^{n-k} \binom{n}{r} x_{r+1}T_{n-r,k} + \sum_{r=0}^{n-k-1} \binom{n}{r} y_{r+1}T_{n-r,k+1}\right]\frac{t^n}{n!}.$$

Equating the coefficients of $t^n/n!$ in both sides of the last expression, (11.53) is deduced.   ∎

11.8 BIBLIOGRAPHIC NOTES

Francesco Faa di Bruno (1855) expressed the derivatives of a composite function as a multivariable polynomial of the derivatives of the component functions. These polynomials are essentially partition polynomials. A history of the problem of deriving a general formula for the derivatives of a composite function was outlined by Eugene Lukacs (1955). The expressions of the cumulants in terms of the (power) moments, and vice versa, were also given in that paper (see Example 11.4). In the classical book of Charles Jordan (1939a) a brief sketch of proof of Faa di Bruno formula, using Taylor series, is quoted. Eric Temple Bell (1927), aiming to the unification of a variety of arithmetical functions, introduced and studied the partition polynomials. Further, Bell (1934a,b) examined additional arithmetical and other properties along with certain particular cases of these polynomials. It is worth noting that R. Frucht (1969), in a combinatorial approach to the Bell polynomials, furnished a nice combinatorial interpretation of these polynomials; Example 11.2 was discussed in that paper. John Riordan (1958, 1968) named the partition polynomials after E. T. Bell and extensively studied them.

The derivation of the expression of the derivatives of a composite function as partition polynomials of the derivatives of the component functions provided by J. Riordan (1958) is based on umbral calculus. A combination of the derivation of Faa di Bruno formula provided by Ch. Jordan (1939a) and J. Riordan (1958) that avoids the use of umbral calculus is adopted in this book (see Theorem 11.4). The reader interested in a rigorous umbral calculus derivation of this formula is referred to the paper of Steven Roman (1980). Louis Comtet (1974) devoted much of the chapter on identities and expansions to a thorough presentation of the Bell polynomials and provided an updated and rich bibliography of the subject.

The exponential, logarithmic and potential polynomials, which are the most striking partition polynomials, have a direct use in combinatorics in expressing the elementary symmetric functions and the homogeneous product sum symmetric functions in terms of the power sum symmetric functions and vice versa. Examples 11.5 and 11.6, in which such expressions are derived, were taken from P. A. MacMahon (1915, 1916). Their use in expressing the probability function and factorial moments of a compound discrete distribution in terms of the probability functions and factorial moments of the component distributions was demonstrated by Ch. A. Charalambides (1977b). Apart from the classical convolution formula, G. P. M. Heselden (1973) derived an interesting convolution formula for the exponential Bell partition polynomials; Exercise 2 is based on this paper. The bipartitional polynomials and their applications in combinatorics and statistics, which appear in the exercises, were examined by Ch. A. Charalambides (1981). The derivatives of a bivariate composite function may be expressed as bipartitional polynomials of the derivatives of the component functions. L. Comtet (1968) expressed the derivatives of an implicit function in terms of certain bipartitional polynomials.

11.9 EXERCISES

1. *Convolution of Bell polynomials.* Let $B_n(x_1, x_2, \ldots, x_n)$ be the exponential Bell partition polynomial. Show that

$$B_n(x_1 + y_1, x_2 + y_2, \ldots, x_n + y_n)$$
$$= \sum_{k=0}^{n} \binom{n}{k} B_k(x_1, x_2, \ldots, x_k) B_{n-k}(y_1, y_2, \ldots, y_{n-k}).$$

2*. *Heselden's convolution of Bell polynomials* (Heselden, 1973). For any real or complex number a such that $a - k \neq 0$, $k = 0, 1, \ldots, n$, show that

$$\sum_{k=0}^{n} \frac{a}{a-k} \binom{n}{k} B_k((k-a)z_1, (k-a)z_2, \ldots, (k-a)z_k)$$
$$\times B_{n-k}((a-k)z_1, (a-k)z_2, \ldots, (a-k)z_{n-k}) = \delta_{n,0}.$$

Using this relation, derive the following convolution formula

$$\sum_{k=0}^{n} \frac{a}{a-k} \binom{n}{k} B_k((k-a)z_1, (k-a)z_2, \ldots, (k-a)z_k)$$
$$\times B_{n-k}(y_1 - kz_1, y_2 - kz_2, \ldots, y_{n-k} - kz_{n-k})$$
$$= B_n(y_1 - az_1, y_2 - az_2, \ldots, y_n - az_n).$$

Further, show that

$$\sum_{k=0}^{n} \frac{a}{a-k} \binom{n}{k} B_k(x_1 + kz_1, x_2 + kz_2, \ldots, x_k + kz_k)$$
$$\times B_{n-k}(y_1 - kz_1, y_2 - kz_2, \ldots, y_{y-k} - kz_{n-k})$$
$$= \sum_{k=0}^{n} \frac{a}{a-k} \binom{n}{k} B_k(x_1 + az_1, x_2 + az_2, \ldots, x_k + az_k)$$
$$\times B_{n-k}(y_1 - az_1, y_2 - az_2, \ldots, y_{n-k} - az_{n-k}).$$

3. *Hermite and Laguerre polynomials.* Show that the exponential Bell partition polynomial $B_n(x_1, x_2, \ldots, x_n)$ in the particular case of $x_1 = x$, $x_2 = 1$, $x_r = 0$, $r = 3, 4, \ldots$, reduces to

$$B_n(x, 1) = i^{-n} H_n(ix), \quad i = \sqrt{-1},$$

where

$$H_n(x) = (-1)^n e^{x^2/2} D_x^n e^{-x^2/2} = n! \sum_{k=0}^{[n/2]} \frac{(-1)^k x^{n-2k}}{(n-2k)!k!2^k}$$

is the Hermite polynomial. Also, show that $B_n(x_1, x_2, \ldots, x_n)$ in the particular case of $x_r = r! x^{-r+1}$, $r = 1, 2, \ldots$, reduces to

$$B_n(1!, 2! x^{-1}, \ldots, n! x^{-n+1}) = (n-1)! x^{-n+1} L_{n-1}^{(1)}(-x),$$

where

$$L_n^{(1)}(x) = \frac{1}{n!} x^{-1} e^x D_x^n (x^{n+1} e^{-x}) = \sum_{k=0}^{n} \binom{n+1}{n-k} \frac{(-x)^k}{k!} \quad n = 0, 1, \ldots$$

is a Laguerre polynomial.

4. *Generalized Hermite polynomials* (Bell, 1934b). Show that the exponential Bell partition polynomial $B_n(x_1, x_2, \ldots, x_n)$ in the particular case of $x_r = (s)_r a x^{s-r}$, $r = 1, 2, \ldots, s$, $x_r = 0$, $r = s+1, s+2, \ldots$, with s a positive integer, reduces to

$$B_n((s)_1 a x^{s-1}, (s)_2 a x^{s-2}, \ldots, (s)_n a x^{s-n}) = H_n(x; a, s),$$

where

$$H_n(x; a, s) = e^{-a x^s} D_x^n e^{a x^s}$$

is a generalized Hermite polynomial. Further, show that

$$H_n(x; a, s) = \sum_{k=0}^{n} C(n, k; s) a^k x^{sk-n},$$

where

$$C(n, k; s) = \frac{1}{k!} \sum_{r=0}^{k} (-1)^r \binom{k}{r} (sr)_n$$

is the generalized factorial coefficient.

5. *Chebyshev polynomials of the first kind.* Show that the logarithmic partition polynomial $L_n(x_1, x_2, \ldots, x_n)$ in the particular case of $x_1 = x$, $x_2 = 1$, $x_r = 0$, $r = 3, 4, \ldots$, reduces to

$$L_n(x, 1) = -T_n(-x)/n,$$

where

$$T_n(x) = n! \sum_{k=0}^{[n/2]} (-1)^k \frac{n}{n-k} \binom{n-k}{k} \frac{x^{n-2k}}{2^k}$$

is the Chebyshev polynomial of the first kind.

6. *Gegenbauer polynomials.* Show that the potential partition polynomial $C_{n,-s}(x_1, x_2, \ldots, x_n)$ in the particular case of $x_1 = x$, $x_2 = 1$, $x_r = 0$, $r = 3, 4, \ldots$, reduces to

$$C_{n,-s}(x, 1) = G_n^{(s)}(-x), \quad -1/2 < s < 0 \ \text{ or } \ 0 < s < \infty,$$

where

$$G_n^{(s)}(x) = n! \sum_{k=0}^{[n/2]} (-1)^k \binom{n-k}{k} \binom{s+n-k-1}{n-k} \frac{x^{n-2k}}{2^k}$$

is the Gegenbauer polynomial. Note that $G_n^{(1/2)}(x) = P_n(x)$ is the Legendre polynomial and $G_n^{(1)}(x) = U_n(x)$ is the Chebyshev polynomial of the second kind. Also $\lim_{s \to 0} s^{-1} n G_n^{(s)}(x) = T_n(x)$ is the Chebyshev polynomial of the first kind.

7. Let $B_{n,k}(x_1, x_2, \ldots, x_n)$ be the partial Bell partition polynomial. Show that, for $r = 1, 2, \ldots$,

$$B_{n,k}(0, \ldots, 0, x_{r+1}, x_{r+2}, \ldots, x_n) = \frac{n!}{(n-rk)!} B_{n-rk,k}(y_1, y_2, \ldots, y_{n-rk}),$$

with $y_i = x_{r+i}/(r+i)_r$, $i = 1, 2, \ldots$. Further, show that

$$B_{n,k}(0, \ldots, 0, x_{r+1}, \ldots, x_n)$$
$$= \sum_{j=0}^{k} (-1)^j \frac{(n)_{rj}}{j!(r!)^j} B_{n-rj,k-j}(0, \ldots, 0, x_r, \ldots, x_{n-rj}) x_r^j$$

and

$$B_{n,k}(0, \ldots, 0, x_r, \ldots, x_n)$$
$$= \sum_{j=0}^{k} \frac{(n)_{rj}}{j!(r!)^j} B_{n-rj,k-j}(0, \ldots, 0, x_{r+1}, \ldots, x_{n-rj}) x_r^j.$$

8. (*Continuation*). Show that

$$B_{n,k}(1!, 2!, \ldots, n!) = \frac{n!}{k!} \binom{n-1}{k-1}$$

and

$$B_{n,k}(1,2,\ldots,n) = \binom{n}{k} k^{n-k}.$$

9. (*Continuation*). Show that

$$B_{2r,k}(0,2!,\ldots,0,(2r)!) = \frac{(2r)!}{k!}\binom{r-1}{k-1},$$

$$B_{2r-1,k}(0,2!,\ldots,(2r-2)!,0) = 0$$

and

$$B_{2r,2s}(1!,0,\ldots,(2r-1)!,0) = \frac{(2r)!}{(2s)!}\binom{r+s-1}{2s-1},$$

$$B_{2r,2s-1}(1!,0,\ldots,(2r-1)!,0) = 0,$$

$$B_{2r-1,2s-1}(1!,0,\ldots,(2r-1)!,0) = \frac{(2r-1)!}{(2s-1)!}\binom{r+s-2}{2s-2},$$

$$B_{2r-1,2s}(1!,0,\ldots,0,(2r-1)!) = 0.$$

10. Let $C_{n,s} \equiv C_{n,s}(x_1,x_2,\ldots,x_n)$ be the potential partition polynomial. Show that

$$\binom{n+s}{n} C_{n,s} = \sum_{r=0}^{n} \binom{n+s}{n+r}\binom{n-s}{n-r}\binom{n+r}{n} C_{n,r}$$

with s any real (or complex) number.

11*. (*Continuation*). Setting $y_n = C_{n,-1}(x_1,x_2,\ldots,x_n)$, show that

$$y_n = (-1)^n n! \begin{vmatrix} \frac{x_1}{1!} & \frac{x_2}{2!} & \frac{x_3}{3!} & \cdots & \frac{x_{n-2}}{(n-2)!} & \frac{x_{n-1}}{(n-1)!} & \frac{x_n}{n!} \\ 1 & \frac{x_1}{1!} & \frac{x_2}{2!} & \cdots & \frac{x_{n-3}}{(n-3)!} & \frac{x_{n-2}}{(n-2)!} & \frac{x_{n-1}}{(n-1)!} \\ 0 & 1 & \frac{x_1}{1!} & \cdots & \frac{x_{n-4}}{(n-4)!} & \frac{x_{n-3}}{(n-3)!} & \frac{x_{n-2}}{(n-2)!} \\ \cdot & \cdot & \cdot & \cdots & \cdot & \cdot & \cdot \\ 0 & 0 & 0 & \cdots & 1 & \frac{x_1}{1!} & \frac{x_2}{2!} \\ 0 & 0 & 0 & \cdots & 0 & 1 & \frac{x_1}{1!} \end{vmatrix}$$

12. Let $\sigma(n) = \sum_{d \mid n} d$ be the sum of the divisors of n, $n = 1, 2, \ldots$. Show that

$$h(t) = \sum_{n=1}^{\infty} \sigma(n) t^n = \sum_{k=1}^{\infty} k t^k (1 - t^k)^{-1}.$$

Further, show that the generating function of the number $p(n)$ of partitions of n, which is given by

$$G(t) = \sum_{n=0}^{\infty} p(n) t^n = \prod_{k=1}^{\infty} (1 - t^k)^{-1},$$

may be expressed as

$$G(t) = \exp \left(\sum_{r=1}^{\infty} \sigma(r) \frac{t^r}{r} \right)$$

and thus conclude that

$$p(n) = B_n(\sigma(1), 1!\sigma(2), \ldots, (n-1)!\sigma(n))/n!.$$

13. Show that the factorial moments $\mu_{(n)}$, $n = 1, 2, \ldots$, of the sequence of probabilities

$$p_k = \frac{1}{k!} \binom{rk}{k-1} p^k (1-p)^{rk-k+1}, \quad k = 1, 2, \ldots,$$

where r is a positive integer and $0 < p < 1/r$ or r is a negative real number and $1/r < p < 0$, are given by

$$\mu_{(n)} = \frac{p^{n-1}}{(1-rp)^n} C_{n-1,-n}(x_1, x_2, \ldots, x_{n-1}), \quad n = 1, 2, \ldots,$$

where

$$x_j = (-r)_j \frac{1 - (n+j)p/(j+1)}{1 - rp}, \quad j = 1, 2, \ldots.$$

14. *Laguerre polynomials.* Show that the Touchard polynomial $T_{n,k} \equiv T_{n,k}(x_1, x_2, \ldots, x_n; y_1, y_2, \ldots, y_n)$ in the particular case of $x_r = r!x^{-r+1}$, $y_r = r!x^{-r+1}$, $r = 1, 2, \ldots$, reduces to

$$T_{n,k}(1!, 2!x^{-1}, \ldots, n!x^{-n+1}; 1!, 2!x^{-1}, \ldots, n!x^{-n+1}) = \frac{n! L_{n-k}^{(k-1)}(-x)}{k! x^{n-k}},$$

for $n = k, k+1, \ldots$, $k = 1, 2, \ldots$, where

$$L_n^{(a)}(x) = \frac{1}{n!} x^{-a} e^x D_x^n (x^{a+n} e^{-x}) = \sum_{r=0}^{n} \binom{n+a}{n-r} \frac{(-x)^r}{r!}.$$

is the Laguerre polynomial.

15. Let $T_{n,k}(x_1, x_2, \ldots, x_n; y_1, y_2, \ldots, y_n)$ be the Touchard polynomial. Show that, for $r = 1, 2, \ldots,$

$$T_{n,k}(0, 0, \ldots, 0, x_{r+1}, x_{r+2}, \ldots, x_n; y_1, y_2, \ldots, y_n)$$

$$= \frac{n!}{(n - rk)!} T_{n-rk,k}(z_1, z_2, \ldots, z_{n-rk}; y_1, y_2, \ldots, y_{n-rk}),$$

with $z_i = x_{r+i}/(r + i)_r$, $i = 1, 2, \ldots.$ Further, show that

$$T_{n,k}(0, \ldots, 0, x_{r+1}, \ldots, x_n,; y_1, y_2, \ldots, y_n)$$

$$= \sum_{j=0}^{k} (-1)^j \frac{(n)_{rj}}{j!(r!)^j} T_{n-rj,k-j}(0, \ldots, 0, x_r, \ldots, x_{n-rj}; y_1, y_2, \ldots, y_{n-rj}) x_r^j$$

and

$$T_{n,k}(0, \ldots, 0, x_r, \ldots, x_n,; y_1, y_2, \ldots, y_n)$$

$$= \sum_{j=0}^{k} \frac{(n)_{rj}}{j!(r!)^j} T_{n-rj,k-j}(0, \ldots, 0, x_{r+1}, \ldots, x_{n-rj}; y_1, y_2, \ldots, y_{n-rj}) x_r^j.$$

16. (*Continuation*). Show that

$$T_{n,k}(1!, 2!, \ldots, n!; 0!r, 1!r, \ldots, (n-1)!r) = \frac{n!}{k!} \binom{n+r-1}{k+r-1}$$

and

$$T_{n,k}(1, 2, \ldots, n; r, 0, \ldots, 0) = \binom{n}{k} (k+r)^{n-k}.$$

17. Let $f(u)$, $g(t)$ and $h(t)$ be functions of real variables for which the derivatives

$$g_r = \left[\frac{d^r g(t)}{dt^r} \right]_{t=a} , \quad h_r = \left[\frac{d^r h(t)}{dt^r} \right]_{t=a} , \quad f_r = \left[\frac{d^r f(u)}{du^r} \right]_{u=g(a)} ,$$

for $r = 0, 1, \ldots,$ exist. Show that the derivatives of the composite function $\psi(t) = f(g(t)) \exp[h(t) - h(a)]$,

$$\psi_n = \left[\frac{d^n \psi(t)}{dt^n} \right]_{t=a} , \quad n = 0, 1, \ldots,$$

are given by

$$\psi_n = \sum_{k=0}^{n} f_k T_{n,k}(g_1, g_2, \ldots, g_n; h_1, h_2, \ldots, h_n)$$

$$= B_n(fg_1 + h_1, fg_2 + h_2, \ldots, fg_n + h_n), \quad f^k \equiv f_k.$$

18*. *Exponential bipartitional polynomials.* The polynomial $A_{m,n} \equiv A_{m,n}(x_{0,1}, x_{1,0}, \dots, x_{m,n})$ in the variables $x_{0,1}, x_{1,0}, \dots, x_{m,n}$ defined by the sum

$$A_{m,n} = \sum \frac{m! n!}{k_{0,1}! k_{1,0}! \cdots k_{m,n}!} \left(\frac{x_{0,1}}{0! 1!}\right)^{k_{0,1}} \left(\frac{x_{1,0}}{1! 0!}\right)^{k_{1,0}} \cdots \left(\frac{x_{m,n}}{m! n!}\right)^{k_{m,n}},$$

where the summation is extended over all partitions of the bipartite number (m, n), that is, over all nonnegative integer solutions $(k_{0,1}, k_{1,0}, \dots, k_{m,n})$ of the equations

$$\sum_{i=1}^{m} i \sum_{j=0}^{n} k_{i,j} = m, \quad \sum_{j=1}^{n} j \sum_{i=0}^{m} k_{i,j} = n,$$

is called *exponential bipartitional polynomial.* The polynomial $A_{m,n;k} \equiv A_{m,n;k}(x_{0,1}, x_{1,0}, \dots, x_{m,n})$ in the variables $x_{0,1}, x_{1,0}, \dots, x_{m,n}$ of degree k, defined by the sum

$$A_{m,n;k} = \sum \frac{m! n!}{k_{0,1}! k_{1,0}! \cdots k_{m,n}!} \left(\frac{x_{0,1}}{0! 1!}\right)^{k_{0,1}} \left(\frac{x_{1,0}}{1! 0!}\right)^{k_{1,0}} \cdots \left(\frac{x_{m,n}}{m! n!}\right)^{k_{m,n}},$$

where the summation is extended over all partitions of the bipartite number (m, n) into k parts, that is, over all nonnegative integer solutions $(k_{0,1}, k_{1,0}, \dots, k_{m,n})$ of the equations

$$\sum_{i=1}^{m} i \sum_{j=0}^{n} k_{i,j} = m, \quad \sum_{j=1}^{n} j \sum_{i=0}^{m} k_{i,j} = n, \quad \sum_{\substack{j=0 \\ i+j \neq 0}}^{n} \sum_{i=0}^{m} k_{i,j} = k,$$

is called *(exponential) partial bipartitional polynomial.* Show that

$$A_{m,n}(b x_{0,1}, a x_{1,0}, \dots, a^m b^n x_{m,n}) = a^m b^n A_{m,n}(x_{0,1}, x_{1,0}, \dots, x_{m,n}),$$

$$A_{m,n;k}(b c x_{0,1}, a c x_{1,0}, \dots, a^m b^n c x_{m,n})$$
$$= a^m b^n c^k A_{m,n;k}(x_{0,1}, x_{1,0}, \dots, x_{m,n}),$$

and

$$A_{m,n}(x_{0,1}, x_{1,0}, \dots, x_{m,n}) = \sum_{k=1}^{m+n} A_{m,n;k}(x_{0,1}, x_{1,0}, \dots, x_{m,n}).$$

Further, show that

$$A(t, u) = \sum_{n=0}^{\infty} \sum_{m=0}^{\infty} A_{m,n} \frac{t^m}{m!} \frac{u^n}{n!} = \exp[g(t, u) - x_{0,0}],$$

$$A(t, u, w) = \sum_{n=0}^{\infty} \sum_{m=0}^{\infty} \sum_{k=0}^{m+n} A_{m,n;k} w^k \frac{t^m}{m!} \frac{u^n}{n!} = \exp\{w[g(t,u) - x_{0,0}]\}$$

and

$$A_k(t, u) = \sum_{n=0}^{\infty} \sum_{m=0}^{\infty} A_{m,n;k} \frac{t^m}{m!} \frac{u^n}{n!} = \frac{[g(t,u) - x_{0,0}]^k}{k!},$$

where

$$g(t, u) = \sum_{n=0}^{\infty} \sum_{m=0}^{\infty} x_{m,n} \frac{t^m}{m!} \frac{u^n}{n!}.$$

19*. (*Continuation*). Show that

$$A_{m+1,n} = \sum_{j=0}^{n} \sum_{i=0}^{m} \binom{m}{i} \binom{n}{j} x_{i+1,j} A_{m-i,n-j}$$

and

$$A_{m+1,n;k+1} = \sum_{j=0}^{n} \sum_{i=0}^{m} \binom{m}{i} \binom{n}{j} x_{i+1,j} A_{m-i,n-j;k}.$$

20*. *General bipartitional polynomials.* The polynomial $P_{m,n} \equiv P_{m,n}(x_{0,1}, x_{1,0}, \ldots, x_{m,n})$ in the variables $x_{0,1}, x_{1,0}, \ldots, x_{m,n}$ and parameters $c_1, c_2, \ldots, c_{m+n}$, defined by the sum

$$P_{m,n} = \sum \frac{m!n!}{k_{0,1}! k_{1,0}! \cdots k_{m,n}!} c_k \left(\frac{x_{0,1}}{0!1!} \right)^{k_{0,1}} \left(\frac{x_{1,0}}{1!0!} \right)^{k_{1,0}} \cdots \left(\frac{x_{m,n}}{m!n!} \right)^{k_{m,n}},$$

where $k = k_{0,1} + k_{1,0} + \cdots + k_{m,n}$ and the summation is extended over all partitions of the bipartite number (m, n), that is, over all nonnegative integer solutions $(k_{0,1}, k_{1,0}, \ldots, k_{m,n})$ of the equations

$$\sum_{i=1}^{m} i \sum_{j=0}^{n} k_{i,j} = m, \quad \sum_{j=1}^{n} j \sum_{i=0}^{m} k_{i,j} = n,$$

is called *bipartitional polynonial.* Note that it can be expressed by the exponential bipartitional polynomial $A_{m,n}(x_{0,1}, x_{1,0}, \ldots, x_{m,n})$ as

$$P_{m,n}(x_{0,1}, x_{1,0}, \ldots, x_{m,n}; c_1, c_2, \ldots, c_{m+n})$$
$$= A_{m,n}(cx_{0,1}, cx_{1,0}, \ldots, cx_{m,n}), \quad c^k \equiv c_k.$$

Show that

$$A_{m,n}(cx_{0,1}, cx_{1,0}, \ldots, cx_{m,n}) = \sum_{k=0}^{m+n} c_k A_{m,n;k}(x_{0,1}, x_{1,0}, \ldots, x_{m,n})$$

and

$$A(t, u; c_1, c_2, \dots) = \sum_{n=0}^{\infty} \sum_{m=0}^{\infty} A_{m,n}(cx_{0,1}, cx_{1,0}, \dots, cx_{m,n}) \frac{t^m}{m!} \frac{u^n}{n!}$$

$$= \sum_{k=0}^{\infty} c_k \frac{[g(t, u) - x_{0,0}]^k}{k!} = \exp\{c[g(t, u) - x_{0,0}]\}, \quad c^k \equiv c_k,$$

where

$$g(t, u) = \sum_{n=0}^{\infty} \sum_{m=0}^{\infty} x_{m,n} \frac{t^m}{m!} \frac{u^n}{n!}.$$

Further, show that

$$h_{m,n} = \sum_{k=0}^{m+n} f_k A_{m,n;k}(g_{0,1}, g_{1,0}, \dots, g_{m,n})$$

$$= A_{m,n}(f g_{0,1}, f g_{1,0}, \dots, f g_{m,n}), \quad f^k \equiv f_k,$$

where

$$h_{m,n} = \left[\frac{\partial^{m+n} h(t, u)}{\partial u^n \partial t^m} \right]_{t=a, u=b}, \quad g_{m,n} = \left[\frac{\partial^{m+n} g(t, u)}{\partial u^n \partial t^m} \right]_{t=a, u=b},$$

and

$$f_k = \left[\frac{d^k f(w)}{dw^k} \right]_{w=g(a,b)}.$$

21*. (*Continuation*). *Compound bivariate discrete distributions.* Consider a discrete distribution with probability function p_k, $k = 0, 1, \dots$ and probability generating function $f(w) = \sum_{k=0}^{\infty} p_k w^k$. Further, consider a bivariate discrete distribution with probability mass function $q_{i,j}$, $i = 0, 1, \dots$, $j = 0, 1, \dots$ and probability generating function $g(t, u) = \sum_{j=0}^{\infty} \sum_{i=0}^{\infty} q_{i,j} t^i u^j$. Then the distribution with probability generating function $h(t, u) = f(g(t, u)) = \sum_{n=0}^{\infty} \sum_{m=0}^{\infty} P_{m,n} t^m u^n$ is called *compound bivariate distribution*. Show that

$$P_{m,n} = \frac{1}{m! n!} \sum_{k=0}^{m+n} f_k A_{m,n;k}(q_{0,1}, q_{1,0}, \dots, q_{m,n})$$

$$= \frac{1}{m! n!} A_{m,n}(f q_{0,1}, f q_{1,0}, \dots, f q_{m,n}), \quad f^k \equiv f_k$$

with $f_k = \sum_{i=k}^{\infty} (i)_k p_i q_{0,0}^{i-k}$, and

$$M_{(m,n)} = \frac{1}{m! n!} \sum_{k=0}^{m+n} \nu_{(k)} A_{m,n;k}(\mu_{(0,1)}, \mu_{(1,0)}, \dots, \mu_{(m,n)})$$

$$= \frac{1}{m! n!} A_{m,n}(\nu \mu_{(0,1)}, \nu \mu_{(1,0)}, \dots, \nu \mu_{(m,n)}), \quad \nu^k \equiv \nu_{(k)}$$

with $\nu_{(k)} = \sum_{i=k}^{\infty} (i)_k p_i$, where

$$M_{(m,n)} = \sum_{j=n}^{\infty} \sum_{i=m}^{\infty} (i)_m (j)_n P_{i,j}, \quad \mu_{(m,n)} = \sum_{j=n}^{\infty} \sum_{i=m}^{\infty} (i)_m (j)_n q_{i,j}.$$

22*. (*Continuation*). *Logarithmic bipartitional polynomials.* These polynomials constitute a particular case of the bipartitional polynomials:

$$L_{m,n}(x_{0,1}, x_{1,0}, x_{1,1}, \ldots, x_{m,n}) = A_{m,n}(cx_{0,1}, cx_{1,0}, \ldots, cx_{m,n}), \quad c^k \equiv c_k,$$

with $c_0 = 0$, $c_k = (-1)^{k-1}(k-1)!$, $k = 1, 2, \ldots$. Show that

$$L(t,u) = \sum_{n=0}^{\infty} \sum_{\substack{m=0 \\ m+n \neq 0}}^{\infty} L_{m,n}(x_{0,1}, x_{1,0}, \ldots, x_{m,n}) \frac{t^m}{m!} \frac{u^n}{n!}$$

$$= \log\{1 + [g(t,u) - x_{0,0}]\},$$

where

$$g(t,u) = \sum_{n=0}^{\infty} \sum_{m=0}^{\infty} x_{m,n} \frac{t^m}{m!} \frac{u^n}{n!}.$$

Further, show that

$$L_{m+1,n} = x_{m+1,n} - \sum_{j=0}^{n} \sum_{\substack{i=0 \\ i+j \neq 0}}^{m} \binom{n}{j} \binom{m}{i} x_{i,j} L_{m-i+1,n-j}, \quad L_{1,0} = x_{1,0}$$

and

$$y_{m,n} = A_{m,n}(x_{0,1}, x_{1,0}, \ldots, x_{m,n}),$$

if and only if

$$x_{m,n} = L_{m,n}(y_{0,1}, y_{1,0}, \ldots, y_{m,n}).$$

23*. (*Continuation*) *Cumulants and moments of bivariate distributions.* Let

$$M(t,u) = \sum_{n=0}^{\infty} \sum_{m=0}^{\infty} \mu'_{m,n} \frac{t^m}{m!} \frac{u^n}{n!}, \quad K(t,u) = \sum_{n=0}^{\infty} \sum_{m=0}^{\infty} \kappa_{m,n} \frac{t^m}{m!} \frac{u^n}{n!}$$

be the generating functions of the moments $\mu'_{m,n}$, $m, n = 0, 1, \ldots$ and the cumulants $\kappa_{m,n}$, $m, n = 0, 1, \ldots$, respectively, of a bivariate distribution. Using the relation $K(t,u) = \log M(t,u)$, show that

$$\kappa_{m,n} = L_{m,n}(\mu'_{0,1}, \mu'_{1,0}, \ldots, \mu'_{m,n})$$

and conclude that

$$\mu'_{m,n} = A_{m,n}(\kappa_{0,1}, \kappa_{1,0}, \ldots, \kappa_{m,n}).$$

24*. (*Continuation*). *Potential bipartitional polynomials.* These polynomials constitute a particular case of the bipartitional polynomials:

$$C_{m,n,s}(x_{0,1}, x_{1,0}, \ldots, x_{n,r}) = A_{m,n}(cx_{0,1}, cx_{1,0}, \ldots, cx_{m,n}), \quad c^k \equiv c_k,$$

with $c_k = (s)_k$, $k = 0, 1, \ldots$. Show that

$$C_s(t, u) = \sum_{n=0}^{\infty} \sum_{m=0}^{\infty} C_{m,n,s}(x_{0,1}, x_{1,0}, \ldots, x_{m,n}) \frac{t^m}{m!} \frac{u^n}{n!}$$

$$= [1 + \{g(t, u) - x_{0,0}\}]^s,$$

where

$$g(t, u) = \sum_{n=0}^{\infty} \sum_{m=0}^{\infty} x_{m,n} \frac{t^m}{m!} \frac{u^n}{n!}.$$

Also, with $C_{m,n,s} \equiv C_{m,n,s}(x_{0,1}, x_{1,0}, \ldots, x_{m,n})$, show that

$$C_{m+1,n,s} = s \sum_{j=0}^{n} \sum_{i=0}^{m} \binom{n}{j} \binom{m}{i} x_{i+1,j} C_{m-i,n-j,s}$$

$$- \sum_{\substack{j=0 \\ i+j \neq 0}}^{n} \sum_{i=0}^{m} \binom{n}{j} \binom{m}{i} x_{i,j} C_{m-i+1,n-j,s}, \quad C_{0,0,s} = 1.$$

25*. *Symmetric functions.* The symmetric functions $a_{k,r} \equiv a_{k,r}(x_1, x_2, \ldots, x_n, y_1, y_2, \ldots, y_n)$, $s_{k,r} \equiv s_{k,r}(x_1, x_2, \ldots, x_n, y_1, y_2, \ldots, y_n)$, $h_{k,r} \equiv h_{k,r}(x_1, x_2, \ldots, x_n, y_1, y_2, \ldots, y_n)$, with respect to the variables $x_1, x_2, \ldots, x_n$ and with respect to the variables $y_1, y_2, \ldots, y_n$, have generating functions:

$$A(t, u) = \sum_{r=0}^{n} \sum_{k=0}^{n-r} a_{k,r} t^k u^j = \prod_{i=1}^{n} (1 + x_i t + y_i u),$$

$$S(t, u) = \sum_{\substack{r=0 \\ k+r \neq 0}}^{\infty} \sum_{k=0}^{\infty} (k + r - 1)! s_{k,r} \frac{t^k}{k!} \frac{u^r}{r!} = -\log \prod_{i=1}^{n} (1 - x_i t - y_i u),$$

$$H(t, u) = \sum_{r=0}^{\infty} \sum_{k=0}^{\infty} h_{k,r} t^k u^r = \prod_{i=1}^{n} (1 - x_i t - y_i u)^{-1}.$$

Show that

$$a_{k,r} = \frac{(-1)^{k+r}}{k! r!} A_{k,r}(-s_{0,1}, -s_{1,0}, \ldots, -(k + r - 1)! s_{k,r})$$

$$= \frac{(-1)^{k+r}}{k! r!} C_{k,r,-1}(h_{0,1}, h_{1,0}, \ldots, k! r! h_{k,r}),$$

$$s_{k,r} = \frac{(-1)^{k+r-1}}{(k+r-1)!} L_{k,r}(a_{0,1}, a_{1,0}, \ldots, k!r!a_{k,r})$$
$$= \frac{1}{(k+r-1)!} L_{k,r}(h_{0,1}, h_{1,0}, \ldots, k!r!h_{k,r})$$

and

$$h_{k,r} = \frac{(-1)^{k+r}}{k!r!} C_{k,r,-1}(a_{0,1}, a_{1,0}, \ldots, k!r!a_{k,r})$$
$$= \frac{1}{k!r!} A_{k,r}(s_{0,1}, s_{1,0}, \ldots, (k+r-1)!s_{k,r}).$$

Chapter 12

CYCLES OF PERMUTATIONS

12.1 INTRODUCTION

The permutations of a finite set of n elements may be distinguished and enumerated according to certain characteristics they may possess. Thus, in Chapter 5, the enumeration of the permutations of n with a given number of fixed points and the permutations of n of a given rank was carried out. Also, the (linear) permutations of n and the circular permutations of n with a given number of successions of points were enumerated. Further, the permutations of a finite set of n elements may be distinguished according to their decomposition into disjoint cycles. A host of interesting enumeration problems emerges from this consideration.

The present chapter is devoted to the enumeration of the permutations of a finite set of n elements that are decomposed into cycles of specified or unspecified number, whose number of cycles of any specific length belongs to a subset of nonnegative integers. Specifically, after the introduction of the necessary basic notions, the permutations that are decomposed into a given number of cycles are enumerated. In particular, a combinatorial interpretation of the signless Stirling numbers of the first kind is provided. Then, the permutations are classified into even and odd, according to the number of transpositions into which they can be decomposed. The enumeration of the even and odd permutations by cycles is reduced to the preceding problem of enumeration of the permutations by cycles.

Also, considering the set of possible orderings of the elements of a cycle, the cycles are distinguished by the subset of orderings allowed for their elements. The enumeration of the permutations with a given number of partially ordered cycles is carried out. It is worth noting that the problem of enumeration of the permutations of a finite set by the number of the totally ordered cycles, into which they can be decomposed, is merely a problem of enumeration of the partitions of a finite set by the number of their subsets.

12.2 PERMUTATIONS WITH A GIVEN NUMBER OF CYCLES

A permutation of a finite set $W_n = \{w_1, w_2, \ldots, w_n\}$ of n elements, as defined in Chapter 2, is an ordered n-tuple $(w_{i_1}, w_{i_2}, \ldots, w_{i_n})$ with $w_{i_r} \in W_n$, $r = 1, 2, \ldots, n$ and $w_{i_r} \neq w_{i_s}$ for $r \neq s$. Such a permutation can be equivalently considered as a rearrangement of a fixed ordering. The replacement of the fixed ordering $(w_1, w_2, \ldots, w_n)$ by $(w_{i_1}, w_{i_2}, \ldots, w_{i_n})$ is denoted by

$$\begin{pmatrix} w_1, & w_2, \ldots, & w_n \\ w_{i_1}, & w_{i_2}, \ldots, & w_{i_n} \end{pmatrix}, \tag{12.1}$$

which defines the rule (mapping) $\sigma(w_r) = w_{i_r}$ of the replacement of the point w_r by the point w_{i_r}, $r = 1, 2, \ldots, n$. Recall that a point w_r for which $w_{i_r} = w_r$ is called *fixed point* of this permutation. The columns in (12.1) corresponding to the fixed points may be omitted. Consider a subset $W_k = \{w_{i_1}, w_{i_2}, \ldots, w_{i_k}\}$ of the set W_n. The permutation

$$\begin{pmatrix} w_{i_1}, & w_{i_2}, \ldots, & w_{i_{k-1}}, & w_{i_k} \\ w_{i_2}, & w_{i_3}, \ldots, & w_{i_k}, & w_{i_1} \end{pmatrix}, \tag{12.2}$$

in which the replacements $\sigma(w_{i_1}) = w_{i_2}, \sigma(w_{i_2}) = w_{i_3}, \ldots, \sigma(w_{i_{k-1}}) = w_{i_k}$, $\sigma(w_{i_k}) = w_{i_1}$ close a cycle is called *cycle*. The notation (12.2) of a cycle may be abbreviated by writing in parentheses the rule of succession of the permuted (non-fixed) points,

$$(w_{i_1}, w_{i_2}, \ldots, w_{i_{k-1}}, w_{i_k}). \tag{12.3}$$

The number k of elements of cycle (12.3) is called *length* of it. The representation (12.3) of a cycle is not unique. The k permutations

$$(w_{i_1}, w_{i_2}, \ldots, w_{i_{k-1}}, w_{i_k}), (w_{i_2}, \ldots, w_{i_k}, w_{i_1}), \ldots, (w_{i_k}, w_{i_1}, \ldots, w_{i_{k-1}}),$$

which are formed by putting as first any of the k elements and keeping the same rule of succession, correspond to (represent) the same cycle. It is conventional to write a cycle with its smallest element in the first position, making its representation (12.3) unique.

Evidently, there are $(n)_k/k$ cycles of length k. A cycle of length n, which includes all the elements of the set W_n, is particularly called *circular (cyclic) permutation*. Thus, there are $(n)_n/n = (n-1)!$ circular permutations of the set W_n. A *transposition* is a cycle of length $k = 2$. The number of transpositions of the set W_n equals $n(n-1)/2$. Clearly, any permutation of the set W_n is either a cycle or can be decomposed into disjoint cycles. This decomposition is unique up to a rearrangement of the cycles.

Example 12.1
Consider the following permutation of the set $W_{10} = \{0, 1, \dots, 9\}$:

$$\begin{pmatrix} 0, 1, 2, 3, 4, 5, 6, 7, 8, 9 \\ 2, 1, 0, 5, 9, 7, 8, 6, 3, 4 \end{pmatrix}.$$

Note that the element 1 is a fixed point of this permutation. Further, $(0, 2)$ and $(4, 9)$ are two cycles of length 2 and $(3, 5, 7, 6, 8)$ is a cycle of length 5. Thus, this permutation may be expressed as a product of cycles as

$$(1)(0, 2)(4, 9)(3, 5, 7, 6, 8).$$

The permutation

$$\begin{pmatrix} 0, 1, 2, 3, 4, 5, 6, 7, 8, 9 \\ 3, 9, 5, 7, 6, 1, 8, 2, 0, 4 \end{pmatrix}$$

is a cycle (cyclic permutation) $(0, 3, 7, 2, 5, 1, 9, 4, 6, 8)$. ☐

The permutations of a finite set of n elements may be distinguished according to their type of decomposition into disjoint cycles. Specifically, we introduce the following definition

DEFINITION 12.1   *A permutation of a finite set W_n, of n elements, which is decomposed into $k_i \geq 0$ cycles of length i, $i = 1, 2, \dots, n$, so that $k_1 + 2k_2 + \cdots + nk_n = n$, is said to be of type $[k_1, k_2, \dots, k_n]$.*

Note that the number of different types in which the $n!$ permutations of n may be classified equals the number $p(n)$ of partitions of n.

Example 12.2
The 24 permutations of the set $W_4 = \{w_1, w_2, w_3, w_4\}$, with $w_1 < w_1 < w_3 < w_4$, written as products of cycles can be classified in the following types:
   (a) The only permutation of the type $[4, 0, 0, 0]$ is

$$(w_1)(w_2)(w_3)(w_4).$$

   (b) The permutations of the type $[2, 1, 0, 0]$ are the following six:

$$(w_1)(w_2)(w_3, w_4), \ (w_1)(w_3)(w_2, w_4), \ (w_1)(w_4)(w_2, w_3),$$

$$(w_2)(w_3)(w_1, w_4), \ (w_2)(w_4)(w_1, w_3), \ (w_3)(w_4)(w_1, w_2).$$

   (c) The permutations of the type $[1, 0, 1, 0]$ are the following eight:

$$(w_1)(w_2, w_3, w_4), \ (w_1)(w_2, w_4, w_3), \ (w_2)(w_1, w_3, w_4), \ (w_2)(w_1, w_4, w_3),$$

$(w_3)(w_1, w_2, w_4)$, $(w_3)(w_1, w_4, w_2)$, $(w_4)(w_1, w_2, w_3)$, $(w_4)(w_1, w_3, w_2)$.

(d) The permutations of the type $[0, 2, 0, 0]$ are the following three:

$$(w_1, w_2)(w_3, w_4), \quad (w_1, w_3)(w_2, w_4), \quad (w_1, w_4)(w_2, w_3).$$

(e) The permutations of the type $[0, 0, 0, 1]$ are the following six:

$$(w_1, w_2, w_3, w_4), \quad (w_1, w_2, w_4, w_3), \quad (w_1, w_3, w_2, w_4),$$

$$(w_1, w_3, w_4, w_2), \quad (w_1, w_4, w_2, w_3), \quad (w_1, w_4, w_3, w_2). \quad \Box$$

The next theorem is concerned with the enumeration of the permutations of a finite set by their decomposition into cycles.

THEOREM 12.1

(a) *The number of permutations of a finite set of n elements that are decomposed into k cycles, among which $k_i \geq 0$ are of length i, $i = 1, 2, \ldots, n$, equals*

$$c_{n,k}(k_1, k_2, \ldots, k_n) = \frac{n!}{k_1! 1^{k_1} k_2! 2^{k_2} \cdots k_n! n^{k_n}}, \tag{12.4}$$

with $k_1 + 2k_2 + \cdots + nk_n = n$ and $k_1 + k_2 + \cdots + k_n = k$.

(b) *The number of permutations of a finite set of n elements that are decomposed into cycles, among which $k_i \geq 0$ are of length i, $i = 1, 2, \ldots, n$, equals*

$$c_n(k_1, k_2, \ldots, k_n) = \frac{n!}{k_1! 1^{k_1} k_2! 2^{k_2} \cdots k_n! n^{k_n}}, \tag{12.5}$$

with $k_1 + 2k_2 + \cdots + nk_n = n$.

PROOF   (a) Consider a partition $\{A_1, A_2, \ldots, A_k\}$ of a finite set W_n, with $N(W_n) = n$, into k subsets, among which $k_i \geq 0$ subsets include i elements each, $i = 1, 2, \ldots, n$, where $k_1 + 2k_2 + \cdots + nk_n = n$ and $k_1 + k_2 + \cdots + k_n = k$. To it there correspond

$$[(1 - 1)!]^{k_1} [(2 - 1)!]^{k_2} \cdots [(n - 1)!]^{k_n}$$

permutations of the set W_n that are decomposed into k cycles, among which $k_i \geq 0$ are of length i, $i = 1, 2, \ldots, n$; these permutations are formed by permuting the $i - 1$ elements of each of the $k_i \geq 0$ subsets with i elements (excluding the smallest element, which, by convention, occupies the first position) in $[(i - 1)!]^{k_i}$ ways, $i = 1, 2, \ldots, n$. Therefore

$$c_{n,k}(k_1, k_2, \ldots, k_n)$$
$$= [(1 - 1)!]^{k_1} [(2 - 1)!]^{k_2} \cdots [(n - 1)!]^{k_n} B(n, k; k_1, k_2, \ldots, k_n),$$

where $B(n, k; k_1, k_2, \ldots, k_n)$ is the number of partitions of the set W_n into k subsets, among which $k_i \geq 0$ subsets include i elements each, $i = 1, 2, \ldots, n$, where $k_1 + 2k_2 + \cdots + nk_n = n$ and $k_1 + k_2 + \cdots + k_n = k$. This number, according to Theorem 2.10, is given by

$$B(n, k; k_1, k_2, \ldots, k_n) = \frac{n!}{k_1!(1!)^{k_1} k_2!(2!)^{k_2} \cdots k_n!(n!)^{k_n}},$$

with $k_1 + 2k_2 + \cdots + nk_n = n$, $k_1 + k_2 + \cdots + k_n = k$ and so (12.4) is deduced.

(b) Clearly, the summation of the numbers $c_{n,k}(k_1, k_2, \ldots, k_n)$ in (12.4) for all $k = 0, 1, \ldots, n$ is equivalent to keep the total number of cycles $k_1 + k_2 + \cdots + k_n$ unspecified and so the number of permutations of a finite set of n elements that are decomposed into cycles, among which $k_i \geq 0$ are of length i, $i = 1, 2, \ldots, n$, is given by (12.5). ∎

The number of permutations of a finite set of n elements that are decomposed into k cycles may be obtained by summing the numbers (12.4) for all $k_i \geq 0$, $i = 1, 2, \ldots, n$, with $k_1 + 2k_2 + \cdots + nk_n = n$, $k_1 + k_2 + \cdots + k_n = k$. Then, by virtue of the expression (8.24) of the signless Stirling numbers of the first kind, we deduce the following corollary of Theorem 12.1.

COROLLARY 12.1

The number of permutations of a finite set of n elements that are decomposed into k cycles equals

$$|s(n, k)|,$$

the signless Stirling number of the first kind.

REMARK 12.1 The sum of the numbers $c_{n,k}(k_1, k_2, \ldots, k_n)$ in (12.5) for all $k_i \geq 0$, $i = 1, 2, \ldots, n$, with $k_1 + 2k_2 + \cdots + nk_n = n$, gives the total number $n!$, of the permutations of a finite set of n elements. Thus, we deduce *Cauchy identity:*

$$\sum \frac{1}{k_1!1^{k_1} k_2!2^{k_2} \cdots k_n!n^{k_n}} = 1,$$

where the summation is extended over all integers $k_i \geq 0$, $i = 1, 2, \ldots, n$ with $k_1 + 2k_2 + \cdots + nk_n = n$. ∎

Example 12.3 Restricted permutations with a given number of cycles

Consider a finite set $W_{n+r} = \{w_1, w_2, \ldots, w_{n+r}\}$, with $w_1 < w_2 < \cdots < w_{n+r}$, and its subset $W_r = \{w_1, w_2, \ldots, w_r\}$. Determine the number of permutations of the set W_{n+r} that are decomposed into $k + r$ cycles such that the r elements of the set W_r belong in r distinct cycles.

The k cycles of such a permutation may contain a total of j elements from the set $W_{n+r} - W_r = \{w_{r+1}, w_{r+2}, \ldots, w_{n+r}\}$, of n elements, for $j = k, k+1, \ldots, n$. The j elements can be chosen in $\binom{n}{j}$ different ways. Further, the number of permutations of a set of j elements that are decomposed into k cycles, according to Corollary 12.1, equals the signless Stirling number of the first kind $|s(j, k)|$. Also, the remaining $n - j$ elements of the set $W_{n+r} - W_r$ can be placed into the r distinct cycles in $(r + n - j - 1)_{n-j} = r(r+1)\cdots(r+n-j-1)$ different ways. Indeed, the first element can be placed in any of the r places between the r elements $w_1, w_2, \ldots, w_r$ or after the last element. After the first element is placed, the second element can be placed in any of the $r + 1$ places between the $r + 1$ elements or after the last element. Finally, after the $n - j - 1$ elements are placed, the last element can be placed in any of the $r + n - j - 1$ places between the $r + n - j - 1$ elements or after the last element. Consequently, the number of permutations of the set W_{n+r} that are decomposed into $k + r$ cycles, such that the r elements of the set W_r belong in r distinct cycles, is given by

$$|s(n, k; r)| = \sum_{j=k}^{n} \binom{n}{j} (r + n - j - 1)_{n-j} |s(j, k)|,$$

which is the *non-central signless Stirling number of the first kind* (see Section 8.5).   □

The multivariate generating functions of the numbers of permutations of a finite set that are decomposed into cycles of unspecified or specified number with respect to cycle-length numbers may be expressed in terms of the exponential and partial Bell partition polynomials. Thus, from Theorem 12.1, using (11.1) and (11.2), we deduce the following corollary.

COROLLARY 12.2

(a) *The generating function of the sequence* $c_n(k_1, k_2, \ldots, k_n)$, $k_i \geq 0$, $i = 1, 2, \ldots, n$ *with* $k_1 + 2k_2 + \cdots + nk_n = n$, *for fixed n, is given by*

$$C_n(x_1, x_2, \ldots, x_n) = \sum c_n(k_1, k_2, \ldots, k_n)x_1^{k_1} x_2^{k_2} \cdots x_n^{k_n}$$
$$= B_n(x_1, 1!x_2, \ldots, (n-1)!x_n), \qquad (12.6)$$

where B_n is the exponential Bell partition polynomial.

(b) *The generating function of the sequence* $c_{n,k}(k_1, k_2, \ldots, k_n)$, $k_i \geq 0$, $i = 1, 2, \ldots, n$ *with* $k_1 + 2k_2 + \cdots + nk_n = n$ *and* $k_1 + k_2 + \cdots + k_n = k$, *for fixed n and k, is given by*

$$C_{n,k}(x_1, x_2, \ldots, x_n) = \sum c_{n,k}(k_1, k_2, \ldots, k_n)x_1^{k_1} x_2^{k_2} \cdots x_n^{k_n}$$
$$= B_{n,k}(x_1, 1!x_2, \ldots, (n-1)!x_n), \qquad (12.7)$$

where $B_{n,k}$ is the partial Bell partition polynomial.

Further, using the generating functions of the Bell and the partial Bell partition polynomials (11.6) and (11.9), we deduce the following corollary.

COROLLARY 12.3

(a) *The generating function of the sequence $c_n(k_1, k_2, \dots, k_n)$, $k_i \geq 0$, $i = 1, 2, \dots, n$, with $k_1 + 2k_2 + \cdots + nk_n = n$, $n = 0, 1, \dots$, is given by*

$$\sum_{n=0}^{\infty} \sum c_n(k_1, k_2, \dots, k_n) x_1^{k_1} x_2^{k_2} \cdots x_n^{k_n} \frac{t^n}{n!}$$

$$= \sum_{n=0}^{\infty} C_n(x_1, x_2, \dots, x_n) \frac{t^n}{n!} = \exp\left(\sum_{r=1}^{\infty} x_r \frac{t^r}{r}\right). \qquad (12.8)$$

(b) *The generating function of the sequence $c_{n,k}(k_1, k_2, \dots, k_n)$, $k_i \geq 0$, $i = 1, 2, \dots, n$, with $k_1 + 2k_2 + \cdots + nk_n = n$ and $k_1 + k_2 + \cdots + k_n = k$, $k = 0, 1, \dots, n$, $n = 0, 1, \dots$, is given by*

$$\sum_{n=0}^{\infty} \sum_{k=0}^{n} \sum c_{n,k}(k_1, k_2, \dots, k_n) x_1^{k_1} x_2^{k_2} \cdots x_n^{k_n} u^k \frac{t^n}{n!}$$

$$= \sum_{n=0}^{\infty} \sum_{k=0}^{n} C_{n,k}(x_1, x_2, \dots, x_n) u^k \frac{t^n}{n!} = \exp\left(u \sum_{r=1}^{\infty} x_r \frac{t^r}{r}\right). \qquad (12.9)$$

Example 12.4

Let $c_n(k; r)$ be the number of permutations of the set W_n for which their decomposition into cycles includes, among other cycles, k cycles of length r. Determine a generating function for the sequence $c_n(k; r)$, $k = 0, 1, \dots, [n/r]$, $n = 0, 1, \dots$, and, expanding it, deduce an explicit expression for these numbers.

The number $c_n(k; r)$ is given by the sum

$$c_n(k; r) = \sum \frac{n!}{k_1! 1^{k_1} k_2! 2^{k_2} \cdots k_n! n^{k_n}},$$

where $k_r = k$ and the summation is extended over all $k_i \geq 0$, $i = 1, 2, \dots, n$, $i \neq r$ with $\sum_{i=1, i\neq r}^{n} ik_i = n - rk$. Consequently, from (12.8), with $x_i = 1$, $i = 1, 2, \dots, i \neq r$, and $x_r = x$, we get the bivariate generating function

$$\sum_{n=0}^{\infty} \sum_{k=0}^{[n/r]} c_n(k; r) x^k \frac{t^n}{n!} = \exp\left\{(x-1)\frac{t^r}{r} + \log(1-t)^{-1}\right\}$$

$$= (1-t)^{-1} \exp\left\{(x-1)\frac{t^r}{r}\right\}.$$

Expanding it into powers of t,

$$\sum_{n=0}^{\infty} \sum_{k=0}^{[n/r]} c_n(k;r)x^k \frac{t^n}{n!} = \left(\sum_{i=0}^{\infty} t^i\right)\left(\sum_{j=0}^{\infty} \frac{(x-1)^j}{j!r^j} t^{rj}\right)$$

$$= \sum_{n=0}^{\infty}\left\{\sum_{j=0}^{[n/r]} \frac{(x-1)^j}{j!r^j}\right\} t^n,$$

we deduce the generating function

$$\sum_{k=0}^{[n/r]} c_n(k;r)x^k = n!\sum_{j=0}^{[n/r]} \frac{(x-1)^j}{j!r^j}.$$

Further, expanding it into powers of x,

$$\sum_{k=0}^{[n/r]} c_n(k;r)x^k = n!\sum_{j=0}^{[n/r]}\sum_{k=0}^{j}(-1)^{j-k}\binom{j}{k}\frac{x^k}{j!r^j}$$

$$= \sum_{k=0}^{[n/r]} \frac{n!}{k!}\left\{\sum_{j=k}^{[n/r]} \frac{(-1)^{j-k}}{(j-k)!r^j}\right\} x^k,$$

we conclude for $c_n(k;r)$ the expression

$$c_n(k;r) = \frac{n!}{k!}\sum_{j=k}^{[n/r]} \frac{(-1)^{j-k}}{(j-k)!r^j}.$$

Note that $c_n(k;1) = D_{n,k}$ is the number of permutations of the set W_n that have k fixed points (see Section 5.2). ☐

12.3   EVEN AND ODD PERMUTATIONS

As previously noted, any permutation of a finite set W_n is either a cycle or can be decomposed into disjoint cycles; this decomposition is unique up to a rearrangement of the cycles. Further, each cycle can be generated by successive transpositions. Indeed, a cycle (w_i) of length 1 (fixed point) can be generated as $(w_i) = (w_i, w_j)(w_j, w_i)$, $i \neq j$, while a cycle $(w_{i_1}, w_{i_2}, \ldots, w_{i_{k-1}}, w_{i_k})$ of length $k \geq 2$ can be generated by $k-1$ successive transpositions as

$$(w_{i_1}, w_{i_2}, \ldots, w_{i_{k-1}}, w_{i_k}) = (w_{i_1}, w_{i_k})(w_{i_1}, w_{i_{k-1}})\cdots(w_{i_1}, w_{i_2}).$$

Consequently, any permutation of a finite set W_n can be decomposed into transpositions. But while the decomposition of a permutation into disjoint cycles is unique up to a rearrangement of the cycles, its decomposition into transpositions, among which two or more have elements in common, may be done in different ways. Nevertheless, in these different decompositions of a permutation, the number of transpositions is either even or odd. This fact is used for the following discrimination of the permutations.

DEFINITION 12.2 *A permutation of a finite set W_n, of n elements, is called even if it can be decomposed into an even number of transpositions, while is called odd if it can be decomposed into an odd number of transpositions.*

The enumeration of the even and odd permutations is facilitated by the following characteristic property.

LEMMA 12.1
A permutation of a finite set W_n, of n elements, is even or odd if and only if, in its decomposition into cycles, the number of cycles of even length is even or odd, respectively.

PROOF   Consider a permutation of a finite set W_n, of n elements, that is decomposed into k cycles. Assume that its decomposition includes r cycles of even lengths, $j_1, j_2, \dots, j_r$, and $k - r$ cycles of odd lengths, $j_{r+1}, j_{r+2}, \dots, j_k$. Since each cycle of length j_s can be decomposed into $j_s - 1$ transpositions, this permutation can be decomposed into

$$n - k = \sum_{s=1}^{k}(j_s - 1) = \sum_{s=1}^{r} j_s + \sum_{s=r+1}^{k}(j_s - 1) - r$$

transpositions. Note that the summands j_s, $s = 1, 2, \dots, r$ and $j_s - 1$, $s = r + 1, r + 2, \dots, k$ in the sums of the right-hand member are even numbers and so the sum

$$\sum_{s=1}^{r} j_s + \sum_{s=r+1}^{k}(j_s - 1)$$

is always an even number. Consequently, the number $n - k$ of transpositions into which the considered permutation can be decomposed is even or odd if and only if the number r of cycles of even length is even or odd, respectively.   ∎

Let us denote by $a_{n,k}(k_1, k_2, \dots, k_n)$ and $b_{n,k}(k_1, k_2, \dots, k_n)$ the number of even and odd permutations of the type $[k_1, k_2, \dots, k_n]$, respectively, so that $k_1 + 2k_2 + \cdots + nk_n = n$ and $k_1 + k_2 + \cdots + k_n = k$, with $k_2 + k_4 + \cdots + k_{2m}$

an even and odd number, respectively, where $m = [n/2]$. Clearly,

$$a_{n,k}(k_1, k_2, \ldots, k_n) = \frac{1 + (-1)^r}{2} c_{n,k}(k_1, k_2, \ldots, k_n) \qquad (12.10)$$

and

$$b_{n,k}(k_1, k_2, \ldots, k_n) = \frac{1 - (-1)^r}{2} c_{n,k}(k_1, k_2, \ldots, k_n), \qquad (12.11)$$

where $r = k_2 + k_4 + \cdots + k_{2m}$, $m = [n/2]$. Utilizing these expressions, generating functions for the numbers $a_{n,k}(k_1, k_2, \ldots, k_n)$ and $b_{n,k}(k_1, k_2, \ldots, k_n)$ may be deduced from the corresponding generating functions for the numbers $c_{n,k}(k_1, k_2, \ldots, k_n)$.

COROLLARY 12.4

(a) *The generating function of the sequence* $a_{n,k}(k_1, k_2, \ldots, k_n)$, $k_i \geq 0$, $i = 1, 2, \ldots, n$, *with* $k_1 + 2k_2 + \cdots + nk_n = n$ *and* $k_1 + k_2 + \cdots + k_n = k$, *for fixed n and k, is given by*

$$\sum a_{n,k}(k_1, k_2, \ldots, k_n) x_1^{k_1} x_2^{k_2} \cdots x_n^{k_n}$$
$$= \frac{1}{2} B_{n,k}(x_1, 1! x_2, \ldots, (n-1)! x_n)$$
$$+ \frac{1}{2} B_{n,k}(x_1, -1! x_2, \ldots, (-1)^{n-1}(n-1)! x_n), \qquad (12.12)$$

where $B_{n,k}$ *is the partial Bell partition polynomial.*

(b) *The generating function of the sequence* $a_{n,k}(k_1, k_2, \ldots, k_n)$, $k_i \geq 0$, $i = 1, 2, \ldots, n$, *with* $k_1 + 2k_2 + \cdots + nk_n = n$ *and* $k_1 + k_2 + \cdots + k_n = k$, $k = 0, 1, \ldots, n$, $n = 0, 1, \ldots$, *is given by*

$$\sum_{n=0}^{\infty} \sum_{k=0}^{n} \sum a_{n,k}(k_1, k_2, \ldots, k_n) x_1^{k_1} x_2^{k_2} \cdots x_n^{k_n} u^k \frac{t^n}{n!}$$
$$= \frac{1}{2} \exp\left(u \sum_{r=1}^{\infty} x_r \frac{t^r}{r} \right) + \frac{1}{2} \exp\left(u \sum_{r=1}^{\infty} (-1)^{r-1} x_r \frac{t^r}{r} \right). \qquad (12.13)$$

PROOF   (a) The required generating function, using (12.12) and since $r = k_2 + k_4 + \cdots + k_{2m}$, $m = [n/2]$, may be expressed as

$$\sum a_{n,k}(k_1, k_2, \ldots, k_n) x_1^{k_1} x_2^{k_2} \cdots x_n^{k_n}$$
$$= \frac{1}{2} \sum c_{n,k}(k_1, k_2, \ldots, k_n) x_2^{k_1} x_2^{k_2} \cdots x_n^{k_n}$$
$$+ \frac{1}{2} \sum c_{n,k}(k_1, k_2, \ldots, k_n)(-1)^{k_2 + k_4 + \cdots + k_{2m}} x_1^{k_1} x_2^{k_2} \cdots x_n^{k_n}.$$

Then, by virtue of (12.7), (12.12) is deduced.

(b) Multiplying (12.12) by $u^k t^n / n!$ and summing for $k = 0, 1, \ldots, n$, $n = 0, 1, \ldots$, upon using the generating function (11.6) of the partial Bell partition polynomials, (12.13) is readily obtained. ∎

Example 12.5

Find the number $a_{n,k}$ of even permutations of a finite set W_n of n elements that are decomposed into k cycles.

Clearly, setting $x_i = 1$, $i = 1, 2, \ldots$ in (12.13), a bivariate generating function of the sequence $a_{n,k}$, $k = 0, 1, \ldots, n$, $n = 0, 1, \ldots$, is obtained as

$$\sum_{n=0}^{\infty} \sum_{k=0}^{n} a_{n,k} u^k \frac{t^n}{n!} = \frac{1}{2} \exp[u \log(1-t)^{-1}] + \frac{1}{2} \exp[u \log(1+t)]$$

$$= \frac{1}{2}(1-t)^{-u} + \frac{1}{2}(1+t)^u.$$

Thus, according to (8.14) and (8.5), it follows that

$$a_{n,k} = \frac{1}{2}|s(n,k)| + \frac{1}{2}s(n,k) = \frac{1 + (-1)^{n-k}}{2} |s(n,k)|,$$

where $|s(n,k)|$ is the *signless Stirling number of the first kind*. ꠸

12.4 PERMUTATIONS WITH PARTIALLY ORDERED CYCLES

In the preceding enumeration of permutations, cycles are distinguished only by their length. In Theorem 12.1, the number $c_{n,k}(k_1, k_2, \ldots, k_n)$ of permutations of a finite set W_n, of n elements, that are decomposed into k cycles, among which $k_i \geq 0$ are of length i, $i = 1, 2, \ldots, n$, has been deduced from the number $B(n, k; k_1, k_2, \ldots, k_n)$ of partitions of the set W_n into k subsets, among which $k_i \geq 0$ include i elements, $i = 1, 2, \ldots, n$. In that case $(i-1)!$ permutations were formed from any subset of i elements by permuting the $i - 1$ elements (excluding the smallest element which, by convention, occupies the first position). It is possible to go further by considering the set of possible orderings of the elements within the cycles. Then the cycles may also be distinguished by the subset of orderings allowed for their elements. Only orderings independent of the particular elements of any cycle are considered. Note that, in the extreme case of (totally) ordered cycles with all their elements in a specific order, for example from the smallest to the largest, only one cycle of length i is formed from a subset of i elements. In the enumeration of permutations by the number of partially ordered cycles, the number a_i of cycles of length i that are formed from a subset of i elements, where $1 \leq a_i \leq (i-1)!$, $i = 1, 2, \ldots, n$, is needed. Then, Theorem 12.1 is readily extended as follows.

THEOREM 12.2

(a) *The number of permutations of a finite set of n elements that are decomposed into k partially ordered cycles, among which $k_i \geq 0$ are of length i and a_i cycles are formed from any subset of i elements, $i = 1, 2, \ldots, n$, equals*

$$c_{n,k}(k_1, k_2, \ldots, k_n; a_1, a_2, \ldots, a_n) = \frac{n! a_1^{k_1} a_2^{k_2} \cdots a_n^{k_n}}{k_1!(1!)^{k_1} k_2!(2!)^{k_2} \cdots k_n!(n!)^{k_n}},$$

(12.14)

with $k_1 + 2k_2 + \cdots + nk_n = n$ and $k_1 + k_2 + \cdots + k_n = k$.

(b) *The number of permutations of a finite set of n elements that are decomposed into partially ordered cycles, among which $k_i \geq 0$ are of length i and a_i cycles are formed from any subset of i elements, $i = 1, 2, \ldots, n$, equals*

$$c_n(k_1, k_2, \ldots, k_n; a_1, a_2, \ldots, a_n) = \frac{n! a_1^{k_1} a_2^{k_2} \cdots a_n^{k_n}}{k_1!(1!)^{k_1} k_2!(2!)^{k_2} \cdots k_n!(n!)^{k_n}},$$

(12.15)

with $k_1 + 2k_2 + \cdots + nk_n = n$.

PROOF   (a) Consider a partition $\{A_1, A_2, \ldots, A_k\}$ of finite set W_n, with $N(W_n) = n$, into k subsets, among which $k_i \geq 0$ subsets include i elements each, $i = 1, 2, \ldots, n$, where $k_1 + 2k_2 + \cdots + nk_n = n$ and $k_1 + k_2 + \cdots + k_n = k$. To it there correspond $a_1^{k_1} a_2^{k_2} \cdots a_n^{k_n}$, permutations of the set W_n that are decomposed into k partially ordered cycles, among which $k_i \geq 0$ are of length i and a_i cycles are formed from each subset of i elements, $i = 1, 2, \ldots, n$; these permutations are formed by permuting the $i - 1$ elements of each of the $k_i \geq 0$ subsets with i elements (excluding the smallest element, which, by convention, occupies the first position) in all $a_i^{k_i}$ permissible ways, for $i = 1, 2, \ldots, n$. Therefore

$$c_{n,k}(k_1, k_2, \ldots, k_n; a_1, a_2, \ldots, a_n) = a_1^{k_1} a_2^{k_2} \cdots a_n^{k_n} B(n, k; k_1, k_2, \ldots, k_n),$$

where $B(n, k; k_1, k_2, \ldots, k_n)$ is the number of partitions of the set W_n into k subsets, among which $k_i \geq 0$ subsets include i elements each, $i = 1, 2, \ldots, n$, where $k_1 + 2k_2 + \cdots + nk_n = n$ and $k_1 + k_2 + \cdots + k_n = k$. Thus, introducing into it the expression of $B(n, k; k_1, k_2, \ldots, k_n)$, (12.14) is obtained.

(b) Clearly, the summation of the numbers (12.14) for all $k = 0, 1, \ldots, n$ is equivalent to keep the total number of cycles $k_1 + k_2 + \cdots + k_n$ unspecified and so the number of permutations of a finite set of n elements that are decomposed into cycles, among which $k_i \geq 0$ are of length i, $i = 1, 2, \ldots, n$, is given by (12.15). ∎

The number of permutations of a finite set of n elements that are decomposed into k partially ordered cycles may be obtained by summing the numbers (12.14) for all $k_i \geq 0$, $i = 1, 2, \ldots, n$ with $k_1 + 2k_2 + \cdots + nk_n = n$ and $k_1 + k_2 + \cdots + k_n = k$. Also, the number of permutations of a finite set of n elements that are decomposed into partially ordered cycles may be obtained by summing the numbers (12.15) for all $k_i \geq 0$, $i = 1, 2, \ldots, n$ with $k_1 + 2k_2 + \cdots + nk_n = n$. These sums, according to Definitions 11.1 and 11.2, are the partial and the exponential Bell partition polynomials. Thus the following corollary is deduced.

COROLLARY 12.5

(a) *The number of permutations of a finite set of n elements that are decomposed into k partially ordered cycles, such that a_i cycles are formed from each subset of i elements, equals*

$$B_{n,k}(a_1, a_2, \ldots, a_n),$$

the partial Bell partition polynomial.

(b) *The total number of permutations of a finite set of n elements that are decomposed into partially ordered cycles, such that a_i cycles are formed from each subset of i elements, equals*

$$B_n(a_1, a_2, \ldots, a_n),$$

the exponential Bell partition polynomial.

Further, using the generating functions of the Bell and the partial Bell partition polynomials (11.6) and (11.9), we deduce the following corollary.

COROLLARY 12.6

(a) *The generating function of the sequence $c_n(k_1, k_2, \ldots, k_n; a_1, a_2, \ldots, a_n)$, $k_i \geq 0$, $i = 1, 2, \ldots, n$, with $k_1 + 2k_2 + \cdots + nk_n = n$, $n = 0, 1, \ldots$, is given by*

$$\sum_{n=0}^{\infty} \sum c_n(k_1, k_2, \ldots, k_n; a_1, a_2, \ldots, a_n) x_1^{k_1} x_2^{k_2} \cdots x_n^{k_n} \frac{t^n}{n!}$$

$$= \exp\left(\sum_{r=1}^{\infty} a_r x_r \frac{t^r}{r!}\right). \qquad (12.16)$$

(b) *The generating function of the sequence $c_{n,k}(k_1, k_2, \ldots, k_n; a_1, a_2, \ldots, a_n)$, $k_i \geq 0$, $i = 1, 2, \ldots, n$, with $k_1 + 2k_2 + \cdots nk_n = n$, $k_1 + k_2 + \cdots + k_n = k$,*

$k = 0, 1, \ldots, n,\; n = 0, 1, \ldots$ *is given by*

$$\sum_{n=0}^{\infty} \sum_{k=0}^{n} \sum c_{n,k}(k_1, k_2, \ldots, k_n; a_1, a_2, \ldots, a_n) x_1^{k_1} x_2^{k_2} \cdots x_n^{k_n} u^k \frac{t^n}{n!}$$

$$= \exp\left(u \sum_{r=1}^{\infty} a_r x_r \frac{t^r}{r!}\right). \quad (12.17)$$

Example 12.6

Find the number $c_{n,k}$ of permutations of a finite set W_n of n elements that are decomposed into k totally ordered cycles.

Clearly, according to the first part of Corollary 12.5, setting $a_i = 1,\, i = 1, 2, \ldots$ and using expression (11.2), the number $c_{n,k}$ is obtained as

$$c_{n,k} = \sum \frac{n!}{k_1!(1!)^{k_1} k_2!(2!)^{k_2} \cdots k_n!(n!)^{k_n}},$$

where the summation is extended over all nonnegative integer solutions $(k_1, k_2, \ldots, k_n)$ of the equations $k_1 + 2k_2 + \cdots + nk_n = n$ and $k_1 + k_2 + \cdots + k_n = k$. Thus, by virtue of (8.25),

$$c_{n,k} = S(n, k),$$

where $S(n, k)$ is the *Stirling number of the second kind*. Further, the total number of permutations of a finite set W_n of n elements that are decomposed into totally ordered cycles is given by

$$B_n = \sum_{k=1}^{n} S(n, k)$$

the *Bell number*. Note that the numbers of permutations of a finite set W_n that are decomposed into a specified or unspecified number of totally ordered cycles coincide with the corresponding numbers of partitions of W_n into a specified or unspecified number of subsets (see Exercises 2.43 and 2.46). This is a consequence of the one-to-one and onto correspondence between the sets of the above permutations and partitions of W_n.   □

Example 12.7 Restricted permutations with a given number of ordered cycles

Consider a finite set $W_{n+r} = \{w_1, w_2, \ldots, w_{n+r}\}$, with $w_1 < w_2 < \cdots < w_{n+r}$, and its subset $W_r = \{w_1, w_2, \ldots, w_r\}$. Determine the number of permutations of the set W_{n+r} that are decomposed into $k + r$ ordered cycles, such that the r elements of the set W_r belong in r distinct cycles.

Note first that the required number equals the number of partitions of the set W_{n+r} into $k + r$ subsets, such that the r elements of the set W_r belong in r

distinct subsets, since each permutation of a finite set that is decomposed into ordered cycles uniquely corresponds to a partition of this set into subsets. Further, the k subsets of such a partition may contain a total of j elements from the set $W_{n+r} - W_r = \{w_{r+1}, w_{r+2}, \ldots, w_{n+r}\}$, of n elements, for $j = k, k+1, \ldots, n$. The j elements can be chosen in $\binom{n}{j}$ different ways. Also, the number of partitions of a set of j elements into k subsets equals the Stirling number of the second kind $S(j, k)$ (see Example 12.6), while the remaining $n - j$ elements of the set $W_{n+r} - W_r$ can be distributed into the r distinct subsets in r^{n-j} different ways. Consequently, the number of partitions of the set W_{n+r} into $k + r$ subsets, such that the r elements of the set W_r belong in r distinct subsets, is given by

$$S(n, k; r) = \sum_{j=k}^{n} \binom{n}{j} r^{n-j} S(j, k),$$

which is the *non-central Stirling number of the second kind* (see Section 8.5). $\square$

The next theorem is concerned with the more general problem of enumeration of permutations of a finite set by partially ordered cycles, where in a specified number of cycles the orderings belong in a given subset and in the other cycles the orderings belong to another given subset.

THEOREM 12.3

(a) *The number of permutations of a finite set W_n of n elements that are decomposed* (i) *into k partially ordered cycles, with elements from a set $U \subseteq W_n$, such that $k_i \geq 0$ are of length i and a_i cycles are formed from any subset of i elements, $i = 1, 2, \ldots, n$, and* (ii) *into other partially ordered cycles, with elements from the set $W_n - U$, such that $r_i \geq 0$ are of length i and b_i cycles are formed from any subset of i elements, $i = 1, 2, \ldots, n$, equals*

$$c_{n,k}(k_1, r_1, k_2, r_2, \ldots, k_n, r_n; a_1, b_1, a_2, b_2, \ldots, a_n b_n)$$
$$= \frac{n! a_1^{k_1} b_1^{r_1} a_2^{k_2} b_2^{r_2} \cdots a_n^{k_n} b_n^{r_n}}{k_1! r_1! (1!)^{k_1+r_1} k_2! r_2! (2!)^{k_2+r_2} \cdots k_n! r_n! (n!)^{k_n+r_n}}, \quad (12.18)$$

with $\sum_{i=1}^{n} i(k_i + r_i) = n$ and $\sum_{i=1}^{n} k_i = k$.

(b) *The number of permutations of a finite set W_n of n elements that are decomposed* (i) *into k partially ordered cycles, with elements from a set $U \subseteq W_n$, such that a_i cycles are formed from any subset of i elements, $i = 1, 2, \ldots, n$, and* (ii) *into other partially ordered cycles, with elements from the set $W_n - U$, such that b_i cycles are formed from any subset of i elements, $i = 1, 2, \ldots, n$, equals*

$$T_{n,k}(a_1, a_2, \ldots, a_n; b_1, b_2, \ldots, b_n),$$

the Touchard polynomial.

PROOF   (a) Consider a partition $\{A_1, A_2, \dots, A_k, B_1, B_2, \dots\}$ of a finite set W_n, of n elements, where $\{A_1, A_2, \dots, A_k\}$ is a partition of a set $U \subseteq W_n$ into k subsets, among which $k_i \geq 0$ include i elements each, $i = 1, 2, \dots, n$, and $\{B_1, B_2, \dots\}$ is a partition of the set $W_n - U$ into subsets among which $r_i \geq 0$ include i elements each, $i = 1, 2, \dots, n$, so that $\sum_{i=1}^{n} i(k_i + r_i) = n$ and $\sum_{i=1}^{n} k_i = k$. To it there correspond $a_1^{k_1} b_1^{r_1} a_2^{k_2} b_2^{r_2} \cdots a_n^{k_n} b_n^{k_n}$, permutations of the set W_n of n elements that are decomposed (i) into k partially ordered cycles, with elements from the set $U \subseteq W_n$, such that $k_i \geq 0$ are of length i and a_i cycles are formed from any subset of i elements, $i = 1, 2, \dots, n$, and (ii) into other partially ordered cycles, with elements from the set $W_n - U$, such that $r_i \geq 0$ are of length i and b_i cycles are formed from any subset of i elements, $i = 1, 2, \dots, n$. Specifically, these permutations are formed by permuting the elements of each of the $k_i \geq 0$ and each of the $r_i \geq 0$ subsets with i elements in all $a_i^{k_i}$ and $b_i^{r_i}$ permissible ways for $i = 1, 2, \dots, n$, respectively. Therefore

$$
\begin{aligned}
c_{n,k}&(k_1, r_1, k_2, r_2, \dots, k_n, r_n; a_1, b_1, a_2, b_2, \dots, a_n, b_n) \\
&= a_1^{k_1} b_1^{r_1} a_2^{k_2} b_2^{r_2} \cdots a_n^{k_n} b_n^{r_n} B(n, k; k_1, r_1, k_2, r_2, \dots, k_n, r_n; \\
&\qquad\qquad\qquad\qquad\qquad\qquad a_1, b_1, a_2, b_2, \dots, a_n, b_n),
\end{aligned}
$$

with $B(n, k; k_1, r_1, k_2, r_2, \dots, k_n, r_n; a_1, b_1, a_2, b_2, \dots, a_n, b_n)$ the number of partitions $\{A_1, A_2, \dots, A_k, B_1, B_2, \dots\}$ of a finite set W_n of n elements; $\{A_1, A_2, \dots, A_k\}$ is a partition of a set $U \subseteq W_n$ into k subsets, among which $k_i \geq 0$ include i elements each, $i = 1, 2, \dots, n$, and $\{B_1, B_2, \dots\}$ is a partition of the set $W_n - U$ into subsets among which $r_i \geq 0$ include i elements each, $i = 1, 2, \dots, n$, so that $\sum_{i=1}^{n} i(k_i + r_i) = n$ and $\sum_{i=1}^{n} k_i = k$. Clearly, this number is given by

$$
\begin{aligned}
B(n&, k; k_1, r_1, k_2, r_2, \dots, k_n, r_n; a_1, b_1, a_2, b_2, \dots, a_n, b_n) \\
&= \frac{n!}{k_1! r_1! (1!)^{k_1 + r_1} k_2! r_2! (2!)^{k_2 + r_2} \cdots k_n! r_n! (n!)^{k_n + r_n}},
\end{aligned}
$$

with $\sum_{i=1}^{n} i(k_i + r_i) = n$, $\sum_{i=1}^{n} k_i = k$ and so (12.18) is deduced.

(b) The number of permutations of a finite set W_n of n elements that are decomposed (i) into k partially ordered cycles, with elements from the set $U \subseteq W_n$, such that a_i cycles are formed from any subset of i elements, $i = 1, 2, \dots, n$, and (ii) into other partially ordered cycles, with elements from a set $W_n - U$, such that b_i cycles are formed from any subset of i elements, $i = 1, 2, \dots, n$, is obtained by summing the numbers (12.18) for all $k_i \geq 0$ and all $r_i \geq 0$, $i = 1, 2, \dots, n$, with $\sum_{i=1}^{n} i(k_i + r_i) = n$ and $\sum_{i=1}^{n} k_i = k$. According to the Definition 11.6, this sum is the Touchard polynomial and so the proof is completed. ∎

Further, using the generating function of the Touchard polynomial (11.50), we deduce the following corollary.

COROLLARY 12.7

*The generating function of the sequence $c_{n,k} \equiv c_{n,k}(k_1, r_1, k_2, r_2, \ldots, k_n, r_n;$
$a_1, b_1, a_2, b_2, \ldots, a_n, b_n)$, for $k_i \geq 0$, $r_i \geq 0$, $i = 1, 2, \ldots, n$, with
$\sum_{i=1}^{n} i(k_i + r_i) = n$, $\sum_{i=1}^{n} k_i = k$, $n = k, k+1, \ldots$, and for fixed k, is
given by*

$$\sum_{n=k}^{\infty} \sum c_{n,k} x_1^{k_1} x_2^{k_2} \cdots x_n^{k_n} y_1^{r_1} y_2^{r_2} \cdots y_n^{r_n} \frac{t^n}{n!}$$

$$= \frac{1}{k!} \left(\sum_{m=1}^{\infty} a_m x_m \frac{t^m}{m!} \right)^k \exp \left(\sum_{m=1}^{\infty} b_m y_m \frac{t^m}{m!} \right). \tag{12.19}$$

Example 12.8

Find the number $R(n, k)$ of permutations of a finite set W_n of n elements that are
decomposed into k (non-ordered) cycles, with elements from a set $U \subseteq W_n$ and
into other (totally) ordered cycles, with elements from the set $W_n - U$.

A generating function for the sequence $R(n, k)$, $n = k, k+1, \ldots$, for fixed k,
may be obtained from (12.19) by setting $a_m = (m-1)!$, $b_m = 1$ and $x_m = 1$,
$y_m = 1$, $m = 1, 2, \ldots$. Then

$$\sum_{n=k}^{\infty} R(n, k) \frac{t^n}{n!} = \frac{[-\log(1-t)]^k}{k!} \exp(e^t - 1).$$

Expanding the right-hand member into powers of t, using the generating function
of the signless Stirling numbers of the first kind $|s(n, k)|$,

$$\sum_{n=k}^{\infty} |s(n, k)| \frac{t^n}{n!} = \frac{[-\log(1-t)]^k}{k!},$$

and the generating function of the Bell numbers $B_n = \sum_{k=0}^{n} S(n, k)$,

$$\sum_{n=0}^{\infty} B_n \frac{t^n}{n!} = \exp(e^t - 1),$$

we get

$$\sum_{n=k}^{\infty} R(n, k) \frac{t^n}{n!} = \left(\sum_{n=k}^{\infty} |s(n, k)| \frac{t^n}{n!} \right) \left(\sum_{j=0}^{\infty} B_j \frac{t^j}{j!} \right)$$

$$= \sum_{n=k}^{\infty} \left\{ \sum_{j=0}^{n} \binom{n}{j} |s(n-j, k)| B_j \right\} \frac{t^n}{n!}.$$

Therefore

$$R(n, k) = \sum_{j=0}^{n} \binom{n}{j} |s(n-j, k)| B_j. \quad \square$$

12.5 BIBLIOGRAPHIC NOTES

A systematic and complete study of the theory of enumeration of permutations by cycles is provided by the elegant paper of Jacques Touchard (1939). In this study several partition polynomials are examined. The Touchard partition polynomials, which were briefly presented in Chapter 11, were so named in recognition of his fundamental contribution to this subject. Credit for its illuminating introduction to the theory of enumeration of permutations by cycles belongs to J. Riordan (1958). Examples 12.3 and 12.7 on the enumeration of restricted permutations by cycles are based on the paper of A. Z. Broder (1984).

12.6 EXERCISES

1. Show that the number of permutations of n that are decomposed into k cycles of length no smaller than 2 is given by

$$|s_2(n,k)| = \sum_{j=0}^{n}(-1)^j \binom{n}{j}|s(n-j,k-j)|,$$

where $|s(n,k)|$ is the *signless Stirling number of the first kind*. The number $|s_2(n,k)|$ is the *signless associated Stirling number of the first kind* (see Exercise 8.10).

2. (*Continuation*) Show that the number of permutations of n that are decomposed into k cycles of length no smaller than r is given by *signless r-associated Stirling number of the first kind* $|s_r(n,k)|$ for which

$$|s_{r+1}(n,k)| = \sum_{j=0}^{n}(-1)^j \frac{(n)_{rj}}{j!r^j}|s_r(n-rj,k-j)|, \quad r = 1,2,\ldots,$$

with $|s_1(n,k)| = |s(n,k)|$ the *signless Stirling number of the first kind* (see Exercise 8.11).

3. Let $c_n(k;r)$ be the number of permutations of a finite set W_n for which their decomposition into cycles includes, among other cycles, k cycles of length r. Show that the generating function

$$C_n(x;r) = \sum_{k=0}^{[n/r]} c_n(k;r)x^k$$

satisfies the recurrence relation

$$C_n(x;r) - nC_{n-1}(x;r) = \delta_{n,rj}\frac{n!(x-1)^j}{j!r^j},$$

and conclude that

$$c_n(k;r) - nc_{n-1}(k;r) = \delta_{n,rj}\frac{n!(-1)^{j-k}}{k!(j-k)!r^j},$$

where $\delta_{n,rj} = 1$ for $n = rj$ and $\delta_{n,rj} = 0$ for $n \neq rj$ is the Kronecker delta.

4. (*Continuation*) Assume that a permutation is randomly chosen from the set of the $n!$ permutations of a finite set of n elements. The probability that its decomposition into cycles includes, among other cycles, k cycles of length r is given by

$$p_k(n;r) = \frac{c_n(k;r)}{n!}, \quad k = 0, 1, \ldots, [n/r].$$

Show that its factorial moments

$$\mu_{(j)}(n;r) = \sum_{k=j}^{[n/r]} (k)_j p_k(n;r), \quad j = 1, 2, \ldots, [n/r]$$

are given by

$$\mu_{(j)}(n;r) = r^{-j}, \quad j = 1, 2, \ldots, [n/r].$$

5. Let $d_n(k;r)$ be the number of permutations of a finite set W_n that are decomposed into k cycles none of which is of length r. Show that the generating function

$$D_n(x;r) = \sum_{k=0}^{[n/r]} d_n(k;r)x^k$$

satisfies the recurrence relation

$$D_{n+1}(x;r) = (n+x)D_n(x;r) - x(n)_{r-1}D_{n-r+1}(x;r) + x(n)_r D_{n-r}(x;r),$$

for $n = r, r+1, \ldots$ and $D_n(x;r) = (x+n-1)_n$, for $n = 0, 1, \ldots, r-1$. Further, conclude the recurrence relation

$$d_{n+1}(k;r) = nd_n(k;r) + d_n(k-1;r) - (n)_{r-1}d_{n-r+1}(k-1;r)$$
$$+ (n)_r d_{n-r}(k-1;r),$$

for $k = 1, 2, \ldots, n+1$, $n = r, r+1, \ldots$, and $d_n(k;r) = |s(n,k)|$, for $k = 0, 1, \ldots, n$, $n = 0, 1, \ldots, r-1$.

6. (*Continuation*) Let $d_n(r)$ be the number of permutations of the set W_n that are decomposed into cycles none of which is of length r. Show that

$$d_n(r) = n! \sum_{j=0}^{[n/r]} \frac{(-1)^j}{j! r^j},$$

and

$$d_n(r) - n d_{n-1}(r) = \delta_{n,rj} \frac{n!(-1)^j}{j! r^j},$$

where $\delta_{n,rj} = 1$ for $n = rj$ and $\delta_{n,rj} = 0$ for $n \neq rj$ is the Kronecker delta.

7. Let $Q(n; k, r)$ be the number of permutations of a finite set W_n for which their decomposition into cycles includes, among other cycles, k cycles of length 1 and r cycles of length 2 and consider the generating function

$$Q_n(x, y) = \sum_{r=0}^{[n/2]} \sum_{k=0}^{n-2r} Q(n; k, r) x^k y^r, \quad n = 1, 2, \dots .$$

Show that

$$\sum_{n=0}^{\infty} Q_n(x, y) \frac{t^n}{n!} = (1 - t)^{-1} \exp[(x - 1)t + (y - 1)t^2/2]$$

and conclude that

$$Q_n(x, y) - n Q_{n-1}(x, y) = \sum_{j=0}^{[n/2]} \binom{n}{j} 2^{-j} (x - 1)^j (y - 1)^{n-2j}.$$

8. (*Continuation*) Show that

$$(1 - t) \sum_{n=0}^{\infty} Q_{n+1}(x, y) \frac{t^n}{n!}$$

$$= \{1 + (x - 1)(1 - t) + (y - 1)(1 - t)t\} \sum_{n=0}^{\infty} Q_n(x, y) \frac{t^n}{n!}$$

and conclude the recurrence relation

$$Q_{n+1}(x, y) = (n + x) Q_n(x, y) + n(y - x) Q_{n-1}(x, y)$$
$$+ n(n - 1)(1 - x) Q_{n-2}(x, y),$$

for $n = 2, 3, \dots$, with initial conditions $Q_0(x, y) = 1$, $Q_1(x, y) = x$ and $Q_2(x, y) = x^2 + y$.

9. *Cayley identity.* Show that

$$\sum \frac{(-1)^{k_1+k_2+\cdots+k_n}}{k_1!1^{k_1}k_2!2^{k_2}\cdots k_n!n^{k_n}} = 0, \quad n = 2, 3, \ldots,$$

where the summation is extended over all integers $k_i \geq 0$, $i = 1, 2, \ldots, n$, with $k_1 + 2k_2 + \cdots + nk_n = n$, and conclude that the number of permutations that are decomposed into an odd number of cycles equals the number of permutations that are decomposed into an even number of cycles.

10. (*Continuation*) Show that

$$\sum \frac{x^{k_1+k_2+\cdots+k_n}}{k_1!1^{k_1}k_2!2^{k_2}\cdots k_n!n^{k_n}} = \binom{x+n-1}{n},$$

where the summation is extended over all integers $k_i \geq 0$, $i = 1, 2, \ldots, n$, with $k_1 + 2k_2 + \cdots + nk_n = n$. Note that, in particular for $x = 1$, it reduces to the Cauchy identity (see Remark 12.1), while for $x = -1$, it reduces to the Cayley identity.

11. Let $a_{n,k}$ and $b_{n,k}$ be the numbers of even and odd permutations, respectively, of a finite set W_n that are decomposed into k cycles and consider the generating functions

$$A_n(u) = \sum_{k=0}^{n} a_{n,k}u^k, \quad B_n(u) = \sum_{k=0}^{n} b_{n,k}u^k, \quad n = 1, 2, \ldots.$$

Show that

$$A_{n+1}(u) = uA_n(u) + nB_n(u), \quad B_{n+1}(u) = uB_n(u) + nA_n(u)$$

and conclude that

$$nA_{n+2}(u) = (2n+1)uA_{n+1}(u) + (n+1)(n^2 - u^2)A_n(u), \quad n = 1, 2, \ldots,$$

with $A_1(u) = u$, $A_2(u) = u^2$ and

$$nB_{n+2}(u) = (2n+1)uB_{n+1}(u) + (n+1)(n^2 - u^2)B_n(u), \quad n = 1, 2, \ldots,$$

with $B_1(u) = 0$, $B_2(u) = u$.

12. Let $g_{n,k}$ and $h_{n,k}$ be the numbers of even and odd permutations, respectively, of a finite set W_n that are decomposed into k cycles of length no smaller than 2 and consider the generating functions

$$G_n(u) = \sum_{k=0}^{n} g_{n,k}u^k, \quad H_n(u) = \sum_{k=0}^{n} h_{n,k}u^k, \quad n = 0, 1, \ldots.$$

Show that

$$\sum_{n=0}^{\infty} G_n(u)\frac{t^n}{n!} = \frac{1}{2}e^{-ut}[(1-t)^{-u} + (1+t)^u],$$

$$\sum_{n=0}^{\infty} H_n(u)\frac{t^n}{n!} = \frac{1}{2}e^{-ut}[(1-t)^{-u} - (1+t)^u]$$

and

$$G_{n+1}(u) = nH_n(u) + nuH_{n-1}(u), \quad n = 1, 2, \dots,$$
$$H_{n+1}(u) = nG_n(u) + nuG_{n-1}(u), \quad n = 1, 2, \dots.$$

Deduce the recurrence relations

$$G_{n+4}(u) = (n+2)(n+3)G_{n+2}(u) + (n+3)(2n+3)uG_{n+1}(u)$$
$$+(n+1)(n+3)u^2 G_n(u),$$

for $n = 0, 1, \dots$, with $G_0(u) = 1$, $G_1(u) = 0$, $G_2(u) = 0$, $G_3(u) = 2u$ and

$$H_{n+4}(u) = (n+2)(n+3)H_{n+2}(u) + (n+3)(2n+3)uH_{n+1}(u)$$
$$+(n+1)(n+3)u^2 H_n(u),$$

for $n = 0, 1, \dots$, with $H_0(u) = 0$, $H_1(u) = 0$, $H_2(u) = u$, $H_3(u) = 0$.

13. (*Continuation*) (a) Show that

$$g_{n,k} = \sum_{j=0}^{k} (-1)^j \binom{n}{j} a_{n-j,k-j}, \quad h_{n,k} = \sum_{j=0}^{k} (-1)^j \binom{n}{j} b_{n-j,k-j},$$

where

$$a_{n,k} = \frac{1 + (-1)^{n-k}}{2}|s(n,k)|, \quad b_{n,k} = \frac{1 - (-1)^{n-k}}{2}|s(n,k)|$$

are the numbers of even and odd permutations, respectively, of a finite set of n elements that are decomposed into k cycles.
   (b) Setting

$$G_n = \sum_{k=0}^{n} g_{n,k}, \quad H_n = \sum_{k=0}^{n} h_{n,k}, \quad n = 0, 1, \dots,$$

show that

$$G_n = nG_{n-1} + (-1)^n \left[1 - \binom{n}{2}\right], \quad n = 2, 3, \dots, \quad G_1 = 0,$$

$$H_n = nH_{n-1} + (-1)^n \binom{n}{2}, \quad n = 2, 3, \dots, \quad H_1 = 0$$

and

$$G_n + H_n = D_n, \quad G_n - H_n = (-1)^{n-1}(n-1), \quad n = 1, 2, \ldots,$$

where D_n is the number of derangements of a finite set of n elements (see Section 5.2).

14. Let $b(n, k)$ be the number of permutations of a finite set of n elements that are decomposed into k odd (of even length) cycles and

$$b_n(u) = \sum_{k=0}^{n} b(n, k) u^k, \quad n = 0, 1, \ldots .$$

Show that

$$\sum_{n=0}^{\infty} b_n(u) \frac{t^n}{n!} = (1 - t^2)^{-u/2}$$

and conclude that

$$b_n(2u) = \begin{cases} \dfrac{(2r)!(u + r - 1)_r}{r!}, & n = 2r, \quad r = 0, 1, \ldots . \\ 0, & n = 2r + 1, r = 0, 1, \ldots, \end{cases}$$

and

$$b(n, k) = \begin{cases} \dfrac{(2r)! |s(r, k)|}{r! 2^k}, & n = 2r, \quad r = 0, 1, \ldots, \\ 0, & n = 2r + 1, r = 0, 1, \ldots, \end{cases}$$

where $|s(r, k)|$ is the *signless Stirling number of the first kind*.

15. (*Continuation*) Let $a(n, k)$ be the number of permutations of a finite set of n elements that are decomposed into k even (of odd length) cycles and

$$a_n(u) = \sum_{k=0}^{n} a(n, k) u^k, \quad n = 0, 1, \ldots .$$

Show that

$$\sum_{n=0}^{\infty} a_n(u) \frac{t^n}{n!} = (1 + t)^{u/2}(1 - t)^{-u/2} = (1 + t)^u(1 - t^2)^{-u/2}$$

and conclude that

$$a_n(u) = \sum_{r=0}^{[n/2]} \binom{n}{2r} (u)_{n-2r} b_{2r}(u)$$

and

$$a(n, k) = \sum_{r=0}^{[n/2]} \sum_{j=0}^{k} (-1)^{n+k-j} \frac{(n)_{2r}}{r! 2^j} |s(n - 2r, k - j)| \cdot |s(r, j)|.$$

16. (*Continuation*) Derive the recurrence relation

$$a_{n+2}(u) = ua_{n+1}(u) + n(n+1)a_n(u), \quad n = 0, 1, \dots,$$

with $a_0(u)$, $a_1(u) = u$, and

$$b_{2r+2}(u) = (2r+1)(2r+u)b_{2r}(u), \quad r = 0, 1, \dots,$$

with $b_0(u) = 1$. Deduce the particular expressions

$$a_2(u) = u^2, \ a_3(u) = 2u + u^3, \ a_4(u) = 8u^2 + u^4, \ a_5(u) = 24u + 20u^3 + u^5$$

and

$$b_2(u) = u, \quad b_4(u) = 6u + 3u^2, \quad b_6(u) = 120u + 90u^2 + 15u^3.$$

17. (*Continuation*) Let $a(n)$ and $b(n)$ be the numbers of permutations of a finite set of n elements that are decomposed into even (of odd length) and into odd (of even length) cycles, respectively. (a) Derive the recurrence relations

$$a(n+2) = a(n+1) + n(n+1)a(n), \quad n = 0, 1, \dots,$$

with $a(0) = 1$, $a(1) = 1$, and

$$b(n+2) = (n+1)^2 b(n), \quad n = 0, 1, \dots,$$

with $b(0) = 1$, $b(1) = 0$. (b) Show that

$$b(2r) = \left[\frac{(2r)!}{r!2^r} \right]^2, \quad b(2r+1) = 0, \quad r = 0, 1, \dots$$

and

$$a(2r) = \left[\frac{(2r)!}{r!2^r} \right]^2, \quad a(2r+1) = (2r+1)\left[\frac{(2r)!}{r!2^r} \right]^2, \quad r = 0, 1, \dots.$$

18. *A universal generating function.* Let $\mathbf{A} = (a_{i,j})$, $i = 1, 2, \dots, j = 0, 1, \dots$ be an infinite matrix with elements $a_{i,j} = 0$ or 1 and $c(n, k; \mathbf{A})$ be the number of permutations of a finite set W_n, of n elements, that are decomposed into k cycles such that the number of cycles of length i belongs to the set $\{j : a_{i,j} = 1\}$. Show that

$$G(t, u; \mathbf{A}) = \sum_{n=0}^{\infty} \sum_{k=0}^{n} c(n, k; \mathbf{A}) u^k \frac{t^n}{n!} = \prod_{i=1}^{\infty} \left[\sum_{j=0}^{\infty} a_{i,j} \frac{u^j}{j!} \left(\frac{t^i}{i} \right)^j \right].$$

19. Let $c_n(r)$ be the number of permutations of n that are decomposed into cycles none of which is of length longer than r. Show that

$$\sum_{n=0}^{\infty} c_n(r)\frac{t^n}{n!} = \exp\left(\sum_{i=1}^{r} \frac{t^i}{i}\right)$$

and conclude the recurrence relation

$$c_{n+1}(r) = \sum_{j=0}^{r-1}(n)_j c_{n-j}(r).$$

20. Let $T(n,k)$ be the number of permutations of n that are decomposed into cycles among which k are of length 2 and the other $n-2k$ are of length 1 (fixed points). Show that the generating function

$$T_n(x) = \sum_{k=0}^{[n/2]} T(n,k)x^k$$

satisfies the recurrence relation

$$T_n(x) = T_{n-1}(x) + (n-1)xT_{n-2}(x), \quad n = 2, 3, \ldots,$$

with $T_0(x) = 1$, $T_1(x) = 1$ and deduce for the numbers $T(n,k)$ the recurrence relation

$$T(n,k) = T(n-1,k) + (n-1)T(n-2,k-1),$$

for $n = 2k, 2k+1, \ldots$, $k = 1, 2, \ldots$, with $T(0,0) = T(1,1) = 1$, $T(n,k) = 0$, $n < 2k$.

Chapter 13

EQUIVALENCE CLASSES

13.1 INTRODUCTION

The enumeration of the number of k-combinations of a finite set $W_n = \{w_1, w_2, \ldots, w_n\}$, of n elements, is usually carried out in two stages. Specifically, if $\mathcal{C}_k(W_n)$ and $\mathcal{P}_k(W_n)$ are the sets of k-combinations and k-permutations of W_n, respectively, then, at the first stage, to each k-combination $A_k = \{a_1, a_2, \ldots, a_k\}$ that belongs to $\mathcal{C}_k(W_n)$ we uniquely correspond the set of its permutations $\mathcal{P}_k(A_k)$, which is a subset of $\mathcal{P}_k(W_n)$. Note that $\mathcal{P}_k(A_k)$ and $\mathcal{P}_k(B_k)$ are disjoint sets for A_k and B_k different k-combinations from $\mathcal{C}_k(W_n)$ and also that each k-permutation from $\mathcal{P}_k(W_n)$ belongs in one of the sets $\mathcal{P}_k(A_k)$, $A_k \in \mathcal{C}_k(W_n)$. Thus, the set $\{\mathcal{P}_k(A_k) \subseteq \mathcal{P}_k(W_n) : A_k \in \mathcal{C}_k(W_n)\}$ constitutes a partition of the set $\mathcal{P}_k(W_n)$. Further, according to the definition of a permutation, $N(\mathcal{P}_k(A_k)) = N(\mathcal{P}_k(W_k))$ for every $A_k \in \mathcal{C}_k(W_n)$ and so, by the addition principle,

$$N(\mathcal{P}_k(W_n)) = \sum_{A_k \in \mathcal{C}_k(W_n)} N(\mathcal{P}_k(A_k)) = N(\mathcal{C}_k(W_n))N(\mathcal{P}_k(W_k)).$$

At the second stage we enumerate the number $P(n, k) = N(\mathcal{P}_k(W_n))$ of k-permutations of the set W_n and conclude the number $P(k) \equiv P(k, k) = N(\mathcal{P}_k(W_k))$. Then the number $C(n, k) = N(\mathcal{C}_k(W_n))$, of k-combinations of the set W_n, is obtained as $C(n, k) = P(n, k)/P(k)$. This technique has been employed for the derivation of this number in Theorem 2.5.

Note that the sets $\mathcal{P}_k(A_k)$, $A_k \in \mathcal{C}_k(W_n)$ constitute equivalence classes of the k-permutations that belong to the set $\mathcal{P}_k(W_n)$. The equivalence relation is defined as follows: A permutation $(a_1, a_2, \ldots, a_k)$ that belongs to $\mathcal{P}_k(W_n)$ is equivalent to a permutation $(b_1, b_2, \ldots, b_k)$ that also belongs to $\mathcal{P}_k(W_n)$ if there exists a permutation $(j_1, j_2, \ldots, j_k)$ of the set of the k indices $\{1, 2, \ldots, k\}$ such that $a_s = b_{j_s}$, $s = 1, 2, \ldots, k$. In other words, two k-permutations of W_n are equivalent if they contain the same k elements. Thus, the enumeration of the number of k-combinations of a finite set of n

elements is merely a problem of enumeration of the equivalence classes into which the k-permutations of the finite set of n elements are distributed.

This chapter is devoted to the study of the general problem of enumerating equivalent classes of a finite set with respect to a group of its permutations. After a brief presentation of groups of permutations and their cycle indicator, we proceed to the enumeration of the orbits (equivalence classes) of the elements of a finite set, under a group of permutations of it. In this respect Burnside's lemma is shown. The study is then focused on the enumeration of models of colorings of the elements of a finite set, under a group of permutations of it; Pólya's fundamental counting theorem is derived.

13.2   CYCLE INDICATOR OF A PERMUTATION GROUP

Consider a finite set $W_n = \{w_1, w_2, \ldots, w_n\}$, of n elements, and let $\mathcal{P}_n \equiv \mathcal{P}_n(W_n)$ be the set of its $n!$ permutations. To each pair of permutations in $\mathcal{P}_n$

$$\sigma = \begin{pmatrix} w_1, & w_2, & \ldots, & w_n \\ w_{i_1}, & w_{i_2}, & \ldots, & w_{i_n} \end{pmatrix}, \quad \tau = \begin{pmatrix} w_{i_1}, & w_{i_2}, & \ldots, & w_{i_n} \\ w_{j_1}, & w_{j_2}, & \ldots, & w_{j_n} \end{pmatrix},$$

there corresponds the permutation $\sigma\tau$ in $\mathcal{P}_n$ defined by

$$\sigma\tau = \begin{pmatrix} w_1, & w_2, & \ldots, & w_n \\ w_{j_1}, & w_{j_2}, & \ldots, & w_{j_n} \end{pmatrix},$$

which is called their product. If

$$\rho = \begin{pmatrix} w_{j_1}, & w_{j_2}, & \ldots, & w_{j_n} \\ w_{r_1}, & w_{r_2}, & \ldots, & w_{r_n} \end{pmatrix}$$

is another permutation in $\mathcal{P}_n$, then

$$(\sigma\tau)\rho = \begin{pmatrix} w_1, & w_2, & \ldots, & w_n \\ w_{r_1}, & w_{r_2}, & \ldots, & w_{r_n} \end{pmatrix}$$

and

$$\tau\rho = \begin{pmatrix} w_{i_1}, & w_{i_2}, & \ldots, & w_{i_n} \\ w_{r_1}, & w_{r_2}, & \ldots, & w_{r_n} \end{pmatrix}, \quad \sigma(\tau\rho) = \begin{pmatrix} w_1, & w_2, & \ldots, & w_n \\ w_{r_1}, & w_{r_2}, & \ldots, & w_{r_n} \end{pmatrix}.$$

Thus, the product of permutations is associative, $(\sigma\rho)\rho = \sigma(\tau\rho)$. Further, the identity permutation

$$\varepsilon = \begin{pmatrix} w_1, w_2, \ldots, w_n \\ w_1, w_2, \ldots, w_n \end{pmatrix}$$

is the neutral element for the product, $\varepsilon\sigma = \sigma\varepsilon = \sigma$. Also, for every permutation σ in $\mathcal{P}_n$, there exists its inverse permutation in $\mathcal{P}_n$,

$$\sigma^{-1} = \begin{pmatrix} w_{i_1}, w_{i_2}, \ldots, w_{i_n} \\ w_1, \ w_2, \ \ldots, \ w_n \end{pmatrix}$$

so that $\sigma\sigma^{-1} = \sigma^{-1}\sigma = \varepsilon$. Consequently, the set $\mathcal{P}_n$ of permutations of W_n equipped with the operation of multiplication constitutes a group called *symmetric group of permutations*. This group is of order $N(\mathcal{P}_n) = n!$ and degree $N(W_n) = n$.

Let us consider a subset $\mathcal{G}_n$ of permutations of W_n that is closed under the operation of multiplication. This set, as a subset of the finite group $\mathcal{P}_n$, is a subgroup and is called *group of permutations* of W_n. Further, consider a permutation σ in $\mathcal{G}_n$, the decomposition of which into disjoint cycles includes $k_i = k_i(\sigma) \geq 0$ cycles of length i, $i = 1, 2, \ldots, n$, so that $k_1(\sigma) + 2k_2(\sigma) + \cdots + nk_n(\sigma) = n$. Let us attach the indicator x_i to any cycle of length i, $i = 1, 2, \ldots, n$. Then the homogeneous product $x_1^{k_1(\sigma)} x_2^{k_2(\sigma)} \cdots x_n^{k_n(\sigma)}$, of degree $k_1(\sigma) + k_2(\sigma) + \cdots + k_n(\sigma) = k(\sigma)$, is the cycle indicator of the permutation σ and the sum

$$C(x_1, x, \ldots, x_n; \mathcal{G}_n) N(\mathcal{G}_n) = \sum_{\sigma \in \mathcal{G}_n} x_1^{k_1(\sigma)} x_2^{k_2(\sigma)} \cdots x_n^{k_n(\sigma)}, \qquad (13.1)$$

where the summation is extended over all permutations σ in $\mathcal{G}_n$, is the cycle indicator of the permutations contained in the group $\mathcal{G}_n$. The polynomial $C(x_1, x_2, \ldots, x; \mathcal{G}_n)$ is called *cycle indicator* of the group of permutation $\mathcal{G}_n$. Introducing the number $c(k_1, k_2, \ldots, k_n; \mathcal{G}_n)$ of permutations contained in $\mathcal{G}_n$, for which the decomposition into cycles includes $k_i \geq 0$ cycles of length i, $i = 1, 2, \ldots, n$, so that $k_1 + 2k_2 + \cdots + nk_n = n$, the cycle indicator may be written as

$$C(x_1, x_2, \ldots, x_n; \mathcal{G}_n) = \frac{1}{N(\mathcal{G}_n)} \sum c(k_1, k_2, \ldots, k_n; \mathcal{G}_n) x_1^{k_1} x_2^{k_2} \cdots x_n^{k_n}, \qquad (13.2)$$

where the summation is extended over all partitions of n, that is, over all nonnegative integer solutions $(k_1, k_2, \ldots, k_n)$ of the equation $k_1 + 2k_2 + \cdots + nk_n = n$.

In the following examples several specific groups of permutations of a finite set are considered and their cycle indicators are calculated.

Example 13.1 Identity group of permutations

The identity permutation $\varepsilon = (w_1)(w_2) \cdots (w_n)$ of the set $W_n = \{w_1, w_2, \ldots, w_n\}$ is clearly a group. Since this permutation includes n cycles of length 1, the cycle indicator of the identity group $\{\varepsilon\}$, according to (13.1), is given by

$$C(x_1, x_2, \ldots, x_n; \{\varepsilon\}) = x_1^n. \qquad \square$$

Example 13.2 Symmetric group of permutations

The number of permutations that are contained in the symmetric group of permutations $\mathcal{G}_n = \mathcal{P}_n$ and are decomposed into $k_i \geq 0$ cycles of length i, $i = 1, 2, \ldots, n$, with $k_1 + 2k_2 + \cdots + nk_n = n$, is given by (see Theorem 12.1)

$$c_{n,k}(k_1, k_2, \ldots, k_n; \mathcal{P}_n) = \frac{n!}{k_1! 1^{k_1} k_2! 2^{k_2} \cdots k_n! n^{k_n}}.$$

Hence, the cycle indicator of the symmetric group of permutations, according to (13.2) and since $N(\mathcal{P}_n) = n!$, is given by

$$C(x_1, x_2, \ldots, x_n; \mathcal{P}_n) = \frac{1}{n!} \sum \frac{n!}{k_1! k_2! \cdots k_n!} \left(\frac{x_1}{1}\right)^{k_1} \left(\frac{x_2}{2}\right)^{k_2} \cdots \left(\frac{x_n}{n}\right)^{k_n},$$

where the summation is extended over all partitions of n, that is over all nonnegative integer solutions $(k_1, k_2, \ldots, k_n)$ of the equation $k_1 + 2k_2 + \cdots + nk_n = n$. This polynomial, using (11.1), may be expressed as

$$C(x_1, x_2, \ldots, x_n; \mathcal{P}_n) = \frac{1}{n!} B_n(x_1, 1! x_2, \ldots, (n-1)! x_n),$$

where $B_n(x_1, x_2, \ldots, x_n)$ is the exponential Bell partition polynomial.   ⬜

Example 13.3 Alternating group of permutations

Consider the set $\mathcal{A}_n \subset \mathcal{P}_n$ of even permutations (which can be decomposed into an even number of transpositions) of the set W_n. Note that the product of two even permutations $\sigma = t_{i_1} t_{i_2} \cdots t_{i_{2m}}$ and $\tau = t_{j_1} t_{j_2} \cdots t_{j_{2r}}$, where $t_{i_1}, t_{i_2}, \ldots, t_{i_{2m}}$ and $t_{j_1}, t_{j_2}, \cdots, t_{j_{2r}}$ are transpositions of elements of W_n, is also an even permutation $\sigma\tau = t_{i_1} t_{i_2} \cdots t_{i_{2m}} t_{j_1} t_{j_2} \cdots t_{j_{2r}}$. Therefore, the set $\mathcal{A}_n$ is closed under multiplication and, as a subset of the finite symmetric group of permutations $\mathcal{P}_n$, is a subgroup called *alternating group of permutations*. Further, the number of permutations that are contained in the alternating group of permutations $\mathcal{G}_n = \mathcal{A}_n$ and are decomposed into $k_i \geq 0$ cycles of length i, $i = 1, 2, \ldots, n$, with $k_1 + 2k_2 + \cdots + nk_n = n$, is given by

$$c_n(k_1, k_2, \ldots, k_n; \mathcal{A}_n) = \frac{1 + (-1)^r}{2} c_n(k_1, k_2, \ldots, k_n; \mathcal{P}_n),$$

where $r = k_2 + k_4 + \cdots + k_{2m}$, $m = [n/2]$. Then, the cycle indicator of the alternating group of permutations, by (13.2) and since $N(\mathcal{A}_n) = n!/2$, is given by

$$C(x_1, x_2, \ldots, x_n; \mathcal{A}_n) = \frac{1}{n!} B_n(x_1 1! x_2, \ldots, (n-1)! x_n)$$

$$+ \frac{1}{n!} B_n(x_1, -1! x_2, \ldots, (-1)^{n-1}(n-1)! x_n),$$

where $B_n(x_1, x_2, \ldots, x_n)$ is the exponential Bell partition polynomial.   ⬜

Example 13.4 **Cyclic group of permutations**

Consider a permutation σ of the symmetric group $\mathcal{P}_n$ of permutations of the set W_n. The powers of σ are recursively defined by the relations $\sigma^j = \sigma\sigma^{j-1}, j = 2, 3, \ldots,$ $\sigma^0 = \varepsilon$ and $\sigma^{-j} = (\sigma^{-1})^j$, $j = 2, 3, \ldots,$ with σ^{-1} the inverse permutation of σ. Since the symmetric group of permutations $\mathcal{P}_n$ has a finite number of elements, $N(\mathcal{P}_n) = n!$, it follows that there exists a positive integer r such that $\sigma^r = \varepsilon$ because, otherwise, $\mathcal{P}_n$ would have an infinite number of elements. The smallest positive integer r with this property is called the *order* of the permutation σ. In this case, the r powers $\sigma_j = \sigma^j$, $j = 1, 2, \ldots, r$ are all different because, if we assume that $\sigma^i = \sigma^j$ for $1 \leq i < j \leq r$, then $\sigma^{j-i} = \varepsilon$ for $1 \leq j - i < r$, which contradicts the hypothesis that r is the smallest positive integer with this property. Also, any positive, zero or negative power σ^k of σ coincides with one of the r different powers $\sigma_j = \sigma^j$, $j = 1, 2, \ldots, r$, since the integer k can be written in the form $k = sr + j$, with $0 \leq j < r$, whence $\sigma^k = \sigma^{sr+j} = (\sigma^r)^s \sigma^j = \varepsilon\sigma^j = \sigma^j$.

The set $\mathcal{G}_{n,r} = \{\varepsilon, \sigma, \sigma^2, \ldots, \sigma^{r-1}\}$, whose elements are powers of a permutation σ of order r, is closed under multiplication and, as a subset of the finite symmetric group of permutations $\mathcal{P}_n$, is a subgroup called *cyclic group of permutations* generated by the permutation σ. Note that, for $j = 1, 2, \ldots, r$ and d the greatest common divisor of j and r, the permutations $\sigma_j = \sigma^j$ and $\sigma_d = \sigma^d$, which belong to the same cyclic group of permutations $\mathcal{G}_{n,r}$, generate the same cyclic subgroup of permutations $\mathcal{G}_{n,r/d}$ of order r/d. The *orbit* of a point w of the set W_n under a permutation σ of order r is the set $\{w, \sigma(w), \sigma^2(w), \ldots, \sigma^{r-1}(w)\} \subseteq W_n$, where the length r is the smallest positive integer that satisfies the relation $\sigma^r(w) = w$.

The cycle indicator of the cyclic group of permutations $\mathcal{G}_{n,r}$ is determined as follows: let d be a divisor of r and $\mathcal{H}_{n,r/d} = \{\sigma_j \in \mathcal{G}_{n,r} : d \text{ the greatest common divisor of } j \text{ and } r\} = \{\sigma_j \in \mathcal{G}_{n,r} : j/d \text{ a positive integer relatively prime to } r/d\}$. Then, if s is a divisor of r different from d, the sets $\mathcal{H}_{n,r/s}$ and $\mathcal{H}_{n,r/d}$ are disjoint. Also, every permutation $\sigma_j = \sigma^j$, which belongs to the cyclic group of permutations $\mathcal{G}_{n,r}$, is contained in one of the sets $\mathcal{H}_{n,r/d}$ and specifically in the set for which j/d is relatively prime to r/d. Therefore the sets $\mathcal{H}_{n,r/d}$ for all divisors d of r constitute a partition of the cyclic group of permutations $\mathcal{G}_{n,r}$,

$$\mathcal{G}_{n,r} = \sum_{d|r} \mathcal{H}_{n,r/d}.$$

Note that the permutation $\sigma_d = \sigma^d$ clearly belongs to the set $\mathcal{H}_{n,r/d}$. Further, for every other permutation $\sigma_j = \sigma^j$ that belongs to $\mathcal{H}_{n,r/d}$, d is the greatest common divisor of j and r, whence the permutations σ_j and σ_d generate the same cyclic subgroup of permutations $\mathcal{G}_{n,r/d}$ of order r/d. Hence, the orbit of a point w of the set W_n under a permutation σ_j that belongs to $\mathcal{H}_{n,r/d}$ is $\{w, \sigma_j(w), \sigma_j^2(w), \ldots, \sigma_j^{r/d-1}(w)\}$, with length r/d. Consequently, the permutation σ_j is decomposed into d cycles, each of length r/d. The number $N(\mathcal{H}_{n,r/d})$, of permutations that belong to the set $\mathcal{H}_{n,r/d}$, which is the number of positive integers less than or equal and relatively prime to r/d, is equal to the Euler's function

$\phi(r/d)$ (see Example 4.4). Thus, by (13.1) and since $N(\mathcal{G}_{n,r}) = r$, the cycle indicator of the cyclic group of permutations $\mathcal{G}_{n,r}$ is given by

$$C(x_1, x_2, \ldots, x_n; \mathcal{G}_n) = \frac{1}{r} \sum_{d|r} \phi(r/d) x_{r/d}^d = \frac{1}{r} \sum_{d|r} \phi(d) x_d^{r/d}. \qquad \square$$

Example 13.5

Let us consider the set $W_6 = \{1, 2, 3, 4, 5, 6\}$ of the six faces of a cube. Its faces are numbered so that $\{1, 6\}$, $\{2, 5\}$ and $\{3, 4\}$ are non-ordered pairs of opposite faces. Let $\mathcal{G}_6$ be the group of permutations of W_6 that are induced by rotations of the cube around its axes of symmetry. Determine the cycle indicator of the group $\mathcal{G}_6$.

The group $\mathcal{G}_6$ contains the following permutations written in the form of product of cycles:

(a) The identity permutation

$$(1)(2)(3)(4)(5)(6).$$

(b) The permutations induced by $\pi/2$, π and $3\pi/2$ rotations of the cube around the axes connecting the centers of opposite faces in the counterclockwise direction:

$$(1)(2, 4, 5, 3)(6), \quad (1)(2, 5)(3, 4)(6), \quad (1)(2, 3, 5, 4)(6),$$
$$(1, 3, 6, 4)(2)(5), \quad (1, 6)(2)(3, 4)(5), \quad (1, 4, 6, 3)(2)(5),$$
$$(1, 5, 6, 2)(3)(4), \quad (1, 6)(2, 5)(3)(4), \quad (1, 2, 6, 5)(3)(4).$$

(c) The permutations induced by π rotations of the cube around the axes connecting opposite edges in the counterclockwise direction:

$$(1, 2)(3, 4)(5, 6), \quad (1, 3)(2, 5)(4, 6), \quad (1, 4)(2, 5)(3, 6),$$
$$(1, 5)(2, 6)(3, 4), \quad (1, 6)(2, 3)(4, 5), \quad (1, 6)(2, 4)(3, 5).$$

(d) The permutations induced by $3\pi/2$ rotations of the cube around the axes connecting opposite vertices in the counterclockwise direction:

$$(1, 2, 3)(4, 5, 6), \quad (1, 2, 4)(3, 6, 5), \quad (1, 3, 2)(4, 5, 6), \quad (1, 3, 5)(2, 6, 4),$$
$$(1, 4, 2)(3, 5, 6), \quad (1, 4, 5)(2, 6, 3), \quad (1, 5, 3)(2, 4, 6), \quad (1, 5, 4)(2, 3, 6).$$

Consequently, the cycle indicator of the group $\mathcal{G}_6$, according to (13.2) and since $N(\mathcal{G}_6) = 24$, is given by

$$C(x_1, x_2, \ldots, x_6; \mathcal{G}_6) = \frac{1}{24}(x_1^6 + 3x_1^2 x_2^2 + 6x_1^2 x_4 + 6x_2^3 + 8x_3^2). \qquad \square$$

13.3 ORBITS OF ELEMENTS OF A FINITE SET

Consider a group $\mathcal{G}_n \subseteq \mathcal{P}_n$, of permutations of a finite set $W_n = \{w_1, w_2, \ldots, w_n\}$ and two elements u and w from W_n. If there exists a permutation τ belonging to $\mathcal{G}_n$ such that $\tau(w) = u$, then the element u is said to be *equivalent* to the element w under the group $\mathcal{G}_n$; this is denoted by $u \equiv w$. Clearly, this relation is (a) *reflexive*: $u \equiv u$, (b) *symmetric*: $u \equiv w$ implies $w \equiv u$ and (c) *transitive*: $u \equiv v$ and $v \equiv w$ imply $u \equiv w$, and so it is an *equivalence relation*.

DEFINITION 13.1   *The set $T_w = \{u \in W_n : \tau(w) = u, \tau \in \mathcal{G}_n\}$, $w \in W_n$, which contains the elements of W_n equivalent to w, constitutes an equivalence class that is called orbit of the element w under $\mathcal{G}_n$.*

Note that, for any two different elements u and w of W_n, the equivalence classes T_u and T_w either coincide (when the two elements u and w are equivalent under the group $\mathcal{G}_n$) or are disjoint. Thus, the set $\mathcal{T} = \{T_{w_{i_1}}, T_{w_{i_2}}, \ldots, T_{w_{i_k}}\}$, of the different orbits of the elements of W_n under the group $\mathcal{G}_n$, constitutes a partition of W_n.

DEFINITION 13.2   *The set $\mathcal{S}_w = \{\sigma \in \mathcal{G}_n : \sigma(w) = w\}$, $w \in W_n$, which contains the permutations of $\mathcal{G}_n$ that keep the element w fixed and is a subgroup of $\mathcal{G}_n$ is called stabilizer of w.*

For each subgroup $\mathcal{S}_w$ of the group $\mathcal{G}_n$ and for each permutation τ belonging to $\mathcal{G}_n$, the *left coset* is defined by $\tau \mathcal{S}_w = \{\tau\sigma : \sigma \in \mathcal{S}_w\}$ and the *right coset* by $\mathcal{S}_w \tau = \{\sigma\tau : \sigma \in \mathcal{S}_w\}$. Clearly, $N(\tau \mathcal{S}_w) = N(\mathcal{S}_w \tau) = N(\mathcal{S}_w)$. Further, for any two different permutations ρ and τ that belong to $\mathcal{G}_n$, the left cosets $\rho \mathcal{S}_w$ and $\tau \mathcal{S}_w$ (as well as the right cosets $\mathcal{S}_w \rho$ and $\mathcal{S}_w \tau$) either coincide or are disjoint. Hence, the set $\{\tau_{j_1} \mathcal{S}_w, \tau_{j_2} \mathcal{S}_w, \ldots, \tau_{j_r} \mathcal{S}_w\}$ of distinct left cosets (as well as the set $\{\mathcal{S}_w \tau_{j_1}, \mathcal{S}_w \tau_{j_2}, \ldots, \mathcal{S}_w \tau_{j_r}\}$ of distinct right cosets) constitutes a partition of $\mathcal{G}_n$.

We are now prepared to prove a theorem concerning the number of elements of the orbit of an arbitrary element of a finite set under a group of permutations of it.

THEOREM 13.1

The number $N(T_w)$, of elements of the orbit T_w of an element w of a finite set W_n, *under a group $\mathcal{G}_n$ of permutations of W_n, is given by*

$$N(T_w) = \frac{N(\mathcal{G}_n)}{N(\mathcal{S}_w)},$$

(13.3)

where $\mathcal{S}_w$ is the stabilizer of the element w.

PROOF   Let u and v be two elements of the orbit T_w of w, under $\mathcal{G}_n$. Then, there exist two permutations τ and ρ that belong to $\mathcal{G}_n$ such that $\tau(w) = u$ and $\rho(w) = v$. Further, $u = v$ if and only if $\rho(w) = \tau(w)$ or, equivalently, $\tau^{-1}(\rho(w)) = w$. The last relation is equivalent to $\sigma = \tau^{-1}\rho$ being in the stabilizer $\mathcal{S}_w$ of w, which in turn is equivalent to $\rho = \tau\sigma$ being in the left coset $\tau\mathcal{S}_w$. Note that the permutation τ belongs to the left coset $\tau\mathcal{S}_w$, since $\tau = \tau\varepsilon$ and the identity permutation ε belongs to the stabilizer $\mathcal{S}_w$. Hence, $u = v$ if and only if the permutations τ and ρ belong to the same left coset $\tau\mathcal{S}_w$. As a consequence, there exist a one-to-one and onto correspondence of the orbit T_w to the set of the left cosets of the stabilizer $\mathcal{S}_w$ in which the element u, with $\tau(w) = u$, corresponds to the left coset $\tau\mathcal{S}_w$. Therefore, the number $N(T_w)$ of elements of the orbit T_w equals the number of the left cosets of the stabilizer $\mathcal{S}_w$. Since, the set of the left cosets of $\mathcal{S}_w$ constitutes a partition of $\mathcal{G}_n$ and $N(\tau\mathcal{S}_w) = N(\mathcal{S}_w)$ for every left coset $\tau\mathcal{S}_w$ of $\mathcal{S}_w$, according to the addition principle, it follows that

$$N(\mathcal{G}_n) = N(T_w)N(\mathcal{S}_w),$$

from which (13.3) is deduced.   ∎

The next theorem, which is concerned with the number of orbits of the elements of a finite set, under a group of permutations of it, is due to W. S. Burnside.

THEOREM 13.2
The number of orbits of the elements of a finite set W_n, under a group $\mathcal{G}_n$ of permutations of it, is given by

$$N(\mathcal{T}) = \frac{1}{N(\mathcal{G}_n)} \sum_{\sigma \in \mathcal{G}_n} k_1(\sigma),$$

(13.4)

where $k_1(\sigma)$ is the number of cycles of length 1 of a permutation σ.

PROOF   The number $k_1(\sigma)$ of cycles of length 1 of a permutation σ and the number $N(\mathcal{S}_u)$ of the elements of the stabilizer $\mathcal{S}_u$ of an element u of the set W_n, on using Kronecker's delta function, $\delta(u, v) = 0$ for $u \neq v$ and $\delta(u, u) = 1$, may be expressed as

$$k_1(\sigma) = \sum_{u \in W_n} \delta(u, \sigma(u)), \quad N(\mathcal{S}_u) = \sum_{\sigma \in \mathcal{G}_n} \delta(u, \sigma(u)).$$

Thus

$$\sum_{\sigma \in \mathcal{G}_n} k_1(\sigma) = \sum_{\sigma \in \mathcal{G}_n} \sum_{u \in W_n} \delta(u, \sigma(u)) = \sum_{u \in W_n} \sum_{\sigma \in \mathcal{G}_n} \delta(u, \sigma(u)) = \sum_{u \in W_n} N(\mathcal{S}_u)$$

and, utilizing the fact that the set $\mathcal{T}$ of the different orbits of the elements of W_n under a group $\mathcal{G}_n$ constitutes a partition of W_n, we get the expression

$$\sum_{\sigma \in \mathcal{G}_n} k_1(\sigma) = \sum_{T_w \in \mathcal{T}} \sum_{u \in T_w} N(\mathcal{S}_u).$$

Further, the orbit T_u of every element $u \in T_w$ coincides with the orbit T_w and, according to (13.3), it follows that

$$N(\mathcal{S}_u) = \frac{N(\mathcal{G}_n)}{N(T_u)} = \frac{N(\mathcal{G}_n)}{N(T_w)} = N(\mathcal{S}_w).$$

Hence

$$\sum_{u \in T_w} N(\mathcal{S}_u) = N(T_w)N(\mathcal{S}_w) = N(\mathcal{G}_n)$$

and

$$\sum_{\sigma \in \mathcal{G}_n} k_1(\sigma) = \sum_{T_w \in \mathcal{T}} N(\mathcal{G}_n) = N(\mathcal{T})N(\mathcal{G}_n).$$

The last expression implies (13.4).   ∎

Example 13.6

The alternating subgroup $\mathcal{A}_4$ of the group $\mathcal{P}_4$ of permutations of the set $W_4 = \{1, 2, 3, 4\}$ contains the following permutations:

$$(1)(2)(3)(4), \ (1,2,3)(4), \ (1,2,4)(3), \ (1,3,2)(4), \ (1,3,4)(2), \ (1,4,2)(3),$$

$$(1,4,3)(2), \ (1)(2,3,4), \ (1)(2,4,3), \ (1,2)(3,4), \ (1,3)(2,4), \ (1,4)(2,3).$$

The orbits of the elements of the set W_4, under the group $\mathcal{A}_4$, are

$$T_1 = T_2 = T_3 = T_4 = W_4 = \{1, 2, 3, 4\}$$

and the stabilizers of the elements of W_4 are

$$\mathcal{S}_1 = \{(1)(2)(3)(4), (1)(2,3,4), (1)(2,4,3)\},$$

$$\mathcal{S}_2 = \{(1)(2)(3)(4), (1,3,4)(2), (1,4,3)(2)\},$$

$$\mathcal{S}_3 = \{(1)(2)(3)(4), (1,2,4)(3), (1,4,2)(3)\},$$

$$\mathcal{S}_4 = \{(1)(2)(3)(4), (1,2,3)(4), (1,3,2)(4)\}.$$

Note that $N(\mathcal{A}_4) = 12$, $N(\mathcal{S}_i) = 3$, $i = 1, 2, 3, 4$ and $N(\mathcal{T}_i) = 4$, $i = 1, 2, 3, 4$, in agreement with Theorem 13.1. Also, $N(\mathcal{T}) = 1$ and

$$\frac{1}{N(\mathcal{A}_4)} \sum_{\sigma \in \mathcal{A}_4} k_1(\sigma) = \frac{1}{12}(4 + 1 + 1 + 1 + 1 + 1 + 1 + 1 + 1) = 1,$$

a result that agrees with Theorem 13.2. $\square$

Example 13.7 Models of coloring the faces of a cube

Calculate the number of models of coloring the six faces of a cube with six different colors, each face with a different color, when two colorings are considered to be of the same model if they can be transformed to each other by rotation of the cube.

Any coloring of the six faces of a cube with six different colors, each face with a different color, constitutes a permutation $w_s = (c_{s_1}, c_{s_2}, \ldots, c_{s_6})$ of the set $C = \{c_1, c_2, \ldots, c_6\}$ of the six colors, where c_{s_i} is the color of the i-th face, $i = 1, 2, \ldots, 6$. Hence, the set W_{720} of colorings contains $6! = 720$ elements. Let $\mathcal{G}_{720}$ be the group of permutations of the set W_{720} that are induced by rotations of the corresponding colored cube around the axes of symmetry. Also, let $\mathcal{H}_6$ be the group of permutations of the set $\mathcal{F}_6 = \{1, 2, \ldots, 6\}$, of the six faces of the cube, which are induced by rotations of the cube around the axes of symmetry. To the permutation $\sigma \in \mathcal{G}_{720}$, in which the coloring $w_s = (c_{s_1}, c_{s_2}, \ldots, c_{s_6})$ is replaced by the coloring $\sigma(w_s) = (c_{s_{r_1}}, c_{s_{r_2}}, \ldots, c_{s_{r_6}})$, $w_s \in W_{720}$, we correspond the permutation $\tau \in \mathcal{H}_6$, in which the face i is replaced by the face $\tau(i) = r_i$, $i = 1, 2, \ldots, 6$.

For example, the identity permutation $\varepsilon = (w_1)(w_2) \cdots (w_{720}) \in W_{720}$, in which the coloring $w_s = (c_{s_1}, c_{s_2}, \ldots, c_{s_6})$ is replaced by the same coloring $\varepsilon(w_s) = w_s = (c_{s_1}, c_{s_2}, \ldots, c_{s_6})$, $w_s \in W_{720}$, we correspond the identity permutation $\varepsilon = (1)(2) \cdots (6) \in \mathcal{H}_6$, in which the face i is replaced by the same face $\varepsilon(i) = i$, $i = 1, 2, \ldots, 6$. The permutation $\sigma \in \mathcal{G}_{720}$, in which the coloring $w_s = (c_{s_1}, c_{s_2}, c_{s_3}, c_{s_4}, c_{s_5}, c_{s_6})$ is replaced by the coloring $\sigma(w_s) = (c_{s_1}, c_{s_3}, c_{s_4}, c_{s_5}, c_{s_2}, c_{s_6})$, we correspond the permutation $\tau = (1)(2, 3, 4, 5)(6) \in \mathcal{H}_6$, in which $\tau(1) = 1$, $\tau(2) = 3$, $\tau(3) = 4$, $\tau(4) = 5$, $\tau(5) = 2$ and $\tau(6) = 6$.

This correspondence of the group of permutations $\mathcal{G}_{720}$ to the group of permutations $\mathcal{H}_6$ is one-to-one and onto and so $N(\mathcal{G}_{720}) = N(\mathcal{H}_6)$. Further, according to the analysis of Example 13.5, $N(\mathcal{G}_{720}) = 24$ permutations, the identity permutation $\varepsilon = (w_1)(w_2) \cdots (w_{720}) \in W_{720}$ retains fixed the $k_1(\varepsilon) = 720$ elements of W_{720}. Also, none of the other 23 permutations retains fixed any of the elements of W_{720}, $k_1(\sigma) = 0$, since no cube whose six faces are colored with six different colors, each face with a different color, is left fixed by rotations around any of the axes of symmetry. The required number of models of colorings is equal to the number $N(\mathcal{T})$ of orbits of the elements of W_{720}, under the group of permutations

$\mathcal{G}_{720}$, which, according to Theorem 13.2, equals

$$N(\mathcal{T}) = \frac{720}{24} = 30. \quad \square$$

A simplified expression for the number of orbits of the elements of a finite set, under a cyclic group of permutations of it, is deduced in the following corollary of Theorem 13.2.

COROLLARY 13.1
The number of orbits of the elements of a finite set W_n, under a cyclic group $\mathcal{G}_{n,r}$
of permutations of it, is given by

$$N(\mathcal{T}) = \frac{1}{r} \sum_{d|r} k_1(\sigma^d)\phi(r/d), \tag{13.5}$$

where $k_1(\sigma^d)$ is the number of cycles of length 1 of a permutation σ^d and $\phi(r/d)$
is Euler's function.

PROOF Consider a divisor d of the order r of the cyclic group of permutations $\mathcal{G}_{n,r} = \{\varepsilon, \sigma, \sigma^2, \dots, \sigma^{r-1}\}$ and let $\mathcal{H}_{n,r/d} = \{\sigma^j \in \mathcal{G}_{n,r} : j/d \text{ a positive integer}$ relatively prime to $r/d\}$. Then, if s is a divisor of r different from d, the sets $\mathcal{H}_{n,r/s}$ and $\mathcal{H}_{n,r/d}$ are disjoint. Also, every permutation σ^j that belongs to the cyclic group of permutations $\mathcal{G}_{n,r}$ is contained in one of the sets $\mathcal{H}_{n,r/d}$ and, specifically, in the set for which j/d is relatively prime to r/d. Therefore, the sets $\mathcal{H}_{n,r/d}$ for all divisors d of r constitute a partition of the cyclic group of permutations $\mathcal{G}_{n,r}$,

$$\mathcal{G}_{n,r} = \sum_{d|r} \mathcal{H}_{n,r/d},$$

and, since $N(\mathcal{G}_{n,r}) = r$, expression (13.4) may be written as

$$N(\mathcal{T}) = \frac{1}{r} \sum_{d|r} \sum_{\sigma^j \in \mathcal{H}_{n,r/d}} k_1(\sigma^j).$$

Note that the permutation σ^d is contained in the set $\mathcal{H}_{n,r/d}$ and, for any other permutation σ^j contained in $\mathcal{H}_{n,r/d}$, d is the greatest common divisor of j and r. Then, the permutations σ^j and σ^d generate the same cyclic subgroup of permutations $\mathcal{G}_{n,r/d}$ of order r/d and so there exist positive integers m and s such that $\sigma^j = \sigma^{dm}$ and $\sigma^d = \sigma^{js}$. Thus, if an element w of W_n is left fixed by the permutation σ^j, $\sigma^j(w) = w$, then

$$\sigma^d(w) = \sigma^{js}(w) = \sigma^{j(s-1)}(\sigma(w)) = \sigma^{j(s-1)}(w) = \cdots = \sigma^j(w) = w$$

and w is also left fixed by σ^d and inversely. Consequently, $k_1(\sigma^j) = k(\sigma^d)$ for every permutation σ^j contained in $\mathcal{H}_{n,r/d}$ and so

$$N(\mathcal{T}) = \frac{1}{r} \sum_{d|r} k_1(\sigma^d) N(\mathcal{H}_{n,r/d}).$$

Since the number $N(\mathcal{H}_{n,r/d})$ of permutations that belong to the set $\mathcal{H}_{n,r/d}$, which is the number of positive integers less than or equal and relatively prime to r/d, is equal to Euler's function $\phi(r/d)$ (see Example 4.4), the required expression (13.5) is established.   ∎

Example 13.8 Models of necklaces

Calculate the number $K_{r,s}$ of models of necklaces with r beads of s different colors when two necklaces are considered to be of the same model if they can be transformed to each other by rotation.

Let $W_n = \{w_1, w_2, \dots, w_n\}$ be the set of necklaces and consider a permutation σ_j of the n necklaces that rotates each necklace by $2\pi j/r$ degrees in the counterclockwise direction, $j = 1, 2, \dots, r$. Then, with $\sigma_1 = \sigma$, we have $\sigma_j = \sigma^j$, $j = 1, 2, \dots, r$, $\sigma^r = \varepsilon$, and the set $\mathcal{G}_{n,r} = \{\varepsilon, \sigma, \sigma^2, \dots, \sigma^{r-1}\}$ constitutes a cyclic subgroup of permutations of W_n of order r generated by σ. Since a permutation σ^d, with d a divisor of r, rotates any necklace by $2\pi d/r$ degrees, and as a result, any d consecutive beads are moved to the positions of the next d consecutive beads, a necklace is left invariant, under σ^d, if and only if any of the r/d blocks of d consecutive beads has the same color arrangement. In other words, a model of necklaces is completely determined once the color of d consecutive beads is specified and so $k_1(\sigma^d) = s^d$. Hence, according to (13.5), the number $K_{r,s} = N(\mathcal{T})$ of models of necklaces is given by

$$K_{r,s} = \frac{1}{r} \sum_{d|r} \phi(r/d) s^d.$$

When the number r of the beads is a prime, this expression reduces to

$$K_{r,s} = \frac{(r-1)s + s^r}{r}.$$

Note that the number $K_{r,s}$ equals the number of cyclic r-permutations of the s colors with repetition. Using Table 4.1 of Euler's function, we get the following few values of the number $K_{r,s}$:

$$K_{2,s} = \frac{s(s+1)}{2}, \quad K_{3,s} = \frac{s(s^2+2)}{3}, \quad K_{4,s} = \frac{s(s^3+s+2)}{4},$$

$$K_{5,s} = \frac{s(s^4+4)}{5}, \quad K_{6,s} = \frac{s(s^5+s^2+2s+2)}{6},$$

$$K_{7,s} = \frac{s(s^6+6)}{7}, \quad K_{8,s} = \frac{s(s^7+s^3+2s+4)}{8}. \qquad □$$

13.4 MODELS OF COLORINGS OF A FINITE SET

Consider a group $\mathcal{G}_n \subseteq \mathcal{P}_n$ of permutations of a finite set $W_n = \{w_1, w_2, \ldots, w_n\}$ and a finite set of colors $Z_r = \{z_1, z_2, \ldots, z_r\}$. The notion of coloring the elements of the set W_n is introduced in the next definition.

DEFINITION 13.3   *A function f defined on the set W_n and taking values from the set Z_r is called a coloring (of the elements) of the set W_n with colors from the set Z_r, in which an element w_i of W_n is painted with color $f(w_i)$ from Z_r, $i = 1, 2, \ldots, n$.*

Note that a coloring f of the set W_n, with $f(w_i) = z_{s_i}$, $i = 1, 2, \ldots, n$, constitutes an n-permutation $(z_{s_1}, z_{s_2}, \ldots, z_{s_n})$ of the set Z_r with repetition. Thus, the set $\mathcal{F}$ of the functions f from W_n to Z_r, called *set of colorings* of the set W_n with colors from the set Z_r, contains $N(\mathcal{F}) = r^n$ functions (colorings). Further, if f is a coloring that belongs to $\mathcal{F}$ and σ is a permutation that belongs to $\mathcal{P}_n$, then $f(\sigma)$ is also a coloring that belongs to $\mathcal{F}$, in which the element w_i is painted with color $f(\sigma(w_i))$. Let f and g be two colorings that belong to $\mathcal{F}$. If there exists a permutation σ that belongs to the group $\mathcal{G}_n$ such that $f(\sigma) = g$, the coloring g is called *equivalent* to the coloring f, under the group $\mathcal{G}_n$, and is denoted by $g \equiv f$. Clearly, this relation is (a) *reflexive*: $f \equiv f$, (b) *symmetric*: $g \equiv f$ implies $f \equiv g$ and (c) *transitive*: $g \equiv h$ and $h \equiv f$ imply $g \equiv f$ and so it is an *equivalence relation*. Using this equivalence relation, the notion of a model of colorings is introduced in the following definition.

DEFINITION 13.4   *The set $\mathcal{M}_f = \{g \in \mathcal{F} : f(\sigma) = g, \ \sigma \in \mathcal{G}_n\}$, $f \in \mathcal{F}$, which contains the colorings in $\mathcal{F}$ that are equivalent to a coloring f, under the group $\mathcal{G}_n$, is called model of f, with respect to the group $\mathcal{G}_n$.*

Note that, for any two different colorings f and g in $\mathcal{F}$, the models $\mathcal{M}_f$ and $\mathcal{M}_g$ either coincide (when the colorings f and g are equivalent under $\mathcal{G}_n$) or are disjoint. Thus the set $\mathcal{M}_{\mathcal{F}} = \{\mathcal{M}_{f_1}, \mathcal{M}_{f_2}, \ldots, \mathcal{M}_{f_k}\}$ of the different models of colorings, under the group $\mathcal{G}_n$, constitutes a partition of $\mathcal{F}$.

As regards the number of models of coloring the elements of a finite set, we first prove the following theorem.

THEOREM 13.3

Let $\mathcal{F}$ be the set of colorings (of the elements) of a finite set W_n with colors from a finite set Z_r. Then the number $N_n(j_1, j_2, \ldots, j_r)$ of models in $\mathcal{M}_{\mathcal{F}}$, under the *group $\mathcal{G}_n$ of permutations of W_n, in which $j_s \geq 0$ elements of W_n are painted with color z_s from Z_r, $s = 1, 2, \ldots, r$, with $j_1 + j_2 + \cdots + j_r = n$, is given by*

$$N_n(j_1, j_2, \ldots, j_r)$$
$$= \frac{1}{N(\mathcal{G}_n)} \sum c(k_1, k_2, \ldots, k_n; \mathcal{G}_n) N_n(j_1, j_2, \ldots, j_r; k_1, k_2, \ldots, k_n),$$
$$\tag{13.6}$$

where the summation is extended over all nonnegative integer solutions $(k_1, k_2, \ldots, k_n)$ of the equation $k_1 + 2k_2 + \cdots + nk_n = n$, $c(k_1, k_2, \ldots, k_n; \mathcal{G}_n)$ is the number of permutations of the type $[k_1, k_2, \ldots, k_n]$ that belong to the group $\mathcal{G}_n$ and $N_n(j_1, j_2, \ldots, j_r; k_1, k_2, \ldots, k_n)$ is the number of colorings in $\mathcal{F}$ for which $j_s \geq 0$ elements of W_n are painted with color z_s from Z_r, $s = 1, 2, \ldots, r$ and are constant on every cycle of the permutations of the type $[k_1, k_2, \ldots, k_n]$ that belong to the group $\mathcal{G}_n$.

PROOF   Let $\mathcal{F}_{j_1,j_2,\ldots,j_r} \subseteq \mathcal{F}$ be the set of colorings in which $j_s \geq 0$ elements of W_n are painted with color z_s from Z_r for $s = 1, 2, \ldots, r$ and let $\mathcal{P}(\mathcal{F}_{j_1,j_2,\ldots,j_r})$ be the symmetric group of permutations of $\mathcal{F}_{j_1,j_2,\ldots,j_r}$. A subgroup of $\mathcal{P}(\mathcal{F}_{j_1,j_2,\ldots,j_r})$ can be constructed as follows. Consider a permutation σ in $\mathcal{G}_n$ and let $\bar{\sigma}$ be a mapping of $\mathcal{F}_{j_1,j_2,\ldots,j_r}$ into itself defined by $\bar{\sigma}(f) = f(\sigma)$. If $f_1 \neq f_2$, then $f_1(\sigma) \neq f_2(\sigma)$, $\bar{\sigma}(f_1) \neq \bar{\sigma}(f_1)$ and so $\bar{\sigma}$ is a one-to-one mapping of $\mathcal{F}_{j_1,j_2,\ldots,j_r}$ onto itself. Also, since the set $\mathcal{F}_{j_1,j_2,\ldots,j_r}$ is finite, $\bar{\sigma}$ is a permutation that belongs to the set $\mathcal{P}(\mathcal{F}_{j_1,j_2,\ldots,j_r})$. The set $\bar{\mathcal{G}}$ of all permutations $\bar{\sigma}$ for σ in $\mathcal{G}_n$, which is a subset of $\mathcal{P}(\mathcal{F}_{j_1,j_2,\ldots,j_r})$, is closed, with respect to the multiplication of permutations. Indeed, if $\bar{\sigma}$ and $\bar{\tau}$ belong to $\bar{\mathcal{G}}$ and $\bar{\rho} = \bar{\sigma}\bar{\tau}$, then $\bar{\sigma}(f) = f(\sigma)$, $\bar{\tau}(f) = f(\tau)$ and $\bar{\rho}(f) = \bar{\sigma}(\bar{\tau}(f)) = \bar{\sigma}(f(\tau)) = f(\tau(\sigma)) = f(\rho)$, with $\rho = \sigma\tau$ a permutation in $\mathcal{G}_n$, whence $\bar{\rho} = \bar{\sigma}\bar{\tau}$ belongs to $\bar{\mathcal{G}}$. Since $\bar{\mathcal{G}}$ is a subset of the symmetric group of permutations $\mathcal{P}(\mathcal{F}_{j_1,j_2,\ldots,j_r})$, it is a subgroup of it. Further, the correspondence of the permutation $\bar{\sigma}$ in $\bar{\mathcal{G}}$ to the permutation σ in $\mathcal{G}_n$ is one-to-one since, for $\sigma \neq \tau$, $f(\sigma) \neq f(\tau)$ and so $\bar{\sigma}(f) \neq \bar{\tau}(f)$, whence $\bar{\sigma} \neq \bar{\tau}$ and inversely. Clearly, this is a correspondence of $\bar{\mathcal{G}}$ onto $\mathcal{G}_n$ and so $N(\bar{\mathcal{G}}) = N(\mathcal{G}_n)$.

Let f be any coloring in $\mathcal{F}_{j_1,j_2,\ldots,j_r}$ and consider its orbit T_f, under the group $\bar{\mathcal{G}}$, and its model $\mathcal{M}_f$, under the group $\mathcal{G}_n$. Then, $T_f = \{g \in \mathcal{F}_{j_1,j_2,\ldots,j_r} : \bar{\sigma}(f) = g, \ \bar{\sigma} \in \mathcal{G}\} = \{g \in \mathcal{F}_{j_1,j_2,\ldots,j_r} : f(\sigma) = g, \ \sigma \in \mathcal{G}_n\} = \mathcal{M}_f$ and so the number $N_n(j_1, j_2, \ldots, j_r)$ of models in $\mathcal{M}_{\mathcal{F}}$, under the group $\mathcal{G}_n$, equals the number $N(\mathcal{T})$, of orbits of the elements of $\mathcal{F}_{j_1,j_2,\ldots,j_r}$, under the group $\bar{\mathcal{G}}$. Thus, according to Theorem 13.2 and since $N(\bar{\mathcal{G}}) = N(\mathcal{G}_n)$, we deduce the expression

$$N_n(j_1, j_2, \ldots, j_r) = \frac{1}{N(\mathcal{G}_n)} \sum_{\bar{\sigma} \in \bar{\mathcal{G}}} k_1(\bar{\sigma}),$$

where $k_1(\bar{\sigma})$ is the number of colorings in $\mathcal{F}_{j_1,j_2,\ldots,j_r}$ which are kept fixed by the permutation $\bar{\sigma}$. Introducing the set $\mathcal{S}_{j_1,j_2,\ldots,j_r}(\sigma) = \{f \in \mathcal{F}_{j_1,j_2,\ldots,j_r} : f(\sigma) = $

$f\}$ for σ in $\mathcal{G}_n$ and using Kronecker's delta function, $\delta(f, g) = 0$ for $f \neq g$ and $\delta(f, f) = 1$, the sum in the right-hand side of the last expression may be written as

$$\sum_{\bar{\sigma} \in \bar{\mathcal{G}}} k_1(\bar{\sigma}) = \sum_{\bar{\sigma} \in \bar{\mathcal{G}}} \sum_{f \in \mathcal{F}_{j_1, j_2, \dots, j_r}} \delta(f, \bar{\sigma}(f))$$

$$= \sum_{\sigma \in \mathcal{G}_n} \sum_{f \in \mathcal{F}_{j_1, j_2, \dots, j_r}} \delta(f, f(\sigma)) = \sum_{\sigma \in \mathcal{G}_n} N(\mathcal{S}_{j_1, j_2, \dots, j_r}(\sigma))$$

and so

$$N_n(j_1, j_2, \dots, j_r) = \frac{1}{N(\mathcal{G}_n)} \sum_{\sigma \in \mathcal{G}_n} N(\mathcal{S}_{j_1, j_2, \dots, j_r}(\sigma)). \qquad (13.7)$$

Further, note that, if w_i is an element of W_n and $\{w_i, \sigma(w_i), \sigma^2(w_i), \dots, \sigma^{m-1}(w_i)\}$, with $\sigma^m(w_i) = w_i$, is its orbit, under the permutation σ, then for any coloring f in $\mathcal{S}_{j_1, j_2, \dots, j_r}(\sigma)$ it holds $f(\sigma^j(w_i)) = f(\sigma(\sigma^{j-1}(w_i))) = f(\sigma^{j-1}(w_i)) = \dots = f(w_i)$, $j = 1, 2, \dots, m - 1$. Hence, every coloring f in $\mathcal{S}_{j_1, j_2, \dots, j_r}(\sigma)$ is constant on every cycle of the permutation σ (painting with the color $f(w_i) = z_{s_i}$ the elements of the cycle of the permutation σ in which w_i is contained). Inversely, every coloring f in $\mathcal{F}_{j_1, j_2, \dots, j_r}(\sigma)$ that is constant on every cycle of the permutation σ belongs to $\mathcal{S}_{j_1, j_2, \dots, j_r}(\sigma)$ since every element w_i of W_n is contained in a cycle of σ in which the element $\sigma(w_i)$ of W_n is also contained and so $f(w_i) = f(\sigma(w_i))$. Consequently, the number $N(\mathcal{S}_{j_1, j_2, \dots, j_r}(\sigma))$ equals the number of colorings in $\mathcal{F}_{j_1, j_2, \dots, j_r}(\sigma)$ that are constant on every cycle of the permutation σ. Partitioning the group $\mathcal{G}_n$, according to the permutations type,

$$\mathcal{G}_n = \sum \mathcal{G}_{n, [k_1, k_2, \dots, k_n]},$$

where the summation is extended over all nonnegative integer solutions $(k_1, k_2, \dots, k_n)$ of the equation $k_1 + 2k_2 + \dots + nk_n = n$ and $\mathcal{G}_{n[k_1, k_2, \dots, k_n]}$ includes all the permutations of the type $[k_1, k_2, \dots, k_n]$ that belong to the group $\mathcal{G}_n$, we deduce from (13.7) the required expression (13.6). $\blacksquare$

The expression (13.6) of the number $N_n(j_1, j_2, \dots, j_r)$ of models in $\mathcal{M}_{\mathcal{F}}$, under the group $\mathcal{G}_n$ of permutations of W_n, in which $j_s \geq 0$ elements of W_n are painted with color z_s from Z_r, $s = 1, 2, \dots, r$, with $j_1 + j_2 + \dots + j_r = n$, is unwieldy. A multivariate generating function of sequence of these numbers, which in general constitutes a powerful tool, is more manageable. In order to construct such a generating function let us assign to each element z_s of the set of colors $Z_r = \{z_1, z_2, \dots, z_r\}$ its *weight* $w(z_s)$, $s = 1, 2, \dots, r$. The *weight* of a coloring f that belongs to $\mathcal{F}$, denoted by $W(f)$, is defined by

$$W(f) = w(f(w_1))w(f(w_2)) \cdots w(f(w_n)).$$

When in this coloring $j_s \geq 0$ elements of W_n are painted with the color z_s, $s = 1, 2, \ldots, r$, its weight takes the form

$$W(f) = [w(z_1)]^{j_1} [w(z_2)]^{j_2} \cdots [w(z_r)]^{j_r}.$$

The multivariate generating function of the numbers $N_n(j_1, j_2, \ldots, j_r)$, $j_s \geq 0$, $s = 1, 2, \ldots, r$, with $j_1 + j_2 + \cdots + j_r = n$, is expressed in terms of the corresponding cycle indicator in the following *Pólya counting theorem*.

THEOREM 13.4
The multivariate generating function of the numbers $N_n(j_1, j_2, \ldots, j_r)$ of models in $\mathcal{M}_{\mathcal{F}}$, under the group $\mathcal{G}_n$ of permutations of W_n, in which $j_s \geq 0$ elements of W_n are painted with color z_s from Z_r, $s = 1, 2, \ldots, r$, with $j_1 + j_2 + \cdots + j_r = n$, is given by

$$\sum N_n(j_1, j_2, \ldots, j_r) t_1^{j_1} t_2^{j_2} \cdots t_r^{j_r} = C(x_1, x_2, \ldots, x_n; \mathcal{G}_n), \quad (13.8)$$

where

$$x_j = x_j(t_1, t_2, \ldots, t_r) = \sum_{s=1}^{r} t_s^j, \quad j = 1, 2, \ldots, n, \quad (13.9)$$

and $C(x_1, x_2, \ldots, x_n; \mathcal{G}_n)$ is the cycle indicator of the group $\mathcal{G}_n$ of permutations of W_n.

PROOF   Expression (13.6), of the numbers $N_n(j_1, j_2, \ldots, j_r)$, $j_s \geq 0$, $s = 1, 2, \ldots, r$, with $j_1 + j_2 + \cdots + j_r = n$, suggests first deriving the generating function

$$g_{k_1, k_2, \ldots, k_n}(t_1, t_2, \ldots, t_r) = \sum N_n(j_1, j_2, \ldots, j_r; k_1, k_2, \ldots, k_n) t_1^{j_1} t_2^{j_2} \cdots t_r^{j_r},$$

in terms of which the required generating function may easily be expressed. Thus, consider a permutation σ of the type $[k_1, k_2, \ldots, k_n]$ that belongs to the group $\mathcal{G}_n$ of permutations of W_n. The set $\{A_1, A_2, \ldots, A_k\}$ of the subsets of W_n that contain the elements of the $k_1 + k_2 + \cdots + k_n = k$ cycles of the permutation σ constitutes a partition of W_n and so, setting $n_i = N(A_i)$, $i = 1, 2, \ldots, k$, we have $n_1 + n_2 + \cdots + n_k = n$. Let $\mathcal{S}(\sigma)$ be the subset of colorings f in $\mathcal{F}$ that are constant on any cycle of the permutation σ, painting with the color $f(a_i) = z_{s_i}$ the n_i elements of the set A_i in which a_i belongs, $i = 1, 2, \ldots, k$. Note that a coloring f in $\mathcal{S}(\sigma)$ is completely determined by the n-permutation $u \in \mathcal{U}_n(Z_r)$, of the set of colors $Z_r = \{z_1, z_2, \ldots, z_r\}$ with repetition, in which the color z_{s_i} appears n_i times, for $i = 1, 2, \ldots, k$, with $n_1 + n_2 + \cdots + n_k = n$. Introducing the weights of the colors by $w(z_s) = t_s$, $s = 1, 2, \ldots, r$, the generating function $g_{k_1, k_2, \ldots, k_n}(t_1, t_2, \ldots, t_r)$ is expressed as

$$g_{k_1, k_2, \ldots, k_n}(t_1, t_2, \ldots, t_r) = \sum_{f \in S(\sigma)} W(f)$$

and since

$$
\sum_{f \in S(\sigma)} W(f) = \sum_{f \in S(\sigma)} w(f(w_1)) w(f(w_2)) \cdots w(f(w_n))
$$

$$
= \sum_{f \in S(\sigma)} [w(f(a_1))]^{n_1} [w(f(a_2))]^{n_2} \cdots [w(f(a_k))]^{n_k}
$$

$$
= \sum_{u \in \mathcal{U}_k(Z_r)} [w(z_{s_1})]^{n_1} [w(z_{s_2})]^{n_2} \cdots [w(z_{s_k})]^{n_k}
$$

$$
= \left(\sum_{s=1}^{r} [w(z_s)]^{n_1} \right) \left(\sum_{s=1}^{r} [w(z_s)]^{n_2} \right) \cdots \left(\sum_{s=1}^{r} [w(z_s)]^{n_k} \right),
$$

it reduces to

$$
g_{k_1, k_2, \ldots, k_n}(t_1, t_2, \ldots, t_r) = \left(\sum_{s=1}^{r} t_s^{n_1} \right) \left(\sum_{s=1}^{r} t_s^{n_2} \right) \cdots \left(\sum_{s=1}^{r} t_s^{n_k} \right).
$$

Note that $k_i \geq 0$ among the exponents $n_1, n_2, \ldots, n_k$ are equal to i, for $i = 1, 2, \ldots, n$, with $k_1 + 2k_2 + \cdots + nk_n = n$, and so

$$
g_{k_1, k_2, \ldots, k_n}(t_1, t_2, \ldots, t_r) = \left(\sum_{s=1}^{r} t_s \right)^{k_1} \left(\sum_{s=1}^{r} t_s^2 \right)^{k_2} \cdots \left(\sum_{s=1}^{r} t_s^n \right)^{k_n}.
$$

Now, multiplying both members of expression (13.6) by $t_1^{j_1} t_2^{j_2} \cdots t_r^{j_r}$ and summing for $j_s = 0, 1, \ldots, s = 1, 2, \ldots, r$, with $j_1 + j_2 + \cdots + j_r = n$, we get

$$
\sum N_n(j_1, j_2, \ldots, j_r) t_1^{t_1} t_2^{j_2} \cdots t_r^{j_r}
$$

$$
= \frac{1}{N(\mathcal{G}_n)} \sum c(k_1, k_2, \ldots, k_n; \mathcal{G}_n) g_{k_1, k_2, \ldots, k_n}(t_1, t_2, \ldots, t_r),
$$

where in the second sum the summation is extended over all partitions of n, that is, over all nonnegative integer solutions $(k_1, k_2, \ldots, k_n)$ of the equation $k_1 + 2k_2 + \cdots + nk_n = n$. Introducing into it the last expression of $g_{k_1, k_2, \ldots, k_n}(t_1, t_2, \ldots, t_r)$ and using the cycle indicator (13.2) of the group of permutations $\mathcal{G}_n$, we deduce (13.8). ∎

The total number of models of coloring the elements of a finite set may be deduced from (13.8) by setting $t_s = 1$, $s = 1, 2, \ldots, r$, whence $x_j = r$, $j = 1, 2, \ldots, n$. This number is given in the following corollary of Theorem 13.4.

COROLLARY 13.2

Let $\mathcal{F}$ be the set of colorings (of the elements) of a finite set W_n with colors from a finite set Z_r. Then the number $N(\mathcal{M}_{\mathcal{F}})$ of models of $\mathcal{F}$, under the group $\mathcal{G}_n$ of *permutations of W_n, is given by*

$$N(\mathcal{M}_{\mathcal{F}}) = C(r, r, \ldots, r; \mathcal{G}_n), \qquad (13.10)$$

where $C(x_1, x_2, \ldots, x_n; \mathcal{G}_n)$ is the cycle indicator of the group $\mathcal{G}_n$ of permutations of W_n.

Example 13.9 Combinations with restricted repetition

The number of n-combinations of r with restricted repetition may be derived, by an application of Theorem 13.4, as follows.

Consider the sets $\mathcal{E}_n(Z_r)$ and $\mathcal{U}_n(Z_r)$ of n-combinations and n-permutations of the set $Z_r = \{z_1, z_2, \ldots, z_r\}$ with repetition, respectively. Let us correspond to each combination $X_n = \{z_{s_1}, z_{s_2}, \ldots, z_{s_n}\}$ that belongs to $\mathcal{E}_n(Z_r)$ the set of its permutations $\mathcal{P}_n(X_n)$, which is a subset of $\mathcal{U}_n(Z_r)$. Note that $\mathcal{P}_n(X_n)$ and $\mathcal{P}_n(Y_n)$ are disjoint sets for X_n and Y_n different combinations from $\mathcal{E}_n(Z_r)$ and also that each permutation from $\mathcal{U}_n(Z_r)$ belongs in one of the sets $\mathcal{P}_n(X_n)$, $X_n \in \mathcal{E}_n(Z_r)$. Thus, the set $\{\mathcal{P}_n(X_n) \subseteq \mathcal{U}_n(Z_r) : X_n \in \mathcal{E}_n(Z_r)\}$ constitutes a partition of the set $\mathcal{U}_n(Z_r)$. Further, each permutation $(z_{s_1}, z_{s_2}, \ldots, z_{s_n})$ that belongs to $\mathcal{U}_n(Z_r)$ completely determines a coloring f of the elements of the set $W_n = \{w_1, w_2, \ldots, w_n\}$ with colors from the set $Z_r = \{z_1, z_2, \ldots, z_r\}$, in which the element w_i of W_n is painted with the color $f(w_i) = z_{s_i}, i = 1, 2, \ldots, n$. Thus, the set $\mathcal{U}_n(Z_r)$ completely determines a set $\mathcal{F}$ of colorings of the set W_n with colors from the set Z_r.

Also, two colorings f and g in $\mathcal{F}$, which are respectively determined by the permutations $(z_{s_1}, z_{s_2}, \ldots, z_{s_n})$ and $(z_{s_{i_1}}, z_{s_{i_2}}, \ldots, z_{s_{i_n}})$ in $\mathcal{P}_n(X_n)$, are equivalent under the permutation σ in the symmetric group $\mathcal{P}_n = \mathcal{P}_n(W_n)$ with $\sigma(w_k) = w_{i_k}$, $k = 1, 2, \ldots, n$. Indeed, from $f(w_i) = z_{s_i}$ and $g(w_k) = z_{s_{i_k}}$, it follows that $f(\sigma(w_k)) = f(w_{i_k}) = g(w_k)$, $k = 1, 2, \ldots, n$, and so $f(\sigma) = g$. Hence the subset $\mathcal{P}_n(X_n)$ of $\mathcal{U}_n(Z_r)$ uniquely corresponds to the model $\mathcal{M}_f$.

Consequently, the number $E_n(j_1, j_2, \ldots, j_r)$ of n-combinations of the set Z_r with repetition that include $j_s \geq 0$ times the element z_s for $s = 1, 2, \ldots, r$, with $j_1 + j_2 + \cdots + j_r = n$, equals the number $N_n(j_1, j_2, \ldots, j_r)$ of models in $\mathcal{M}_{\mathcal{F}}$, under the symmetric group of permutations $\mathcal{P}_n = \mathcal{P}_n(W_n)$, in which $j_s \geq 0$ elements of W_n are painted with color z_s from Z_r for $s = 1, 2, \ldots, r$, with $j_1 + j_2 + \cdots + j_r = n$. The multivariate generating function of this sequence of numbers, according to Theorem 13.4 and using the cycle indicator of the symmetric group derived in Example 13.2, is given by

$$\sum E_n(j_1, j_2, \ldots, j_r) t_1^{j_1} t_2^{j_2} \cdots t_r^{j_r} = \frac{1}{n!} B_n(x_1, 1! x_2, \ldots, (n-1)! x_n),$$

where

$$x_j = x_j(t_1, t_2, \ldots, t_r) = \sum_{s=1}^{r} t_s^j, \quad j = 1, 2, \ldots, n,$$

and $B_n(x_1, x_2, \dots, x_n)$ is the exponential Bell partition polynomial.

The number $E(r, n)$ of n-combinations of r with repetition equals the number of models in $\mathcal{M}_{\mathcal{F}}$ under the symmetric group of permutations $\mathcal{P}_n = \mathcal{P}_n(W_n)$ and, according to Corollary 13.2, is given by

$$E(r, n) = \frac{1}{n!} B_n(r, 1!r, \dots, (n-1)!r).$$

The generating function of the exponential Bell partition polynomials (see Corollary 11.1),

$$\sum_{n=0}^{\infty} B_n(x_1, x_2, \dots, x_n) \frac{t^n}{n!} = \exp\left(\sum_{j=1}^{\infty} x_j \frac{t^j}{j!}\right),$$

for $x_j = (j-1)!r$ yields

$$\sum_{n=0}^{\infty} B_n(r, 1!r, \dots, (n-1)!r) \frac{t^n}{n!} = \exp[r \log(1-t)^{-1}] = (1-t)^{-r}.$$

Therefore

$$\sum_{n=0}^{\infty} B_n(r, 1!r, \dots, (n-1)!r) \frac{t^n}{n!} = \sum_{n=0}^{\infty} \binom{r+n-1}{n} t^n$$

and

$$E(r, n) = \frac{1}{n!} B_n(r, 1!r, \dots, (n-1)!r) = \binom{r+n-1}{n}. \qquad \square$$

Example 13.10 *Cyclic permutations with restricted repetition*

The number of cyclic n-permutations of r with restricted repetition may be derived, by using Theorem 13.4, as follows. Note that an n-permutation $(z_{s_1}, z_{s_2}, \dots, z_{s_n})$ of the set $Z_r = \{z_1, z_2, \dots, z_r\}$ with repetition, in which $z_{s_{i+1}}$ follows z_{s_i}, $i = 1, 2, \dots, n-1$, and z_{s_1} follows z_{s_n}, is called a cyclic n-permutation of r with repetition.

Consider the sets $\mathcal{K}_n(Z_r)$ and $\mathcal{U}_n(Z_r)$ of the cyclic and the linear n-permutations, respectively, of the set Z_r. Let us correspond to each cyclic permutation $c_n = (z_{s_1}, z_{s_2}, \dots, z_{s_n})$ that belongs to $\mathcal{K}_n(Z_r)$ the set of the corresponding linear permutations $\mathcal{L}_n(c_n)$, which is a subset of $\mathcal{U}_n(Z_n)$. Clearly, the set $\{\mathcal{L}_n(c_n) \subseteq \mathcal{U}_n(Z_r) : c_n \in \mathcal{K}_n(Z_r)\}$ constitutes a partition of the set $\mathcal{U}_n(Z_r)$. Further, each (linear) permutation $(z_{s_1}, z_{s_2}, \dots, z_{s_n})$ that belongs to $\mathcal{U}_n(Z_r)$ completely determines a coloring f of the elements of the set $W_n = \{w_1, w_2, \dots, w_n\}$ with colors from the set $Z_r = \{z_1, z_2, \dots, z_r\}$, in which the element w_i of W_n is painted with the color $f(w_i) = z_{s_i}$, $i = 1, 2, \dots, n$. Thus, the set $\mathcal{U}_n(Z_r)$ completely determines a set $\mathcal{F}$ of colorings of the set W_n with colors from the set Z_r.

Also, two colorings f and g in $\mathcal{F}$, which are respectively determined by the permutations $(z_{s_1}, z_{s_2}, \ldots, z_{s_n})$ and $(z_{s_{m+1}}, z_{s_{m+2}}, \ldots, z_{s_n}, z_{s_1}, \ldots, z_{s_m})$ in $\mathcal{L}_n(c_n)$, are equivalent under the permutation σ_m in the cyclic group $\mathcal{G}_{n,n}$, of order n and degree n, with $\sigma_m(w_i) = w_{m+i}$, $i = 1, 2, \ldots, n - m$ and $\sigma_m(w_i) = w_{m-n+i}$, $i = n - m + 1, n - m + 2, \ldots, n$. Indeed, from $f(w_i) = z_{s_i}$, $i = 1, 2, \ldots, n$ and $g(w_i) = z_{s_{m+i}}$, $i = 1, 2, \ldots, n - m$, $g(w_i) = z_{s_{m-n+i}}$, $i = n - m + 1, n - m + 2, \ldots, n$, it follows that $f(\sigma_m(w_i)) = f(w_{m+i}) = g(w_i)$, $i = 1, 2, \ldots, n - m$, $f(\sigma_m(w_i)) = f(w_{m-n+i}) = f(w_i)$, $i = n - m + 1$, $n - m + 2, \ldots, n$ and so $f(\sigma) = g$. Hence the subset $\mathcal{L}_n(c_n)$ of $\mathcal{U}_n(Z_r)$ uniquely corresponds to the model $\mathcal{M}_f$.

Consequently, the number $K_n(j_1, j_2, \ldots, j_r)$ of cyclic n-permutations of the set Z_r with repetition that include $j_s \geq 0$ times the element z_s for $s = 1, 2, \ldots, r$, with $j_1 + j_2 + \cdots + j_r = n$, equals the number $N_n(j_1, j_2, \ldots, j_r)$ of models in $\mathcal{M}_{\mathcal{F}}$, under the cyclic group $\mathcal{G}_{n,n}$, of order n and degree n, in which $j_s \geq 0$ elements of W_n are painted with color z_s from Z_r for $s = 1, 2, \ldots, r$, with $j_1 + j_2 + \cdots + j_r = n$. The multivariate generating function of this sequence of numbers, according to Theorem 13.4 and using the cycle indicator of the cyclic group of permutations derived in Example 13.4, is obtained as

$$\sum K_n(j_1, j_2, \ldots, j_r) t_1^{j_1} t_2^{j_2} \cdots t_r^{j_r} = \frac{1}{n} \sum_{d|n} \phi(d)(t_1^d + t_2^d + \cdots + t_r^d)^{n/d}$$

and so

$$K_n(j_1, j_2, \ldots, j_r) = \frac{1}{n} \sum_{d|j} \phi(d) \frac{(n/d)!}{(j_1/d)!(j_2/d)! \cdots (j_r/d)!},$$

where j is the greatest common divisor of the nonzero numbers among the r numbers $j_1, j_2, \ldots, j_r$ and $\phi(d)$ is Euler's function.

The number $K_{n,r}$ of cyclic n-permutations of r with repetition equals the number of models in $\mathcal{M}_{\mathcal{F}}$, under the cyclic group of permutations $\mathcal{G}_{n,n}$, of order n and degree n and, according to Corollary 13.2, is given by

$$K_{n,r} = \frac{1}{n} \sum_{d|n} \phi(d) r^{n/d}. \qquad \square$$

Example 13.11 Models of colorings of a cube with r colors

Calculate the number of models of coloring the six faces of a cube with r different colors, when two colorings are considered to be of the same model if they can be transformed to each other by rotation of the cube.

Let $W_6 = \{1, 2, 3, 4, 5, 6\}$ be the set of the six faces of a cube and $Z_r = \{z_1, z_2, \ldots, z_r\}$ the set of the r different colors. Note that the six faces of a usual cube are numbered so that $\{1, 6\}$, $\{2, 5\}$ and $\{3, 4\}$ are non-ordered pairs of opposite faces. Further, let $\mathcal{G}_6$ be the group of permutations of W_6 that are induced by rotations of the cube around its axes of symmetry. The cycle indicator of the group $\mathcal{G}_6$, derived in Example 13.5, is given by

$$C(x_1, x_2, \ldots, x_6; \mathcal{G}_6) = \frac{1}{24}(x_1^6 + 3x_1^2 x_2^2 + 6x_1^2 x_4 + 6x_2^3 + 8x_3^2).$$

The set $\mathcal{F}$ of colorings of the set $W_6 = \{1, 2, 3, 4, 5, 6\}$, of the six faces of a cube, with colors from the set Z_r is partitioned, under the group $\mathcal{G}_6$, into a set of models $\mathcal{M}_\mathcal{F}$. The number $C_{6,r} \equiv N(\mathcal{M}_\mathcal{F})$ of models of coloring the six faces of a cube with r colors, according to Corollary 13.2, is given by the cycle indicator $C(x_1, x_2, \ldots, x_6; \mathcal{G}_6)$ for $x_j = r, j = 1, 2, \ldots, 6$ and so

$$C_{6,r} = \frac{1}{24}(r^6 + 3r^4 + 12r^3 + 8r^2).$$

The following few values of the number $C_{6,r}$ may help to get an idea of its magnitude:

$$C_{6,2} = 10, \quad C_{6,3} = 57, \quad C_{6,4} = 234, \quad C_{6,5} = 800,$$

$$C_{6,6} = 2226, \quad C_{6,7} = 5390, \quad C_{6,8} = 11{,}712. \qquad \Box$$

13.5 BIBLIOGRAPHIC NOTES

W. S. Burnside, in 1897, essentially enumerated the orbits of the elements of a finite set, under a group of permutations of it. This result, known as Burnside's lemma, is included in Burnside (1911). The great Hungarian mathematician George Pólya provided fundamental techniques for counting equivalence classes. The famous Pólya's counting theorem was derived in G. Pólya (1937) and its combinatorial significance was demonstrated by J. Riordan (1958) and R. C. Read (1987). An elegant and useful generalization of it is given by N. G. DeBruijn (1964). The article of A. Tucker (1974) provides an elementary presentation of Pólya's counting method through examples. A rigorous treatment of this method can be found in C. Berge (1971) and L. Comtet (1974).

13.6 EXERCISES

1. Let $W_3 = \{1, 2, 3\}$ be the set of the corners of an equilateral triangle numbered circularly in the counterclockwise direction. (a) Determine the group $\mathcal{G}_3$, of permutations of W_3, that is generated by rotation or reflection of the triangle around its axes of symmetry and show that its cycle indicator is given by

$$C(x_1, x_2, x_3; \mathcal{G}_3) = \frac{1}{6}(x_1^3 + 3x_1 x_2 + 2x_3).$$

(b) Calculate the number of models of coloring the three corners of the triangle with a set of r colors when two colorings are considered to be of the same model if they can be transformed to each other by rotation or reflection of the triangle around any of its axes of symmetry. (c) Show that, if r is a positive integer, then the positive integer $r(r^2 + 3r + 2)$ is divisible by 6.

2. Let $W_4 = \{1, 2, 3, 4\}$ be the set of the four corners of a square. Determine the dihedral group $\mathcal{G}_4$, of permutations of W_4, that is generated by rotation or reflection of the square around its axes of symmetry and show that its cycle indicator is given by

$$C(x_1, x_2, x_3, x_4; \mathcal{G}_4) = \frac{1}{8}(x_1^4 + 2x_1^2 x_2 + 3x_2^2 + 2x_4).$$

3. (*Continuation*) Calculate (a) the number $K_4(j)$ of models of coloring the four corners of a square with j corners red and the other $4 - j$ corners black, for $j = 0, 1, 2, 3, 4$ and (b) the number $K_{4,r}$ of models of coloring the four corners of a square with a set of r colors when two colorings are considered to belong to the same model if they can be transformed to each other by rotation or reflection of the square around any of its axes of symmetry.

4. Let $W_4 = \{1, 2, 3, 4\}$ be the set of the sides of a pyramid (tetrahedron). The base bears number 1 and the other sides are numbered circularly 2, 3 and 4 in the counterclockwise direction. (a) Determine the group $\mathcal{G}_4$, of permutations of W_4, which is generated by rotation of the pyramid around the vertical axis that connects the apex with the center of the base and show that its cycle indicator is given by

$$C(x_1, x_2, x_3, x_4; \mathcal{G}_4) = \frac{1}{3}(x_1^4 + 2x_1 x_3).$$

(b) Calculate the number of models of coloring the four sides of the pyramid with a set of r colors when two colorings are considered to belong to the same model if they can be transformed to each other by rotation of the pyramid around its axis of symmetry.

5. (*Continuation*) Calculate (a) the number $K_4(j)$ of models of coloring the four sides of a pyramid with j sides red and the other $4 - j$ sides black, for $j = 0, 1, 2, 3, 4$ and (b) the number $K_4(1, 1, 1)$ of models of coloring the four sides of a pyramid with a set of four colors when in both cases two colorings are considered to belong to the same model if they can be transformed to each other by rotation of the pyramid around its axis of symmetry.

6. Let $W_4 = \{1, 2, 3, 4\}$ be the set of the sides of a regular tetrahedron. (a) Determine the group $\mathcal{G}_4$, of permutations of W_4, which is generated by rotation of the tetrahedron around its axes of symmetry and show that its cycle indicator is given by

$$C(x_1, x_2, x_3, x_4; \mathcal{G}_4) = \frac{1}{12}(x_1^4 + 8x_1 x_3 + 3x_2^2).$$

7. (*Continuation*) Calculate (a) the number $K_4(j)$ of models of coloring the four sides of a regular tetrahedron with j sides red and the other $4 - j$ sides black, for $j = 0, 1, 2, 3, 4$ and (b) the number $K_{4,r}$ of models of coloring the four sides of a regular tetrahedron with a set of r colors when, in both cases, two colorings are considered to belong to the same model if they can be transformed to each other by rotation of the tetrahedron around any of its axes of symmetry.

8. (*Continuation*) Calculate the number of models of coloring the four sides of a regular tetrahedron with four different colors, each side with a different color, when two colorings are considered to belong to the same model if they can be transformed to each other by rotation of the tetrahedron around any of its axes of symmetry.

9. Let $W_8 = \{1, 2, \ldots, 8\}$ be the set of the eight vertices of a cube. Determine the group $\mathcal{G}_8$, of permutations of W_8, that is generated by rotation of the cube around its axes of symmetry and show that its cycle indicator is given by

$$C(x_1, x_2, \ldots, x_8; \mathcal{G}_8) = \frac{1}{24}(x_1^8 + 8x_1^2 x_3^2 + 9x_2^4 + 6x_4^2).$$

10. (*Continuation*) Let $N_8(i, j, k)$ be the number of models of coloring the eight vertices of a cube with i red, j green and $k = 8 - i - j$ black vertices, when two colorings are considered to be of the same model if they can be transformed to each other by rotation of the cube. Determine the generating function

$$g_8(t, u, w) = \sum N_8(i, j, k) t^i u^j w^k$$

and show that $N_8(1, 3, 4) = 13$, $N_8(2, 2, 2) = 22$, $N_8(2, 3, 3) = 24$ and $N_{8,3} = \sum N_8(i, j, k) = 333$.

11. Let $W_{12} = \{1, 2, \ldots, 12\}$ be the set of the 12 edges of a cube. Determine the group $\mathcal{G}_{12}$, of permutations of W_{12}, that is generated by rotation of the cube around its axes of symmetry and show that its cycle indicator is given by

$$C(x_1, x_2, \ldots, x_{12}; \mathcal{G}_{12}) = \frac{1}{24}(x_1^{12} + 6x_1^2 x_2^5 + 3x_2^6 + 8x_3^4 + 6x_4^3).$$

12. (*Continuation*) Calculate (a) the number of models of coloring the 12 edges of a cube with 12 different colors, each edge with a different color and (b) the number of models of coloring the 12 edges of a cube with a set of r colors, when in both cases two colorings are considered to belong to the same model if they can be transformed to each other by rotation of the cube.

13. Let $W_6 = \{1, 2, \ldots, 6\}$ be the set of the six vertices of a regular octahedron. Determine the group $\mathcal{G}_6$, of permutations of W_6, that is generated by rotation of the octahedron around its axes of symmetry and show that its cycle indicator is given by

$$C(x_1, x_2, \ldots, x_6; \mathcal{G}_6) = \frac{1}{24}(x_1^6 + 6x_1^2 x_4 + 3x_1^2 x_2^2 + 6x_2^3 + 8x_3^2).$$

14. (*Continuation*) Calculate (a) the number $K_6(j)$ of models of coloring the six vertices of an octahedron with j vertices red and the other $6 - j$ vertices black, for $j = 0, 1, \ldots, 6$ and (b) the number $K_{6,r}$ of models of coloring the six vertices of an octahedron with a set of r colors when in both cases two colorings are considered to belong to the same model if they can be transformed to each other by rotation or reflection of the octahedron.

15. *Models of coloring a roulette.* The disc of roulette is divided into n equal sectors and can be rotated around its axis. Evaluate the number of models of coloring the n sectors of the disc of the roulette with a set of r colors.

16. (*Continuation*) Let $K_n(j)$ be the number of models of coloring the n sectors of the disc of roulette with j red and $n - j$ black sectors. Show that

$$K_n(j) = \frac{1}{n} \sum_{d \mid m_{j,n}} \phi(d) \binom{n/d}{j/d}, \quad j = 0, 1, \ldots, n, \quad n = 1, 2, \ldots,$$

where $m_{0,n} = n$, $m_{j,n}$, $j = 1, 2, \ldots, n$, $n = 1, 2, \ldots$, is the greatest common divisor of j and n, and $\phi(d)$ is Euler's function.

17*. Let $W_n = \{1, 2, \ldots, n\}$ be the set of the n vertices of a regular n-gon. Determine the dihedral group $\mathcal{G}_n$, of permutations of W_n, that is generated by rotation or reflection of the n-gon around its axes of symmetry and show that its cycle indicator is given by

$$C(x_1, x_2, \ldots, x_n; \mathcal{G}_n) = \frac{1}{2n} \sum_{d|n} \phi(d) x_d^{n/d} + \frac{1}{2} x_1 x_2^m,$$

for $n = 2m + 1$ and

$$C(x_1, x_2, \ldots, x_n; \mathcal{G}_n) = \frac{1}{2n} \sum_{d|n} \phi(d) x_d^{n/d} + \frac{1}{4} (x_1^2 x_2^{m-1} + x_2^m),$$

for $n = 2m$, where $\phi(d)$ is Euler's function.

18. *Models of necklaces.* Evaluate the number of models of necklaces with n beads of r different colors, when two necklaces are considered to be of the same model if they can be transformed to each other by rotation or reflection with respect to the diameter.

19*. Let $\mathcal{H}_n = \{h_1, h_2, \ldots, h_n\}$ be a finite group of order $N(\mathcal{H}_n) = n$ and consider a fixed point h in $\mathcal{H}_n$. Show that (a) the mapping σ_h of $\mathcal{H}_n$ into itself, defined by $\sigma_h(h_r) = hh_r$, $r = 1, 2, \ldots, n$, constitutes a permutation of the elements of $\mathcal{H}_n$, and (b) the group $\mathcal{G}_n$ of the permutations σ_h for all h in $\mathcal{H}_n$ is isomorphic to $\mathcal{H}_n$, that is, $\sigma_{gh} = \sigma_g \sigma_h$ for any g and h in $\mathcal{H}_n$. Also show that (c) the cycle indicator of the group $\mathcal{G}_n$ is given by

$$C(x_1, x_2, \ldots, x_n; \mathcal{G}_n) = \frac{1}{n} \sum_{d|n} N(\mathcal{H}_{n;d}) x_d^{n/d},$$

where $N(\mathcal{H}_{n;d})$ is the number of elements h in $\mathcal{H}_n$ of order d.

20*. *Partitions and compositions of integers.* Let $\mathcal{P}_{n,k}$ be the set of partitions of n into at most k parts and $\mathcal{C}_{n,k}$ the set of compositions (ordered partitions) of n into at most k parts. Let us correspond to each partition $R_k = \{r_1, r_2, \ldots, r_k\}$, $r_1 \geq r_2 \geq \cdots \geq r_k \geq 0$, that belongs to $\mathcal{P}_{n,k}$ the set of its permutations $\mathcal{P}_k(R_k)$, which is a subset of $\mathcal{C}_{n,k}$. The set $\{\mathcal{P}_k(R_k) : R_k \in \mathcal{P}_{n,k}\}$ constitutes a partition of the set $\mathcal{C}_{n,k}$. Using Pólya's counting theorem, show that the generating function of the sequence of numbers $P(n, k) = N(\mathcal{P}_{n,k})$, $n = 0, 1, \ldots$, for fixed k, is given by

$$F_k(t) = \sum_{n=0}^{\infty} P(n, k) t^n = C_k((1 - t)^{-1}, (1 - t^2)^{-1}, \ldots, (1 - t^k)^{-1}; \mathcal{P}_k),$$

where $C(x_1, x_2, \ldots, x_k; \mathcal{P}_k)$ is the cycle indicator of the symmetric group $\mathcal{P}_k$ of order $k!$ and conclude the identity

$$C_k((1-t)^{-1}, (1-t^2)^{-1}, \ldots, (1-t^k)^{-1}; \mathcal{P}_k) = \prod_{i=1}^{k}(1-t^i)^{-1}.$$

Chapter 14

RUNS OF PERMUTATIONS AND EULERIAN NUMBERS

14.1 INTRODUCTION

In the classical Newcomb's problem, cards are successively drawn one after the other from a deck of m numbered cards of a general specification. The cards are placed on a pile as long as the number x_j of the card drawn at the j-th drawing is greater than or equal to the number x_{j-1} of the card drawn in the previous drawing; a new pile is started whenever the number x_j is less than x_{j-1}, for $j = 1, 2, \ldots, m$. The enumeration of the number of different arrangements of the cards that lead to a given number of k piles is of interest. Clearly, this is a problem of counting the permutations of n kinds of elements, with $s_1, s_2, \ldots, s_n$ elements respectively, which include k non-decreasing sequences of consecutive elements.

This chapter is devoted to the enumeration of the permutations of the set $\{1, 2, \ldots, n\}$ that have k increasing sequences of consecutive elements and the n-permutations of the set $\{1, 2, \ldots, s\}$ with repetition that have k non-decreasing sequences of consecutive elements. The Eulerian and the Carlitz numbers, which enumerate these permutations, are presented in the first two sections.

14.2 EULERIAN NUMBERS

The n-th power may be expressed as a sum of binomials (binomial coefficients) of the n-th order. Specifically, we successively get the expressions

$$t^0 = \binom{t}{0}, \quad t^1 = \binom{t}{1}, \quad t^2 = t\frac{(t+1) + (t-1)}{2} = \binom{t+1}{2} + \binom{t}{2},$$

$$t^3 = \binom{t+1}{2} \frac{(t+2) + 2(t-1)}{3} + \binom{t}{2} \frac{2(t+1) + (t-1)}{3}$$

$$= \binom{t+2}{3} + 4\binom{t+1}{3} + \binom{t}{3}$$

and, generally,

$$t^n = \sum_{k=0}^{n} A(n,k)\binom{t+n-k}{n}, \quad n = 0, 1, \ldots . \tag{14.1}$$

Then, the following definition is introduced.

DEFINITION 14.1   *The coefficient $A(n,k)$ of expansion (14.1), of the n-th power into binomials of the n-th order, is called Eulerian number.*

Clearly, this definition implies $A(n,k) = 0$, $k > n$. The numbers $A(n,k)$, $k = 0, 1, \ldots, n$, $n = 0, 1, \ldots$, attributed to Euler are called Eulerian to distinguish them from the Euler numbers E_{2n} (see Exercise 18).

REMARK 14.1   Expansion (14.1), upon replacing t by $-t$ and using the relation

$$(-1)^n \binom{-t+n-k}{n} = \binom{t+k-1}{n},$$

is transformed to

$$t^n = \sum_{k=0}^{n} A(n,k)\binom{t+k-1}{n}, \quad n = 0, 1, \ldots,$$

or equivalently to

$$t^n = \sum_{j=0}^{n} A(n, n-j+1)\binom{t+n-j}{n}, \quad n = 0, 1, \ldots .$$

Consequently

$$A(n,k) = A(n, n-k+1), \quad k = 0, 1, \ldots, n, \quad n = 0, 1, \ldots .$$

The number

$$\bar{A}(n,k) = A(n+k+1, k+1), \quad k = 0, 1, \ldots, n, \quad n = 0, 1, \ldots, \tag{14.2}$$

which satisfies the symmetric property $\bar{A}(n,k) = \bar{A}(k,n)$ is called *symmetric Eulerian number*.   ∎

An explicit expression and a recurrence relation for the Eulerian numbers are derived in the following two theorems.

THEOREM 14.1
The Eulerian number $A(n,k)$, $k = 0, 1, \ldots, n$, $n = 0, 1, \ldots$, is given by the sum

$$A(n,k) = \sum_{r=0}^{k} (-1)^r \binom{n+1}{r} (k-r)^n. \tag{14.3}$$

PROOF   From expansion (14.1), replacing the variable k by j and putting $t = k - r$, for $r = 0, 1, \ldots, k$, we deduce the expression

$$(k-r)^n = \sum_{j=0}^{k-r} A(n,j) \binom{n+k-r-j}{n}, \quad r = 0, 1, \ldots, k,$$

which, upon using the relation

$$\binom{n+k-r-j}{n} = \binom{n+k-r-j}{k-r-j} = (-1)^{k-r-j} \binom{-n-1}{k-r-j},$$

may be written as

$$(k-r)^n = \sum_{j=0}^{k-r} A(n,j)(-1)^{k-r-j} \binom{-n-1}{k-r-j}, \quad r = 0, 1, \ldots, k.$$

Multiplying both sides of the last expression by $(-1)^r \binom{n+1}{r}$ and summing for $r = 0, 1, \ldots, k$, we get

$$\sum_{r=0}^{k} (-1)^r \binom{n+1}{r} (k-r)^n = \sum_{r=0}^{k} \sum_{j=0}^{k-r} A(n,j)(-1)^{k-j} \binom{n+1}{r} \binom{-n-1}{k-r-j}$$

$$= \sum_{j=0}^{k} A(n,j)(-1)^{k-j} \sum_{r=0}^{k-j} \binom{n+1}{r} \binom{-n-1}{k-j-r}$$

and since, by Cauchy's formula,

$$\sum_{r=0}^{k-j} \binom{n+1}{r} \binom{-n-1}{k-j-r} = \binom{0}{k-j} = \delta_{k,j},$$

expression (14.3) is deduced.   ∎

THEOREM 14.2
The Eulerian numbers $A(n, k)$, $k = 0, 1, \ldots, n$, $n = 0, 1, \ldots$, satisfy the triangular recurrence relation

$$A(n + 1, k) = kA(n, k) + (n - k + 2)A(n, k - 1), \qquad (14.4)$$

for $k = 1, 2, \ldots, n + 1$, $n = 0, 1, \ldots$, with initial conditions

$$A(0, 0) = 1, \quad A(n, 0) = 0, \quad n > 0, \quad A(n, k) = 0, \quad k > n.$$

PROOF   Expanding both members of the identity $t^{n+1} = tt^n$ into binomial coefficients, by the using (14.1),

$$\sum_{k=0}^{n+1} A(n + 1, k)\binom{t + n - k + 1}{n + 1} = \sum_{k=0}^{n} A(n, k)\binom{t + n - k}{n}t$$

and, since

$$\binom{t + n - k}{n}t = \binom{t + n - k}{n}\frac{k(t + n - k + 1) + (n - k + 1)(t - k)}{n + 1}$$

$$= k\binom{t + n - k + 1}{n + 1} + (n - k + 1)\binom{t + n - k}{n + 1},$$

we get the expression

$$\sum_{k=0}^{n+1} A(n + 1, k)\binom{t + n - k + 1}{n + 1} = \sum_{k=0}^{n} kA(n, k)\binom{t + n - k + 1}{n + 1}$$

$$+ \sum_{k=0}^{n}(n - k + 1)A(n, k)\binom{t + n - k}{n + 1}.$$

Equating the coefficients of binomials $\binom{t+n-k+1}{n+1}$ in both sides of this expression we deduce (14.4). The initial conditions follow directly from (14.1). Note that the initial conditions may be replaced by $A(n, 1) = 1$, $n > 0$, $A(n, k) = 0$, $k > n$, which can be checked in the applications.   ∎

The Eulerian numbers can be tabulated by using recurrence relation (14.4) and its initial conditions. Table 14.1 gives the numbers $A(n, k)$, $k = 1, 2, \ldots, n$, $n = 1, 2, \ldots, 9$.

The derivation of a bivariate generating function of the Eulerian numbers is facilitated by an expression of the *Eulerian polynomial,*

$$A_n(t) = \sum_{k=0}^{n} A(n, k)t^k, \quad n = 0, 1, \ldots, \qquad (14.5)$$

deduced in the following lemma.

Table 14.1 **Eulerian Numbers $A(n, k)$**

k / n	1	2	3	4	5	6	7	8	9
1	1								
2	1	1							
3	1	4	1						
4	1	11	11	1					
5	1	26	66	26	1				
6	1	57	302	302	57	1			
7	1	120	1191	2416	1191	120	1		
8	1	247	4293	15619	15619	4293	247	1	
9	1	502	14608	88234	156190	88234	14608	502	1

LEMMA 14.1

The Eulerian polynomial $A_n(t)$, $n = 0, 1, \ldots$, is alternatively expressed as

$$A_n(t) = (1 - t)^{n+1} \sum_{j=0}^{\infty} j^n t^j, \quad n = 0, 1, \ldots . \tag{14.6}$$

PROOF Expression (14.5) of the Eulerian polynomial, upon introducing the explicit expression (14.3) of the Eulerian numbers, may be written as

$$A_n(t) = \sum_{k=0}^{n} \sum_{r=0}^{k} (-1)^r \binom{n+1}{r} (k - r)^n t^k, \quad n = 0, 1; \ldots .$$

Note that the inner sum, in agreement with $A(n, k) = 0, k > n$, vanishes for $k = n + 1, n + 2, \ldots$, and so in the outer sum the summation may be extended to infinity without altering its value. Thus

$$A_n(t) = \sum_{k=0}^{\infty} \sum_{r=0}^{k} (-1)^r \binom{n+1}{r} (k - r)^n t^k$$

$$= \sum_{r=0}^{\infty} (-1)^r \binom{n+1}{r} t^r \sum_{k=r}^{\infty} (k - r)^n t^{k-r}$$

and, since

$$\sum_{r=0}^{\infty} (-1)^r \binom{n+1}{r} t^r = (1 - t)^{n+1},$$

expression (14.5) is deduced. ∎

THEOREM 14.3
*The bivariate generating function of the Eulerian numbers $A(n, k)$, $k = 0, 1$,
... , n, $n = 0, 1, \ldots$, is given by*

$$g(t, u) = \sum_{n=0}^{\infty} \sum_{k=0}^{n} A(n, k) t^k \frac{u^n}{n!} = \frac{1 - t}{1 - t e^{u(1-t)}}. \tag{14.7}$$

PROOF   The bivariate generating function of the Eulerian numbers, using
expression (14.6) of the Eulerian polynomials, may be expressed as

$$g(t, u) = \sum_{n=0}^{\infty} (1 - t)^{n+1} \sum_{j=0}^{\infty} j^n t^j \frac{u^n}{n!} = (1 - t) \sum_{j=0}^{\infty} t^j \sum_{n=0}^{\infty} \frac{[ju(1 - t)]^n}{n!}$$

and, since

$$\sum_{n=0}^{\infty} \frac{[ju(1 - t)]^n}{n!} = e^{ju(1-t)}, \quad \sum_{j=0}^{\infty} [t e^{u(1-t)}]^j = \frac{1}{1 - t e^{u(1-t)}},$$

expression (14.7) is readily deduced.   ∎

REMARK 14.2   The bivariate generating function of the shifted Eulerian numbers $A(n, k + 1)$, $k = 0, 1, \ldots, n - 1$, $n = 1, 2, \ldots$,

$$h(t, u) = 1 + \sum_{n=1}^{\infty} \sum_{k=0}^{n-1} A(n, k + 1) t^k \frac{u^n}{n!},$$

which emerges in their applications, is closely connected with generating function
(14.7). Specifically, since $A(n, 0) = 0$, $n > 0$, it follows that

$$h(t, u) = 1 + \frac{g(t, u) - 1}{t}$$

and so

$$h(t, u) = 1 + \sum_{n=1}^{\infty} \sum_{k=1}^{n} A(n, k) t^{k-1} \frac{u^n}{n!} = \frac{1 - t}{e^{-u(1-t)} - t}, \tag{14.8}$$

which is the required generating function.   ∎

The Eulerian numbers are connected with the Stirling numbers of the
second kind as shown in the following theorem.

THEOREM 14.4
The Eulerian numbers $A(n, k)$, $k = 0, 1, \ldots, n$, $n = 0, 1, \ldots$, are expressed in terms of the Stirling numbers of the second kind $S(n, k)$, $k = 0, 1, \ldots, n$, $n = 0, 1, \ldots$, by

$$A(n, k) = \sum_{r=0}^{k} (-1)^{k-r} \binom{n-r}{k-r} r! S(n, r) \tag{14.9}$$

and inversely

$$S(n, r) = \frac{1}{r!} \sum_{k=0}^{r} \binom{n-k}{r-k} A(n, k). \tag{14.10}$$

PROOF Expression (14.6) of the Eulerian polynomials, upon expanding the powers of j into factorials of it by using (8.3), may be written as

$$A_n(t) = (1-t)^{n+1} \sum_{j=0}^{\infty} \sum_{r=0}^{n} S(n, r)(j)_r t^j = (1-t)^{n+1} \sum_{r=0}^{n} r! S(n, r) \sum_{j=0}^{\infty} \binom{j}{r} t^j,$$

and, since

$$\sum_{j=0}^{\infty} \binom{j}{r} t^j = \sum_{j=r}^{\infty} \binom{j}{r} t^j = t^r \sum_{i=0}^{\infty} \binom{r+i}{i} t^i = t^r (1-t)^{-r-1},$$

it reduces to

$$A_n(t) = \sum_{r=0}^{n} r! S(n, r) t^r (1-t)^{n-r}. \tag{14.11}$$

Thus, by virtue of (14.5),

$$\sum_{k=0}^{n} A(n, k) t^k = \sum_{r=0}^{n} r! S(n, r) t^r \sum_{k=r}^{n} (-1)^{k-r} \binom{n-r}{k-r} t^{k-r}$$

$$= \sum_{k=0}^{n} \left\{ \sum_{r=0}^{k} (-1)^{k-r} \binom{n-r}{k-r} r! S(n, r) \right\} t^k.$$

Equating the coefficients of t^k in the first and last member of this relation we conclude (14.9). Again, from (14.11) and setting $u = t/(1-t)$, whence $t = u/(1+u)$, we get, by virtue of (14.5),

$$\sum_{r=0}^{n} r! S(n, r) u^r = \sum_{k=0}^{n} A(n, k) u^k (1+u)^{n-k}.$$

Thus

$$\sum_{r=0}^{n} r! S(n, r) u^r = \sum_{k=0}^{n} A(n, k) u^k \sum_{r=k}^{n} \binom{n-k}{r-k} u^{r-k}$$

$$= \sum_{r=0}^{n} \left\{ \sum_{k=0}^{r} \binom{n-k}{r-k} A(n, k) \right\} u^r$$

and, equating the coefficients of u^r in the first and last member of this relation. we conclude (14.10).   ∎

Example 14.1 Power series with coefficients powers of fixed order
 The first application of the Eulerian polynomials is in the expression of the power series with coefficients powers of fixed order. Specifically, solving (14.6) with respect to the power series, we get the expression

$$\sum_{j=0}^{\infty} j^n t^j = \frac{A_n(t)}{(1-t)^{n+1}}, \quad n = 0, 1, \ldots .$$

In particular, for $n = 1, 2, 3, 4$ and using Table 14.1, we get the expressions

$$\sum_{j=1}^{\infty} j t^j = \frac{t}{(1-t)^2}, \quad \sum_{j=1}^{\infty} j^2 t^j = \frac{t(1+t)}{(1-t)^2},$$

$$\sum_{j=1}^{\infty} j^3 t^j = \frac{t(1 + 4t + t^2)}{(1-t)^3}, \quad \sum_{j=1}^{\infty} j^4 t^j = \frac{t(1 + 11t + 11t^2 + t^3)}{(1-t)^4}.$$

A probabilistic problem that requires the evaluation of such a power series is the following. Consider a sequence of throws of a coin and assume that at any throw the probability of falling heads is p and so the probability of falling tails is $q = 1 - p$. Then the probability that j tails preceded the first appearance of heads is given by

$$p_j = pq^j, \quad j = 0, 1, \ldots .$$

This is the probability function of a *geometric distribution*. Calculate the moments $\mu'_r, r = 1, 2, \ldots$, of the sequence of probabilities $p_j, j = 0, 1, \ldots$.

The r-th order moment is given by

$$\mu'_r = \sum_{j=1}^{\infty} j^r p_j = p \sum_{j=1}^{\infty} j^r q^j, \quad r = 1, 2, \ldots$$

and so

$$\mu'_r = \frac{A_r(q)}{p^r}, \quad r = 1, 2, \ldots .$$

In particular, the mean and the variance are obtained as

$$\mu = \mu_1' = \frac{q}{p}, \quad \sigma^2 = \mu_2' - \mu^2 = \frac{q}{p^2}. \quad \square$$

Example 14.2 Sample moments

Consider a sample $\{i_1, i_2, \ldots, i_N\}$ of N observations from a population (set) of non-negative integers and let N_j be the frequency (number of appearances) of the number j for $j = 0, 1, \ldots, n$, so that $N_0 + N_1 + \cdots + N_n = N$. The r-th order sample moment

$$m_r = \frac{1}{N} \sum_{j=0}^{n} j^r N_j$$

can be expressed in terms of the successive partial sums,

$$N_{r,k} = \sum_{j=k}^{n} N_{r-1,j}, \quad r = 1, 2, \ldots, \quad k = 0, 1, \ldots, n,$$

where $N_{0,j} = N_j$, $j = 0, 1, \ldots, n$. Clearly, the double-index sequence $N_{r,k}$, $r = 0, 1, \ldots, k = 0, 1, \ldots, n$, satisfies the triangular recurrence relation

$$N_{r,k} = N_{r,k-1} - N_{r-1,k-1},$$

for $r = 1, 2, \ldots, k = 1, 2, \ldots, n$, with $N_{0,k} = N_k$, $k = 0, 1, \ldots, n$. Setting $r = i$, $k = j + 1$ and multiplying the resulting relation by

$$\binom{j+r-k-i+1}{r-i+1} = \binom{j+r-k-i}{r-i} + \binom{j+r-k-i+1}{r-i+1},$$

for $i = 1, 2, \ldots, r$, we get

$$\binom{j+r-k-i+1}{r-i+1} N_{i-1,j} = \binom{j+r-k-i}{r-i} N_{i,j} + \binom{j+r-k-i}{r-i+1} N_{i,j}$$
$$- \binom{j+r-k-i+1}{r-i+1} N_{i,j+1},$$

for $i = 1, 2, \ldots, r$. Summing this expression for $j = k, k+1, \ldots, n$ and since $\binom{r-i}{r-i+1} = 0$, $N_{i,n+1} = 0$, we get

$$\sum_{j=k}^{n} \binom{j+r-k-i+1}{r-i+1} N_{i-1,j} = \sum_{j=k}^{n} \binom{j+r-k-i}{r-i} N_{i,j}$$
$$+ \sum_{j=k+1}^{n} \binom{j+r-k-i}{r-i+1} N_{i,j} - \sum_{j=k}^{n-1} \binom{j+r-k-i+1}{r-i+1} N_{i,j+1},$$

for $i = 1, 2, \ldots, r$ and so

$$\sum_{j=k}^{n} \binom{j+r-k-i+1}{r-i+1} N_{i-1,j} = \sum_{j=k}^{n} \binom{j+r-k-i}{r-i} N_{i,j}, \quad i = 1, 2, \ldots, r.$$

Applying it successively, we get

$$\sum_{j=k}^{n} \binom{j+r-k-i}{r-i} N_{i,j} = \sum_{j=k}^{n} \binom{j+r-k}{r} N_{0,j} = \sum_{j=k}^{n} \binom{j+r-k}{r} N_{j}.$$

In particular, for $i = r$ we deduce the expression

$$N_{r+1,k} = \sum_{j=k}^{n} \binom{j+r-k}{r} N_{j}.$$

Using (14.1), the r-th order sample moment m_r, is expressed as

$$m_r = \frac{1}{N} \sum_{j=0}^{n} j^r N_j = \frac{1}{N} \sum_{j=0}^{n} \sum_{k=0}^{r} A(r,k) \binom{j+r-k}{r} N_j$$

$$= \frac{1}{N} \sum_{k=0}^{r} A(r,k) \sum_{j=k}^{n} \binom{j+r-k}{r} N_j$$

and so

$$m_r = \frac{1}{N} \sum_{k=0}^{r} A(r,k) N_{r+1,k}. \quad \square$$

14.3   CARLITZ NUMBERS

Consider the generalized binomial of t of order n and scale parameter s,

$$\binom{st}{n} = \frac{(st)_n}{n!} = \frac{st(st-1)\cdots(st-n+1)}{n!}, \quad n = 1, 2, \ldots, \quad \binom{st}{0} = 1,$$

with s a real number. It can be expressed as a sum of binomials of $t+n-k$ of order n for $k = 0, 1, \ldots, n$ with coefficients depending on the nonnegative integers n and k and the parameter s. Specifically, we get successively the expressions

$$\binom{st}{0} = \binom{t}{0} = 1, \quad \binom{st}{1} = \binom{s}{1}\binom{t}{1}, \quad \binom{st}{2} = \binom{s}{2}\binom{t+1}{2} + \binom{s+1}{2}\binom{t}{2}$$

and, generally,

$$\binom{st}{n} = \sum_{k=0}^{n} B(n, k; s) \binom{t + n - k}{n}, \quad n = 0, 1, \ldots \quad (14.12)$$

Then, the following definition is introduced.

DEFINITION 14.2   *The coefficient $B(n, k; s)$ of expansion (14.12), of the n-th order generalized binomial of t with scale parameter s into binomials of the n-th order, is called Carlitz number.*

Clearly, this definition implies $B(n, k; s) = 0$, $k > n$.

REMARK 14.3   Expansion (14.12), upon replacing t by $-t$ and s by $-s$ and using the relation

$$(-1)^n \binom{-t + n - k}{n} = \binom{t + k - 1}{n},$$

is transformed to

$$\binom{st}{n} = (-1)^n \sum_{k=0}^{n} B(n, k; -s) \binom{t + k - 1}{n}, \quad n = 0, 1, \ldots,$$

or, equivalently, to

$$\binom{st}{n} = (-1)^n \sum_{k=0}^{n} B(n, n - k + 1; -s) \binom{t + n - k}{n}, \quad n = 0, 1, \ldots.$$

Consequently,

$$B(n, k; s) = (-1)^n B(n, n - k + 1; -s) \quad (14.13)$$

for $k = 0, 1, \ldots, n$ and $n = 0, 1, \ldots$.   ∎

An explicit expression and a recurrence relation for the Carlitz numbers are derived in the following two theorems.

THEOREM 14.5
The Carlitz number $B(n, k; s)$, $k = 0, 1, \ldots, n$, $n = 0, 1, \ldots$, is given by the sum

$$B(n, k; s) = \sum_{r=0}^{k} (-1)^r \binom{n + 1}{r} \binom{s(k - r)}{n}. \quad (14.14)$$

PROOF   From expansion (14.12), replacing the variable k by j and putting $t = k - r$, for $r = 0, 1, \ldots, k$, we deduce the expression

$$\binom{s(k-r)}{n} = \sum_{j=0}^{k-r} B(n,j;s) \binom{n+k-r-j}{n}, \quad r = 0, 1, \ldots, k$$

which, upon using the relation

$$\binom{n+k-r-j}{n} = \binom{n+k-r-j}{k-r-j} = (-1)^{k-r-j} \binom{-n-1}{k-r-j},$$

may be written as

$$\binom{s(k-r)}{n} = \sum_{j=0}^{k-r} B(n,j;s)(-1)^{k-r-j} \binom{-n-1}{k-r-j}, \quad r = 0, 1, \ldots, k.$$

Multiplying both sides of the last expression by $(-1)^r \binom{n+1}{r}$ and summing for $r = 0, 1, \ldots, k$, we get

$$\sum_{r=0}^{k} (-1)^r \binom{n+1}{r} \binom{s(k-r)}{n}$$

$$= \sum_{r=0}^{k} \sum_{j=0}^{k-r} B(n,j;s)(-1)^{k-j} \binom{n+1}{r} \binom{-n-1}{k-r-j}$$

$$= \sum_{j=0}^{k} B(n,j;s)(-1)^{k-j} \sum_{r=0}^{k-j} \binom{n+1}{r} \binom{-n-1}{k-j-r}$$

and since, by Cauchy's formula,

$$\sum_{r=0}^{k-j} \binom{n+1}{r} \binom{-n-1}{k-j-r} = \binom{0}{k-j} = \delta_{k,j},$$

expression (14.14) is deduced.   ∎

THEOREM 14.6
The Carlitz numbers $B(n, k; s)$, $k = 0, 1, \ldots, n$, $n = 0, 1, \ldots$, satisfy the triangular recurrence relation

$$(n+1)B(n+1, k; s) = (sk - n)B(n, k; s)$$
$$+[s(n-k+2)+n]B(n, k-1; s), \quad (14.15)$$

for $k = 1, 2, \ldots, n+1$, $n = 0, 1, \ldots$, with initial conditions

$$B(0, 0; s) = 1, \quad B(n, 0; s) = 0, \quad n > 0, \quad B(n, k; s) = 0, \quad k > n.$$

PROOF   Expanding the generalized binomials in both sides of the identity

$$(n+1)\binom{st}{n+1} = (st-n)\binom{st}{n},$$

by using (14.12),

$$\sum_{k=0}^{n+1}(n+1)B(n+1,k;s)\binom{t+n-k+1}{n+1}$$

$$= \sum_{k=0}^{n}B(n,k;s)\binom{t+n-k}{n}(st-n)$$

and, since

$$\binom{t+n-k}{n}(st-n) = (sk-n)\binom{t+n-k+1}{n+1}$$

$$+[s(n-k+1)+n]\binom{t+n-k}{n+1},$$

we get the expression

$$\sum_{k=0}^{n+1}(n+1)B(n+1,k;s)\binom{t+n-k+1}{n+1}$$

$$= \sum_{k=0}^{n}(sk-n)B(n,k;s)\binom{t+n-k+1}{n+1}$$

$$+ \sum_{k=0}^{n}[s(n-k+1)+n]B(n,k;s)\binom{t+n-k}{n+1}.$$

Equating the coefficients of binomials $\binom{t+n-k+1}{n+1}$ in both sides of this expression we deduce (14.15). The initial conditions follow from (14.12).   ∎

The Carlitz numbers can be tabulated by using the recurrence relation (14.15) and its initial conditions. Table 14.2 gives the numbers $B(n,k;s)$, $k=1,2,\ldots,n$, $n=1,2,\ldots,4$.

The derivation of a bivariate generating function of the Carlitz numbers is facilitated by an expression of the *Carlitz polynomial*,

$$B_n(t;s) = \sum_{k=0}^{n}B(n,k;s)t^k, \quad n=0,1,\ldots, \tag{14.16}$$

deduced in the following lemma.

Table 14.2   *Carlitz Numbers $B(n, k; s)$*

$\ \ \ \ \ \ \ \ k$ n	1	2	3	4
1	$\binom{s}{1}$			
2	$\binom{s}{2}$	$\binom{s+1}{2}$		
3	$\binom{s}{3}$	$4\binom{s+1}{3}$	$\binom{s+2}{3}$	
4	$\binom{s}{4}$	$\binom{s+2}{4} + 10\binom{s+1}{4}$	$10\binom{s+2}{4} + \binom{s+1}{4}$	$\binom{s+3}{4}$

LEMMA 14.2

The Carlitz polynomial $B_n(t; s)$, $n = 0, 1, 2, \ldots$, is alternatively expressed as

$$B_n(t; s) = (1 - t)^{n+1} \sum_{j=0}^{\infty} \binom{sj}{n} t^j, \quad n = 0, 1, \ldots . \qquad (14.17)$$

PROOF   Introducing in polynomial (14.16) the explicit expression (14.14) of the Carlitz numbers, we get

$$B_n(t; s) = \sum_{k=0}^{n} \sum_{r=0}^{k} (-1)^r \binom{n+1}{r} \binom{s(k-r)}{n} t^k, \quad n = 0, 1, \ldots .$$

Note that the inner sum, in agreement with $B(n, k; s) = 0$, $k > n$, vanishes for $k = n + 1, n + 2, \ldots$ and so in the outer sum the summation may be extended to infinity without altering its value. Thus

$$B_n(t; s) = \sum_{k=0}^{\infty} \sum_{r=0}^{k} (-1)^r \binom{n+1}{r} \binom{s(k-r)}{n} t^k$$

$$= \sum_{r=0}^{\infty} (-1)^r \binom{n+1}{r} t^r \sum_{k=r}^{\infty} \binom{s(k-r)}{n} t^{k-r}$$

and, since

$$\sum_{r=0}^{\infty}(-1)^r\binom{n+1}{r}t^r = (1-t)^{n+1},$$

expression (14.17) is deduced.   ∎

THEOREM 14.7
The bivariate generating function of the Carlitz numbers $B(n,k;s)$, $k = 0,1,$
$\ldots,n$, $n = 0,1,\ldots$, *is given by*

$$g(t,u;s) = \sum_{n=0}^{\infty}\sum_{k=0}^{n}B(n,k;s)t^k u^n = \frac{1-t}{1-t[1+u(1-t)]^s}. \qquad (14.18)$$

PROOF   The bivariate generating function of the Carlitz numbers, using expression (14.17) of the polynomials (14.16), may be expressed as

$$g(t,u;s) = \sum_{n=0}^{\infty}(1-t)^{n+1}\sum_{j=0}^{\infty}\binom{sj}{n}t^j u^n$$

$$= (1-t)\sum_{j=0}^{\infty}t^j\sum_{n=0}^{\infty}\binom{sj}{n}[u(1-t)]^n$$

and, since

$$\sum_{n=0}^{\infty}\binom{sj}{n}[u(1-t)]^n = [1+u(1-t)]^{sj},$$

$$\sum_{j=0}^{\infty}(t[1+u(1-t)]^s)^j = \frac{1}{1-t[1+u(1-t)]^s},$$

expression (14.18) is readily deduced.   ∎

REMARK 14.4   The bivariate generating function of the shifted Carlitz numbers $B(n,k+1;s)$, $k = 0,1,\ldots,n-1$, $n = 1,2,\ldots$,

$$h(t,u;s) = 1 + \sum_{n=1}^{\infty}\sum_{k=0}^{n-1}B(n,k+1;s)t^k\frac{u^n}{n!},$$

which emerges in their applications, is closely connected with the generating function (14.18). Specifically, since $B(n,0;s) = 0$, $n > 0$, it follows that

$$h(t,u;s) = 1 + \frac{g(t,u;s)-1}{t}$$

and so

$$h(t, u; s) = 1 + \sum_{n=1}^{\infty} \sum_{k=1}^{n} B(n, k; s) t^{k-1} \frac{u^n}{n!} = \frac{1 - t}{[1 + u(1 - t)]^{-s} - t},$$

(14.19)

which is the required generating function. ∎

The Carlitz numbers are connected with the generalized factorial coefficients (see Section 8.4) as shown in the following theorem.

THEOREM 14.8
The Carlitz numbers $B(n, k; s)$, $k = 0, 1, \ldots, n$, $n = 0, 1, \ldots$, *are expressed in terms of the generalized factorial coefficients* $C(n, k; s)$, $k = 0, 1, \ldots, n$, $n = 0, 1, \ldots$, *by*

$$B(n, k; s) = \sum_{r=0}^{k} (-1)^{k-r} \frac{r!}{n!} \binom{n - r}{k - r} C(n, r; s)$$

(14.20)

and inversely

$$C(n, r; s) = \frac{n!}{r!} \sum_{k=0}^{r} \binom{n - k}{r - k} B(n, k; s).$$

(14.21)

PROOF   Expression (14.17) of the polynomial (14.16), upon expanding the generalized binomial of j of order n and scale parameter s into factorials of j by using (8.39), may be written as

$$B_n(t; s) = (1 - t)^{n+1} \sum_{j=0}^{\infty} \sum_{r=0}^{n} \frac{r!}{n!} C(n, r; s) \binom{j}{r} t^j$$

$$= (1 - t)^{n+1} \sum_{r=0}^{n} \frac{r!}{n!} C(n, r; s) \sum_{j=0}^{\infty} \binom{j}{r} t^j,$$

and, since

$$\sum_{j=0}^{\infty} \binom{j}{r} t^j = \sum_{j=r}^{\infty} \binom{j}{r} t^j = t^r \sum_{i=0}^{\infty} \binom{r + i}{i} t^i = t^r (1 - t)^{-r-1},$$

it reduces to

$$B_n(t; s) = \sum_{r=0}^{n} \frac{r!}{n!} C(n, r; s) t^r (1 - t)^{n-r}.$$

(14.22)

Thus, by virtue of (14.16),

$$\sum_{k=0}^{n} B(n,k;s)t^k = \sum_{r=0}^{n} \frac{r!}{n!}C(n,r;s)t^r \sum_{k=r}^{n}(-1)^{k-r}\binom{n-r}{k-r}t^{k-r}$$

$$= \sum_{k=0}^{n}\left\{\sum_{r=0}^{k}(-1)^{k-r}\frac{r!}{n!}\binom{n-r}{k-r}C(n,r;s)\right\}t^k$$

and, equating the coefficients of t^k in the first and last member of this relation, we conclude (14.20). Again, from (14.22) and setting $u = t/(1-t)$, whence $t = u/(1+u)$, we get, by virtue of (14.16),

$$\sum_{r=0}^{n}\frac{r!}{n!}C(n,r;s)u^r = \sum_{k=0}^{n}B(n,k;s)u^k(1+u)^{n-k}.$$

Thus

$$\sum_{r=0}^{n}\frac{r!}{n!}C(n,r;s)u^r = \sum_{k=0}^{n}B(n,k;s)u^k\sum_{r=k}^{n}\binom{n-k}{r-k}u^{r-k}$$

$$= \sum_{r=0}^{n}\left\{\sum_{k=0}^{r}\binom{n-k}{r-k}B(n,k;s)\right\}u^r$$

and, equating the coefficients of u^r in the first and last member of this relation, we conclude (14.21). ∎

A limiting expression of the Carlitz numbers as $s \to \pm\infty$ is derived in the following theorem.

THEOREM 14.9
Let $B(n,k;s)$ be the Carlitz number. Then

$$\lim_{s\to\pm\infty} s^{-n}n!B(n,k;s) = A(n,k), \tag{14.23}$$

where $A(n,k)$ is the Eulerian number.

PROOF Multiplying both members of expression (14.14) by $s^{-n}n!$, taking the limit as $s \to \pm\infty$ and since $\lim_{s\to\pm\infty} s^{-n}(s(k-r))_n = (k-r)^n$, we get

$$\lim_{s\to\pm\infty} s^{-n}n!B(n,k;s) = \sum_{r=0}^{k}(-1)^r\binom{n+1}{r}[\lim_{s\to\pm\infty} s^{-n}(s(k-r))_n]$$

$$= \sum_{r=0}^{k}(-1)^r\binom{n+1}{r}(k-r)^n.$$

Thus, by virtue of (14.3), the limiting expression (14.23) is established. ∎

14.4 PERMUTATIONS WITH A GIVEN NUMBER OF RUNS

The permutations of a finite set can be classified according to the number and the length of the increasing (or decreasing) sequences of consecutive elements they include. In this respect, the following definition is introduced.

DEFINITION 14.3   *Let $(j_1, j_2, \ldots, j_n)$ be a permutation of the set $\{1, 2, \ldots, n\}$. The sequence of consecutive elements*

$$(j_m, j_{m+1}, \ldots, j_{m+r-1}), \quad \text{with} \quad j_m < j_{m+1} < \cdots < j_{m+r-1},$$

is called ascending run of length r, for $2 \le m \le n - 1$, $1 \le r \le n - 2$, if $j_{m-1} > j_m$ and $j_{m+r-1} > j_{m+r}$; in particular, for $m = 1$, $1 \le r \le n - 1$, if $j_r > j_{r+1}$, for $m = n - r + 1$, $1 \le r \le n - 1$, if $j_{n-r} > j_{n-r+1}$, and for $m = 1$, $r = n$, without any additional condition.

This sequence is called descending run of length r if the corresponding conditions are satisfied with the reverse inequalities.

Example 14.3

Consider the permutations (j_1, j_2, j_3, j_4) of the set $\{1, 2, 3, 4\}$. Clearly these 24 permutations can be classified according to the number of their ascending runs as follows:

(a) The permutation

$$(1, 2, 3, 4)$$

is the only one that has one ascending run of length 4.

(b) The permutations with two ascending runs are the following 11:

$$(1, 2, 4, 3), \; (1, 3, 2, 4), \; (1, 3, 4, 2), \; (1, 4, 2, 3), \; (2, 1, 3, 4), \; (2, 3, 1, 4),$$

$$(2, 3, 4, 1), \; (2, 4, 1, 3), \; (3, 1, 2, 4), \; (3, 4, 1, 2), \; (4, 1, 2, 3).$$

(c) The permutations with three ascending runs are the following 11:

$$(1, 4, 3, 2), \; (2, 1, 4, 3), \; (2, 4, 3, 1), \; (3, 1, 4, 2), \; (3, 2, 1, 4), \; (3, 2, 4, 1),$$

$$(3, 4, 2, 1), \; (4, 1, 3, 2), \; (4, 2, 1, 3), \; (4, 2, 3, 1), \; (4, 3, 1, 2).$$

(d) The permutation

$$(4, 3, 2, 1)$$

is the only one that has four ascending runs, each of length 1.   ▯

THEOREM 14.10

The number of permutations of the set $\{1, 2, \ldots, n\}$ with k ascending runs equals $A(n, k)$, the Eulerian number.

PROOF Note that, from any permutation of the set $\{1, 2, \ldots, n\}$, by attaching the element $n + 1$ in any of the $n + 1$ possible positions (one before the first element, $n - 1$ between the n elements and one after the last element) $n + 1$ permutations of the set $\{1, 2, \ldots, n+1\}$ are constructed. Further, this attachment either keeps unchanged or increases by one the number of runs. Specifically, (a) from each permutation of the set $\{1, 2, \ldots, n\}$ with k ascending runs, by attaching the element $n + 1$ after the last element of any of these runs, k permutations of the set $\{1, 2, \ldots, n + 1\}$ with k ascending runs are constructed. Also, (b) from each permutation of the set $\{1, 2, \ldots, n\}$ with $k - 1$ ascending runs, by attaching the element $n + 1$ in any of the possible positions except after the last element of any of these runs $(n + 1) - (k - 1) = n - k + 2$ permutations of the set $\{1, 2, \ldots, n+1\}$ with k ascending runs are constructed. Consequently, the number $a(n, k)$ of permutations of the set $\{1, 2, \ldots, n\}$ with k ascending runs satisfies the recurrence relation

$$a(n + 1, k) = ka(n, k) + (n - k + 2)a(n, k - 1),$$

for $k = 2, 3, \ldots, n + 1, n = 1, 2, \ldots$, with

$$a(n, 1) = 1, \quad n > 0, \quad a(n, k) = 0, \quad k > n.$$

Comparing this recurrence relation and its initial conditions with the recurrence relation (14.4) of the Eulerian numbers and its initial conditions, we conclude that $a(n, k) = A(n, k)$. ∎

REMARK 14.5 Consider a permutation $(j_1, j_2, \ldots, j_n)$ of the set $\{1, 2, \ldots, n\}$ with k ascending runs. Then, the permutation $(i_1, i_2, \ldots, i_n)$ of the set $\{1, 2, \ldots, n\}$, with $i_m = j_{n-m+1}, m = 1, 2, \ldots, n$, which is the reverse permutation, has k descending runs. Since this correspondence is one-to-one, it follows from Theorem 14.10 that the number of permutations of the set $\{1, 2, \ldots, n\}$ with k descending runs equals the Eulerian number $A(n, k)$. ∎

REMARK 14.6 *Rises and falls of permutations.* The ascending or descending runs of a permutation of the set $\{1, 2, \ldots, n\}$ are connected with its rises or falls. Specifically, consider a permutation $(j_1, j_2, \ldots, j_n)$ of the set $\{1, 2, \ldots, n\}$. The pair (j_r, j_{r+1}) is called a *rise* if $j_r < j_{r+1}$ and a *fall* if $j_r > j_{r+1}, r = 1, 2, \ldots, n - 1$. If the permutation $(j_1, j_2, \ldots, j_n)$ has k rises (or k falls) $\{(j_{r_1}, j_{r_1+1}), (j_{r_2}, j_{r_2+1}), \ldots, (j_{r_k}, j_{r_k+1})\}$, then it has $k + 1$ descending runs (or $k+1$ ascending runs) $\{(j_1, j_2, \ldots, j_{r_1}), (j_{r_1+1}, j_{r_2+2}, \ldots, j_{r_2}), (j_{r_k+1}, j_{r_k+2}, \ldots, j_n)\}$ and inversely. Hence, by Theorem 14.10, the number of permutations of the set $\{1, 2, \ldots, n\}$ with k rises (or k falls) equals the shifted Eulerian number $A(n, k + 1)$. ∎

Example 14.4

Consider an urn containing n numbered balls $\{1, 2, \ldots, n\}$ and suppose that the balls are randomly drawn one after the other, without replacement. If j_m is the number drawn at the m-th drawing for $m = 1, 2, \ldots, n$, then $(j_1, j_2, \ldots, j_n)$ represents a drawing of the n balls. Determine (a) the probability $p(k; n)$ that the drawing of the n balls has k ascending runs, for $k = 1, 2, \ldots, n$, and (b) the r-th factorial moment $\mu_{(r)}$ of the sequence of probabilities $p(k; n)$, $k = 1, 2, \ldots, n$, for $r = 1, 2, \ldots$.

(a) The number of permutations $(j_1, j_2, \ldots, j_n)$ of the set $\{1, 2, \ldots, n\}$ with k ascending runs, according to Theorem 14.10, is given by the Eulerian number $A(n, k)$. Thus, the probability $p(k; n)$ that the drawing of the n balls has k ascending runs, on using the Laplace's classical definition of probability, is given by

$$p(k; n) = \frac{A(n, k)}{n!}, \quad k = 1, 2, \ldots, n.$$

(b) The generating function of the sequence of probabilities $p(k; n)$, $k = 1, 2, \ldots, n$, by (14.5), is given by

$$P(t) = \sum_{k=1}^{n} p(k; n) t^k = \frac{A_n(t)}{n!}$$

and so the factorial moment generating function, $B(t) = P(t + 1)$, (see Section 6.4) is

$$B(t) = \sum_{r=0}^{\infty} \mu_{(r)} \frac{t^r}{r!} = \frac{A_n(t + 1)}{n!}.$$

Further, by (14.7), the exponential generating of the Eulerian polynomial $A_n(t)$, $n = 0, 1, \ldots$, may be written as

$$\sum_{n=0}^{\infty} A_n(t) \frac{u^n}{n!} = 1 - t + \frac{t}{1 - [e^{u(t-1)} - 1]/(1 - t)}.$$

Expanding it into a geometric series of $z = (e^{u(t-1)} - 1)/(1-t)$ and then expanding the powers of $e^{u(t-1)} - 1$ into a series of $u(t-1)$, by using the generating function of the Stirling numbers of the second kind, (8.17), we get

$$\sum_{n=0}^{\infty} A_n(t) \frac{u^n}{n!} = 1 + t \sum_{k=1}^{\infty} [e^{u(t-1)} - 1]^k (t - 1)^{-k}$$

$$= 1 + t \sum_{k=1}^{\infty} \sum_{n=k}^{\infty} k! S(n, k)(t - 1)^{n-k} \frac{u^n}{n!}$$

$$= 1 + \sum_{n=1}^{\infty} \left\{ t \sum_{k=1}^{n} k! S(n, k)(t - 1)^{n-k} \right\} \frac{u^n}{n!}$$

and so

$$A_n(t) = t \sum_{k=1}^{n} k! S(n, k)(t - 1)^{n-k}$$

$$= \sum_{k=0}^{n} k!\{(k + 1)S(n, k + 1) + S(n, k)\}(t - 1)^{n-k}.$$

Thus, on using the triangular recurrence relation of the Stirling numbers of the second kind, (8.27), we deduce the relation

$$A_n(t) = \sum_{k=0}^{n} k! S(n + 1, k + 1)(t - 1)^{n-k}$$

whence

$$B(t) = \sum_{r=0}^{\infty} \mu_{(r)} \frac{t^r}{r!} = \sum_{r=0}^{n} \frac{(n - r)!}{n!} S(n + 1, n - r + 1) t^r.$$

Consequently,

$$\mu_{(r)} = S(n + 1, n - r + 1) / \binom{n}{r}, \quad r = 1, 2, \ldots, n,$$

and $\mu_{(r)} = 0$ for $r > n$. ▯

14.5 PERMUTATIONS WITH REPETITION AND A GIVEN NUMBER OF RUNS

The permutations of a finite set with repetition can be classified according to the number and the length of the non-decreasing (or non-increasing) sequences of consecutive elements they include.

DEFINITION 14.4   *Let $(j_1, j_2, \ldots, j_n)$ be an n-permutation of the set $\{1, 2, \ldots, s\}$ with repetition. The sequence of consecutive elements*

$$(j_m, j_{m+1}, \ldots, j_{m+r-1}), \quad \text{with} \quad j_m \leq j_{m+1} \leq \cdots \leq j_{m+r-1},$$

is called non-descending run of length r, for $2 \leq m \leq n - 1$, $1 \leq r \leq n - 2$, if $j_{m-1} > j_m$ and $j_{m+r-1} > j_{m+r}$; in particular, for $m = 1$, $1 \leq r \leq n - 1$, if $j_r > j_{r+1}$, for $m = n - r + 1$, $1 \leq r \leq n - 1$, if $j_{n-r} > j_{n-r+1}$, and for $m = 1$, $r = n$, without any additional condition.

This sequence is called non-ascending run of length r if the corresponding conditions are satisfied with the reverse inequalities.

Example 14.5

Consider the permutations (j_1, j_2, j_3) of the set $\{1, 2, 3\}$ with repetition. Clearly these 27 permutations can be classified according to the number of their ascending runs as follows.

(a) The permutations with one non-descending run are the following 10:

$$(1,1,1), \quad (1,1,2), \quad (1,1,3), \quad (1,2,2), \quad (1,2,3),$$

$$(1,3,3), \quad (2,2,2), \quad (2,2,3), \quad (2,3,3), \quad (3,3,3).$$

(b) The permutations with two non-descending runs are the following 16:

$$(1,2,1), \ (2,1,1), \ (1,3,1), \ (3,1,1), \ (2,1,2), \ (2,2,1), \ (3,1,3), \ (3,3,1),$$

$$(1,3,2), \ (2,1,3), \ (2,3,1), \ (3,1,2), \ (2,3,2), \ (3,2,2), \ (3,2,3), \ (3,3,2).$$

(c) The permutation

$$(3,2,1)$$

is the only one that has three non-descending runs. Each of these runs is of length one. ▯

The enumeration of the permutations of a finite set with repetition and a given number of non-descending runs is facilitated by distinguishing them according to the last element they include. A bivariate generating function of the number of such permutations is derived in the following theorem.

THEOREM 14.11

Let $Q(n, k; s, r)$ be the number of n-permutations of the set $\{1, 2, \ldots, s\}$ with repetition and last element r, which have k non-descending runs. Then

$$\sum_{n=1}^{\infty} \sum_{k=1}^{n} Q(n, k; s, r) t^k u^n = \frac{ut(1-t)[1 - u(1-t)]^{-r}}{1 - t[1 - u(1-t)]^{-s}}. \qquad (14.24)$$

PROOF   Distinguishing the n-permutations of the set $\{1, 2, \ldots, s\}$ with repetition and last element r that have k non-descending runs, according to the element before the last, we conclude the expressions

$$Q(n, k; s, r) = \sum_{j=1}^{r} Q(n-1, k; s, j) + \sum_{j=r+1}^{s} Q(n-1, k-1; s, j),$$

for $r = 1, 2, \ldots, s - 1$, $k = 1, 2, \ldots, n$, $n = 2, 3, \ldots$, and

$$Q(n, k; s, s) = \sum_{j=1}^{s} Q(n - 1, k; s, j),$$

for $k = 1, 2, \ldots, n$, $n = 2, 3, \ldots$, with $Q(n, 0; s, r) = 0$, $n = 1, 2, \ldots$, $Q(1, 1; s, r) = 1$ and $Q(1, k; s, r) = 0$, $k = 2, 3, \ldots$. Multiplying these expressions by t^k and summing for $k = 1, 2, \ldots, n$, we get for the generating function

$$f_n(t; s, r) = \sum_{k=1}^{n} Q(n, k; s, r) t^k, \quad n = 1, 2, \ldots, \quad r = 1, 2, \ldots, s,$$

the recurrence relations

$$f_n(t; s, r) = \sum_{j=1}^{r} f_{n-1}(t; s, j) + t \sum_{j=s+1}^{s} f_{n-1}(t; s, j),$$

for $r = 1, 2, \ldots, s - 1$, $n = 2, 3, \ldots$, and

$$f_n(t; s, s) = \sum_{j=1}^{s} f_{n-1}(t; s, j),$$

for $n = 2, 3, \ldots$, with $f_1(t; s, r) = t$. Taking the difference of these relations with respect to r, we deduce the recurrence relation

$$f_n(t; s, r - 1) = f_n(t; s, r) + (t - 1) f_{n-1}(t; s, r),$$

for $r = 2, 3, \ldots, s$, $n = 2, 3, \ldots$, with $f_1(t; s, r) = t$. Replacing n by $n - j$, multiplying by

$$\binom{i-1}{j}(t-1)^j = \binom{i}{j}(t-1)^j - \binom{i-1}{j-1}(t-1)^j$$

and summing the resulting expression for $j = 0, 1, \ldots, i$, we get

$$\sum_{j=0}^{i-1} \binom{i-1}{j}(t-1)^j f_{n-j}(t; s, r-1) = \sum_{j=0}^{i} \binom{i}{j}(t-1)^j f_{n-j}(t; s, r)$$

$$- \sum_{j=1}^{i} \binom{i-1}{j-1}(t-1)^j f_{n-j}(t; s, r) + \sum_{j=0}^{i-1} \binom{i-1}{j}(t-1)^{j+1} f_{n-j-1}(t; s, r)$$

and so

$$\sum_{j=0}^{i} \binom{i}{j}(t-1)^j f_{n-j}(t; s, r) = \sum_{j=0}^{i-1} \binom{i-1}{j}(t-1)^j f_{n-j}(t; s, r-1).$$

Applying successively the last relation, we deduce the expression

$$f_n(t; s, r - i) = \sum_{j=0}^{i} \binom{i}{j} (t - 1)^j f_{n-j}(t; s, r)$$

for $n = 1, 2, \ldots, i = 0, 1, \ldots, r - 1, r = 1, 2, \ldots, s$. Multiplying both members of this expression by u^n and summing for $n = 1, 2, \ldots$, we find for the bivariate generating function

$$f(t, u; s, r) = \sum_{n=1}^{\infty} \sum_{k=1}^{n} Q(n, k; s, r) t^k u^n = \sum_{n=1}^{\infty} f_n(t; s, r) u^n,$$

the relation

$$f(t, u; s, r - i) = [1 - u(1 - t)]^i f(t, u; s, r),$$

for $i = 0, \ldots, r - 1, r = 1, 2, \ldots, s$. In particular, for $r = s$,

$$f(t, u; s, s - i) = [1 - u(1 - t)]^i f(t, u; s, s), \quad i = 0, 1, \ldots, s - 1. \quad (14.25)$$

Summing it for $i = 0, 1, \ldots, s - 1$, we get

$$\sum_{i=0}^{s-1} f(t, u; s, s - i) = \frac{1 - [1 - u(1 - t)]^s}{u(1 - t)} f(t, u; s, s).$$

Also

$$\sum_{i=0}^{s-1} f(t, u; s, s - i) = \sum_{i=0}^{s-1} \sum_{n=1}^{\infty} f_n(t; s, s - i) u^n = \sum_{n=1}^{\infty} \left\{ \sum_{i=0}^{s-1} f_n(t; s, s - i) \right\} u^n$$

and, since $f_{n+1}(t; s, s) = \sum_{j=1}^{s} f_n(t; s, j)$,

$$\sum_{i=0}^{s-1} f(t, u; s, s - i) = \frac{f(t, u; s, s) - ut}{u}.$$

Therefore

$$(1 - t) f(t, u; s, s) - ut(1 - t) = \{1 - [1 - u(1 - t)^s]\} f(t, u; s, s)$$

and

$$f(t, u; s, s) = \frac{ut(1 - t)}{[1 - ut(1 - t)]^s - t}.$$

Introducing the last expression into (14.25) and setting $r = s - i$, whence $i = s - r$, we conclude that

$$f(t, u; s, r) = \frac{ut(1 - t)[1 - u(1 - t)]^{-r}}{1 - t[1 - u(1 - t)]^{-s}}$$

and the derivation of generating function (14.24) is completed. ∎

COROLLARY 14.1
*Let $Q(n, k; s)$ be the number of n-permutations of the set $\{1, 2, \ldots, s\}$ with
repetition, which have k non-descending runs. Then*

$$\sum_{n=1}^{\infty} \sum_{k=1}^{n} Q(n, k; s) t^{k-1} u^n = \frac{(1 - t)}{[1 - u(1 - t)]^s - t}. \tag{14.26}$$

PROOF   Notice that the attachment of the element s after the last element
of an n-permutation of the set $\{1, 2, \ldots, s\}$ with repetition, which has k non-
descending runs uniquely yields an $(n + 1)$-permutation of the set $\{1, 2, \ldots, s\}$
with repetition and last element s, which has k non-descending runs. Consequently,
$Q(n, k; s) = Q(n + 1, k; s, s)$ and introducing this expression into (14.24), we
deduce (14.26). ∎

Clearly, comparing the generating function (14.26) with the generating
function (14.19) of the Carlitz numbers we conclude the following corollary.

COROLLARY 14.2
*The number of n-permutations of the set $\{1, 2, \ldots, s\}$ with repetition, which have
k non-descending runs is given by*

$$Q(n, k; s) = |B(n, k; -s)| = (-1)^n B(n, k; -s),$$

where $B(n, k; s)$ is the Carlitz number.

14.6   BIBLIOGRAPHIC NOTES

The problem of evaluating the power series with coefficients powers of
fixed order led Euler to the introduction of Eulerian polynomials and numbers (see Example 14.1). Expression (14.1), of the n-th power of a number
as sum of binomial coefficients of the n-th order with coefficients the Eulerian numbers, is due to J. Worpitzky (1883). Other properties of the
Eulerian numbers and polynomials can be found in L. Carlitz (1959). Motivated by the problem of expressing the power moments of a frequency
distribution in terms of cumulative totals (see Example 14.2), P. S. Dwyer
(1938, 1940) introduced and studied the cumulative numbers, which are
non-central Eulerian numbers. The Carlitz numbers were so named in
honor of Leonard Carlitz for his stimulating contribution on the Eulerian

numbers, their generalizations and combinatorial applications. These numbers were first studied in L. Carlitz (1979) as degenerate Eulerian numbers. Ch. A. Charalambides (1982) further examined them as composition numbers. The general Simon Newcomb's problem was studied through generating functions by P. A. MacMahon (1915, 1916). J. Riordan (1958) and Ch. A. Charalambides (2002) examined it as a problem of enumeration of permutations with restricted positions by using power and factorial rook polynomials. An interesting combinatorial treatment of the same problem was provided by J. E. Dillon and D. P. Roselle (1969). The presentation of the section on the enumeration of permutations with repetition and a given number of runs (or rises) is based on the paper of L. Carlitz, D. P. Roselle and R. A. Scoville (1966).

14.7 EXERCISES

1. *Non-central Eulerian numbers.* Consider the expansion

$$(t+r)^n = \sum_{k=0}^{n} A(n,k;r) \binom{t+n-k}{n}, \quad n = 0, 1, \ldots .$$

The coefficient $A(n,k;r)$ is called non-central Eulerian number or Dwyer number. Derive (a) the explicit expression

$$A(n,k;r) = \sum_{j=0}^{k} (-1)^j \binom{n+1}{j} (k+r-j)^n$$

and (b) the triangular recurrence relation

$$A(n+1,k;r) = (k+r)A(n,k;r) + (n-k-r+2)A(n,k-1;r),$$

for $k = 1, 2, \ldots, n+1$, $n = 0, 1, \ldots$, with initial conditions

$$A(0,0;r) = 1, \quad A(n,0;r) = r^n, \quad n > 0, \quad A(n,k;r) = 0, \quad k > n.$$

2. (*Continuation*). Show that

$$A_n(t;r) = \sum_{k=0}^{n} A(n,k;r)t^k = (1-t)^{n+1} \sum_{i=0}^{\infty} (i+r)^n t^i$$

and

$$g(t,u;r) = \sum_{n=0}^{\infty} \sum_{k=0}^{n} A(n,k;r)t^k \frac{u^n}{n!} = \frac{(1-t)e^{ru(1-t)}}{1 - te^{u(1-t)}}.$$

3. (*Continuation*). Show that the non-central Eulerian numbers $A(n, k; r)$, $k = 0, 1, \ldots, n$, $n = 0, 1, \ldots$ are connected with the non-central Stirling numbers of the second kind $S(n, k; r)$, $k = 0, 1, \ldots, n$, $n = 0, 1, \ldots$ by

$$A(n, k; r) = \sum_{j=0}^{k} (-1)^{k-j} \binom{n-j}{k-j} j! S(n, j; r)$$

and

$$S(n, j; r) = \frac{1}{j!} \sum_{k=0}^{j} \binom{n-k}{j-k} A(n, k; r).$$

4. *Non-central Carlitz numbers.* Consider the expansion

$$\binom{st + r}{n} = \sum_{k=0}^{n} B(n, k; s, r) \binom{t + n - k}{n}, \quad n = 0, 1, \ldots .$$

The coefficient $B(n, k; s, r)$ is called non-central Carlitz number. Derive (a) the explicit expression

$$B(n, k; s, r) = \sum_{j=0}^{k} (-1)^j \binom{n+1}{j} \binom{s(k-j) + r}{n}$$

and (b) the triangular recurrence relation

$$(n + 1)B(n + 1, k; s, r) = (sk + r - n)B(n, k; s, r)$$
$$+ [s(n - k + 2) + n - r]B(n, k - 1; s, r),$$

for $k = 1, 2, \ldots, n + 1$, $n = 0, 1, \ldots$, with initial conditions

$$B(0, 0; s, r) = 1, \quad B(n, 0; s, r) = \binom{r}{n}, \quad n > 0, \quad B(n, k; s, r) = 0, \quad k > n.$$

5. (*Continuation*). Show that

$$B_n(t; s, r) = \sum_{k=0}^{n} B(n, k; s, r) t^k = (1 - t)^{n+1} \sum_{j=0}^{\infty} \binom{sj + r}{n} t^j$$

and

$$g(t, u; s, r) = \sum_{n=0}^{\infty} \sum_{k=0}^{n} B(n, k; s, r) t^k u^n = \frac{(1 - t)[1 + u(1 - t)]^r}{1 - t[1 + u(1 - t)]^s}.$$

6. (*Continuation*). Show that

$$\lim_{s \to \pm \infty} s^{-n} n! B(n, k; s, r) = A(n, k; \rho), \quad \text{if} \quad \lim_{s \pm \infty} r/s = \rho$$

and

$$\sum_{k=0}^{n} B(n, k; s, r) = s^n.$$

7. (*Continuation*). Show that the non-central Carlitz numbers $B(n, k; s, r)$, $k = 0, 1, \ldots, n$, $n = 0, 1, \ldots$ are connected with the non-central generalized factorial coefficients $C(n, k; s, r)$, $k = 0, 1, \ldots, n$, $n = 0, 1, \ldots$ by

$$B(n, k; s, r) = \sum_{j=0}^{k} (-1)^{k-j} \binom{n-j}{k-j} \frac{j!}{n!} C(n, j; s, r)$$

and

$$C(n, j; s, r) = \frac{n!}{j!} \sum_{k=0}^{j} \binom{n-k}{j-k} B(n, k; s, r).$$

8*. *q-Eulerian numbers.* Consider the expansion of the n-th power of the q-number $[t]_q$ into q-binomial coefficients of order n:

$$[t]_q^n = \sum_{k=0}^{n} q^{k(k-1)/2} A(n, k|q) \begin{bmatrix} t + n - k \\ n \end{bmatrix}_q, \quad n = 0, 1, \ldots.$$

The coefficient $A(n, k|q)$ is called q-Eulerian number. (a) Show that

$$A(n, k|q) = A(n, n - k + 1|q), \quad k = 0, 1, \ldots, n, \quad n = 0, 1, \ldots$$

and (b) derive the explicit expression

$$A(n, k|q) = q^{-k(k-1)/2} \sum_{r=0}^{n} (-1)^r q^{r(r-1)/2} \begin{bmatrix} n + 1 \\ r \end{bmatrix}_q [k - r]_q^n,$$

for $k = 0, 1, \ldots, n$, $n = 0, 1, \ldots$.

9*. (*Continuation*). Show that the q-Eulerian numbers $A(n, k|q)$, $k = 0, 1, \ldots, n$, $n = 0, 1, \ldots$, satisfy the triangular recurrence relation

$$A(n + 1, k|q) = [k]_q A(n, k|q) + [n - k + 1]_q A(n, k|q),$$

for $k = 1, 2, \ldots, n$, $n = 0, 1, \ldots$, with initial conditions

$$A(0, 0|q) = 1, \quad A(n, 0|q) = 0, \quad n > 0 \quad A(n, k|q) = 0, \quad k > n.$$

10*. *Records.* Let $(j_1, j_2, \ldots, j_n)$ be a permutation of $\{1, 2, \ldots, n\}$. The element j_r, for $2 \leq r \leq n$, is called a *record* of this permutation if $j_r > j_s$, $s = 1, 2, \ldots, r - 1$. Especially, the first element j_1 is regarded, by convention, as a record. Show that the number of permutations of the set $\{1, 2, \ldots, n\}$ that have k records equals $|s(n, k)|$, the signless (absolute) Stirling number of the second kind.

11. *Inversions of permutations.* Let $(j_1, j_2, \ldots, j_n)$ be a permutation of the set $\{1, 2, \ldots, n\}$. If $j_r > j_s$ for $1 \leq r < s \leq n$, this permutation has an inversion at the pair of points (r, s). Show that the number $b(n, k)$ of permutations of the set $\{1, 2, \ldots, n\}$ with k inversions satisfies the horizontal recurrence relation

$$b(n, k) = \sum_{j=m}^{k} b(n-1, j), \quad m = \max\{0, k-n+1\},$$

for $k = 0, 1, \ldots, n(n-1)/2$, $n = 1, 2, \ldots$, with initial conditions

$$b(n, 0) = 1, \quad b(0, k) = 0, \quad k \geq 1, \quad b(n, k) = 0, \quad k > n(n-1)/2$$

and conclude that

$$b(n, k) = b(n, k-1) + b(n-1, k),$$

for $k = 1, 2, \ldots, n-1$, $n = 1, 2, \ldots$ and

$$b(n, k) = b(n, k-1) + b(n-1, k) - b(n-1, k-n),$$

for $k = n, n+1, \ldots, n(n-1)/2$, $n = 3, 4, \ldots$.

12. (*Continuation*). Show that

$$\phi_n(t) = \sum_{k=0}^{n(n-1)/2} b(n, k) t^k = \prod_{j=1}^{n} \frac{1 - t^j}{1 - t}, \quad n = 1, 2, \ldots$$

and conclude that

$$b(n, n(n-1)/2 - k) = b(n, k),$$

$$\sum_{k=0}^{n(n-1)/2} b(n, k) = n!, \quad \sum_{k=0}^{n(n-1)/2} (-1)^k b(n, k) = 0.$$

13. *Local maxima and minima.* Let $(j_1, j_2, \ldots, j_n)$ be a permutation of the set $\{1, 2, \ldots, n\}$. The element j_r, for $2 \leq r \leq n-1$, is called a *local maximum* or a *peak* at the point r of this permutation if $j_r > j_{r-1}$ and $j_r > j_{r+1}$. Especially, the element j_1 is a peak if $j_1 > j_2$, while the element j_n is a peak if $j_n > j_{n-1}$. The element j_r is a *local minimum* if the conditions are satisfied with the reverse inequalities. Show that the number $T_1(n, k)$ of permutations of the set $\{1, 2, \ldots, n\}$ with k local maxima (minima) satisfies the triangular recurrence relation

$$T_1(n+1, k) = (2k+2)T_1(n, k) + (n - 2k + 1)T_1(n, k-1),$$

for $k = 1, 2, \ldots, n = 2, 3, \ldots$, with initial conditions

$$T_1(n, 0) = 2^{n-1}, \quad n = 2, 3, \ldots, \quad T_1(2, k) = 0, \quad k = 1, 2, \ldots.$$

14. Show that the number $T_2(n, k)$ of permutations of the set $\{1, 2, \ldots, n\}$ with k ascending runs, each of length greater than one, satisfies the triangular recurrence relation

$$T_2(n+1, k) = (2k+1)T_2(n, k) + (n - 2k + 2)T_2(n, k-1),$$

for $k = 1, 2, \ldots, n = 2, 3, \ldots$, with initial conditions

$$T_2(n, 0) = 1, \quad n = 2, 3, \ldots, \quad T_2(2, 1) = 1 \quad T_2(2, k) = 0, \quad k = 2, 3, \ldots.$$

15*. Consider the double sequence of numbers $T(n, k)$, $k = 0, 1, \ldots$, $n = 2, 3, \ldots$, where $T_1(n, k) = T(n, 2k + 1)$ and $T_2(n, k) = T(n, 2k)$ are the numbers of permutations of the set $\{1, 2, \ldots, n\}$ with k local maxima (minima) and with k ascending runs of length greater than one, respectively. (a) Show that

$$T(n+1, k) = (k+1)T(n, k) + (n - k + 2)T(n, k-2),$$

for $k = 2, 3, \ldots, n = 2, 3, \ldots$, with

$$T(n, 0) = 1, \quad T(n, 1) = 2^{n-1}, \quad n = 2, 3, \ldots,$$

$$T(2, 2) = 1, \quad T(2, k) = 1, \quad k = 3, 4, \ldots.$$

(b) Setting $T(1, k) = 1$, $k = 0, 1$ and $T(1, k) = 0$, $k = 2, 3, \ldots$, derive the generating function

$$g(t, u) = \sum_{n=1}^{\infty} \sum_{k=1}^{\infty} T(n, k) t^k \frac{u^{n-1}}{(n-1)!} = \frac{1 - t^2}{t(\cosh z - 1)},$$

with $z = u(1 - t^2)^{1/2} + \text{arccosh}(1/t)$, where $\cosh z = (e^z + e^{-z})/2$ is the hyperbolic cosine and $\text{arccosh}w$ is the arch of the hyperbolic cosine of w.

16*. Show that the number $R(n, k)$ of permutations of $\{1, 2, \ldots, n\}$ with k (ascending or descending) runs of length greater than one satisfies the recurrence relation

$$R(n+1, k) = kR(n, k) + 2R(n, k-1) + (n - k + 1)R(n, k-2),$$

for $k = 2, 3, \ldots, n = 2, 3, \ldots$, with initial conditions

$$R(n, 0) = 0, \quad R(n, 1) = 2, \quad n = 2, 3, \ldots, \quad R(2, k) = 0, \quad k = 2, 3, \ldots.$$

17*. *Up-down and down-up permutations.* A permutation $(j_1, j_2, \ldots, j_n)$ of the set $\{1, 2, \ldots, n\}$ is called *up-down permutation* if $j_{2r-1} < j_{2r}$ for $r = 1, 2, \ldots, [n/2]$ and $j_{2r} > j_{2r+1}$ for $r = 1, 2, \ldots, [n/2] - 1$, while it is called *down-up permutation* if $j_{2r-1} > j_{2r}$ for $r = 1, 2, \ldots, [n/2]$ and $j_{2r} < j_{2r+1}$ for $r = 1, 2, \ldots, [n/2] - 1$. (a) Show that the number A_n of up-down permutations of the set $\{1, 2, \ldots, n\}$, which is equal to the number of down-up permutations of the same set, satisfies the recurrence relation

$$2A_{n+1} = \sum_{k=0}^{n} \binom{n}{k} A_k A_{n-k}, \quad n = 2, 3, \ldots,$$

with $A_n = 1$, $n = 0, 1, 2$ and (b) deduce the generating function

$$A(t) = \sum_{n=0}^{\infty} A_n \frac{t^n}{n!} = \tan\left(\frac{t}{2} + \frac{\pi}{4}\right).$$

18*. (*Continuation*). *Euler numbers.* The sequence of Euler numbers E_{2n}, $n = 0, 1, \ldots$, ($E_{2n+1} = 0$, $n = 0, 1, \ldots$) has generating function

$$E(t) = \sum_{n=0}^{\infty} E_{2n} \frac{t^{2n}}{(2n)!} = \frac{2e^t}{1 + e^{2t}}.$$

(a) Show that

$$|E_{2n}| = (-1)^n E_{2n} = A_{2n},$$

where A_{2n} is the number of up-down permutations of the set $\{1, 2, \ldots, 2n\}$. (b) Derive the expression

$$|E_{2n}| = \sum_{k=1}^{n} (-1)^{n-k} \frac{(2k)!}{2^k} R(2n, 2k)$$

where

$$R(n, k) = \frac{1}{k!} \sum_{j=0}^{k} (-1)^j \binom{k}{j} \left(\frac{k}{2} - j\right)^n$$

is the Carlitz-Riordan number of the second kind (see Exercise 8.22).

19*. (*Continuation*). *Tangent coefficients.* Consider the sequence T_{2n+1}, $n = 0, 1, \ldots$, ($T_{2n} = 0$, $n = 0, 1, \ldots$) with generating function

$$T(t) = \sum_{n=0}^{\infty} T_{2n+1} \frac{t^{2n+1}}{(2n+1)!} = \frac{1 - e^{2t}}{1 + e^{2t}} = -\tanh t.$$

(a) Show that

$$|T_{2n+1}| = (-1)^n T_{2n+1} = A_{2n+1},$$

where A_{2n+1} is the number of up-down permutations of the set $\{1, 2, \ldots, 2n + 1\}$. (b) Derive the expression

$$|T_{2n+1}| = \sum_{k=1}^{n} (-1)^{n-k} \frac{(k+1)(2k)!}{2^k} R(2n, 2k),$$

where $R(n, k)$ is the Carlitz-Riordan number of the second kind.

20*. (*Continuation*). *Genocchi numbers.* The sequence of Genocchi numbers G_n, $n = 1, 2, \ldots$, has generating function

$$G(t) = \sum_{n=1}^{\infty} G_n \frac{t^n}{n!} = \frac{2t}{1 + e^t}$$

and so $G_1 = 1$, $G_{2n+1} = 0$, $n = 1, 2, \ldots$. (a) Show that

$$|G_{2n}| = (-1)^n G_{2n} = \frac{n}{2^{2(n-1)}} A_{2n-1},$$

where A_{2n-1} is the number of up-down permutations of the set $\{1, 2, \ldots, 2n - 1\}$. (b) Derive the expression

$$|G_{2n}| = 2n \sum_{k=1}^{2n-1} (-1)^{n-k} \frac{k!}{2^k} S(2n - 1, k),$$

where $S(n, k)$ is the Stirling number of the second kind.

HINTS AND ANSWERS TO EXERCISES

CHAPTER 1

1. $S_3 = A^3 = \{(t,t,t),\ (t,t,h),\ (t,h,t),\ (h,t,t),\ (t,h,h),\ (h,t,h),$
$$\qquad\qquad\qquad\qquad\qquad (h,h,t),\ (h,h,h)\},$$
$\quad F((t,t,t)) = 0,\quad F((t,t,h)) = F((t,h,t)) = F((h,t,t)) = 1,$
$\quad F((t,h,h)) = F((h,t,h)) = F((h,h,t)) = 2,\ F((h,h,h)) = 3.$

2. (a) $N(S_2) = 36$,

(b) $A_2 = \{(1,1)\}$, $\quad A_3 = \{(1,2),\ (2,1)\}$, $\quad A_4 = \{(1,3),\ (2,2),\ (3,1)\}$,
$A_5 = \{(1,4),\ (2,3),\ (3,2),\ (4,1)\}$, $\quad A_6 = \{(1,5),\ (2,4),\ (3,3),\ (4,2),\ (5,1)\}$,
$A_7 = \{(1,6),\ (2,5),\ (3,4),\ (5,2),\ (6,1)\}$,
$A_8 = \{(2,6),\ (3,5),\ (4,4),\ (5,3),\ (6,2)\}$, $\quad A_9 = \{(3,6),\ (4,5),\ (5,4),\ (6,3)\}$,
$A_{10} = \{(4,6),\ (5,5),\ (6,4)\}$, $\quad A_{11} = \{(5,6),\ (6,5)\}$, $\quad A_{12} = \{(6,6)\}$.

3. $W = \{1,2,3,4,5,6\}$, $S_3 = W^3$, $N(S_3) = [N(W)]^3 = 6^3 = 216$, $A = \{(w,b,r) \in S_3 : w+b+r > 10\}$, $A' = \{(w',b',r') \in S_3 : w'+b'+r' \leq 10\}$. The element (w,b,r) in A uniquely corresponds to the element (w',b',r') in A', with $w' = 7-w$, $b' = 7-b$, $r' = 7-r$. Indeed, $w'+b'+r' = 21-(w+b+r) \leq 10$ if and only if $w+b+r > 10$.

4. $N(L_5 \times R_5) = N(L_5)N(R_5) = 25$, $\quad N(L_5 \times R_4) = N(L_5)N(R_4) = 20$.

5. The set B_n of n-digit binary sequences $(1, a_1, a_2, \ldots, a_{n-1})$ is equivalent to the Cartesian product $A_1 \times A_2 \times \cdots \times A_{n-1}$, with $A_i = \{0,1\}$, $i = 1, 2, \ldots, n-1$, and so $N(B_n) = N(A_1)N(A_2) \cdots N(A_{n-1}) = 2^{n-1}$, $n = 2, 3, \ldots$, $N(B_1) = 2$. Also, $N(T_n) = \sum_{r=1}^{n} N(B_r) = 2 + \sum_{r=2}^{n-1} 2^{r-1} = 2^n$. Alternatively, the set T_n of binary sequences of at most n digits is equivalent to the Cartesian product $A_1 \times A_2 \times \cdots \times A_n$, with $A_i = \{0,1\}$, $i = 1, 2, \ldots, n$ and so $N(T_n) = 2^n$.

6. The set B_n of sequences of n symbols $(a_1, a_2, \ldots, a_n)$ is equivalent to the Cartesian product $A_1 \times A_2 \times \cdots \times A_n$, with $A_i = \{0,1\}$, $i = 1, 2, \ldots, n$, where 0 corresponds to a "dot" and 1 to a "dash." Thus $N(B_n) = 2^n$, $n = 1, 2, \ldots$, and the required number is $\sum_{r=1}^{n} N(B_r) = 2^{n+1}$.

7. (a) $8 + 9 = 17$, (b) $2 \cdot 8 \cdot 9 + 9 \cdot 9 = 225$, (c) $1 + 17 + 225 = 243$.

8. (a) $9 \cdot 5 = 45$, (b) $9 \cdot 10 \cdot 5 = 450$, (c) $5 + 45 + 450 = 500$.

9. (a) n^3, (b) $n(n-1)(n-2)$, (c) $3 \cdot n^2$.

10. (a) $4 \cdot 3 \cdot 2 = 24$, (b) $3^3 = 27$, (c) $3 \cdot 2^2 = 12$.

11. Note that any factor of the number N is of the form $p_1^{i_1} p_2^{i_2} \cdots p_n^{i_n}$, where $i_r \in A_r = \{0, 1, \ldots, k_r\}$, $r = 1, 2, \ldots, n$, and apply the multiplication principle.

12. 4^n.

13. (a) $3^{13} = 1{,}594{,}323$, (b) $2^5 \cdot 3^2 = 288$.

14. $14^3 \cdot 9 \cdot 10^3 = 24{,}696{,}000$.

15. k^n.

16. Apply the multiplication principle (a) with $A_r = \{1, 2, \}$ and (b) with $A_r = \{1, 2, \ldots, k\}$ for $r = 1, 2, \ldots, n$. Then the number of divisions (a) in two subsets is 2^n and (b) in k subsets is k^n.

17. Note that a map f from X into Y corresponds to an ordered n-tuple $(y_{j_1}, y_{j_2}, \ldots, y_{j_n})$ with $y_{j_r} = f(x_r)$, $r = 1, 2, \ldots, n$ and apply the multiplication principle.

18. Suppose that each cell contains at most one object and conclude that this hypothesis leads to a contradiction. Use the fact that each integer $k_i \in K$ can be written in the form $k_i = 2^{r_i} \cdot s_i$, where r_i is a nonnegative integer and s_i is a positive odd integer less than $2n$, and apply the pigeonhole principle.

19. Consider the partial sums $s_1 = k_1$, $s_2 = k_1 + k_2, \ldots, s_n = k_1 + k_2 + \cdots + k_n$ and apply the pigeonhole principle.

20. Each integer i from the set $\{1, 2, \ldots, 10\}$ can be written in the form $i = 2^{r_i} \cdot s_i$, where r_i is a nonnegative integer and s_i belongs to the set $\{1, 3, 5, 7, 9\}$ of five positive odd integers. Apply the pigeonhole principle and conclude that between the six chosen numbers there are two having the same s.

CHAPTER 2

1. (a) Let the four boys $\{b_1, b_2, b_3, b_4\}$ seat first. Then the three girls $\{g_1, g_2, g_3\}$ may take any three among the five permissible positions

$\{w_1, w_2, w_3, w_4, w_5\}$, where w_i is the position before the i-th boy for $i = 1, 2, 3, 4$ and w_5 is the position after fourth boy. Apply the multiplication principle and conclude that the required number is $4!(5)_3 = 480$. (b) $7! = 5040$.

2. $16 \cdot 15 = (2 \cdot 15) \cdot 8 = 240$.

3. $4!(51 \cdot 3! \cdot 1 \cdot 1) = 17{,}280$.

4. $4^3 = 64$, $3 \cdot 3^2 = 27$.

5. $(7)_4 = 840$.

6. Consider the set A_j that contains the permutations of $\{j, j+1, j+2\}$, $j = 1, 2, 3, 4$, and conclude that the required number is $N(A_1) + N(A_2) + N(A_3) + N(A_4) = 4 \cdot 3! = 24$.

7. (a) $2 \cdot 3^3 = 54$, (b) $9 \cdot 10^2 \cdot 5 = 4500$, (c) $9 \cdot 10^2 \cdot 5 = 4500$.

8. (a) $(10)_4 = 5040$, (b) $10^4 = 10{,}000$.

9. Consider the two like elements of each of the 2 kinds $\{w_1, w_2\}$ as two single elements and conclude that the required number is
$$\frac{(2 \cdot 3 + 2)!}{2^3 \cdot 1^2} = \frac{8!}{2^3} = 7! = 5040.$$

10. $10!/2^2 = 907{,}200$.

11. $N(A) = 3 \cdot 3! + 2 \cdot 3 + 1 = 25$, $\quad N(B) = 3 \cdot 3! + 3 \cdot 3 = 27$, $N(\Omega) = 6^3 = 216$.

12. Note that each circular permutation $(i_1, i_2, \dots, i_n)$ of the set $\{1, 2, \dots, n\}$ corresponds to n (linear) permutations of the same set and conclude the required number.

13. (a) $2 \cdot (5)_3 \cdot 2 \cdot 3 = 720$, (b) $(5)_3 \cdot 2 = 120$

14. (a) $(8 + 3)! \cdot 2^3 = 319{,}334{,}400$, (b) $(8 + 3 - 1)! \cdot 2^3 = 29{,}030{,}400$

15. (a) $N(A_4) = 4^2 = 16$, (b) $N(A_3) = 3^2 = 9$,
(c) $N(B_4) = N(A_4 - A_3) = N(A_4) - N(A_3) = 4^2 - 3^2 = 7$.

16. $N(D_4) = N(C_4 - C_5) = N(C_4) - N(C_5)$
$= (6 - 4 + 1)^2 - (6 - 5 + 1)^2 = 5$.

17. $\binom{7}{2} + 7 = \binom{7 + 2 - 1}{2} = 28$.

18. $2 \binom{6}{1}\binom{6}{5} + \binom{6}{3}^2 = 472$.

19. (a) $\binom{5}{2} = 10$, (b) $\binom{5}{3} = 10$.

20. (a) $\binom{n}{2}\binom{n}{3}$, (b) $\binom{n}{5}2^5$, (c) $\binom{2n}{5}$.

21. (a) $5^{10} = 9,765,625$, (b) $\binom{5 + 10 - 1}{10} = 1001$.

22. $\binom{7}{2}\binom{5}{2} + \binom{7}{2}\binom{5}{3} = 420$.

23. (a) $\binom{10}{2}8! = \frac{10!}{2} = 1,814,400$, (b) $\binom{10}{3}7! = \frac{10!}{3!} = 604,800$,

(c) $\binom{10}{2}\binom{8}{2}6! = \frac{10!}{2^2} = 907,200$.

24. $\dfrac{10!}{1!2!3!4!} = 12,600$.

25. (a) $\dfrac{6!}{(2!)^3} = 90$, (b) $\dfrac{6!}{3!(2!)^3} = 15$.

26. (a) $\dfrac{10!}{2^5} = 113,400$, $(5!)^2 = 14,400$, (b) $\dfrac{10!}{5!2^5} = 945$, $5! = 120$.

27. (a) $3^{11} = 177,147$, (b) $\displaystyle\sum_{j=6}^{11} 3 \cdot 1^j 2^{11-j} = 3(2^6 - 1) = 189$.

28. The required number is the number of paths from the point $(0,0)$ to the point $(5,3)$ and equals $\binom{8}{3} = 56$.

29. Consider the inverse procedure starting with n one-element sets and, at each stage merging together any two subsets derive, by applying the multiplication principle, the required number.

30. (a) The least number of lockers needed is the number of different "restricted" lockers needed and equals $a_n = \binom{n}{k}$, which is the number of k-combinations of the n generals, with $k = [n/2]$ the integer part of $n/2$. (b) The number of keys a general should have equals $b_n = \binom{n-1}{k}$, which is the number of k-combinations of $n - 1$ generals.

31. (a) The number of n-combinations of a set W_{r+s}, of $r + s$ elements, which equals $\binom{r+s}{n}$, may equivalently be evaluated by selecting k elements from a subset $A_r \subseteq W_{r+s}$, of r elements, and $n - k$ elements from the complementary set $W_{r+s} - A_r$ of s elements, for $k = 0, 1, \ldots, n$, and using the principles of multiplication and addition. (b) Similarly, evaluate the number of n-combinations of W_{r+s} with repetitions.

32. (a) Use the expression $\binom{n}{k} = \dfrac{n!}{k!(n-k)!}$.

(b) The pair of sets (A, B), with $A \subseteq B \subseteq W$ and $N(A) = k$, $N(B) = s$, $N(W) = n$, uniquely corresponds to the pair of sets (A, C), where $C =$

$B - A$, with $A \subseteq \Omega$ and $C \subseteq A' = W - A$. Also, the same pair of sets (A, B) uniquely corresponds to the pair of sets (A, D), where $D = B - A$, with $D \subseteq \Omega$ and $A \subseteq \Omega - D$. Evaluate the pairs of sets (A, B), (A, C) and (A, D) and conclude the first two equalities. The other two equalities are similarly deduced with $s = r + k$.

33. Multiply the expression $\dbinom{n}{k} = \dfrac{n!}{k!(n-k)!}$ by $n = (n - k) + k$ and deduce the first relation. Express $\dbinom{n}{k} = \dbinom{s + (n + s)}{k}$ as a sum of binomial coefficients, by using the first expression in Exercise 31; using the second part of Exercise 32, deduce the required formula.

34. Multiply the expression $\dbinom{n}{k} = \dfrac{n!}{k!(n-k)!}$ by $n = \dfrac{(n+1)k + (n-k)}{k+1}$ and deduce the first relation. Work as in the second part of Exercise 33.

35. Use Pascal's triangle and, splitting the sum into two sums, conclude the first formula. Using the first part of Exercise 32 and the first part of this exercise, derive the second formula.

36. The use of Pascal's triangle splits the sum into two sums, which reduce to the required expression.

37. Verify that each injective map f of X into Y, with $f(x_r) = y_{j_r}$, $r = 1, 2, \ldots, k$, uniquely corresponds to a k-permutation $(y_{j_1}, y_{j_2}, \ldots, y_{j_k})$ of the set Y.

38. Verify that each map f of $\{1, 2, \ldots, k\}$ into $\{1, 2, \ldots, n\}$ uniquely corresponds to a k-combination (a) without repetition if f is strictly increasing and (b) with repetition if f is increasing.

39. Consider the set $\mathcal{C}$ of k-combinations of the set W_{n+1} in which the element w_0 appears at most s times and the subset $\mathcal{C}_j \subseteq \mathcal{C}$ of these combinations in which the element w_0 appears exactly j times, for $j = 0, 1, \ldots, m$, with $m = \min\{k, s\}$, and apply the addition principle.

40. Consider the set $\mathcal{E}$ of k-combinations of n, with repetition, in which each element appears at most twice and the subset $\mathcal{E}_j \subseteq \mathcal{E}$ of these combinations in which exactly j elements appear twice, for $j = 0, 1, \ldots, m$, with $m = \min\{n, [k/2]\}$ and apply the addition principle.

41. Consider the set $\mathcal{A}$ of k-combinations of $n + r$ with repetition, in which each of n specified elements appears at most twice, while each of the other r elements appears at most once. Further, consider (a) the subset $\mathcal{A}_j \subseteq \mathcal{A}$ of these combinations in which exactly j of the n specified elements appear twice, for $j = 0, 1, \ldots, m$, with $m = \min\{n, [k/2]\}$, and apply the addition principle. Also, consider (b) the subset $\mathcal{B}_j \subseteq \mathcal{A}$ of these combinations in which exactly i of the r elements appear, for $i = 0, 1, \ldots, s$, with $s = \min\{r, k\}$, and apply the addition principle.   •

42. Consider the set $\mathcal{K}$ of the k-combinations of n elements belonging in r kinds with $k_1, k_2, \ldots, k_r$ elements, respectively. Further, consider the subset $\mathcal{K}_j \subseteq \mathcal{K}$ of these combinations which contain exactly j of the k_1 elements of the first kind, for $j = 0, 1, \ldots, k_1$ and apply the addition principle.

43. (a) Use Theorem 2.10. (b) Consider an element w of the set W_n, of n elements, and enumerate the partitions of W_n in k subsets in which w belongs in a unit set and the partitions of W_n in k subsets in which w belongs in a subset that is not a unit set. Applying the addition principle, conclude the required recurrence relation. (c) Enumerate the partitions of W_n in k subsets in which w belong in a subset with a total of j elements, for $j = 1, 2, \ldots, n$ and apply the addition principle.

44. (a) Show that the required number equals the product of the number of partitions of X in k subsets and the number of k-permutations of Y. (b) Use the result of Exercise 1.17. (c) Use the first part of Exercise 32.

45. Note that, to each partition of a set W_n in k (nonempty) subsets, there correspond $k!$ divisions of W_n in k nonempty subsets, and use the result (c) of Exercise 44.

46. (a) Use Theorem 2.10. (b) Consider an element w of the set W_n, of n elements, and enumerate the partitions of W_n in which w belongs in a subset with a total of j elements, for $j = 1, 2, \ldots, n$, and apply the addition principle.

47. Consider the set $\mathcal{P}$ of partitions of the set W_n, of n elements, and the subset $\mathcal{P}_k \subseteq \mathcal{P}$ of partitions of W_n in k subsets, for $k = 1, 2, \ldots, n$, and apply the addition principle and the result (a) of Exercise 43. Further, introduce the expression (c) of Exercise 44 into the preceding expression.

48. The required number equals $\binom{k-1}{n-1}$, the number of positive integer solutions of the linear equation $x_1 + x_2 + \cdots + x_n = k$.

49. Select first the s positions for the s runs of zeros among the $n - k + 1$ positions between the $n - k$ ones or before the first and after the last one. Then (a) allocate the k zeros in the selected s positions, placing at least one zero in each of these positions. (b) Choose the s_r positions of the runs of length r, for $r = 1, 2, \ldots, k$, among the s selected positions.

50. To each k-combination $\{i_1, i_2, \ldots, i_k\}$ of the set $\{1, 2, \ldots, n\}$ correspond the permutation $(a_1, a_2, \ldots, a_n)$ of k zeros and $n - k$ ones in which the element $a_{i_1}, a_{i_2}, \ldots, a_{i_k}$ are zeros. Verify that this correspondence is one-to-one and use the result of Exercise 49.

51. To each (circular) k-combination of the first n integral numbers displayed on a circle, correspond a circular permutation of k zeros and $n-k$ ones in which one of the n elements is marked with a star corresponding to number 1.

52. (a) Use the one-to-one transformation $y_i = x_i - s$, $i = 1, 2, \ldots, n$, and conclude that the required numbers equals $\binom{n+k-sn-1}{k-sn}$. (b) Use the one to one transformation $z_i = m - x_i$, $i = 1, 2, \ldots, n$ and conclude that the required number equals $\binom{n+mn-k-1}{mn-k}$.

53. To each partial derivative of order k of an analytic function of n variables, correspond a nonnegative integer solutions of a linear equation $r_1 + r_2 + \cdots + r_n = k$ and conclude that the required number is $\binom{n+k-1}{k}$.

54. Verify that the number of monomials in the most general polynomial in n variables $x_1, x_2, \ldots, x_n$ of degree k equals the number of nonnegative integer solutions of a linear inequality $r_1 + r_2 + \cdots + r_n \leq k$ and conclude that $a_{n,k} = \binom{n+k}{k}$.

55. Correspond to each k-combination of $\{1, 2, \ldots, n\}$, satisfying the required property, an integer solution of linear equation $j_1 + j_2 + \cdots + j_k + j_{k+1} = n$, satisfying the restrictions $j_1 \geq 1$, $j_2 \geq s + 1$, $j_3 \geq s + 1, \ldots,$ $j_k \geq s + 1$, $j_{k+1} \geq 0$.

56. Using the result of Exercise 55 enumerate the k-combinations of $\{1, 2, \ldots, n\}$, displayed on a circle, which satisfy the required property and (a) do not contain the elements $n - s + 1, n - s + 2, \ldots, n$ and (b) contain at least one of the s elements $n - s + 1, n - s + 2, \ldots, n$ and apply the addition principle.

57. (a) The evaluation of the number $C(n, k; s_{k-1})$ may be reduced to the enumeration of the k-combinations $\{j_1, j_2, \ldots, j_k\}$ of $\{1, 2, \ldots, n - s_{k-1}\}$ by using the transformation $j_1 = i_1$, $j_m = i_m - s_{m-1}$, $m = 2, 3, \ldots, k$. (b) Using the same transformation, the evaluation of the number $B(n, k; s_{k-1}, r)$ may be reduced to the enumeration of the k-combinations $\{j_1, j_2, \ldots, j_k\}$ of the set $\{1, 2, \ldots, n - s_{k-1}\}$, which satisfy the restriction $j_k \leq n - s_{k-1} + (j_1 - r - 1)$.

58. The number $Q_{n,k}$ equals the number of ways of selecting k positions for the zeros among the $(n-k-1)+2 = n-k+1$ positions between the $n-k$ ones or before the first and after the last one. Apply the addition principle to deduce Q_n. Use Pascal's triangle to deduce the recurrence relation.

59. The required number is given by $C_{n+1} = N(A-B) = N(A) - N(B)$, where A is the set of lattice paths from the point $(0,0)$ to the point (n,n) and B is the set of lattice paths from the point $(0,0)$ to the point (n,n) that do not touch or intersect the straight line $x = y - 1$.

60. (a) Note that each lattice path from $(0,0)$ to (n,k) with r diagonal steps requires, in addition, $k-r$ vertical and $n-r$ horizontal steps. Select the r diagonal and $k-r$ vertical steps among the total of $n+k-r$ steps. (b) Exclude the r diagonal steps and enumerate the number of lattice paths from $(0,0)$ to $(n-r,k-r)$ that do not touch or intersect the straight line $x=y$. Such a path passes through $n+k-2r$ points (not including the origin). Select the r starting points of the diagonal steps among the $n+k-2r$ points.

61. Use Laplace's classical definition of probability and the results of Exercise 60.

62. The probability that number r is drawn at the k-th drawing is given by $p_{n,k} = (n-1)_{k-1}/(n)_k$.

63. Consider the set Ω_k of possible outcomes of k drawings and the subset A_k of these outcomes in which a white ball is drawn for the first time at the k-th drawing. Then $p_{n,r,k} = P(A_k)/P(\Omega_k)$, $k = 1, 2, \ldots, r+1$. Use the fact that $A_1 + A_2 + \cdots + A_{r+1} = \Omega_k$ to conclude the required expression.

64. (a) $P(A_k) = (n-1)^{k-1}/n^k$,   (b) $P(B_k) = (n)_k/n^k$.

65. Use Laplace's classical definition of probability and the result of Exercise 58.

CHAPTER 3

1. Use Vandermonde's formula to get $s_{n,r} = (n)_r(a)_r(a+b-r)_{n-r}$.

2. Express $(x+y+n)_n$ into factorials of x and $y+n$, using Vandermonde's formula, and then multiply both members by $(y)_{-n} = 1/(y+n)_n$.

3. Express $(y)_n = (-1)^n(x+(-x-y-1+n))_n$ into factorials of x and $(-x-y-1+n)$ and then divide both members by $(x+y)_n$.

4. Rewrite the expression to be shown as

$$\sum_{j=n}^{n+s} \binom{j-1}{n-1}\binom{r+s-j}{r-n} = \binom{r+s}{r},$$

which follows from the expansion of the identity

$$[t^n(1-t)^{-n}][t^{r-n}(1-t)^{-r+n-1}] = t^r(1-t)^{-r-1}$$

into powers of t.

5. Use the result of the first part of Exercise 4 to get $s_n = ns/(r+1)_{n+1}$ and $s_{n,m} = (n+m-1)_m(s)_m/(r+m)_{n+m}$.

6. Use Newton's binomial formula to get $s_{n,r} = (n)_r p^r$.

7. Use Newton's negative binomial formula to get $s_{n,r} = (n+r-1)_r \frac{q^r}{p^r}$.

8. Differentiate a suitable expression of Newton's binomial formula (see Example 3.4).

9. Integrate a suitable expression of Newton's binomial formula (see Example 3.5).

10. Use the result of Exercise 2.32 and Newton's binomial formula.

11. Use Pascal's triangle to split the sum and derive the recurrence relation.

12. Use Newton's binomial formula (3.10) with $x = -t$, $y = 1$ and formula $\sum_{k=1}^{n} u^{k-1} = (1-u)^{-1}(1-u^n)$ with $u = 1-t$.

13. Integrate, with respect to t, in the interval $[0, u]$ the identity derived in the first part of Exercise 12 and then integrate, with respect to u, in the interval $[0, 1]$ the resulting expression to deduce the required relation.

14. Use Pascal's triangle to derive the recurrence relation.

15. Use the result of Exercise 14 with $z = 1/2$ and $z = -1/2$.

16. Use the relation

$$\frac{1}{r-k+1} = \frac{1}{r+1}\left(1 + \frac{k}{r-k+1}\right)$$

to derive the recurrence relation and iterate it to get the required expression.

17. Express $\binom{x+k}{n}$ into binomials of x and k, using Cauchy's formula, and then use the result of Exercise 10.

18. Use Cauchy's formula with $x = y = n = 2r$ and $x = y = n = 2r+1$ to derive the first and the second sums, respectively.

19. Express $\binom{n+r}{n+k}$ into binomials of n and r, using Cauchy's formula and then use the result of Exercise 10.

20. Use the relation $k\binom{r}{k} = r\binom{r-1}{k-1}$ and evaluate the resulting expression by expanding the identity $(x+1)^{r-1}(x+1)^s = (x+1)^{r+s-1}$ into powers of x.

21. Expand the identity $(1-t)^r(1-t)^{-s-1} = (1-t)^{-s+r-1}$ into powers of t and equate the coefficients of t^n in both sides of the resulting expression.

22. Use the relation

$$\binom{n}{k}\binom{n+r}{k}^{-1} = \binom{r+n-k}{n-k}\binom{n+r}{n}^{-1}$$

and then the result of Exercise 21.

23. Expand into powers of t the identity $(1 - t)^{-r-1}(1 - t)^{-(s-n)-1} = (1 - t)^{-(r+s-n+1)-1}$ and equate the coefficients of t^n in both sides of the resulting expression.

24. Using Newton's binomial formula, show that

$$t^r(1 - t)^{-r-1} = \sum_{k=0}^{r-s}(-1)^{r-s-k}\binom{r-k}{k}t^s(1 - t)^{-s-k-1}.$$

Expand this identity into powers of t and equate the coefficients of t^n in both sides of the resulting expression.

25. Expand into powers of t the identity $(1 + t)^r(1 + t)^{-s} = (1 + t)^{r-s}$ and equate the coefficients of t^n in both sides of the resulting expression.

26. Use for part (a) the expression $\binom{x}{k} = \dfrac{x(x - 1)\cdots(x - k + 1)}{k!}$, and for parts (b) and (c) the "triangular" recurrence relation.

27. Expand into powers of t and u the function $(1 + t + u + tu)^x = (1 + [t(1+u) + u])^x$, using Newton's general binomial formula. Also, expand into powers of t and u the product of functions $(1 + t)^x(1 + u)^x$. Equate the coefficients of $t^n u^k$ in the resulting expansion of the identity.

28. Use the expression $\binom{x}{k} = \dfrac{x(x - 1)\cdots(x - k + 1)}{k!}$, with $x = -1/2$, for the derivation of the first relation and expand into powers of u the identity $(1 - 4u)^{-1/2}(1 - 4u)^{-1/2} = (1 - 4u)^{-1}$ to conclude the second relation.

29. Use Pascal's triangle to derive the relations

$$\binom{x}{k}\binom{-x}{n - k} = \frac{n - k}{n}\binom{x - 1}{k}\binom{-x}{n - k} - \frac{n - k + 1}{n}\binom{x - 1}{k - 1}\binom{-x}{n - k + 1}$$

and

$$\binom{x}{k}\binom{1 - x}{n - k + 1} = \frac{n(1 - x) - k}{n(n + 1)}\binom{x - 1}{k}\binom{-x}{n - k}$$
$$- \frac{n(1 - x) - (k - 1)}{n(n + 1)}\binom{x - 1}{k - 1}\binom{-x}{n - k + 1}.$$

Summing these relations for $k = 0, 1, \ldots, r$, deduce the required expressions.

30. Expand into powers of t the identity

$$(1 + t)^{x_1 + x_2 + \cdots + x_r} = (1 + t)^{x_1}(1 + t)^{x_2}\cdots(1 + t)^{x_r}$$

and equate the coefficients of t^n in both sides of the resulting expression.

CHAPTER 4

1. $N(A_1' A_2' A_3') = 10^4 - 3 \cdot 9^4 + 3 \cdot 8^4 - 7^4 = 204.$

2. $N(A'B') = 1000 - 31 - 10 + 3 = 962.$

3. $N(A_1' A_2' A_3' A_4' A_5') = 25^5 - 25 \cdot 24^4 + 10 \cdot 20 \cdot 23^3 - 10 \cdot 60 \cdot 22^2$
$\qquad\qquad + 5 \cdot 120 \cdot 21 - 120.$

4. $N(A_1 \cup A_2 \cup A_3 \cup A_4 \cup A_5) = \binom{5}{1}\binom{8}{2} - \binom{5}{2} = 130.$

5. $\binom{9}{3} - \left\{ \binom{7}{3} + \binom{6}{3} + \binom{5}{3} + \binom{4}{3} \right\} - \left\{ \binom{4}{3} + \binom{3}{3} \right\} = 20.$

6. $N_{5,2} = \binom{5}{2} \sum_{j=0}^{3} (-1)^j \binom{3}{j} (3-j)^4 = 360.$

7. Apply the inclusion and exclusion principle with Ω the set of positive integers less than or equal to $70n$ and A_i its subset that contains the integers divisible by the prime s_i, $i = 1, 2, 3$, with $s_1 = 2$, $s_2 = 5$ and $s_3 = 7$.

8. Apply the inclusion and exclusion principle with Ω the set of positive integers less than or equal to 100 and A_i its subset that contains the multiples of s_i^2, $i = 1, 2, 3, 4$, with $s_1 = 2$, $s_2 = 3$, $s_3 = 5$ and $s_4 = 7$.

9. Apply the inclusion and exclusion principle with Ω the set of positive integers less than or equal to n and A_i its subset that contains the integers divisible by the prime s_i, $i = 1, 2, \ldots, r$.

10. Consider the set $\Omega = \{2, 3, \ldots, r\}$ and its subset A_i that contains the multiples of a_i, $i = 1, 2, \ldots, n$, with $n = E(\sqrt{r})$, and show that $E(r) - E(\sqrt{r}) = N(A_1' A_2' \cdots A_n')$. Apply the inclusion and exclusion principle to establish the required expression.

11. Apply the result of Exercise 10.

12. Consider the set Ω of nonnegative integer solutions of the linear equation $x_1 + x_2 + \cdots + x_n = k$, with k a nonnegative integer, and the set $A_i \subseteq \Omega$ that contains the solutions for which $x_i \geq 2$, $i = 1, 2, \ldots, n$. Evaluate the number $N(A_1' A_2' \cdots A_n')$ (a) directly and (b) applying the inclusion and exclusion principle to establish the required identity.

13. Consider the set Ω of nonnegative integer solutions of the linear equation $x_1 + x_2 + \cdots + x_n = k$, with k a nonnegative integer, and the set $A_i \subseteq \Omega$ that contains the solutions for which $x_i = 0$, $i = 1, 2, \ldots, n$. Evaluate the number $N(A_1' A_2' \cdots A_n')$ (a) directly and (b) applying the inclusion and exclusion principle to establish the required identity.

14. Consider the set Ω of positive integer solutions of the linear equation $x_1 + x_2 + \cdots + x_n = k$, with k a positive integer, and the set $A_i \subseteq \Omega$ that contains the solutions for which $x_i > 6$, $i = 1, 2, \ldots, n$. Then apply the inclusion and exclusion principle to evaluate the required number $N(A_1' A_2' \cdots A_n')$.

15. Note that the number $U(n, k, r)$ is equal to the number of positive integer solutions $(j_1, j_2, \ldots, j_k)$ of the linear inequality $j_1 + j + 2 + \cdots + j_k \leq r$ which satisfy the restrictions $j_s \leq n$, $s = 1, 2, \ldots, k$. Use the inclusion and exclusion principle to derive the required expression.

16. Note that the number $C(n, k, s)$ is equal to the number of nonnegative integer solutions of the linear equation $x_1 + x_2 + \cdots + x_{n-k+1} = k$ satisfying the restrictions $x_i \leq s - 1$, for $i = 1, 2, \ldots, n$. Use the inclusion and exclusion principle to derive the required expression.

17. Consider the number $B_u(n, k, s)$ of $(n - k)$-combinations $\{j_1, j_2, \ldots, j_{n-k}\}$ of $\{1, 2, \ldots, n\}$, $1 \leq j_1 < j_2 < \cdots < j_{n-k} \leq n$, for which $0 < j_{m+1} - j_m \leq s$, $m = 1, 2, \ldots, n - k - 1$ and $n - (j_{n-k} - j_1) > u$, and show that $B(n, k, s) = B_0(n, k, s) - B_s(n, k, s)$. Further, consider the set Ω of $(n - k)$-combinations $\{j_1, j_2, \ldots, j_{n-k}\}$ of $\{1, 2, \ldots, n\}$, with $1 \leq j_1 < j_2 < \cdots < j_{n-k} \leq n$, $n - (j_{n-k} - j_1) > u$ and its subset A_m that contains those combinations for which $j_{m+1} - j_m > s$, $m = 1, 2, \ldots, n-k-1$. Verify that $B_u(n, k, s) = N(A_1' A_2' \cdots A_{n-k-1}')$ and use the inclusion and exclusion principle to evaluate this number.

18. Consider the set Ω of k-combinations $\{i_1, i_2, \ldots, i_k\}$, $1 \leq i_1 < i_2 < \cdots < i_k \leq n$, for which $d_m = i_{m+1} - i_m > a_m$, $m = 1, 2, \ldots, k - 1$ and its subset A_m that contains those combinations for which $d_m > b_m$, $m = 1, 2, \ldots, k - 1$. Verify that the required number equals $N(A_1' A_2' \cdots A_{k-1}')$ and evaluate it by using the inclusion and exclusion principle and the result of Exercise 2.57.

19. Consider the number $B_u(n, k; a_1, b_1, \ldots, a_{k-1}, b_{k-1})$ of k-combinations $\{i_1, i_2, \ldots, i_k\}$, $1 \leq i_1 < i_2 < \cdots < i_k \leq n$, with $a_m < d_m < b_m$, $m = 1, 2, \ldots, k - 1$ and $n - d > u$, and show that

$$B(n, k; a_1, b_1, \ldots, a_k, b_k) = B_{a_k}(n, k; a_1, b_1, \ldots, a_{k-1}, b_{k-1})$$
$$- B_{b_k}(n, k; a_1, b_1, \ldots, a_{k-1}, b_{k-1}).$$

Further, consider the set Ω of k-combinations $\{i_1, i_2, \ldots, i_k\}$, $1 \leq i_1 < i_2 < \cdots < i_k \leq n$, with $d_m > a_m$, $m = 1, 2, \ldots, k-1$, $n-d > u$ and its subset A_m that contains those combinations for which $d_m > b_m$, $m = 1, 2, \ldots, k - 1$. Verify that $B_u(n, k; a_1, b_1, \ldots, a_{k-1}, b_{k-1}) = N(A_1' A_2' \cdots A_{k-1}')$ and evaluate it by using the inclusion and exclusion principle and the result of Exercise 2.57.

20. Work as in Example 4.5 by considering the subsets $A_1, A_2, \ldots, A_{2n-1}$ of the set Ω of the permutations of $\{1, 2, \ldots, n\}$.

21. Use the explicit expressions of M_n and L_n to express M_n in terms of L_n. Then express M_n and L_n in terms of the auxiliary numbers

$$K_n = \sum_{k=0}^{n} (-1)^k \binom{2n-k+1}{k} (n-k)!$$

and, crossing out K_n, deduce the required recurrence relations for the M_n and L_n.

22. Consider the set Ω of seating n married couples around a circular table and its subset A_j which contains the seating of the j-th man next to his wife, $j = 1, 2, \ldots, n$. Using the inclusion and exclusion principle, evaluate the number $H_n = N(A_1' A_2' \cdots A_n')$.

23. Work as in Exercise 22.

24. Consider the set Ω of the terms in the development of a determinant of order n and its subset A_i that includes the terms containing the element a_i, $i = 1, 2, \ldots, n$. Using the inclusion and exclusion principle, evaluate $D_{n,k}$.

25. Consider the set Ω of the different outcomes of a series of k throws of a die and its subset A_i of outcomes in which the i-th face does not appear, $i = 1, 2, \ldots, 6$. Using the inclusion and exclusion principle, evaluate the number $Q(k, r) = N(A_{i_1}' A_{i_2}' \cdots A_{i_r}')$.

26. Consider the subset B_i of Ω_k that contains the different outcomes of a series of k throws of a pair of distinguishable dice in which the pair (i, i) does not appear, $i = 1, 2, \ldots, 6$. Using the inclusion and exclusion principle, evaluate the number $N(A_k) = N(B_1' B_2' \cdots B_6')$.

27. Consider the set Ω of k-permutations $(a_1, a_2, \ldots, a_k)$ of $W_n = \{w_1, w_2, \ldots, w_n\}$ with repetition and its subset A_i which includes those permutations that do not contain the element w_i, $i = 1, 2, \ldots, n$. Evaluate the number $U_1(n, k) = N(A_1' A_2' \cdots A_n')$ by applying the inclusion and exclusion principle.

28. Work as in Example 4.7.

29. Consider the set Ω of the permutations of the $2n$ elements and its subset A_j of those permutations in which the two like elements of the j-th kind are consecutive, $j = 1, 2, \ldots, n$, and apply the inclusion and exclusion principle to evaluate the number $Q_n = N(A_1' A_2' \cdots A_n')$.

30. Consider the set Ω of the s-tuple drawings and its subset A_i that contains the s-tuple drawings in which number i is not drawn, $i = 1, 2, \ldots, n$. Then apply the inclusion and exclusion principle to evaluate the number $L(n, r, s) = N(A_1' A_2' \cdots A_n')$.

31. Consider the set Ω of divisions $(A_1, A_2, \ldots, A_r)$ of W, with $N(W) = n$ and $N(A_i) \geq s$, $i = 1, 2, \ldots, n$, and its subset D_i that contains those divisions for which $N(A_i) = s$, $i = 1, 2, \ldots, r$. Note that $D(n, r, s + 1) = N(D_1' D_2' \cdots D_r')$ and use the inclusion and exclusion principle to evaluate this number in terms of the numbers $N(D_{i_1} D_{i_2} \cdots D_{i_j})$, $\{i_1, i_2, \ldots, i_j\} \subseteq \{1, 2, \ldots, r\}$.

32. Work as in Example 4.10 and use the result of Exercise 31.

33. Express the product sum of binomial coefficients, $r! C(n, r; s)/n!$, as a coupon collector's number (see Example 4.11) and using the inclusion and exclusion principle, evaluate it.

34. Work as in Exercise 33.

35. Use expression (4.20) and the result of Exercise 2.32.

36. Use expression (4.21) and the result of Exercise 2.32.

37. (a) The necessity of the condition follows from $f(x, x)g(x, x) = 1$ for every x in P. The sufficiency is shown by induction starting from $g(x, x) = [f(x, x)]^{-1}$ for every x in P since P is locally finite. (b) Introducing into the expression $h(x, y) = \sum_{x \leq w \leq y} h(x, w)\delta(w, y)$ the function $g(x, y)$, through the result of part (a), and interchanging the order of summation, conclude that $h(x, y) = g(x, y)$. (c) Set into the expression of part (a) $f(x, y) = \zeta(x, y)$ and $g(x, y) = h(x, y) = \mu(x, y)$.

38. Use the zeta function and the fact that $\sum_{z \leq y \leq x} \zeta(z, x)\mu(y, x) = \delta(z, x)$.

39. Use the expression of $\mu(X, Y)$ in part (c) of Exercise 37 to show that the relation holds for $N(Y) - N(X) = 0, 1$ and, assuming that it holds for $N(Y) - N(X) \leq r - 1$, show that it also holds for $N(Y) - N(X) = r$.

40. Use the results of Exercises 38 and 39.

CHAPTER 5

1. Use the explicit expression of D_n and the inequality connecting the remainder of an alternating series with the first omitted term.

2. Use the explicit expression of D_n to derive the first recurrence and using it deduce the second recurrence relation.

3. Note that, in a permutation $(j_1, j_2, \ldots, j_n)$ of the set $\{1, 2, \ldots, n\}$, which is of rank k, there are r elements among the $n - k$ elements $(j_{k+1}, j_{k+2},$

..., j_n) that remain fixed, for $r = 0, 1, \ldots, n - k$. Then use the addition principle to deduce the required expression.

4. Summing the recurrence relation for $k = 2, 3, \ldots, n$, conclude that $R_{n-1} = (n-1)!$. Multiplying the recurrence relation by k and summing for $k = 2, 3, \ldots, n$, conclude that $E_{n-1} = n! - D_n$.

5. Using the relation $n - s - 1 = (n - k) + (k - s - 1)$, split the explicit expression of $R_{n,k}$.

6. Consider the set Ω of permutations $(j_1, j_2, \ldots, j_n)$ of $\{1, 2, \ldots, n\}$ and its subset A_s that contains the permutations in which j_s is a fixed point, $s = 1, 2, \ldots, n$. Use the inclusion and exclusion principle to evaluate the number $E_{n,r} = N(A'_{i_1} A'_{i_2} \cdots A'_{i_r})$. Use Pascal's triangle to split the explicit expression of $E_{n,r}$ and then the relation $n - j = (n - r) + (r - j)$, derive the first and second recurrences, respectively.

7. Consider the subsets $A_{i_1}, A_{i_2}, \ldots, A_{i_r}$ of Ω defined in Exercise 6 and evaluate the number $E_{n,r,k}$ of elements of Ω that are contained in k among the r sets $A_{i_1}, A_{i_2}, \ldots, A_{i_r}$ by applying the inclusion and exclusion principle.

8. Using the relation $n - k - j = (n - k) - j$, split the explicit expression of $E_{n,r,k}$.

9. Utilize the fact that there are s fixed elements among the $n - r$ non-specified positions for $s = 0, 1, \ldots, n - r$.

10. Consider the set Ω of allocations of n letters and n invoices into n envelopes and its subset A_r that contains the allocations in which both the r-th letter and the r-th invoice are placed in the r-th envelope, for $r = 1, 2, \ldots, n$. Use the inclusion and exclusion principle to evaluate the number $Q_n = N(A'_1 A'_2 \cdots A'_n)$.

11. Utilize the explicit expression of Q_n.

12. Consider the subsets $A_1, A_2, \ldots, A_n$ of Ω defined in Exercise 10 and evaluate the number $Q_{n,k}$ of elements of Ω that are contained in k among the n sets by applying the inclusion and exclusion principle.

13. Consider the set Ω of the different drawings of the balls from the r urns and its subset A_j that contains the drawings in which the j-th ball is drawn from each urn at the j-th drawing, $j = 1, 2, \ldots, n$. Evaluate the number $B_{n,r} = N(A'_1 A'_2 \cdots A'_n)$ and the number $B_{n,r,k}$ of elements of Ω that are contained in k among the n sets $A_1, A_2, \ldots, A_n$ by applying the inclusion and exclusion principle.

14. Use the explicit expression of $B_{n,r}$.

15. Consider the set Ω of the different drawings of the sn balls from the urn and its subset A_i that contains the drawings in which a ball bearing the number i is drawn at the i-th drawing, $i = 1, 2, \ldots, n$. Evaluate the number $G_{n,s} = N(A'_1 A'_2 \cdots A'_n)$ and the number $G_{n,k,s}$ of elements of Ω that are contained in k among the n sets $A_1, A_2, \ldots, A_n$ by applying the inclusion and exclusion principle.

16. Consider the subsets $A_1, A_2, \ldots, A_n$ of Ω that are defined in Exercise 15. Then $H_{n,k,s}$ is the number of elements of Ω of rank k with respect to the sets $A_1, A_2, \ldots, A_n$.

17. Note that the permutations of $\{1, 2, \ldots, n, n+1\}$ may be constructed from the permutations of $\{1, 2, \ldots, n\}$ by attaching the element $n + 1$ in one of the $n + 1$ possible positions. Further, this attachment either alters (reduces or increases by one) or keeps unchanged the number of successions.

18. Use the expressions connecting $C_{n,k}$ and $D_{n,k}$ with C_n and D_n, respectively, along with the expressions (5.12) and part (a) of Exercise 2.

19. Consider the set $A_{r,s}$ of permutations of $\{1, 2, \ldots, n\}$ that include the transposition (r, s), for $r, s = 1, 2, \ldots, n$. Note that there are $\binom{n}{2}$ different such sets and a permutation of $\{1, 2, \ldots, n\}$ is contained in at most $m = [n/2]$ among these sets. Using the inclusion and exclusion principle, evaluate the number $T_{n,k}$ of permutations of $\{1, 2, \ldots, n\}$ that are contained in exactly k among these sets.

20. Use the explicit expression of $T_{n,k}$ for $k = 2r$ and $k = 2r + 1$.

CHAPTER 6

1. $A(t) = \displaystyle\sum_{k=7}^{\infty} a_k t^k = t^7 (1-t)^{-5}, \quad a_k = \binom{k-3}{4}, \quad k = 7, 8, \ldots.$

2. $A(t) = \displaystyle\sum_{k=3}^{18} a_k t^k = (t + t^2 + \cdots + t^6)^3 = t^3 (1 - t^6)^3 (1 - t)^{-3},$

$a_9 = \binom{8}{2} - \binom{3}{1}\binom{2}{2} = 25, \quad a_{11} = \binom{9}{2} - \binom{3}{1}\binom{3}{2} = 27.$

3. $A(t) = \displaystyle\sum_{k=1}^{17} a_k t^k = (1 + t^2 + \cdots + t^8)(t + t^3 + \cdots + t^9)$

$= t(1 - t^{10})^2 (1 - t^2)^{-2}, \quad a_7 = 4, \quad a_9 = 5, \quad a_{11} = 4.$

4. $A(t) = (1 + t + \cdots + t^{13})^4 = (1 - t^{14})^4 (1 - t)^{-4}$,

$$a_{13} = \binom{16}{3} = 280, \quad a_{26} = \binom{29}{3} - \binom{4}{1}\binom{15}{3} = 1834.$$

5. (a) $E(t) = \sum_{k=0}^{\infty} a_k \frac{t^k}{k!} = (1 + t)^5$,

(b) $F(t) = \sum_{k=0}^{\infty} b_k \frac{t^k}{k!} = \left(1 + t + \frac{t^2}{2!} + \cdots\right)^5 = e^{5t}$.

6. (a) $E(t) = \sum_{k=0}^{\infty} a_k \frac{t^k}{k!} = \left(1 + t + \frac{t^2}{2!} + \cdots\right)^8 \left(t + \frac{t^3}{3!} + \cdots\right)^2$

$$= (e^{10t} - 2e^{8t} + e^{6t})/4, \quad a_k = (10^4 - 2 \cdot 8^k + 6^k)/4,$$

(b) $F(t) = \sum_{k=0}^{\infty} b_k \frac{t^k}{k!} = \left(1 + t + \frac{t^2}{2!} + \cdots\right)^8 \left(1 + \frac{t^2}{2!} + \frac{t^4}{4!} + \cdots\right)^2$

$$= (e^{10t} + 2e^{8t} + e^{6t})/4, \quad b_k = (10^k + 2 \cdot 8^k + 6^k)/4.$$

7. (a) $E(t) = \sum_{k=0}^{\infty} a_k \frac{t^k}{k!} = \frac{1}{4}(e^{8t} - e^{6t}), \quad a_k = (8^k - 6^k)/4,$

(b) $F(t) = \sum_{k=0}^{\infty} b_k \frac{t^k}{k!} = \frac{1}{2}(e^{10t} - e^{bt}), \quad b_k = (10^k - 6^k)/2.$

8. (a) $E(t) = \sum_{k=0}^{\infty} a_k \frac{t^k}{k!} = e^{4t} + e^{2t}, \quad a_k = 4^k + 2^k,$

(b) $F(t) = \sum_{k=0}^{\infty} b_k \frac{t^k}{k!} = e^{4t} + e^{3t} - 1, \quad b_k = 4^k + 3^k.$

9. $T_n(t) = \sum_{k=0}^{2n} T(n, k)t^k = (1 + t + t^2)^n$. Use first Newton's binomial formula to expand $T_n(t) = ((1 + t) + t^2)^n$. Then, expand the resulting expression into powers of t to deduce the first expression. In order to derive the second expression, expand each factor of $T_n(x) = (1 - t^3)^n (1 - t)^{-n}$ into powers of t and form the Cauchy product of series.

10. Expand each member of $T_n(t) = T_{n-1}(t) + tT_{n-1}(t) + t^2 T_{n-1}(t)$ into powers of t and equate the coefficients of t^k in both members of the resulting expression.

11. Differentiate the generating function $T_n(t)$ to get the expressions $T'_n(t) = nT_{n-1}(t) + 2nT_{n-1}(t)$, $T''_n(t) = 2n(2n-1)T_{n-1}(t) - 3n(n-1)T_{n-2}(t)$, which, expanded into powers of t, imply the required recurrence relations.

12. Cross out the term $T(n-1, k-2)$ between the two recurrence relations of Exercise 11 to deduce the recurrence relation for $T_n \equiv T(n, n)$. Multiplying the recurrence relation by t^{n-1} and summing for $n = 2, 3, \ldots$, derive the first order differential equation $(1 - 2t - et^2)T'(t) = (1 + et)T(t)$, with $T(0) = 1$. Integrate it to obtain the required expression.

13. Use Lagrange formula with $f(t) = 1$, $g(t) = 1 + t + t^2$ to show that $T(n) = (1 - u - 2ut)^{-1}$, with $t = h(u)$ the inverse of $u = h^{-1}(t) = t/(1 + t + t^2)$, and thus conclude that $T(u) = \mp(1 - 2u - 3u^2)^{-1/2}$, where the case $-(1 - 2u - 3u^2)^{-1/2}$ is rejected.

14. $A_{n,r}(t) = \sum_{k=0}^{2n+r} A(n, r, k)t^k = (1 + t + t^2)^n(1 + t)^r$. Expand first the term $((1 + t) + t^2)^n$, using Newton's binomial formula, and then expand the resulting expression of $A_{n,r}(t)$ into powers of t to obtain the first expression. In order to derive the second expression, expand each member of $A_{n,r}(t) = T_n(t)(1 + t)^n$ into powers of t and form the Cauchy product of series.

15. $A_n(t) = \sum_{k=0}^{\infty} G(n, k)t^k = (1 + t^2 + t^4 + \cdots)^n = (1 - t^2)^{-n}$.

16. $B_n(t) = \sum_{k=0}^{\infty} H(n, k)t^k = (t + t^3 + t^5 + \cdots)^n = t^n(1 - t^2)^{-n}$.

17. $B_{n,s}(t) = \sum_{k=0}^{n+s} B(n, k, s)t^k = (1 + t + t^2 + \cdots + t^s)(1 + t)^n$. Expand the second factor into powers of t, using Newton's binomial formula, and in the resulting expansion of $B_{n,s}(t)$, equate the coefficients of t^k. Further, derive the relation $(1 - t)B_{n,s} = (1 - t^{s+1})(1 + t)^n$, which, expanded into powers of t, yields the required expression.

18. $E_{n,s}(t) = \sum_{k=0}^{sn} E_s(n, k)t^k = (1 + t + \cdots + t^s)^n$. Expand each member of $E_{n,s}(t) = E_{n-1,s}(t) + tE_{n-1,s}(t) + \cdots + t^s E_{n-1,s}(t)$ into powers of t and equate the coefficients of t^k in both sides of the resulting expression.

19. Expand the identity $A_{n,s}(t) = A_{r,s}(t)A_{n-r,s}(t)$ into powers of t. Also, expand $A_{n,s}(t) = (1 + t(1 + t + \cdots + t^{s-1}))^n$, using Newton's binomial formula, and then expand the resulting expression into powers of t.

20. Use the exponential and the geometric series to deduce the expression of the generating function $P(t)$, which can be written in the form $P(t) = tP(t) + e^t$. Expanding both members of the last relation into powers of t, deduce the required recurrence relation.

21. $A_n(t) = \sum_{k=0}^{\infty} A(n, k)\dfrac{t^k}{k!} = \left(1 + \dfrac{t^2}{2!} + \dfrac{t^4}{4!} + \cdots\right)^n = \left(\dfrac{e^t + e^{-t}}{2}\right)^n$.

Expand the generating function to get the explicit expression. Differentiate twice the generating function to find the differential equation $A_n''(t) = n^2 A_n(t) + n(n-1)A_{n-2}(t)$, which implies the required recurrence relations.

22. $B_n(t) = \sum_{k=0}^{\infty} B(n, k)\dfrac{t^k}{k!} = \left(t + \dfrac{t^3}{3!} + \dfrac{t^5}{5!} + \cdots\right)^n = \left(\dfrac{e^t - e^{-t}}{2}\right)^n$.

Expand the generating function to find the required expression.

23. Differentiate the generating function $B_n(t)$ to get the differential equation $B_n''(t) = n^2 B_n(t) + n(n-1)B_{n-2}(t)$, which implies the required recurrence relations.

24. $R_n(t) = \sum_{k=0}^{2n} R(n,k)t^k/k! = (1 + t + t^2/2)^n$. This generating function satisfies the relation $R_{n+1}(t) = R_n(t)+tR_n(t)+(t^2/2)R_n(t)$, which implies the first recurrence relation. Further, differentiate $R_n(t)$ to find the differential equations $R_n'(t) = nR_n(t) - (nt^2/2)R_{n-1}(t)$ and $R_n''(t) = n(2n-1)R_{n-1}(t) - n(n-1)R_{n-2}(t)$, which, expanded into powers of t, imply the other two recurrence relations.

25. Sum for all $k = 0, 1, \ldots, 2n - 2$ the third recurrence relation of Exercise 24 to find the required recurrence relation.

26. Use the fact that the stick is initially broken into two parts of length j and $n-j$ units, respectively, to derive the recurrence relation. Note that the recurrence relation and its initial conditions coincide with the recurrence relation and initial conditions of the Catalan numbers.

27. Consider the vertices $v_1, v_2, \ldots, v_n$ of the convex n-gon circularly ordered and note that $v_n v_1$ is a side of the triangle $v_n v_1 v_k$, for $k = 3, 4,$ $\ldots, n-2$, obtained by ruling the diagonals $v_n v_k$ and $v_k v_1$, or the triangle $v_n v_1 v_2$ obtained by ruling the diagonal $v_n v_2$, or the triangle $v_n v_1 v_{n-1}$ obtained by ruling the diagonal $v_{n-1}v_1$. Use that the ruling of these diagonals divides the convex n-gon into a convex k-gon and a convex $(n-k+1)$-gon, for $k = 3, 4, \ldots, n-2$, or into a triangle and a convex $(n-1)$-gon, or into a convex $(n-1)$-gon and a triangle, to establish the recurrence relation. Multiplying it by t^n and summing for $n = 3, 4, \ldots$, derive the generating function which implies the required expression.

28. (a) Multiplying the recurrence relation by t^n and summing for $n = 1, 2, \ldots$, derive the recurrence $C_k(t) = C_1(t)C_{k-1}(t)$, $k = 2, 3, \ldots$, and iterate it to find the required expression. (b) Show that $C_{k+1}(t) = C_k(t) - tC_{k-1}(t)$ and deduce the required recurrence relation. (c) Expand the generating function $C_k(t)$ to get the explicit expression of the numbers $C_n^{(k)}$.

29. Work as in Example 6.4.

30. Use the symbolic calculus to show that $S(t,u) = Se^{Su+tu}$, $S^r \equiv S_r$. Then use the result of Exercise 29 to deduce the required expression of the generating function.

31. Determine the moment generating function $B(t) = \sum_{r=0}^{\infty} \mu_{(r)}t^r/r!$ and from it deduce the probability generating function $P(t) = \sum_{k=0}^{\infty} p_k t^k$. Expand it to find the probabilities $p_k = e^{-\lambda}\lambda^k/k!$, $k = 0, 1, \ldots$.

32. Use the expansion of the exponential function and the geometric series.

33. Use the general binomial formula and the geometric series.

34. In order to evaluate the series $A_n(t) = \sum_{k=0}^{\infty} a_{n,k} t^k$, split it for $k \leq n-1$ and $k \geq n$.

35. Use the negative binomial formula.

CHAPTER 7

1. $a_n = (1+r)^n \cdot 10,000.$

2. $a_n = 10 \cdot 9^{n-1}.$

3. $(b+1)a_n = \sum_{k=0}^{n} a_k + a$, $n = 2, 3, \ldots$, $(b+1)a_1 = a_1 + a$. Take the difference $(b+1)a_n - (b+1)a_{n-1}$ to establish the required recurrence relation. Iterate it to get $a_n = (a/b)(1+1/b)^{n-1}.$

4. $a_n = \binom{n}{2}.$

5. Consider a plane divided into R_{n-1} regions by $n-1$ lines, each two of them having a point in common but no three of them having a point in common. Rule the n-th line and establish the recurrence relation. Iterate it to find the unique solution.

6. Consider also the number b_n of n-permutations of the set $\{0, 1, 2\}$ with repetition that include an odd number of zeros and establish the relation $a_{n+1} = 2a_n + b_n$, which, along with the relation $a_n + b_n = 3^n$, implies the required recurrence relation. Its solution is $a_n = 2 + 3(3^{n-1} - 1)/2.$

7. Iterate the recurrence to get $a_n = (2n-1)(2n-3)\cdots 3 \cdot 1/(2^n \cdot n!)$ and multiply it by $2n(2n-2)\cdots 4 \cdot 2 = 2^n n!$ to establish the required expression.

8. Derive the auxiliary relations $a_{n+1} = 9a_n + b_n$ and $b_{n+1} = a_n + 9b_n$ and deduce the required recurrence relations.

9. Use the fact that the last two elements of an n-digit nonnegative integer are either different or like to derive the required recurrence relation.

10. Use the fact that the first position of such a permutation may be occupied either by any of the numbers $\{1, 2\}$ or by 0 to establish the required recurrence relation.

11. Use the fact that the last two elements of such a permutation are either two different elements of the set $\{0, 1, 2\}$ or two 2s to establish the required recurrence relation.

12. Use Pascal's triangle to derive the relations $a_n = b_n - a_{n-1}$ and $b_n = 4a_{n-1} - b_{n-1}$.

13. $a_n = \dfrac{1}{2\sqrt{5}} \left\{ \left(2 + \sqrt{5}\right)^{n+1} - \left(2 - \sqrt{5}\right)^{n+1} \right\}$.

14. $a_n = c_1 + c_2(-2)^n + 2^{n-2}$.

15. $a_n = c_1 + c_2 n + 3^{n-1}/4$.

16. Use the fact that in such a permutation either the last element is different from its previous one or the last j elements are like elements different from their previous element, for $j = 2, 3, \dots, r - 1$.

17. (a) Utilize the fact that the first position of such a permutation may be occupied either by any of the numbers $\{1, 2, \dots, s\}$ or by 0. (b) Note that such a permutation contains at least k elements from the set $\{1, 2, \dots, s\}$, with $k = n/2$ if n is even and $k = (n-1)/2$ if n is odd. (c) Multiply the recurrence relation by t^{n+1} and sum for $n = 1, 2, \dots$.

18. Distinguish these permutations according to the number of zeros each permutation contains and show that the number of those permutations that contain k zeros equals $\binom{n-k}{k} s^{n-k-1}$.

19. The first expression may be derived either by the method of characteristic roots or by the method of generating function. Expand this expression, using Newton's binomial formula, to establish the second expression.

20. Use the fact that the last two elements of such a permutation are either different elements of the set Ω or like elements of Ω_1 to establish the recurrence relation.

21. Distinguish these permutations according to the number of 0s that they include before the first 1 in order to get the recurrence relation.

22. Distinguish these permutations according to the number of 0s each permutation contains and show that the number of those permutations that contain k 0s equals $E_{s-1}(n - k, k)$.

23. Work as in Example 7.8.

24. Distinguish these cyclic permutations according to the number of 0s that follow the marked element, if this is 1, or the first 1 after the marked element.

25. Use the result of Exercise 2.55 to find the explicit expression

$$f_{n+s,s} = \sum_{k=0}^{m} \binom{n+s-ks}{k}, \quad m = [(n+s)/(s-1)]$$

and, using Pascal's triangle, deduce the recurrence relation.

26. Use the result of Exercise 2.56 to find the explicit expression

$$g_{n,s} = \sum_{k=0}^{m} \frac{n}{n-ks} \binom{n-ks}{k}, \quad m = [n/(s+1)]$$

and, using Pascal's triangle, deduce the recurrence relation.

27. (a) Multiply the general recurrence relation by t^n and sum for $n = 0, 1, \ldots$ to get for the generating function $F_k(t)$ the recurrence relation $F_k(t) = F_1(t)F_{k-1}(t)$, $k = 2, 3, \ldots$. Iterate it to find the explicit expression for $F_k(t)$. (b) Expand $F_k(t)$ into powers of t to obtain $f_n^{(k)}$. (c) Multiply $F_k(t)$ by u^k and sum for $k = 0, 1, \ldots$ to derive $F(t, u)$.

28. Consider a plane divided in $R_{n-1,k-1}$ regions by $n - 1$ lines, $k - 1$ of them being parallel to each other but no three of them having a point in common. Rule the n-th line and establish the recurrence relation. Iterate it to find the unique solution.

29. Multiplying the recurrence relation by $t^{n-2}/(n - 2)!$ and summing for $n = 3, 4, \ldots$, derive the differential equation $(1 - t)S'(t) = (1 + t)S(t)$; integrating it, deduce the required expression of $S(t)$.

30. (a) Show first that

$$\binom{n}{k}^{-1} = \frac{n+1}{n}\binom{n-1}{k}^{-1} - \binom{n}{k+1}^{-1}$$

and then, introducing it into the expression of S_n, conclude the recurrence relation. (b) Multiplying the recurrence relation by t^{n-1} and summing for $n = 1, 2, \ldots$, derive the differential equation $(1-t/2)S'(t) - S(t) = (1-t)^{-2}$ and, solving it, find the generating function. (c) Expand the generating function to get the required expression of S_n.

31. Use Pascal's triangle to derive the relations $a_n = a_{n-1} + b_n$ and $b_n = b_{n-1} + a_{n-1}$, which imply the generating function relations $A(t) - 1 = tA(t) + B(t)$ and $B(t) = tB(t) + tA(t)$. Solve this system to get the expression of $A(t)$ and $B(t)$.

32. The number B_n, $n = 0, 1, \ldots$, satisfy the recurrence relation (see Exercise 2.46)

$$B_n = \sum_{k=1}^{n} \binom{n-1}{k-1} B_{k-1}, \quad B_0 = 1.$$

Multiplying it by $t^{n-1}/(n - 1)!$ and summing for $n = 1, 2, \ldots$, derive the differential equation $B'(t) = B(t)e^t$. Integrate it to get $B(t)$ and expand it to find the required expression of B_n.

33. Consider the set $\mathcal{P}_0$ of partitions of a finite set W_n of n elements into subsets with even number of elements and its subset $\mathcal{B}_k$ that contains the partitions in which the element $w_n \in W_n$ belongs in a subset with $2k$

elements, for $k = 1, 2, \ldots, m$, $m = [n/2]$. Then, $\mathcal{P}_0 = \mathcal{B}_1 + \mathcal{B}_2 + \cdots + \mathcal{B}_m$. Applying the addition principle, conclude the recurrence relation. Multiplying it by $t^{n-1}/(n-1)!$ and summing for $n = 1, 2, \ldots$, derive the differential equation $A'(t) = A(t) \sinh t$, the solution of which gives the required expression of the generating function $A(t)$.

34. Expand the generating function $A(t)$ into powers of t and use the generating function constructed in Exercise 6.15.

35. Work as in Exercise 33.

36. Expand the generating function $E(t)$ into powers of t and use the generating function constructed in Exercise 6.16.

37. Use the fact that the counting of the $n + k$ votes with $x_r \geq y_x$, $r = 1, 2, \ldots, n + k$ is terminated either with a vote for $\mathcal{N}$ or a vote for $\mathcal{K}$ to derive the recurrence relation.

38. (a) Use the fact that the counting of the votes is terminated with a vote for $\mathcal{N}$ or with a vote for $\mathcal{K}$ or with a vote by both $\mathcal{N}$ and $\mathcal{K}$ to establish the recurrence relation. (b) Rewrite the generating function as $G(t, u) = (1 - u)^{-1}[1 + t(1 + u)(1 - u)^{-1}]^{-1}$ and, expanding it, conclude the first expression of $D(n, k)$. Further, rewrite the generating function as $G(t) = (1 - t)^{-1}(1 - u)^{-1}[1 - 2t(1 - t)^{-1}u(1 - u)^{-1}]^{-1}$ and, expanding it, conclude the second expression of $D(n, k)$.

39. Rewrite the bivariate generating function $G(t, u)$ of the Delannoy numbers (see Exercise 38) in the form $G(t, u) = (1-t)^{-1}[1-(1+t)(1-t)u]^{-1}$ and, expanding it, into powers of u, derive the generating function $G_n(t) = \sum_{k=0}^{n} D(n, k)t^k = (1+t)^n(1-t)^{-n-1}$. Further, use Lagrange formula with $f(t) = (1 - t)^{-1}$, $g(t) = (1 + t)(1 - t)^{-1}$ to find the expression of $G(u)$. Differentiate it to deduce the differential equation $(1 - bu + u^2)G'(u) - (3u - 2)G(u) = 0$, which implies the required recurrence relation.

40. Show first that

$$\binom{n}{k}^{-1} = \frac{n+1}{n}\binom{n-1}{k}^{-1} - \binom{n}{k+1}^{-1}$$

and then, introducing it into the expression of $L(n, k)$, deduce the recurrence relation.

CHAPTER 8

1. (a) Expand the identity $(t)_n = [(t)_{n+1}/t] \cdot [1/(1 - n/t)]$ into powers of t. (b) The generating function $\phi_k(u) = \sum_{n=k}^{\infty} S(n, k)u^k$ satisfies the recurrence relation $\phi_k(u) = u(1 - ku)^{-1}\phi_{k-1}(u)$. Expand it into powers of u.

2. Multiply the relation $\sum_{k=0}^{r} s(r+1, k+1)t^k = (t-1)_r$ by $S(n, r)$ and the relation $\sum_{k=0}^{r} s(r, k)t^k = (t)_r$ by $S(n+1, r+1)$ and sum the resulting expressions for $r = 0, 1, \ldots, n$.

3. Use Lagrange inversion formula with $t = \phi^{-1}(u) = e^u - 1$ and $u = \phi(t) = \log(1 + t)$ (see Theorem 8.5) to find the expression

$$S(n, j) = \sum_{r=0}^{n-j}(-1)^r \binom{n+r-1}{j-1}\binom{2n-j}{n-j-r} s(n-j+r, r),$$

which, for $j = n - k$, reduces to the required expression.

4. Expand the function $(t+u)_n$ into powers of $t+u$ and into factorials of t and u. Further, expand both sides of the resulting expression into powers of t and u.

5. Expand the function $(t + u)^n$ into factorials of $t + u$ and into powers of t and u. Further, expand both sides of the resulting expression into factorials of t and u.

6. Use Lagrange inversion formula with $t = \phi^{-1}(u) = (1 + u)^{1/s} - 1$ and $u = \phi(t) = (1 + t)^s - 1$ (see Theorem 8.5) to find the expression

$$C(n, j; s^{-1}) = \sum_{r=0}^{n-j}(-1)^r \binom{n+r-1}{j-1}\binom{2n-j}{n-j-r} C(n-j+r, r; s),$$

which, for $j = n - k$, reduces to the required expression.

7. Expand the function $(st + su)_n$ into factorials of $t + u$ and into factorials of st and su. Further, expand both sides of the resulting expression into factorials of t and u.

8. Using the expansion which defines the signless Stirling numbers of the first kind, derive the relation

$$\sum_{k=0}^{n} |s(n+1, k+1)|t^k = n!\exp\left\{\sum_{r=1}^{\infty}(-1)^{r-1}\zeta_n(r)t^r/r\right\}.$$

Expanding the right-hand side into powers of t, deduce the required expression.

9. Use the expression of Exercise 8 and $\lim_{n\to\infty} \zeta_n(1)/\log(n+1) = 1$, $\lim_{n\to\infty}[\log(1 + n)]^j/n = 0$, $\zeta_n(s) \leq \pi^2/6$, $s = 2, 3, \ldots$, to show that $\lim_{n\to\infty}|s(n+1, k+1)|[\log(n+1)]^{-k}/n! = 1/k!$. Use the explicit expressions of $S(n, k)$ and $C(n, k; s)$ to derive the limits $\lim_{n\to\infty} S(n, k)/k^n = 1/k!$ and $\lim_{n\to\infty} C(n, k; s)/(sn)_n = 1/k!$.

10. Setting successively $k = n + 1, n, n - 1$ in the recurrence relation, conclude the required expressions. Multiplying the relation $s(n, k) =$

$\sum_{j=k}^{n} s_2(n-j, k-j)\binom{n}{j}$ successively by t^k and $u^n/n!$ and summing for $k = 0, 1, \ldots, [n/2]$ and $n = 0, 1, \ldots$, derive the bivariate generating function. Rewrite it as $g_2(t, u) = \exp\{t[\log(1 + u) - u]\}$ and expand it into powers of t to find $f_{k,2}(u)$. Differentiate either the generating $f_{k,2}(u)$ or the bivariate generating function $g_2(t, u)$, with respect to u, to derive the recurrence relation. Expand the bivariate generating function $g_2(t, u)$, first into powers of u and then into powers of t, get the required expression.

11. Differentiate the generating function to get the differential equation $(1 + u)f'_{k,r}(u) = (-1)^{r-1}u^{r-1}f_{k-1,r}(u)$, which, expanded into powers of u, yields the required recurrence relation. Use Newton's binomial formula to express $f_{k,r+1}(u)$ as a sum of $f_{k-j,r}(u)$, $j = 0, 1, \ldots, k$. Also, express $f_{k,r}(u)$ as a sum of $f_{k-j,r+1}(u)$, $j = 0, 1, \ldots, k$.

12. Work as in Exercise 10.

13. Work as in Exercise 11.

14. Work as in Exercise 10.

15. Work as in Exercise 11.

16. Expand both members of the relation $t[f(t) - 1] = [t - \log(1+t)]f(t)$ into powers of t to derive the recurrence relation. Rewrite the generating function as $f(t) = (e^{\log(1+t)} - 1)/\log(1+t)$ and expand it, first into powers of $u = \log(1 + t)$, and then into powers of t to get the expression of the Cauchy numbers in terms of the Stirling numbers of the first kind. Multiply it by $S(r, n)$ and sum for $n = 0, 1, \ldots, r$.

17. Expand both members of the relation $t[g(t) - 1] = -(e^t - 1 - t)g(t)$ into powers of t to derive the recurrence relation. Rewrite the generating function as $g(t) = \log[1 + (e^t - 1)]/(e^t - 1)$ and expand it, first into powers of $u = e^t - 1$, and then into powers of t to get the expression of the Bernoulli numbers in terms of the Stirling numbers of the second kind. Multiply it by $s(r, n)$ and sum for $n = 0, 1, \ldots, r$.

18. Show that the generating function $h(t) = \sum_{n=2}^{\infty} B_n t^n/n! = -1 + t/2 + t/(e^t - 1)$ is an even function and conclude that $B_{2r+1} = 0$, $r = 1, 2, \ldots$. Further, show that $1 + \sum_{r=1}^{\infty} B_{2r}t^{2r}/(2r)! = (t/2)(e^{t/2} + e^{-t/2})/(e^{t/2} - e^{-t/2})$ and, replacing t by $2it$, with $i = \sqrt{-1}$, conclude the required expression. (b) Show that $t \cot t = 1 - 2\sum_{r=1}^{\infty} \pi^{-2r}\zeta(2r)t^{2r}$ and deduce the required expressions.

19. Differentiate the generating function to get the differential equation $g_{r+1}(t) = (1 - t)g_r(t) - (t/r)g'_r(t)$, which, expanded into powers of t, yields the recurrence relation.

20. Expand both members of the relation
$$st[g(t; s) - 1/s] = -[(1 + t)^s - 1 - st]g(t; s)$$

into powers of t to derive the recurrence relation. Rewrite the generating function in the form

$$g(t; s) = \{[1 + [(1 + t)^s - 1]]^{1/s} - 1\}/[(1 + t)^s - 1]$$

and expand it, first into powers of $u = (1 + t)^s - 1$, and then into powers of t to get the required expression. Take the limits of the corresponding generating function.

21. Use the relation

$$\sum_{n=0}^{\infty} \frac{a}{a + bn} \binom{a + bn}{n} u^n = y^a, \quad u = (y - 1)y^{-b},$$

with $a = t$, $b = 1/2$, and then follow the steps of the derivation of the corresponding relations for the Stirling number of the first kind.

22. Follow the steps of the derivation of the corresponding relations for the Stirling numbers.

23. Follow the steps of the derivation of Theorems 8.7(b) and 8.10.

24. Work as in Theorem 8.3(a).

25. Expand both members of $(t+r+n)_{n+1} = (t+r+n)(t+r+n-1)_n$ into powers of t to derive the triangular recurrence relation. Express the rising factorial $(t + r + n - 1)_n$ into a sum of rising factorials $(t + r - m + j - 1)_j$, $j = 0, 1, \ldots, n$, using Vandermonde's formula, and then expand both members of the resulting expression into powers of t to find the required expression.

26. Work in analogy to Example 8.1.

27. Work as in Theorem 8.3(b) and 8.4.

28. Work as in Theorem 8.7(b) to derive the recurrence triangular recurrence relation. Further, express the power $(t + r)^n$ into a sum of powers $(t + r - m)^j$, $j = 0, 1, \ldots, n$, using Newton's binomial formula, and then expand both numbers of the resulting expression into factorials of t to get the required expression.

29. Express $(t+r_2-r_1)_n$ into powers of $t+r_2$ and then express the powers $(t + r_2)^j$ into factorials of t. In addition, express $(t + r_2 - r_1)_n$ directly into factorials of t, using Vandermonde's formula, and equate the coefficients of $(t)_k$ in the resulting expression. Also, express $(t - r_2 + r_1)^n$ into factorials of $t - r_2$ and then express the factorials $(t - r_2)_j$ into powers of t. Further, express $(t + r_1 - r_2)^n$ directly into powers of t, using Newton's binomial formula, and equate the coefficients of t^k in the resulting expression.

30. Work with the function $(t + u - r_1 - r_2)_n$ as in Exercise 4.

31. Work with the function $(t + u + r_1 + r_2)^n$ as in Exercise 5.

32. Follow the steps of the derivation of Theorems 8.10 and 8.12.

33. Work in analogy to Example 8.7.

34. Work as in Theorem 8.14.

35. Work as in Theorem 8.15.

36. Work as in Theorem 8.19.

37. Work as in Theorem 8.17.

38. Expand the non-central factorial $(-(t-r))_n = (-1)^n(t-r+n-1)_n$ into factorials of t, using Vandermonde's formula, and conclude the expression $L(n, k; r) = (-1)^n \binom{n}{k}(n - r - 1)_{n-k}$, which can be transformed into the required expression.

39. Expand the factorial $(s(t + \rho) - r)_n$ into powers of $s(t + \rho)$ and then expand the powers $(t + \rho)^j$ into factorials of t. Also, expand the same factorial $(st + (s\rho - r))_n$ directly into factorials of t and equate the coefficients of $(t)_k$ in the resulting expression.

40. Expand the factorial $(s_1(s_2t + r_2) + r_1)_n$ into factorials of $s_2t + r_2$ and then expand the factorials $(s_2t + r_2)_j$ into factorials of t. Also, expand the same factorial $(s_1s_2t + (r_1 + r_2s_1))_n$ directly into factorials of t and equate the coefficients of $(t)_k$ in the resulting expression.

41. Use the triangular recurrence relation for the numbers $C(n, k; s, r)$, derived in Exercise 36, and $1/(u)_n = (u-n)/(u)_{n+1}$ to deduce the required recurrence relation for $\phi_k(u; s, r)$.

42. Work with the function $(st + su + r_1 + r_2)_n$ as in Exercise 7.

43. Work as in Example 8.4.

44. Work as in Example 8.8.

45. Work as in Example 8.8.

CHAPTER 9

1. Consider the set Ω of distributions of n distinguishable balls into k distinguishable urns and its subset A_i of these distributions in which j balls are allocated in the i-th urn, for $i = 1, 2, \ldots, k$, and apply the inclusion and exclusion principle.

2. (a) Use the enumerator (9.22). (b) Deduce the factorial moment generating function from the generating function of the sequence a_r, $r = 0, 1, \ldots, k$.

3. Use Corollary 9.2.

4. Consider the set Ω of distributions of n indistinguishable balls into k distinguishable urns and its subset A_i of these distributions in which j balls are placed in the i-th urns, for $i = 1, 2, \ldots, k$, and apply the inclusion and exclusion principle.

5. (a) Use the enumerator (9.27). (b) Deduce the binomial moment generating function from the generating function of the sequence a_r, $r = 0, 1, \ldots, k$.

6. Consider the set Ω of distributions of n indistinguishable balls into k distinguishable urns (of unlimited capacity) and its subset A_i of these distributions in which at least $s + 1$ balls are placed in the i-th urn, for $i = 1, 2, \ldots, k$, and apply the inclusion and exclusion principle to find $L(n, k, s)$. Further, consider the set W of distributions of n indistinguishable balls into k distinguishable urns, each of capacity limited to s balls, and its subset B_i of these distributions in which the i-th urn remains empty, for $i = 1, 2, \ldots, k$, and apply the inclusion and exclusion principle to get the expression of $L(n, k, s, r)$.

7. Work as in the second part of Exercise 6 replacing the subset B_i by the subset C_i of those distributions in which j balls are placed in the i-th urn, for $i = 1, 2, \ldots, k$. For the derivation of the second part, use the first part of Exercise 6.

8. Work as in Exercise 5.

9. Note that these distributions may be carried out in two consecutive stages. Initially kj indistinguishable balls are distributed into the k distinguishable urns with j balls in each urn and then the remaining $n - kj$ indistinguishable balls are distributed into the k distinguishable urns, each of capacity limited to $s-j$ balls. Enumerate the distributions of both stages, using the result of the first part of Exercise 6, and by the multiplication principle deduce the required number.

10. Note that when the interest is only on the occupancy of a single urn, say the i-th urn, and not on the individual occupancy of the n urns, the corresponding enumerator follows from the enumerator (9.24) by setting $x_i = x$ and $x_j = 1$, $j \neq i$. Expand the enumerator $H(t; x)$ into powers of x and t and into powers of $(x - 1)$ and t to find the two expressions of $h_n(x)$. These expressions imply the required expressions for $B(n, k, r)$ and $b_{(s)}(n, k)$.

11. Work as in the first part of Exercise 10.

12. (a) Show that the enumerator for occupancy of the s distinguishable cells of the j-th urn is $(1 + x_{1,j}t)(1 + x_{2,j}t) \cdots (1 + x_{s,j}t)$ and, putting $x_{i,j} = x_j$, $i = 1, 2, \ldots, s$, deduce the enumerator for occupancy of the j-th urn. (b) Use this result to find the enumerator for occupancy of the k distinguishable urns and expand it into powers of t to establish the expression of $h_{n,s}(x_1, x_2, \ldots, x_k)$.

13. Work as in Exercise 12.

14. Consider the set Ω of distributions of n indistinguishable balls into the k distinguishable urns and its subset A_i of these distributions in which j balls are placed in the i-th urn, for $i = 1, 2, \ldots, k$, and apply the inclusion and exclusion principle.

15. Use the result of Exercise 12.

16. Work as in Exercise 14.

17. Use the result of Exercise 13.

18. Consider the set Ω of distributions of n indistinguishable balls into the kr distinguishable cells and its subset A_i of these distributions in which j balls are placed in the i-th column of cells, for $i = 1, 2, \ldots, s$, and apply the inclusion and exclusion principle.

19. Consider the set W of distributions of n indistinguishable balls into the kr distinguishable cells in which at least one ball is placed in each column of cells. Further, consider the subset B_s of these distributions that leave the s-th row of cells empty, for $s = 1, 2, \ldots, r$, and apply the inclusion and exclusion principle.

20. Work as in Exercise 18.

21. Work as in Exercise 19.

22. Use the enumerator (9.22) to find the generating function $M_k(t)$. Further, differentiate it to get the differential equation $M_k'(t) = k\{M_k(t) - (t^s/s!)M_{k-1}(t)\}$, which, expanded into powers of t, yields the first recurrence relation. Finally, derive the expression $M_k(t) = \sum_{j=0}^{s} M_{k-1}(t)t^j/j!$, which implies the second recurrence relation.

23. Work as in Exercise 22.

24. Properly modify the enumerator for occupancy of the j-th urn, for $j = 1, 2, \ldots, k$, obtained in Exercise 12, to establish the generating function $G_k(t)$ and continue as in Exercise 22.

25. Work as in Exercise 24.

26. Properly modify the enumerator for occupancy of the j-th urn, for $j = 1, 2, \ldots, k$, obtained in Exercise 13, to establish the generating function $H_k(t)$ and then continue as in Exercise 22.

27. Work as in Exercise 26.

28. Consider the set Ω of distributions of the $n + r$ balls into the k distinguishable urns and its subset A_i of these distributions which leave the i-th urn empty, for $i = 1, 2, \ldots, k$, and apply the inclusion and exclusion principle to find the expression of the number $W(n, r, k)$.

29. Consider the set A of the distributions of the $n + r$ balls into the k distinguishable urns that do not leave any of the urns empty. Further, consider the subset B_j of these distributions in which $k - j$ of the urns contain none of the n distinguishable balls, for $j = 1, 2, \ldots, m$, $m = \min\{n, k\}$, whence $A = B_1 + B_2 + \cdots + B_m$. Apply the addition principle to conclude the required expression of the number $W(n, r, k)$. In order to derive the recurrence relation, consider the set $B \subseteq A$ of the distributions in which a specified ball b, among the n distinguishable balls, is placed in an urn with at least one other ball and the set $C \subseteq A$ of the distributions in which ball b is placed in an urn with no other ball, whence $A = B + C$ and $N(A) = N(B) + N(C)$.

30. Use the enumerators (9.21) and (9.26) to form the enumerator for occupancy of the k distinguishable urns by distinguishable and indistinguishable balls. For the other generating function, consider a suitable modification of this enumerator.

CHAPTER 10

1. Consider the following division in two subsets $(\mathcal{P}_1, \mathcal{P}_2)$ of the set $\mathcal{P}$ of partitions of n into k parts: $\mathcal{P}_1$ is the set of partitions of n into k parts with at least one part of length 1 and $\mathcal{P}_2$ is the set of partitions of n into k parts with no part of length 1. Apply the addition principle.

2. Consider the following division into two subsets $(\mathcal{P}_1, \mathcal{P}_2)$ of the set $\mathcal{P}$ of partitions of n into at most k parts: $\mathcal{P}_1$ is the set of partitions of n into at most $k - 1$ parts and $\mathcal{P}_2$ is the set of partitions of n into exactly k parts. Apply the addition principle.

3. Deduce the bivariate generating function $F_r(t, u)$ from the corresponding universal generating function. For the derivation of the generating function $F_{r,k}(t)$, note that $(1 - ut^{r+1})F_r(t, ut) = (1 - u)F_r(t, u)$ and work as in the second part of Theorem 10.2.

4. Deduce the generating function $\psi_k(t)$ from the corresponding universal generating function. Note that $\psi_k(t) = t(1 - t^{k-1})\psi_{k-1}(t)$ and establish the recurrence relation for $r(n, k)$.

5. Consider the following division in two subsets (Q_1, Q_2) of the set Q of partitions of n into k unequal parts: Q_1 is the set of partitions of n into k unequal parts with one part of length 1 and Q_2 is the set of partitions of n into k unequal parts with no part of length 1. Apply the addition principle.

6. Deduce the bivariate generating function $F(t, u)$ from the corresponding universal generating function. Note that the generating function $F(t, u)$ satisfies the relation $(1-u)F(t, u) = (1-u^2 t^2)F(t, ut)$, which implies the required recurrence relation for the generating function $F_k(t)$.

7. The bivariate generating functions $H_0(t, u)$ and $H_1(t, u)$ may be deduced for the corresponding universal generating function. For the derivation of the generating functions $H_{0,k}(t)$ and $H_{1,k}(t)$, note that $H_0(t, u) = (1 + ut^2)H_0(t, ut^2)$ and $H_1(t, u) = (1 + ut)H_1(t, ut^2)$ and work as in the second part of Theorem 10.2.

8. Deduce the bivariate generating function $G_r(t, u)$ from the corresponding universal generating function. Note that $(1 - ut^{r+1})G_r(t, ut) = (1 - ut)G_r(t, u)$ and work as in the second part of Theorem 10.2.

9. Using the relations $(1 - ut)G(t, u) = G(t, ut)$, $(1 - ut)\{G_r(t, u) + ut^{r+1}G_{r+1}(t, u)\} = G_r(t, ut)$, derive the recurrence relation $(1 - t^r)g_r(t) = t^{2r-1}g_{r-1}(t)$, $r = 1, 2, \ldots$. Iterate it to get the explicit expression for $g_r(t)$. In order to derive the required identity, use the expression connecting the generating functions $G(t, u)$, $G_r(t, u)$ and $g_r(t)$ along with the first part of Theorem 10.2 and the result of the first part of Exercise 8.

10. Use the identity $(1+t^{2^i}) = (1-t^{2^{i+1}})/(1-t^{2^i})$ to derive the expression $\prod_{i=0}^{k-1}(1 + t^{2^i}) = (1 - t^{2^k})/(1 - t)$ and, for $|t| < 1$, conclude the expression $\prod_{i=0}^{\infty}(1 + t^{2^i}) = (1 - t)^{-1}$, which implies the required expression for n.

11. From the corresponding universal generating function deduce the generating function $\sum_{n=0}^{\infty} R(n)t^n = \prod_{r=0}^{\infty}(1 - t^{2^r})^{-1}$, where $R(n) = R_1(n)$. Use this generating function to show that $\sum_{n=0}^{\infty} R_0(n)t^n - (1 - t)/2 = \sum_{n=0}^{\infty} R_1(n)t^n + (1-t)/2$ and conclude that $R_0(n) = R_1(n)$, for $n = 2, 3, \ldots$.

12. Use the relation $\sum_{r_i=0}^{i} t^{i!r_i} = (1 - t^{(i+1)!})/(1 - t^{i!})$ to derive the expression $\prod_{i=0}^{k-1}\left(\sum_{r_i=0}^{i} t^{i!r_i}\right) = (1 - t^{k!})/(1 - t)$, and for $|t| < 1$, conclude the expression $\prod_{i=1}^{\infty}\left(\sum_{r_i=0}^{i} t^{i!r_i}\right) = (1 - t)^{-1}$, which implies the required expression for n.

13. Deduce the bivariate generating function $H_r(t, u)$ from the corresponding universal generating function. Note that $(1 + ut)H_r(t, ut) = (1 + ut^{r+1})H_r(t, u)$ and work as in the second part of Theorem 10.2.

14. (a) Correspond to each class of multinomial coefficients a nonnegative integer solution of the linear equation $x_1 + x_2 + \cdots + x_r = n$, with $x_1 \geq x_2 \geq \cdots \geq x_r \geq 0$. (b) Note that

$$M(n) = \sum \frac{n!}{(1!)^{r_1} (2!)^{r_2} \cdots (n!)^{r_n}},$$

where the summation is extended over all nonnegative integer solutions of the equation $r_1 + 2r_2 + \cdots + nr_n = n$, and deduce the required generating function.

15. Note first that, in a perfect partition of n, the number of unit parts is $y_1 \geq 1$. Consider a perfect partition of n that includes $y_1 = r_1 - 1$ unit parts. Then, every positive integer $r < r_1$ has exactly one partition and r_1 is the next part of the perfect partition. The number of parts that are equal to r_1 is $y_{r_1} \geq 1$, whence, with $y_{r_1} = r_2 - 1$, every positive integer $r < r_1 r_2$ has exactly one partition with parts 1 and r_1. Continuing this step-by-step procedure, conclude that $n = (r_1 - 1) + r_1(r_2 - 1) + r_1 r_2(r_3 - 1) + \cdots + (r_1 r_2 \cdots r_{k-1})(r_k - 1)$ and $n + 1 = r_1 r_2 \cdots r_k$.

16. Note that the number $c(n, k)$ of compositions of n into k parts equals the number of nonnegative integer solutions of the linear equation $z_1 + z_2 + \cdots + z_k = n - k$, which is equal to the number of distributions of $n - k$ indistinguishable balls (the units) into k distinguishable urns (the summands). Use the expression of $c(n, k)$ to evaluate the generating function $G(t, u)$.

17. Evaluate the expression

$$\sum_{n=k}^{\infty} c(n, k; S) t^n = \sum_{n=k}^{\infty} \left(\sum t^{x_1 + x_2 + \cdots + x_k} \right),$$

where, in the inner sum, the summation is extended over all solutions of the linear equation $x_1 + x_2 + \cdots + x_k = n$, which $x_i \in S$, $i = 1, 2, \ldots, k$.

18. Use the result of Exercise 17, with $S = \{1, 2, \}$, and the relation $c_2(n) = \sum_{k=1}^{n} c_2(n, k)$, with $c_2(n, k) = c(n, k; \{1, 2\})$.

19. Use the result of Exercise 17, with $S = \{1, 2, \ldots, r\}$, and the relation $c_r(n) = \sum_{k=1}^{n} c_r(n, k)$, with $c_r(n, k) = c(n, k; \{1, 2, \ldots, r\})$, to find the generating function $C_r(t)$. Show that $C_r(t) = 1 + \sum_{j=1}^{r} t^j C_r(t)$ and deduce the required recurrence relation.

20. Consider the set $\mathcal{P}_{n,k} = \{X_k = \{x_1, x_2, \ldots, x_k\} : x_1 + x_2 + \cdots + x_k = n, \ x_1 \geq x_2 \geq \cdots \geq x_k \geq 1\}$, of partitions of n into k parts, and the set $\mathcal{C}_{n,k} = \{(x_1, x_2, \ldots, x_k) : x_1 + x_2 + \cdots + x_k = n, \ x_i \geq 1, i = 1, 2, \ldots, k\}$, of compositions of n into k parts. Correspond to each partition $X_k \in \mathcal{P}_{n,k}$ the set $\mathcal{P}(X_k) \subseteq \mathcal{P}_{n,k}$ of compositions of n into k parts that are composed of the same parts (in different orderings). Show that the set $\{\mathcal{P}(X_k) : X_k \in \mathcal{P}_{n,k}\}$ is a partition of $\mathcal{C}_{n,k}$ and, applying the addition principle, conclude that

$\binom{n-1}{k-1} \leq k!p(n,k)$. Further, consider the $Q_{n,k} = \{Y_k = \{x_1, x_2, \ldots, x_k\} :$ $x_1 + x_2 + \cdots + x_k = n$, $x_1 > x_2 > \cdots < x_k \geq 1\}$, of partitions of n into k unequal parts, and correspond to each partition Y_k the set $\mathcal{P}(Y_k)$ of permutations of Y_k. Since $\mathcal{P}(Y_k) \subseteq C_{n,k}$, conclude that $k!q(n,k) \leq \binom{n-1}{k-1}$.

21. In the Gauss-Jacobi identity, replace (a) q by q^r, $r \geq 0$ and x by q^s, $s \geq 0$ and (b) q by q^r, $r \geq 0$ and x by $-q^s$, $s \geq 0$ and deduce the required identities.

22. Use the relation

$$\sum_{n=0}^{\infty} \sum_{\mathcal{P}_n} \prod_{i=1}^{\infty} u_i(k_i)t^{ik_i} = \prod_{i=1}^{\infty} \sum_{k_i=0}^{\infty} u_i(k_i)t^{ik_i}.$$

23. Use the result of Exercise 22 for $a(n) \equiv a(n; u_j)$, with $u_j(k) = u_j k$ and $u_i(k) = 1$, $i \neq j$.

24. Work as in the proof of the first part of Theorem 10.2.

25. Use the expression

$$\begin{bmatrix} x \\ k \end{bmatrix}_q = \frac{(1 - q^x)(1 - q^{x-1}) \cdots (1 - q^{x-k+1})}{(1 - q)(1 - q^2) \cdots (1 - q^k)}.$$

26. Use the expression

$$\begin{bmatrix} n \\ k \end{bmatrix}_q = \frac{[n]_q!}{[k]_q![n - k]_q!}.$$

27. Use the triangular recurrence relation

$$\begin{bmatrix} r \\ k \end{bmatrix}_q = \begin{bmatrix} r - 1 \\ k - 1 \end{bmatrix}_q + q^k \begin{bmatrix} r - 1 \\ k \end{bmatrix}_q$$

to split the left-hand side sum into a difference of two sums.

28. Work as in Exercise 27.

29. In the generating functions $H_r(t, u)$ and $H_{r,k}(t)$ of Exercise 13, which are connected by $H_r(t, u) = \sum_{k=0}^{\infty} H_{r,k}(t)u^k$, replace t by q, ut by x and r by n to obtain the q-binomial formula. Also, in the generating functions $G_r(t, u)$ and $G_{r,k}(t)$ of Exercise 8, which are connected by $G_r(t, u) = \sum_{k=0}^{\infty} G_{r,k}(t)u^k$, replace t by q, ut by x, and r by n to obtain the q-negative binomial formula. These formulae may also be derived algebraically. For the derivation of the third formula, consider the function $F_n(q, x) = q^{xn}$ and its expansion $F_n(q, x) = \sum_{k=0}^{\infty} F_{n,k}(q)[x]_{k,q}$ and deduce the recurrence relation $[k]_q F_{n,k}(q) = (q-1)q^{k-1}[n-k+1]_q F_{n,k-1}(q)$, $k = 1, 2, \ldots, n$, with $F_{n,0}(q) = 1$, which implies $F_{n,k}(q) = q^{k(k-1)/2}(q - 1)^k \begin{bmatrix} n \\ k \end{bmatrix}_q$.

30. Expand the identities

$$\prod_{i=1}^{r+s}(1+xq^{i-1}) = \prod_{i=1}^{r}(1+xq^{i-1})\prod_{j=1}^{s}(1+xq^{r+j-1})$$

and

$$\prod_{i=1}^{r+s}(1-xq^{i-1})^{-1} = \prod_{i=1}^{r}(1-xq^{i-1})^{-1}\prod_{j=1}^{s}(1-xq^{r+j-1})^{-1},$$

using the q-binomial and the q-negative binomial formulae of Exercise 29, respectively, to derive the first two formulae. Show the third formula by induction on n.

31. Utilize the results of Exercise 26(b) together with the q-binomial and negative binomial formulae of Exercise 29.

32. (a) Note that

$$\prod_{i=1}^{n-1}([t]_q - [i]_q) = \sum_{k=1}^{n}s(n,k|q)[t]_q^{k-1}$$

and work as in Theorem 8.1 to get the required expression. (b) Expand both members of the recurrence relation

$$\prod_{i=1}^{n}([t]_q - [i]_q) = ([t]_q - [n]_q)\prod_{i=1}^{n-1}([t]_q - [i]_q)$$

into powers of $[t]_q$ and conclude the required triangular recurrence relation. Further, use the triangular recurrence relation to obtain the vertical recurrence relation.

33. (a) Note first that

$$[t]_q^n = \sum_{k=0}^{n}S(n,k|q)\prod_{i=0}^{k-1}([t]_q - [i]_q)$$

and, using this expression, expand both members of the identity $[t]_q^{n+1} = [t]_q[t]_q^n$. Transform the resulting expression by introducing the expression $[t]_q\prod_{i=0}^{r-1}([t]_q - [i]_q) = \prod_{i=0}^{r}([t]_q - [i]_q) + [r]_q\prod_{i=0}^{r-1}([t]_q - [i]_q)$ and conclude the triangular recurrence relation. Further, use the triangular recurrence relation to get the recurrence relation $\phi_k(u|q) = u(1 - [k]_q u)^{-1}\phi_{k-1}(u|q)$, $k = 1, 2, \ldots$, $\phi_0(u|q) = 1$ and iterate it to obtain the required expression. (b) Expand both members of the recurrence relation for $\phi_k(u|q)$ to get the vertical recurrence relation. Further, expand the generating function $\phi_k(u|q)$ into powers of u and equate the coefficients of u^n in the resulting expression. (c) Expand the n-th order q-factorial $[t]_{n,q}$ into powers of $[t]_q$ and, in the resulting expression, expand the powers $[t]_q^r$ into q-factorials $[t]_{k,q}$ and equate the coefficients of $[t]_{k,q}$ to deduce the first orthogonal relation. The second relation may be similarly derived.

34. Show first that

$$\prod_{j=0}^{k}(t - [j]_q)^{-1} = \sum_{r=0}^{k} c_r \cdot (t - [r]_q)^{-1},$$

where

$$c_r^{-1} = \frac{d}{dt}\left[\prod_{j=0}^{k}(t - [j]_q)\right]_{t=[r]_q} = (-1)^{k-r}q^{r(2k-r-1)/2}[r]_q![k-r]_q!$$

and, using this formula, expand the generating function $\phi_k(u|q)$ into powers of u to find the required expression.

35. Using the last expression in Exercise 30, with $x = n - 1$ and $y = t$, expand the q-factorial $[t + n - 1]_{n,q}$ into a sum of q-factorial $[t]_{t,q}$, and deduce the expression $|L(n, k|q)| = q^{-n(n-1)/2+k(k-1)/2}\begin{bmatrix} n \\ k \end{bmatrix}_q [n - 1]_{n-k,q}$, which can be transformed into the required expression.

CHAPTER 11

1. Use the generating function of the exponential Bell partition polynomials.

2. For $n > 0$ and $k_1, k_2, \ldots, k_n$ nonnegative integers satisfying the equation $k_1 + 2k_2 + \cdots + nk_n = n$, consider the function

$$C_{k_1, k_2, \ldots, k_n}(a) = \sum \left[\prod_{i=1}^{n}(-1)^{r_i}\binom{k_i}{r_i}\right]\left(a - \sum_{j=1}^{n} jr_j\right)^{k_1+k_2+\cdots+k_n-1},$$

where the summation is extended over all $r_i = 0, 1, \ldots, k_i$, $i = 1, 2, \ldots, n$ and show that $C_{k_1, k_2, \ldots, k_n}(a) = 0$. Multiply it by $\prod_{i=1}^{n}(z_i/i!)^{k_i}/k_i!$ and sum for all nonnegative integer solutions of the equation $k_1 + 2k_2 + \cdots + nk_n = n$, to get the relation

$$\sum_{k=0}^{n}\frac{1}{a - k}\left[\sum\prod_{i=1}^{k}\frac{1}{r_i!}\left(\frac{(k-a)z_i}{i!}\right)^{r_i}\right] \cdot \left[\sum\prod_{j=1}^{n-k}\frac{1}{m_j!}\left(\frac{(a-k)z_j}{j!}\right)^{m_j}\right] = 0,$$

where, in the inner sums, the summation is extended over all nonnegative integer solutions of the equations $r_1 + 2r_2 + \cdots + kr_k = k$ and $m_1 + 2m_2 + \cdots + (n - k)m_{n-k} = n - k$, respectively, and thus conclude the required formula. Further, replace n by s, multiply both sides by $B_{n-s}(y_1 - az_1,$

$y_2 - az_2, \ldots, y_{n-s} - az_{n-s})$ and sum for $s = 0, 1, \ldots, n$ to get the convolution formula. Finally, replace $n - k$ by k, a by $n - a$ and $y_i - nz_i$ by x_i. In the resulting expression, replace n by s, multiply both sides by $B_{n-s}(y_1 - az_1, y_2 - az_2, \ldots, y_{n-s} - az_{n-s})$ and sum for $s = 0, 1, \ldots, n$ to get the required expression.

3. Set $x_1 = x$, $x_2 = 1$, $x_r = 0$, $r = 3, 4, \ldots$, in the generating function of the exponential Bell partition polynomials to find the generating function of $B_n(x, 1)$. Deduce from it the generating function of $H_n(x)$ and conclude the required expression. Similarly, find the generating function of $B_n(1!, 2!x^{-1}, \ldots, n!x^{-n+1})$ and deduce the generating function of $L_n^{(1)}(-x)$, which implies the required expression.

4. Set $x_r = (s)_r a x^{s-r}$, $r = 1, 2, \ldots, s$, $x_r = 0$, $r = s + 1, s + 2, \ldots$, in the generating function of the exponential Bell partition polynomials to find the generating function of $H_n(x; a, s)$. Deduce from it the required expressions.

5. Set $x_1 = x$, $x_2 = 1$ and $x_r = 0$, $r = 3, 4, \ldots$, in the generating function of the Logarithmic polynomials and find the generating function of $L_n(x, 1)$, which, expanded into powers of t, yields the required expression.

6. Set $x_1 = x$, $x_2 = 1$ and $x_r = 0$, $r = 3, 4, \ldots$, in the generating function of the potential polynomials to find the generating function of $C_{n,-s}(x, 1) = G_n^{(s)}(-x)$, which, expanded into powers of t, yields the required expression.

7. The vertical generating function of the partial Bell polynomials for $x_i = 0$, $i = 1, 2, \ldots, r$, is given by $B_{k,r+1}(t) = \left(\sum_{i=r+1}^{\infty} x_i t^i / i!\right)^k / k!$. Set $x_i = (i)_r y_{i-r}$, $i = r + 1, r + 2, \ldots$, and conclude the first relation. Use Newton's binomial formula to express the generating function $B_{k,r+1}(t)$ in terms of the generating function $B_{k-j,r}(t)$; expanding into powers of t both sides of the resulting expression, deduce the second relation. The third relation is similarly derived.

8. Set (a) $x_r = r!$, $r = 1, 2, \ldots$, and (b) $x_r = r$, $r = 1, 2, \ldots$, in the vertical generating function of the partial Bell partition polynomials and deduce the required expression.

9. Set (a) $x_{2j-1} = 0$, $x_{2j} = (2j)!$, $j = 1, 2, \ldots$, and (b) $x_{2j-1} = (2j-1)!$, $x_{2j} = 0$, $j = 1, 2, \ldots$, in the vertical generating function of the partial Bell partition polynomials and deduce the required expressions.

10. Multiply both members of (11.34) by $\binom{n+s}{n}$ and use the relation $\binom{n+s}{n}\binom{s}{r} = \binom{n+s}{n+r}\binom{n+r}{r}$.

11. From the generating function of the potential polynomials, with $s = -1$ and putting $a_r = x_r / r!$, $r = 1, 2, \ldots$, $a_0 = 1$, derive the linear system of equations $\sum_{j=1}^{n} a_{k-j} y_j = -a_k$, $k = 1, 2, \ldots, n$ and solve it to find the required expression.

12. In order to evaluate the generating function $h(t) = \sum_{n=1}^{\infty} \sum_{d|n} dt^n$, use the transformation $k = d$, $r = n/k$, with inverse $d = k$, $n = rk$, and sum first for $r = 1, 2, \ldots$ and then for $k = 1, 2, \ldots$. Further, differentiate the function $\log G(t)$ and conclude that $G'(t)/G(t) = h(t)/t$. Finally, integrate this differential equation to get the required expression of $G(t)$.

13. Work as in Example 11.8 and use the result of Example 11.7(b).

14. Set $x_r = y_r = r! x^{-r+1}$, $r = 1, 2, \ldots$, in the generating function of the Touchard polynomials to find the generating function of these particular polynomials. Deduce from it the generating function of $L_n^{(k-1)}(x)$, $n = 1, 2, \ldots$.

15. Work as in Exercise 7.

16. Work as in Exercise 8.

17. Use Leibnitz formula for the derivatives of a product of functions and Faa di Bruno formula for the derivatives of a composite function.

18. Follow the steps of the derivation of the corresponding properties of the exponential partition polynomials in Section 11.2.

19. Take the partial derivatives of the generating functions $A(t, u)$ and $A_k(t, u)$ with respect to t and, expanding the resulting partial differential equations, conclude the corresponding recurrence relations.

20. Follow the steps of the derivation of the corresponding properties of the general partition polynomials in Section 11.3.

21. Work as in Example 11.3, using the last part of Exercise 20.

22. In order to derive the recurrence relation, take the partial derivative of the generating function $L(t, u)$ with respect to t and deduce the relation

$$\frac{\partial L(t, u)}{\partial t} = \frac{\partial g(t, u)}{\partial t} - [g(t, u) - x_{0,0}] \frac{\partial L(t, u)}{\partial t}.$$

23. Use the results of Exercise 22.

24. In order to derive the recurrence relation, take the partial derivative of the generating function $C_s(t, u)$ with respect to t and deduce the relation

$$\frac{\partial C_s(t, u)}{\partial t} = s C_s(t, u) \frac{\partial g(t, u)}{\partial t} - [g(t) - x_{0,0}] \frac{\partial C_s(t, u)}{\partial t}.$$

25. Use the generating functions of exponential, logarithmic and potential bipartitional polynomials (Exercises 18, 22, 24) and conclude the required expressions.

CHAPTER 12

1. Set $x_1 = 0$, $x_j = 1$, $j = 2, 3, \ldots$ in (12.9) to find the bivariate generating function of the required number of permutations and, expanding it into powers of u and t, conclude the required expression.

2. Set $x_j = 0$, $j = 1, 2, \ldots, r - 1$, $x_j = 1$, $j = r, r + 1, \ldots$ into (12.9) to find the bivariate generating function of the required number of permutations and use the results of Exercise 8.11 to conclude the required expression.

3. Set $x_j = 1$, $j = 1, 2, \ldots, j \neq r$ and $x_r = x$ in (12.8) to find the expression generating function $C(x, t; r) = \sum_{n=0}^{\infty} C_n(x; r) t^n / n!$. Rewrite this expression as $(1 - t) C(x, t; r) = \exp[(x - 1) t^r / r]$ and expand it into powers of t to get the required recurrence relation. Further, expanding it into powers of x, deduce the recurrence relation for the numbers $C_n(k; r)$.

4. Use the relation connecting the factorial moment generating function with the probability generating function.

5. Set $x_j = x$, $j = 1, 2, \ldots, j \neq r$ and $x_r = 0$ in (12.8) to get the generating function $D(x, t; r) = \sum_{n=0}^{\infty} D_n(x; r) t^n / n!$. Differentiate it with respect to t and deduce the partial differential equation

$$(1 - t) \frac{\partial D(x, t; r)}{\partial t} = x D(x, t; r) - (t^{r-1} - t^r) D(x, t; r),$$

which, expanded into powers of t, yields for $D_n(x; r)$ the required recurrence relation. Further, expand it into powers of x to deduce the recurrence relation for the numbers $d_n(k; r)$.

6. Note that $d_n(r) = \sum_{k=0}^{[n/r]} d_n(k; r)$ and from the expression of $D(x, t; r)$ deduce the generating function $D(t; r) = \sum_{n=0}^{\infty} d_n(r) t^n / n!$, which, expanded into powers of t, yields the explicit expression of $d_n(t)$. Further, rewrite the expression of $D(t; r)$ as $(1 - t) D(t; r) = \exp(-t^r / r)$ and expand it into powers of t to find the recurrence relation for $d_n(r)$.

7. Set $x_1 = x$, $x_2 = y$ and $x_i = 1$, $i = 3, 4, \ldots$, in (12.8) to get the generating function $Q(x, y, t) = \sum_{n=0}^{\infty} Q_n(x, y) t^n / n!$. Multiply both members of this expression by $(1 - t)$ and, expanding it into powers of t, conclude the required recurrence relation.

8. Differentiate, with respect to t, the generating function $Q(x, y, t)$, which is derived in Exercise 7, and then equate the coefficients of $t^n / n!$ in the resulting expression.

9. Denote by C_n the multiple sum in the left-hand side of the required identity. Put $x_i = -1$, $i = 1, 2, \ldots$, in (12.8) and deduce the generating function, $\sum_{n=0}^{\infty} C_n t^n / n! = 1 - t$, which implies $C_n = 0$, $n = 2, 3, \ldots$.

10. Denote by $C_n(x)$ the multiple sum in the left-hand side of the required identity. Set $x_i = x$, $i = 1, 2, \ldots$, in (12.8) and deduce the generating function $\sum_{n=0}^{\infty} C_n(x) t^n / n! = (1 - t)^{-x}$, which, expanded into powers of t, yields the required expression for $C_n(x)$.

11. Deduce from (12.13) the generating function $A(t, u) = \sum A_n(u) t^n / n!$. Similarly find the generating function $B(t, u) = \sum B_n(u) t^n / n!$. Differentiate these functions with respect to t and get the differential equations

$$\frac{\partial A(t, u)}{\partial t} - t \frac{\partial B(t, u)}{\partial t} = u A(t, u), \quad \frac{\partial B(t, u)}{\partial t} - t \frac{\partial A(t, u)}{\partial t} = u B(t, u).$$

Expand both sides of these equations into powers of t to establish the required recurrence relations.

12. Work as in Exercise 11.

13. (a) Use the results of Exercises 10 and 11 to show that

$$G_n(u) = \sum_{j=0}^{n} (-1)^j \binom{n}{j} u^j A_{n-j}(u), \quad H_n(u) = \sum_{j=0}^{n} (-1)^j \binom{n}{j} u^j B_{n-j}(u)$$

and conclude the required expressions. (b) Note that $G_n = G_n(1)$ and $H_n = H_n(1)$ and, from the first part of Exercise 12, conclude the generating functions $\sum_{n=0}^{\infty} G_n t^n / n!$ and $\sum_{n=0}^{\infty} H_n t^n / n!$. Multiply each by $(1 - t)$ and expand the resulting expression into powers of t to get the recurrence relations. For the derivation of $G_n + H_n = D_n$, use the generating function $\sum_{n=0}^{\infty} D_n t^n / n! = (1 - t)^{-1} e^{-t}$.

14. Set $x_{2i-1} = 0$, $x_{2i} = 1$, $i = 1, 2, \ldots$, in (12.9) to get the required generating function. Expand it into powers of t and conclude the generating function $b_n(2u)$, which implies the required expression for the number $b(n, k)$.

15. Work as in Example 14.

16. Differentiate the generating functions $\sum_{n=0}^{\infty} b_n(u) t^n / n!$ and $\sum_{n=0}^{\infty} a_n(u) t^n / n!$, with respect to t, and multiply each of the resulting expressions by $1 - t^2$. Expand the final expressions into powers of t to deduce the recurrence relations.

17. Note that $a(n) = a_n(1)$ and $b(n) = b_n(1)$ and, from the first part of Exercises 14 and 15, conclude the generating functions $\sum_{n=0}^{\infty} a(n) t^n / n!$ and $\sum_{n=0}^{\infty} b(n) t^n / n!$. Use these generating functions to derive the required recurrence relations.

18. Use the expression

$$c(n, k; \mathbf{A}) = \sum \frac{n!}{k_1! k_2! \cdots k_n! 1^{k_1} 2^{k_2} \cdots n^{k_n}},$$

where the summation is extended over all nonnegative integer solutions of the equations $k_1 + 2k_2 + \cdots + nk_n = n$ and $k_1 + k_2 + \cdots + k_n = k$, with $k_i \in \{j : a_{i,j} = 1\}$, for $i = 1, 2, \ldots, n$, and proceed as in Theorem 10.3.

19. Put $a_{i,0} = 1$, $i = 1, 2, \ldots$, $a_{i,j} = 1$, $i = 1, 2, \ldots, r$, $a_{i,j} = 0$, $i = r+1, r+2, \ldots$, $j = 1, 2, \ldots$ and $u = 1$ in the universal generating function of Exercise 18 to deduce the generating function $C(t; r) = \sum_{n=0}^{\infty} c_n(r) t^n / n!$. Differentiate it to derive the differential equation $C'(t; r) = C(t; r) \sum_{j=0}^{r-1} t^j$, which implies the recurrence relation.

20. Set $x_1 = 1$, $x_2 = x$, $x_i = 0$, $i = 3, 4, \ldots$, in (12.8) to deduce the generating function $C(x, t) = \sum_{n=0}^{\infty} T_n(x) t^n / n! = \exp(t + xt^2/2)$. Differentiate it with respect to t to find the differential equation $\partial C(x, t)/\partial t = (1 + xt)C(x, t)$, which implies the required recurrence relation for $T_n(x)$.

CHAPTER 13

1. (a) $\mathcal{G}_3 = \{(1)(2)(3), (1, 2, 3), (1, 3, 2), (1)(2, 3), (1, 3)(2), (1, 2)(3)\}$,

   (b) $K_{3,r} = r(r^2 + 3r + 2)/6$.

2. $\mathcal{G}_4 = \{(1)(2)(3)(4), (1, 2, 3, 4), (1, 3)(2, 4), (1, 4, 3, 2),$
$(1)(2, 4)(3), (1, 3)(2)(4), (1, 4)(2, 3), (1, 2)(3, 4)\}$.

3. (a) Use Pólya's theorem and the cycle indicator of Exercise 2 to get $\sum_{j=0}^{4} K_4(j) t^j = 1 + t + 2t^2 + t^3 + t^4$ and conclude that $K_4(j) = 1$, $j = 0, 1, 3, 4$ and $K_4(2) = 2$. (b) $K_{4,r} = r(r^3 + 2r^2 + 3r + 2)/8$.

4. (a) $\mathcal{G}_4 = \{(1)(2)(3)(4), (1)(2, 3, 4), (1)(2, 4, 3)\}$,

   (b) $K_{4,r} = r^2(r^2 + 2)/3$.

5. (a) Use Pólya's theorem and the cycle indicator of Exercise 4 to get $\sum_{j=0}^{4} K_4(j) t^j = 1 + 2t + 2t^2 + 3t^3 + t^4$ and conclude that $K_4(j) = 1$, $j = 0, 4$ and $K_4(j) = 2$, $j = 1, 2, 3$. (b) Similarly $K_4(1, 1, 1) = 8$.

6. $\mathcal{G}_4 = \{(1)(2)(3)(4), (1)(2, 3, 4), (1)(2, 4, 3), (1, 3, 4)(2),$
$(1, 4, 3)(2), (1, 2, 4)(3), (1, 4, 2)(3), (1, 2, 3)(4),$
$(1, 3, 2)(4), (1, 2)(3, 4), (1, 3)(2, 4), (1, 4)(2, 3)\}$.

7. (a) Use Pólya's theorem and the cycle indicator of Exercise 6 to find $\sum_{j=0}^{4} K_4(j)t^j = 1 + t + t^2 + t^3 + t^4$ and conclude that $K_4(j) = 1$, $j = 0, 1, \ldots, 4$. (b) $K_{4,j} = r^2(r^2 + 11)/4$.

8. Consider the set W_{24} of the $4! = 24$ colorings and the group $\mathcal{G}_{24}$ of the permutations that are induced by rotations of the corresponding colored tetrahedron around its axes of symmetry. Show that $N(\mathcal{G}_{24}) = N(\mathcal{H}_4) = 12$, where $\mathcal{H}_4$ is the group $\mathcal{G}_4$ of Exercise 6, and apply Burnside's Theorem to conclude that the required number is $N(\mathcal{T}) = 24/12 = 2$. This number may also by obtained by using the cycle indicator of Exercise 6 and applying Pólya's theorem.

9. $\mathcal{G}_8 = \{(1)(2)(3)(4)(5)(6)(7)(8), (1,2,3,4)(5,6,7,8), (1,4,3,2)(5,8,7,6),$
$(1,4,8,5)(2,3,7,6), (1,5,8,4)(2,6,7,3), (1,5,6,2)(3,4,8,7),$
$(1,2,6,5)(3,7,8,4), (1,3)(2,4)(5,7)(6,8), (1,8)(4,5)(2,7)(3,6),$
$(1,6)(2,5)(3,8)(4,7), (1,2)(3,5)(4,6)(7,8), (1,4)(2,8)(3,5)(6,7),$
$(1,5)(2,8)(3,7)(4,6), (1,7)(2,3)(4,6)(5,8), (1,7)(2,6)(3,5)(4,8),$
$(1,7)(2,8)(3,4)(5,6), (1)(2,4,5)(3,8,6)(7), (1)(2,5,4)(3,6,8)(7),$
$(1,6,3)(2)(4,5,7)(8), (1,3,6)(2)(4,7,5)(8), (1,6,8)(2,7,4)(3)(5),$
$(1,8,6)(2,4,7)(3)(5), (1,3,8)(2,7,5)(4)(6), (1,8,3)(2,5,7)(4)(6)\}$

10. Use Pólya's theorem and the cycle indicator of Exercise 9 to get
$$g_8(t, u, w) = [(t + u + w)^8 + 8(t + u + w)^2(t^3 + u^3 + w^3)^2$$
$$+ 9(t^2 + u^2 + w^2)^4 + 6(t^4 + u^4 + 2^4)]/24$$
and conclude the required values. Further,
$$N_{8,3} = g_{8,3}(1,1,1) = (3^8 + 17 \cdot 3^4 + 6 \cdot 3^2)/24.$$

11. The group $\mathcal{G}_{12}$ is

$\{(1)(2)(3)(4)(5)(6)(7)(8)(9)(10)(11)(12), (1,8,2,5)(3,6,4,7)(9,10,11,12),$
$(1,2)(5,8)(3,4)(6,7)(9,11)(10,12), (1,5,2,8)(3,7,4,6)(9,12,11,10),$
$(1,9,4,11)(2,12,3,11)(5,6,7,8), (1,4)(2,3)(5,7)(6,8)(9,10)(11,12),$
$(1,10,4,9)(2,11,3,12)(5,8,7,6), (1,2,3,4)(5,12,6,9)(7,10,8,11),$
$(1,3)(2,4)(5,6)(7,8)(9,12)(10,11), (1,4,3,2)(5,9,6,12)(7,11,8,10),$
$(1)(2,4)(3)(5,10)(6,11)(7,12)(8,9), (1,3)(2)(4)(5,11)(6,10)(7,9)(8,12),$
$(1,12)(2,9)(3,10)(4,11)(5)(6,8)(7), (1,11)(2,10)(3,9)(4,12)(5,7)(6)(8),$
$(1,6)(2,7)(3,8)(4,5)(9)(10,12)(11), (1,7)(2,6)(3,5)(4,8)(9,11)(10)(12),$
$(1,5,9)(2,6,10)(3,7,11)(4,8,12), (1,9,5)(2,10,6)(3,11,7)(4,12,8),$
$(1,10,8)(2,9,7)(3,12,6)(4,11,5), (1,8,10)(2,7,9)(3,6,12)(4,5,11),$
$(1,7,12)(2,8,11)(3,5,10)(4,6,9), (1,12,7)(2,11,8)(3,10,5)(4,9,6),$
$(1,11,6)(2,12,5)(3,9,8)(4,10,7), (1,6,11)(2,5,12)(3,8,9)(4,7,10)\}.$

12. (a) Use Pólya's theorem and the cycle indicator of Exercise 11 to get

$$\sum N_{12}(j_1, j_2, \ldots, j_{12}) t_1^{j_1} t_2^{j_2} \cdots t_{12}^{j_{12}}$$
$$= \frac{1}{24}[(t_1 + t_2 + \cdots + t_{12})^{12} + 6(t_1 + t_2 + \cdots + t_{12})^2 (t_1^2 + t_2^2 + \cdots + t_{12}^2)^5$$
$$+ 3(t_1^2 + t_2^2 + \cdots + t_{12}^2)^6 + 8(t_1^3 + t_2^3 + \cdots + t_{12}^3)^4 + 6(t_1^4 + t_2^4 + \cdots + t_{12}^4)]$$

and conclude that the required number is

$$N_{12}(1, 1, \ldots, 1) = 12!/24 = 19,958,400.$$

(b) $K_{12,r} = r^3(r^9 + 6r^4 + 3r^3 + 8r + 6)/24.$

13. $\mathcal{G}_6 = \{(1)(2)(3)(4)(5)(6), \ (1)(2,3,5,4)(6), \ (1)(2,5)(3,4)(6),$
$(1)(2,4,5,3)(6), \ (1,3,6,4)(2)(5), \ (1,6)(2)(3,4)(5),$
$(1,4,6,3)(2)(5), \ (1,2,6,5)(3)(4), \ (1,6)(2,5)(3)(4),$
$(1,5,6,2)(3)(4), \ (1,2)(3,4)(5,6), \ (1,3)(2,5)(4,6),$
$(1,4)(2,5)(3,6), \ (1,5)(2,6)(3,4), \ (1,6)(2,3)(4,5),$
$(1,6)(2,4)(3,5), \ (1,2,3)(4,5,6), \ (1,3,2)(4,5,6),$
$(1,2,4)(3,6,5), \ (1,4,2)(3,5,6), \ (1,3,5)(2,6,4),$
$(1,5,3)(2,4,6), \ (1,4,5)(2,6,3), \ (1,5,4)(2,3,6)\}.$

14. (a) Use Pólya's theorem and the cycle indicator of Exercise 13 to get
$\sum K_6(j) t^j = 1 + t + 2t^2 + 2t^3 + 2t^4 + t^5 + t^6$ and conclude that $K_6(j) = 1$,
$j = 0, 1, 5, 6$ and $K_6(j) = 2$, $j = 2, 3, 4$. (b) $K_{6,r} = r^2(r^4 + 3r^2 + 12r + 8)/24.$

15. Use Pólya's theorem and the cycle indicator of the symmetric group
$\mathcal{G}_n$ of order n, derived in Example 13.4, to get

$$K_{n,r} = \frac{1}{n} \sum_{d|n} \phi(d) r^{n/d}.$$

16. Use Pólya's theorem and the cycle indicator of the symmetric group
$\mathcal{G}_n$ of order n, derived in Example 13.4, to get

$$\sum_{j=0}^{n} K_n(j) t^j = \frac{1}{n} \sum_{d|n} \phi(d)(1 + t^d)^{n/d}$$

and, expanding the right-hand side into powers of t, deduce the required
expression of $K_n(j)$.

17. Show that $\mathcal{G}_n = \{\epsilon, \sigma, \sigma^2, \ldots, \sigma^{n-1}, \tau, \tau\sigma, \tau\sigma^2, \ldots, \tau\sigma^{n-1}\}$ for $n = 2m + 1$ and $\mathcal{G}_n = \{\epsilon, \sigma, \sigma^2, \ldots, \sigma^{n-1}, \rho, \rho\sigma, \rho\sigma^2, \ldots, \rho\sigma^{n-1}\}$ for $n = 2m$, where the permutations σ, τ and ρ are defined as follows: $\sigma(j) = j + 1$, $j = 2, 3, \ldots, n - 1$, $\sigma(n) = 1$ (rotation of the n-gon by $2\pi/n$ degrees), $\tau(1) = 1$, $\tau(j) = 2m + 3 - j$, $j = 2, 3, \ldots, 2m + 1$ (reflection of the n-gon, for $n = 2m + 1$, around the axis connecting vertex 1 and the middle of the side connecting vertices $m + 1$ and $m + 2$) and $\rho(1) = 1$, $\rho(j) = 2m + 2 - j$, $j = 2, 3, \ldots, 2m$ (reflection of the n-gon, for $n = 2m$, around the axis connecting vertices 1 and $m + 1$).

18. Set $x_i = r$, $i = 1, 2, \ldots, n$, in the cycle indicator derived in Exercise 17 to get, according to Pólya's theorem,

$$K_{n,r} = \frac{1}{2n} \sum_{d|n} \phi(d) r^{n/d} + \frac{1}{2} r^{m+1},$$

for $n = 2m + 1$ and

$$K_{n,r} = \frac{1}{2n} \sum_{d|n} \phi(d) r^{n/d} + \frac{1}{4} (r + 1) r^m$$

for $n = 2m$.

19. (c) Show that the order $k = k(\sigma_h)$ of the permutation σ_h in $\mathcal{G}_n$ equals the order $k(h)$ of the element h in $\mathcal{H}_n$ and, since the orbit $\{h_r, \sigma_h(h_r), \ldots, \sigma_h^{k-1}(h_r)\}$ of any element h_r in $\mathcal{H}_n$ under the permutation σ_h is of length k, conclude that σ_h is decomposed in n/k cycles of length k and so

$$C(x_1, x_2, \ldots, x_n; \mathcal{G}_n) = \frac{1}{n} \sum_{h \in \mathcal{H}_n} x_{k(h)}^{n/k(h)}.$$

Further, use the fact that $\mathcal{H}_n = \sum_{d|n} \mathcal{H}_{n;d}$ to conclude the required expression.

20. Consider the set $\mathcal{F}$ of colorings of the set $W_n = \{1, 2, \ldots, k\}$, of the k parts of a composition $(r_1, r_2, \ldots, r_k)$ in $C_{n,k}$, (zero parts are allowed) with colors from $Z_{n+1} = \{0, 1, \ldots, n\}$. Show that the number $P_k(j_0, j_1, \ldots, j_n)$, $j_0 + j_1 + \cdots + j_n = k$, of partitions of n into at most k parts, in which $j_s \geq 0$ parts are equal to s, $s = 0, 1, \ldots, n$, equals the number of models $M_{\mathcal{F}}$, under the symmetric group $\mathcal{P}_k$ of W_k, in which $j_s \geq 0$ elements of W_k are painted with color z_s from Z_{n+1}, $s = 0, 1, \ldots, n$. For derivation of the required identity, use the vertical generating function of the number of partitions of n into at most k parts obtained in Corollary 10.3.

CHAPTER 14

1. (a) Use the definition of the non-central Eulerian numbers to get the relation

$$(r + k - j)^n = \sum_{i=0}^{k-j} A(n, i; r)(-1)^{k-j-i} \binom{-n-1}{k-j-i}, \quad j = 0, 1, \ldots, k.$$

Multiply it by $(-1)^j \binom{n+1}{j}$ and sum for $j = 0, 1, \ldots, k$ to deduce the expression of $A(n, k; r)$. (b) Expand both members of the identity $(t + r)^{n+1} = (t + r)(t + r)^n$ into binomial coefficients of order $n + 1$ and conclude the recurrence relation.

2. Work as in Lemma 14.1 and Theorem 14.3.

3. Express the coefficients of the power series in the expression of $A_n(t; r)$ (see Exercise 2) into factorials of i, using the non-central Stirling numbers of the second kind, and deduce the relation

$$A_n(t; r) = \sum_{j=0}^{n} j! S(n, j; r) t^j (1 - t)^{n-j}.$$

Expand both sides of it into powers of t and conclude the first expression. Also, set $t = u/(1 + u)$ and expand the resulting relation into powers of u and conclude the second expression.

4. Work as in Exercise 1.

5. Work as in Lemma 14.2 and Theorem 14.7.

6. Work as in Theorem 14.9.

7. Work as in Theorem 14.8.

8. (a) Replace t by $-t$ in the definition of q-Eulerian numbers and use the relations $[-t]_q = -q^{-t}[t]_q$ and

$$\begin{bmatrix} -t + n - k \\ n \end{bmatrix}_q = (-1)^n q^{-nt+n(n-k)-n(n-1)/2} \begin{bmatrix} t + k - 1 \\ k \end{bmatrix}_q.$$

(b) Use the definition of the q-Eulerian numbers to get the relation

$$[k - r]_q^n = \sum_{j=0}^{k-r} q^{j(j-1)/2} A(n, j|q) \begin{bmatrix} n + k - r - j \\ k - r - j \end{bmatrix}_q.$$

Multiply it by $(-1)^r q^{r(r-1)/2} \begin{bmatrix} n + 1 \\ r \end{bmatrix}_q$, sum for $r = 0, 1, \ldots, k$ and conclude the required expression for $A(n, k|q)$, since

$$\sum_{r=0}^{k-j} (-1)^{k-j-r} q^{r(r-1)/2} \begin{bmatrix} n + 1 \\ r \end{bmatrix}_q \begin{bmatrix} n + k - j - r \\ k - j - r \end{bmatrix}_q = \delta_{k,j}.$$

The last relation may be deduced by using the q-binomial and q-negative binomial formulae (see Exercise 10.29).

9. Expand into q-binomial coefficients, using the q-Eulerian numbers, both members of the identity $[t]_q^{n+1} = [t]_q [t]_q^n$ and then use the relation

$$\begin{bmatrix} t + n - k \\ n \end{bmatrix}_q [t]_q = [k]_q \begin{bmatrix} t + n - k + 1 \\ n + 1 \end{bmatrix}_q + q^k [n - k + 1]_q \begin{bmatrix} t + n - k \\ n + 1 \end{bmatrix}_q.$$

10. Show first that the number of permutations $(j_1, j_2, \ldots, j_n)$ of the set $\{1, 2, \ldots, n\}$ that have k records $j_{r_1}, j_{r_2}, \ldots, j_{r_k}$ at the points $r_1, r_2, \ldots, r_k$, with $r_1 = 1$ and $j_{r_k} = n$ is given by $(n - r_k)! \prod_{i=0}^{k-1} (j_{r_i} - r_i)_{r_{i+1} - r_i - 1}$. Summing for $1 \leq j_{r_1} < j_{r_2} < \cdots < j_{r_{k-1}} < j_{r_k} = n$, derive the number of permutations of $\{1, 2, \ldots, n\}$ that have k records at the points $r_1, r_2, \ldots, r_k$ as

$$\frac{(n-1)!}{(r_2 - 1)(r_3 - 1) \cdots (r_k - 1)}.$$

Summing for $2 \leq r_2 < r_3 < \cdots < r_k \leq n$ and using (8.12), conclude that the number of permutations of the set $\{1, 2, \ldots, n\}$ that have k records equals $|s(n, k)|$.

11. Consider the set Ω of permutations $(j_1, j_2, \ldots, j_n)$ of $\{1, 2, \ldots, n\}$ that have k inversions and its subset A_i of the permutations in which $j_1 = i$, $i = 1, 2, \ldots, n$, and use the relation $\Omega = A_1 + A_2 + \cdots + A_n$ to deduce the horizontal recurrence relation.

12. Use the horizontal recurrence relation of the numbers $b(n, k)$, derived in Exercise 11, to get for the generating function $\phi_n(t)$ the recurrence relation $\phi_n(t) = (1 - t)^{-1}(1 - t^n)\phi_{n-1}(t)$, $n = 1, 2, \ldots$, $\phi_0(t) = 1$. Iterate it to deduce the required expression.

13. Note that the permutations of $\{1, 2, \ldots, n + 1\}$ that have k local maxima can be deduced from the permutations of $\{1, 2, \ldots, n\}$ that have k or $k - 1$ local maxima by placing the element $n + 1$ in one of the possible positions. Use it to derive the required recurrence relation.

14. Work as in Exercise 13.

15. (b) Multiply both members of the recurrence by $t^k u^{n-1}/(n - 1)!$ and sum for $k = 0, 1, \ldots$, $n = 1, 2, \ldots$, to deduce the partial differential equation

$$t(t^2 - 1)\frac{\partial g(t, u)}{\partial t} + (1 - t^2 u)\frac{\partial g(t, u)}{\partial u} = (t^2 + 1)g(t, u),$$

the solution of which, with $g(t, 0) = 1 + t$, gives the required expression of $g(t, u)$.

16. Work as in Exercise 13.

17. (a) Note that removing the element $n + 1$ from any up-down permutation $(j_1, j_2, \ldots, j_{n+1})$ of $\{1, 2, \ldots, n + 1\}$ then we get two up-down permutations $(j_1, j_2, \ldots, j_{2r-1})$ and $(j_{2r+1}, j_{2r+2}, \ldots, j_{n+1})$ of two disjoint subsets of $\{1, 2, \ldots, n\}$ with $2r - 1$ and $n - 2r + 1$. Use it to derive the required recurrence relation. (b) Multiply the recurrence relation by $t^n/n!$ and sum for $n = 0, 1, \ldots$ to get the differential equation $2A'(t) = 1 + A^2(t)$, the solution of which gives the required expression of $A(t)$.

18. (a) Use the relation $\tan(t/2 + \pi/4) = (1/\cos t) + \tan t$ to show that $A_0(t) = \sum_{n=0}^{\infty} A_{2n} t^{2n}/(2n)! = 1/\cos t$ and, since $A_0(t) = E(it)$, $i = \sqrt{-1}$, conclude the required expression. (b) Use the result of Exercise 8.22 to derive the required expression.

19. Work as in Exercise 18.

20. (a) In the generating function of G_{2n}, $n = 1, 2, \ldots$, replace t by $2it$, $i = \sqrt{-1}$, to get the generating function $\sum_{n=1}^{\infty} (-1)^n 2^{2n-1} G_{2n} t^{2n-1}/(2n)! = \tan t$ and, since $\sum_{n=1}^{\infty} A_{2n-1} t^{2n-1}/(2n-1)! = \tan t$, conclude the required expression. (b) Rewrite the generating function of G_{2n}, $n = 1, 2, \ldots$, as $\sum_{n=1}^{\infty} G_{2n} t^{2n}/(2n)! = [(e^{-t} - 1)/2]/[1 + (e^{-t} - 1)/2]$ and expand it into powers of $u = (e^{-t} - 1)/2$. Using the generating function (8.17) of the Stirling numbers of the second kind, deduce the required expression of the Genocchi numbers.

BIBLIOGRAPHY

Abel, N. H. (1826) Beweis eines Ausdruckes, von welchem die Binomial-Formel ein einzelner Fall ist, *Journal für die reine und angewandte Mathematik*, **1**, 159-160.

Abramovitz, M. and Stegun, I. A. (1965) *Handbook of Mathematical Functions*, Dover, New York.

Andersen, E. S. (1953) Two summation formulae for product sums of binomial coefficients, *Math. Scand.* **1**, 261-262.

André, D. (1887) Solution directe du probléme résolu par M. Bertrand, *Comptes Rendus de l' Académie des Sciences*, Paris, **105**, 436-437.

Andrews, G. E. (1976) *The Theory of Partitions*, Encyclopedia of Mathematics and its Applications, Vol. 2, Addison-Wesley, Reading, MA.

Apostol, T. M. (1962) *Calculus*, Vol. II, Blaisdell, Publishing Company, New York.

Barton, D. E. (1958) The matching distributions: Poisson limiting forms and derived methods of approximation, *Journal of the Royal Statistical Society, Series B*, **20**, 73-92.

Barton, D. E. and David, F.N. (1959) Contagious occupancy, *Journal of the Royal Statistical Society, Series B*, **21**, 120-133.

Bell, E. T. (1927) Partition polynomials, *Annals of Mathematics*, **24**, 38-46.

Bell, E. T. (1934a) Exponential numbers, *American Mathematical Monthly*, **41**, 411-419.

Bell, E. T. (1934b) Exponential polynomials, *Annals of Mathematics*, **35**, 258-277.

Bell, E. T. (1940) Postulational bases for the umbral calculus, *American Journal of Mathematics*, **62**, 717-724.

Berge, C. (1971) *Principles of Combinatorics*, Academic Press, New York.

Bernoulli, Jacobus (1713) *Ars Conjectandi*, (Opus Posth.) Basileae.

Bernstein, L. (1965) Rational approximations of algebraic irrationals by means of a modified Jacobi-Perron algorithm, *Duke Mathematical Journal*, **32**, 161-176.

Bertrand, J. (1887) Solution d' un problème, *Comptes Rendus de l' Académie des Sciences*, Paris, **105**, 369.

Blissard, J. (1867) On the properties of $\Delta^n 0^n$ class numbers and of other analogous to them, as investigated by means of representative notation, *Quarterly Journal of Pure and Applied Mathematics*, **8**, 85-110.

Bonferroni, C. E. (1936) Teorie statistica delle classi e calcolo delle probabilità, Publicazioni del R. Instituto Superiore di Scienze Economiche e Commerciali di Firenze, **8**, 1-62.

Boole, G. (1860) *A Treatise on the Calculus of Finite Differences*, Stechert, New York. (Reprinted by Chelsea, New York, 1950.)

Broder, A. Z. (1984) The r-Stirling numbers, *Discrete Mathematics*, **49**, 241-259.

Burnside, W. S. (1911) *Theory of Groups of Finite Order*, 2nd ed., Cambridge University Press, Cambridge (Reprinted by Dover, New York, 1955.)

Carlitz, L. (1932) On arrays of numbers, *American Journal of Mathematics*, **54**, 739-752.

Carlitz, L. (1959) Eulerian numbers and polynomials, *Mathematics Magazine*, **32**, 247-260.

Carlitz, L. (1960) Note on Norlund's polynomials $B_n^{(z)}$, *Proceedings of the American Mathematical Society*, **11**, 447-451.

Carlitz, L. (1965) Note on a paper of L. Bernstein, *Duke Mathematical Journal*, **32**, 177-180.

Carlitz, L. (1971) Note on the numbers of Jordan and Ward, *Duke Mathematical Journal*, **38**, 783-790.

Carlitz, L. (1979) Degenerate Stirling, Bernoulli and Eulerian numbers, *Utilitas Mathematica*, **15**, 51-88.

Carlitz, L. (1980a) Weighted Stirling numbers of the first and second kind-I, *Fibonacci Quarterly*, **18**, 147-162.

Carlitz, L. (1980b) Weighted Stirling numbers of the first and second kind-II, *Fibonacci Quarterly*, **18**, 242-257.

Carlitz, L., Roselle, D. P., and Scoville, R. A. (1966) Permutations and sequences with repetition by number of increases, *Journal of Combinatorial Theory*, **1**, 350-374.

Cauchy, A. (1833) *Resumes Analytiques*, Turin.

Cayley, A. (1878a) On a problem of arrangements, *Proceedings of the Royal Society of Edinburgh*, **9**, 338-342.

Cayley, A. (1878b) Note on Mr. Muir's solution of a problem of arrangement, *Proceedings of the Royal Society of Edinburgh*, **9**, 388-391.

Cayley, A. (1887) Note on a formula for $\Delta^n 0^i / n^i$ when n and i are very large numbers, *Proceedings of the Royal Society of Edinburg*, **14**, 149-153.

Charalambides, Ch. A. (1974) The generalized Stirling and C-numbers, *Sankhyā, Series A*, **36**, 397-418.

Charalambides, Ch. A. (1976) The asymptotic normality of certain combinatorial distributions, *Annals of Institute of Statistical Mathematics*, **28**, 499-506.

Charalambides, Ch. A. (1977a) A new kind of numbers appearing in the n-fold convolution of truncated binomial and negative binomial distributions, *SIAM Journal on Applied Mathematics*, **33**, 279-288.

Charalambides, Ch. A. (1977b) On the generalized discrete distributions and the Bell polynomials, *Sankhyā, Series B*, **39**, 36-44.

Charalambides, Ch. A. (1979a) Some properties and applications of the differences of the generalized factorial, *SIAM Journal on Applied Mathematics*, **36**, 273-280.

Charalambides, Ch. A. (1979b) Bernoulli related polynomials and numbers, *Mathematics of Computation*, **33**, 794-804.

Charalambides, Ch. A. (1981) Bipartitional polynomials and their applications in combinatorics and statistics, *Discrete Mathematics*, **34**, 81-84.

Charalambides, Ch. A. (1982) On the enumeration of certain compositions and related sequences of numbers, *Fibonacci Quarterly*, **20**, 132-146.

Charalambides, Ch. A. (2002) The rook numbers of Ferrers boards and the related restricted permutation numbers, *Journal of Statistical Planning and Inference*, **101**, 33-48.

Charalambides, Ch. A. and Koutras, M. (1983) On the differences of the generalized factorials at an arbitrary point and their combinatorial applications, *Discrete Mathematics*, **47**, 183-201.

Charalambides, Ch. A. and Singh, J. (1988) A review of the Stirling numbers, their generalizations and statistical applications, *Communications in Statistics - Theory and Methods*, **17**, 2533-2595.

Chung, K. L. (1941) On the probability of the occurrence of at least m events among n arbitrary events, *Annals of Mathematical Statistics*, **12**, 328-338.

Chung, K. L. (1943a) Generalization of Poincaré's formula in the theory of probability, *Annals of Mathematical Statistics*, **14**, 63-65.

Chung, K. L. (1943b) On fundamental systems of probabilities of a finite number of events, *Annals of Mathematical Statistics*, **14**, 123-133.

Chung, K. L. (1943c) Further results on probabilities of a finite number of events, *Annals of Mathematical Statistics*, **14**, 234-237.

Comtet, L. (1968) Polynômes de Bell et formule explicite de dérivées successives d' une fontion implicite, *Comptes Rendus de l' Académie des Sciences*, Paris, **267**, 457-460.

Comtet, L. (1973) Une formule explicite pour les puissances successives de l' opérateur de derivation de Lie, *Comptes Rendus de l' Académie des Sciences*, Paris, **276**, 165-168.

Comtet, L. (1974) *Advanced Combinatorics*, Reidel, Dordrecht-Holland.

David, F. N. (1962) *Games, God, and Gambling*, Hafner Press, New York.

David, F. N. and Barton, D. E. (1962) *Combinatorial Chance*, Griffins, London.

David, F. N., Kendall, M. G., and Barton, D. E. (1966) *Symmetric Functions and Allied Tables*, Cambridge University Press, Cambridge.

De Moivre, A. (1718) *The Doctrine of Chances*, Pearson, London (2nd ed. 1738 and 3rd ed. 1756).

De Moivre, A. (1730) *Miscellanea Analytica de Seriebus et Quadraturis*, J. Tonson and J. Watts, London.

DeBruijn, N. G. (1964) Polya's theory of counting, in *Applied Combinatorics* as Chapter 5, E. F. Beckenbach (ed.), John Wiley & Sons, New York.

Dickson, L. E. (1920) *History of the Theory of Numbers*, Vol. II, Carnegie Institute, Washington. (Reprinted by Chelsea, New York, 1952.)

Dillon, J. F. and Roselle, D. P. (1969) Simon Newcomb's problem, *SIAM Journal on Applied Mathematics*, **17**, 1086-1093.

Dwyer, P. S. (1938) The calculation of moments with the use of cumulative totals, *Annals of Mathematical Statistics*, **9**, 288-304.

Dwyer, P. S. (1940) The cumulative numbers and their polynomials, *Annals of Mathematical Statistics*, **11**, 66-71.

Euler, L. (1746) *Introductio in analysin infinitorum*.

Euler, L. (1751) Calcul de la probabilité dans le jeu de rencontre, *Mémoires de l' Academie des Sciences de Berlin*, **7**, 255-270 (Opera Omnia I-7, 11-25.)

Euler, L. (1760) Theoremata arithmetica nova methodo demonstrata, *Novi Commentarii Academiae Petropolitanae*, **8**, 74-104 (Opera Omnia I-2, 531-535.)

Faà di Bruno, F. (1855), Sullo sviluppo delle funzioni, *Annali di Scienze Mathematiche et Fisiche di Tortolini*, **6**, 479-480.

Faà di Bruno, F. (1857), Note sur un nouvelle formule de calcul différentiel, *Quarterly Journal of Mathematics*, **1**, 359-60.

Feller, W. (1968) *An Introduction to Probability Theory and Its Applications*, Vol. 1, 3rd ed. John Wiley & Sons, New York.

Fisher, R. A. and Yates, F. (1953) *Statistical Tables*, Oliver Boyd, London.

Franklin, F. (1881) Sur le développement du produit infini $(1-x)(1-x^2)$ $(1-x^3)\cdots$, *Comptes Rendus de l' Acadèmie des Sciences*, Paris, **92**, 448-450.

Frechet, M. (1940) *Le Probabilités associées à un système d' événements compatibles et dépendants, I: Événements en nombre fini fixe*, Actualités Scientifiques et Industrielles, No. 859, Hermann, Paris.

Frechet, M. (1943) *Le Probabilités associées à un système d' événements compatibles et dépendants II: Cas particuliers et applications*, Actualités Scientifiques et Industrielles, No. 942, Hermann, Paris.

Frucht, R. (1969) A combinatorial approach to the Bell polynomials and their generalizations, in *Recent Progress in Combinatorics*, pp. 69-74, Academic Press.

Gould, H. W. (1956) Some generalizations of Vandermonde's convolution, *American Mathematical Monthly*, **63**, 84-91.

Gould, H. W. (1958) A theorem concerning the Bernstein polynomial, *Mathematics Magazine*, **31**, 259-264.

Gould, H. W. (1964) The operator $(a^x D)^n$ and the Stirling numbers of the first kind, *American Mathematical Monthly*, **71**, 850-858.

Grunert, J. A. (1822) *Mathematische Abhandlungen*, Altona.

Grunert, J. A. (1843) Ueber die Summirung der Reihen von der Form $A_0 \phi(0)$, $A_1 \phi(1)x$, $A_2 \phi(2)x^2, \ldots, A_n \phi(n)x^n, \ldots$ wo A eine beliebige constante Gröfse, A_n eine beliebige und $\phi(n)$ eine ganze rationale algebraische Function der positiven ganzen Zahl n bezeichnet, *Journal für die reine und angewandte Mathematik*, **25**, 240-279.

Gupta, H. (1950) Tables of distributions, *Research Bulletin of East Panjab University*, 13-44.

Hardy, G. H. and Ramanujan, S. (1917) Une formule asymptotique pour le nombre des partitions de n, *Comptes Rendus de l' Académie des Sciences*, Paris, **164**, 35-38.

Hardy, G. H. and Ramanujan, S. (1918) Asymptotic formulae in combinatory analysis, *Proceedings of the London Mathematical Society*, *Series 2*, **17**, 75-115.

Heselden, G. P. M. (1973) A convolution involving Bell polynomials, *Proceedings of the Cambridge Philosophical Society*, **74**, 97-106.

Hoggat, V. E. (1969) *Fibonacci and Lucas numbers*, Houghton Mifflin, Boston, MA.

Johnson, N. L. and Kotz, S. (1977) *Urn Models and Their Application*, Wiley, New York.

Jordan, C. (1867) De quelques formules de probabilité, *Comptes Rendus de l' Académie des Sciences*, **65**, 993-994.

Jordan, Ch. (1926) Sur la probabilité des épreuves répétées, le théorème de Bernoulli et son inversion, *Bulletin de la Société Mathématique de France*, **54**, 101-137.

Jordan, Ch. (1927a) Sur un cas généralisé de la probabilité des épreuves répétées, *Comptes Rendus de l' Académie des Sciences*, Paris, **184**, 315-317.

Jordan, Ch. (1927b) Sur un cas généralisé de la probabilité des épreuves répétées, *Acta Scientiarum Mathematicarum*, Szeged, **3**, 193-210.

Jordan, Ch. (1933) On Stirling numbers, *Tôhoku Mathematical Journal*, **37**, 254-278.

Jordan, Ch. (1934) Le thèorém de probabilité de Poincaré généralisé au cas de plusiers variables indépendantes, *Acta Scientiarum Mathematicarum*, Szeged **7**, 103-111.

Jordan, Ch. (1939a) *Calculus of Finite Differences*, Rotting and Romwalter, Sorron, Hungary. (Reprinted Chelsea, New York, 1947, 1960, 1965.)

Jordan, Ch. (1939b) Problémes de la probabilité des épreuves répétées dans le cas général, *Bulletin de la Société Mathématique de France*, **67**, 223-242.

Koutras, M. (1982) Non-central Stirling numbers and some applications, *Discrete Mathematics*, **42**, 73-89.

Lagrange, L. (1770) Nouvelle methode pour resoudre les equations litterales par le moyen de series, *Mem. Acad. Roy. Sci. Belles-Lettres de Berlin*, **24**.

Lah, I. (1955) Eine neue Art von Zahlen, ihre Eigenschaften und Anwendung in der mathematischen Statistik, *Mitteilungsbl. Math. Statist.*, **7**, 203-212.

Laplace, P. S. (1812) *Théorie Analytique des Probabilités*, Courvier, Paris.

Lucas, E. (1891) *Théorié de nombres*, Paris.

Lukacs, E. (1955) Applications of Faà di Bruno's formula in mathematical statistics, *American Mathematical Monthly*, **62**, 340-349.

MacMahon, P. A. (1902) The problem of derangement in the theory of permutations, *Transactions of the Cambridge Philosophical Society*, **21**, 467-481.

MacMahon, P. A. (1915) *Combinatory Analysis*, Vol. 1, Cambridge University Press, London, (Reprinted by Chelsea, New York, 1960.)

MacMahon, P. A. (1916) *Combinatory Analysis*, Vol. 2, Cambridge University Press, London. (Reprinted by Chelsea, New York, 1960.)

Milne-Thomson, L. M. (1933) *The Calculus of Finite Differences*, Mac-Millan, London.

Mohanty, S. G. (1979) *Lattice Path Counting and Applications*, Academic Press, New York.

Montmort, P. R. (1708) *Essai d' Analyse sur les Jeux de Hazards*, Paris.

Muir, T. (1878) On Professor Tait's problem of arrangement, *Proceedings of the Royal Society of Edinburgh*, **9**, 382-387.

Narayana, T. V. (1979) *Lattice Path Combinatorics with Statistical Applications*, University of Toronto Press, Toronto.

Netto, E. (1927) *Lehrbuch der Kombinatorik*, 2nd ed., Teubner. (Reprinted by Chelsea, 1958.)

Nielsen, N. (1906) *Hambuch der Theorie des Gammafunktion*, Leipsig. (Reprinted by Chelsea, 1966.)

Nielsen, N. (1923) *Traite Elementaire des Nombres de Bernoulli*, Gauthier-Villars, Paris.

Nörlund, N. E. (1924) *Vorlesungen über Differenzenrechnung*, Berlin. (Reprinted by Chelsea, 1954.)

Pólya, G. (1937) Kombinatorische Anzahlbestimmungen für Gruppen, Graphen und chemische Verbindungen, *Acta Mathematica*, **68**, 145-254.

Poincaré, H. (1896) *Calcul des Probabilités*, Gauthier-Villars, Paris.

Raabe, J. L. (1851) *Journal für reine und angewandte Mathematik* **42**, 350.

Rademacher, H. (1937a) On the partition function $p(n)$, *Proceedings of the London Mathematical Society*, Series 2, **43**, 241-254.

Rademacher, H. (1937b) A convergent series for the partition function, *Proceedings of the National Academy of Sciences*, U.S.A., **23**, 78-84.

Rademacher, H. (1940) Fourier expansions of modular forms and problems of partition, *Bulletin of the American Mathematical Society*, **46**, 59-73.

Rademacher, H. (1943) On the expansion of the partition function in a series, *Annals of Mathematics*, **44**, 416-422.

Read, R. C. (1987) Pólya's theorem and its progeny, *Mathematics Magazine*, **60**, 275-282.

Riordan, J. (1937) Moment recurrence relations for binomial, Poisson and hypergeometric frequency distributions, *Annals of Mathematical Statistics*, **8**, 103-111.

Riordan, J. (1958). *An Introduction to Combinatorial Analysis*, John Wiley & Sons, New York.

Riordan, J. (1968) *Combinatorial Identities*, John Wiley & Sons, New York.

Roman, S. (1980) The formula of Faà di Bruno, *American Mathematical Monthly*, **87**, 805-809.

Schlafli, L. (1867) Erganzung der abhandlung uber die entwickelung des products, *Journal für die reine und angewandte Mathematik*, **67**, 179-182.

Schlömilch, O. (1852) Recherches sur les coefficients des faculties analytiques, *Journal für die reine und angewandte Mathematik*, **44**, 344-355.

Schlömilch, O. (1895) *Compendium der Hoheren Analysis*, Braunschweig.

Seal, H. L. (1949) The historical development of the use of generating functions in probability theory, *Bulletin de l' Association des Actuaires Suisses*, **49**, 209-228.

Shanmugan, R. (1984) On central versus factorial moments, *South African Statistical Journal*, **18**, 97-110.

Shumway, R. and Gurland, J. (1960) A fitting procedure for some generalized Poisson distributions, *Skandinavisk Aktuarietidskrift*, **43**, 87-108.

Spitzer, F. (1964) *Principles of Random Walks*, Van Nostrand, New York.

Stanley, R. P. (1978) Generating functions, in *Studies in Combinatorics*, Vol. 17, G.-C. Rota (ed.), pp. 100-141, Mathematical Association of America, Washington, D.C.

Steffensen, J. F. (1927) *Interpolation*, Williams and Wilkins, Baltimore. (Reprinted by Chelsea, New York, 1950.)

Stirling, J. (1730) *Methodulus Differentialis sine Tractatus de Summatione et Interpolatione Serierum Infinitarum*, Londini. (English translation by F. Holliday with title: *The Differential Method*, London, 1749.)

Sylvester, J. (1883) Note sur le théorèm de Legendre cité dans une Note insérée dans les Comptes Rendus, *Comptes Rendus de l' Académie des Sciences*, Paris, **96**, 463-465.

Takacs, L. (1967a) *Combinatorial Methods in the Theory of Stochastic Processes*, John Wiley & Sons, New York.

Takacs, L. (1967b) On the method of inclusion and exclusion, *Journal of the American Statistical Association*, **62**, 102-113.

Takacs, L. (1991) A moment convergence theorem, *American Mathematical Monthly*, **98**, 742-746.

Touchard, J. (1934) Sur un problème de permutations, *Comptes Rendus de l' Académie des Sciences*, Paris, **198**, 631-633.

Touchard, J. (1939) Sur les cycles des substitutions, *Acta Mathematica*, **70**, 243-279.

Touchard, J. (1953) Permutations discordant with two given permutations, *Scripta Mathematica*, **19**, 108-119.

Tucker, A. (1974) Pólya's enumeration formula by example, *Mathematics Magazine*, **47**, 248-256.

Vandermonde, N. (1772) Mémoire sur des Irrationneles de différens ordres avec application au cercle, *Histoire de l' Academie Royale des Sciences*, Part 1, 489-498.

Vilenkin, N. Y. (1971) *Combinatorics*, Academic Press, New York.

Ward, W. (1934) The representation of the Stirling's numbers and the Stirling's polynomials as sums of factorials, *American Journal of Mathematics*, **56**, 87-95.

Whitworth, W. A. (1867) *Choice and Chance*, Hafner Press, New York.

Worpitzky, J. (1883) Studien über die Bernoullischen and Eulerschen Zahlen, *Journal für die reine und angewandte Mathematik*, **94**, 203-232.

Index